This timely volume provides the first comprehensive review and synthesis of the current understanding of the origin, evolution, and effects of magnetic fields in the Sun and other cool stars. Magnetic activity results in a wealth of phenomena – including starspots, nonradiatively heated outer atmospheres, activity cycles, deceleration of rotation rates, and even, in close binaries, stellar cannibalism – all of which are covered clearly and authoritatively.

This book brings together for the first time recent results in solar studies, with their wealth of observational detail, and stellar studies, which allow the study of how activity evolves and depends on the mass, age, and chemical composition of stars. The result is an illuminating and comprehensive view of stellar magnetic activity. Observational data are interpreted by using the latest models in convective simulations, dynamo theory, outer-atmospheric heating, stellar winds, and angular momentum loss.

Researchers are provided with a state-of-the-art review of this exciting field, and the pedagogical style and introductory material make the book an ideal and welcome introduction for graduate students.

SOLAR AND STELLAR MAGNETIC ACTIVITY

Cambridge astrophysics series

Series editors

Andrew King, Douglas Lin, Stephen Maran, Jim Pringle and Martin Ward

SOLAR AND STELLAR MAGNETIC ACTIVITY

C. J. SCHRIJVER

Stanford-Lockhead Institute for Space Research, Palo Alto

C. ZWAAN

Astronomical Institute, University of Utrecht

CAMBRIDGE
UNIVERSITY PRESS

CAMBRIDGE UNIVERSITY PRESS
Cambridge, New York, Melbourne, Madrid, Cape Town, Singapore, São Paulo, Delhi

Cambridge University Press
The Edinburgh Building, Cambridge CB2 8RU, UK

Published in the United States of America by Cambridge University Press, New York

www.cambridge.org
Information on this title: www.cambridge.org/9780521582865

First published 2000
This digitally printed version 2008

A catalogue record for this publication is available from the British Library

Library of Congress Cataloguing in Publication data
Schrijver, Carolus J.
Solar and stellar magnetic activity / Carolus J. Schrijver,
Cornelis Zwaan.
        p.   cm. – (Cambridge astrophysics series ; 34)
ISBN 0-521-58286-5 (hc.)
1. Solar magnetic fields.   2. Stars – Magnetic fields.   I. Zwaan,
Cornelis.   II. Title.   III. Series.
QB539.M23S37   1999
523.7′2 – dc21   99-21364
                                                          CIP

ISBN 978-0-521-58286-5 hardback
ISBN 978-0-521-73986-3 paperback

Die Sonne tönt nach alter Weise
In Brudersphären Wettgesang,
Und ihre vorgeschriebene Reise
Vollendet sie mit Donnergang.

Ihr Anblick gibt den Engeln Stärke,
Wenn Keiner Sie ergründen mag.
Die unbegreiflich hohen Werke
Sind herrlich wie am ersten Tag.

Johann Wolfgang von Goethe

Contents

Image taken with *TRACE* in its 171-Å passband on 26 July 1998, at 15:50:23 UT of Active Region 8,272 at the southwest limb, rotated over $-90°$. High-arching loops are filled with plasma at ~ 1 MK up to the top. Most of the material is concentrated near the lower ends under the influence of gravity. Hotter 3–5 MK loops, at which the bulk of the radiative losses from the corona occur, do not show up at this wavelength. Their existence can be inferred from the emission from the top of the conductively heated transition region, however, where the temperature transits the 1-MK range, as seen in the low-lying bright patches of "moss." A filament-prominence configuration causes extinction of the extreme-ultraviolet radiation.

Preface

This book is the first comprehensive review and synthesis of our understanding of the origin, evolution, and effects of magnetic fields in stars that, like the Sun, have convective envelopes immediately below their photospheres. The resulting magnetic activity includes a variety of phenomena that include starspots, nonradiatively heated outer atmospheres, activity cycles, the deceleration of rotation rates, and – in close binaries – even stellar cannibalism. Our aim is to relate the magnetohydrodynamic processes in the various domains of stellar atmospheres to processes in the interior. We do so by exploiting the complementarity of solar studies, with their wealth of observational detail, and stellar studies, which allow us to study the evolutionary history of activity and the dependence of activity on fundamental parameters such as stellar mass, age, and chemical composition. We focus on observational studies and their immediate interpretation, in which results from theoretical studies and numerical simulations are included. We do not dwell on instrumentation and details in the data analysis, although we do try to bring out the scope and limitations of key observational methods.

This book is intended for astrophysicists who are seeking an introduction to the physics of magnetic activity of the Sun and of other cool stars, and for students at the graduate level. The topics include a variety of specialties, such as radiative transfer, convective simulations, dynamo theory, outer-atmospheric heating, stellar winds, and angular momentum loss, which are all discussed in the context of observational data on the Sun and on cool stars throughout the cool part of the Hertzsprung–Russell diagram. Although we do assume a graduate level of knowledge of physics, we do not expect specialized knowledge of either solar physics or of stellar physics. Basic notions of astrophysical terms and processes are introduced, ranging from the elementary fundamentals of radiative transfer and of magnetohydrodynamics to stellar evolution theory and dynamo theory.

The study of the magnetic activity of stars remains inspired by the phenomena of solar magnetic activity. Consequently, we begin in Chapter 1 with a brief introduction of the main observational features of the Sun. The solar terminology is used throughout this book, as it is in stellar astrophysics in general.

Chapter 2 summarizes the internal and atmospheric structure of stars with convective envelopes, as if magnetic fields were absent. It also summarizes standard stellar terminology and aspects of stellar evolution as far as needed in the context of this monograph.

The Sun forms the paradigm, touchstone, and source of inspiration for much of stellar astrophysics, particularly in the field of stellar magnetic activity. Thus, having introduced the basics of nonmagnetic solar and stellar "classical" astrophysics in the first two chapters, we discuss solar properties in Chapters 3–8. This monograph is based on the premise that the phenomena of magnetic activity and outer-atmospheric heating are governed by processes in the convective envelope below the atmosphere and its interface with the atmosphere. Consequently, in the discussion of solar phenomena, much attention is given to the deepest part of the atmosphere, the photosphere, where the magnetic structure dominating the outer atmosphere is rooted. There we see the emergence of magnetic flux, its transport across the photospheric surface, and its ultimate removal from the atmosphere. We concentrate on the systematic patterns in the dynamics of magnetic structure, at the expense of very local phenomena (such as the dynamics in sunspot penumbrae) or transient phenomena (such as solar flares), however fascinating these are. Page limitations do not permit a discussion of heliospheric physics and solar–terrestrial relationships.

Chapter 3 discusses the solar rotation and large-scale flows in the Sun. Chapters 4–8 cover solar magnetic structure and activity. Chapter 4 deals with fundamental aspects of magnetic structure in the solar envelope, which forms the foundation for our studies of fields in stellar envelopes in general. Chapter 5 discusses time-dependent configurations in magnetic structure, namely the active regions and the magnetic networks. Chapter 6 addresses the global properties of the solar magnetic field, and Chapter 7 deals with the solar dynamo and starts the discussion of dynamos in other stars. Chapter 8 discusses the solar outer atmosphere.

Chapters 9 and 11–14 deal with magnetic activity in stars and binary systems. This set of chapters is self-contained, although there are many references to the chapters on solar activity. Chapter 9 discusses observational magnetic-field parameters and various radiative activity diagnostics, and their relationships; stellar and solar data are compared. Chapter 11 relates magnetic activity with other stellar properties. Chapter 12 reviews spatial and temporal patterns in the magnetic structure on stars and Chapter 13 discusses the dependence of magnetic activity on stellar age through the evolution of the stellar rotation rate. Chapter 14 addresses the magnetic activity of components in binary systems with tidal interaction, and effects of magnetic activity on the evolution of such interacting binaries.

Two integrating chapters, 10 and 15, are dedicated to the two great problems in magnetic activity that still require concerted observational and theoretical studies of the Sun and the stars: the heating of stellar outer atmospheres, and the dynamo action in stars with convective envelopes.

We use Gaussian cgs units because these are (still) commonly used in astrophysics. Relevant conversions between cgs and SI units are given in Appendix I.

We limited the number of references in order not to overwhelm the reader seeking an introduction to the field. Consequently, we tried to restrict ourselves to both historical, pioneering papers and recent reviews. In some domains this is not yet possible, so there we refer to sets of recent research papers.

We would appreciate your comments on and corrections for this text, which we intend to collect and eventually post on a web site. Domain and computer names are, however, notoriously unstable. Hence, instead of listing such a URL here, we ask that you send e-mail to kschrijver at solar.stanford.edu with either your remarks or a request to let you know where corrections, notes, and additions will be posted.

In the process of selecting, describing, and integrating the data and notions presented in this book, we have greatly profited from lively interactions with many colleagues by reading, correspondence, and discussions, from our student years, through collaboration with then-Ph.D. students in Utrecht, until the present day. It is impossible to do justice to these experiences here. We can explicitly thank the colleagues who critically commented on specific chapters: V. Gaizauskas (Chapters 1, 3, 5, 6, and 8), H. C. Spruit (Chapters 2 and 4), R. J. Rutten (Chapter 2), F. Moreno-Insertis (Chapters 4 and 5), J. W. Harvey (Chapter 5), A. M. Title (Chapters 5 and 6), N. R. Sheeley (Chapters 5 and 6), P. Hoyng (Chapters 7 and 15), B. R. Durney (Chapters 7 and 15), G. H. J. van den Oord (Chapters 8 and 9), P. Charbonneau (Chapters 8 and 13), J. L. Linsky (Chapter 9), R. B. Noyes (Chapter 11), R. G. M. Rutten (Chapter 11), A. A. van Ballegooijen (Chapter 10), K. G. Strassmeier (Chapter 12), and F. Verbunt (Chapters 2 and 14). These reviewers have provided many comments and asked thought-provoking questions, which have greatly helped to improve the text. We also thank L. Strous and R. Nightingale for their help in proof reading the manuscript. It should be clear, however, that any remaining errors and omissions are the responsibility of the authors.

The origin of the figures is acknowledged in the captions; special thanks are given to T. E. Berger, L. Golub and K. L. Harvey for their efforts in providing some key figures. C. Zwaan thanks E. Landré and S. J. Hogeveen for their help with figure production and with LaTeX problems.

Kees Zwaan died of cancer on 16 June 1999, shortly after the manuscript of this book had been finalized. Despite his illness in the final year of writing this book, he continued to work on this topic that was so dear to him. Kees' research initially focused on the Sun, but he reached out towards the stars already in 1977. During the past two decades he investigated solar as well as stellar magnetic activity, by exploiting the complementarity of the two fields. His interests ranged from sunspot models to stellar dynamos, and from intrinsically weak magnetic fields in the solar photosphere to the merging of binary systems caused by magnetic braking. His very careful observations, analyses, solar studies, and extrapolations of solar phenomena to stars have greatly advanced our understanding of the sun and of other cool stars: he was directly involved in the development of the flux-tube model for the solar magnetic field, he stimulated discussions of flux storage and emergence in a boundary-layer dynamo, lead the study of sunspot nests, and stimulated the study of stellar chromospheric activity. And Kees always loved to teach. That was one of the main reasons for him to undertake the writing of this book.

Kees Zwaan (24 July 1928–16 June 1999)

1

Introduction: solar features and terminology

The Sun serves as the source of inspiration and the touchstone in the study of stellar magnetic activity. The terminology developed in observational solar physics is also used in stellar studies of magnetic activity. Consequently, this first chapter provides a brief illustrated glossary of nonmagnetic and magnetic features, as they are visible on the Sun in various parts of the electromagnetic spectrum. For more illustrations and detailed descriptions, we refer to Bruzek and Durrant (1977), Foukal (1990), Golub and Pasachoff (1997), and Zirin (1988).

The *photosphere* is the deepest layer in the solar atmosphere that is visible in "white light" and in continuum windows in the visible spectrum. Conspicuous features of the photosphere are the *limb darkening* (Fig. 1.1a) and the *granulation* (Fig. 2.12), a time-dependent pattern of bright *granules* surrounded by darker *intergranular lanes*. These nonmagnetic phenomena are discussed in Sections 2.3.1 and 2.5.

The magnetic structure that stands out in the photosphere comprises dark *sunspots* and bright *faculae* (Figs. 1.1a and 1.2b). A large sunspot consists of a particularly dark *umbra*, which is (maybe only partly) surrounded by a less dark *penumbra*. Small sunspots without a penumbral structure are called *pores*. Photospheric faculae are visible in white light as brighter specks close to the limb.

The *chromosphere* is the intricately structured layer on top of the photosphere; it is transparent in the optical continuum spectrum, but it is optically thick in strong spectral lines. It is seen as a brilliantly purplish-red crescent during the first and the last few seconds of a total solar eclipse, when the moon just covers the photosphere. Its color is dominated by the hydrogen Balmer spectrum in emission. *Spicules* are rapidly changing, spikelike structures in the chromosphere observed beyond the limb (Fig. 4.7 in Bruzek and Durrant, 1977, or Fig. 9-1 in Foukal, 1990).

Chromospheric structure can always be seen, even against the solar disk, by means of monochromatic filters operating in the core of a strong spectral line in the visible spectrum or in a continuum or line window in the ultraviolet (see Figs. 1.1b, 1.1c, 1.2c and 1.3). In particular, filtergrams recorded in the red Balmer line Hα display a wealth of structure (Fig. 1.3). *Mottle* is the general term for a (relatively bright or dark) detail in such a monochromatic image. A strongly elongated mottle is usually called a *fibril*.

The photospheric granulation is a convective phenomenon; most other features observed in the photosphere and chromosphere are magnetic in nature. Sunspots, pores, and faculae are threaded by strong magnetic fields, as appears by comparing the magnetograms in Figs. 1.1 and 1.2 to other panels in those figures. On top of the photospheric

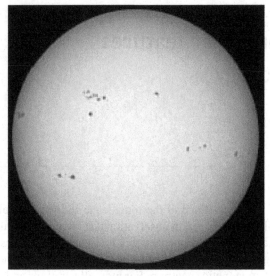

1.1 *a*

Fig. 1.1. Four faces of the Sun and a magnetogram, all recorded on 7 December 1991. North is to the top; West is to the right. Panel *a*: solar disk in white light; note the limb darkening. Dark sunspots are visible in the sunspot belt; the bright specks close to the solar limb are the photospheric *faculae* (NSO-Kitt Peak). Panel *b*: solar disk recorded in the Ca II K line core. Only the largest sunspots remain visible; bright *chromospheric faculae* stand out throughout the activity belt, also near the center of the disk. Faculae cluster in *plages*. In addition, bright specks are seen in the *chromospheric network*, which covers the Sun everywhere outside sunspots and plages (NSO-Sacramento Peak). Panel *c*: solar disk recorded in the H α line core. The plages are bright, covering also the sunspots, except the largest. The dark ribbons are called *filaments* (Observatoire de Paris-Meudon). Panel *d*: the solar *corona* recorded in soft X-rays. The bright *coronal condensations* cover the active regions consisting of sunspot groups and faculae. Note the intricate structure, with loops. Panel *e*: magnetogram showing the longitudinal (line-of-sight) component of the magnetic field in the photosphere; light gray to white patches indicate positive (northern) polarity, and dark gray to black ones represent negative (southern) polarity. Note that the longitudinal magnetic signal in plages and network decreases toward the limb (NSO-Kitt Peak).

faculae are the *chromospheric faculae*, which are well visible as bright fine mottles in filtergrams obtained in the Ca II H or K line (Fig. 1.1*b*) and in the ultraviolet continuum around 1,600 Å (Fig. 1.2*c*). Whereas the faculae in "white light" are hard to see near the center of the disk,* the chromospheric faculae stand out all over the disk.

The magnetic features are often found in specific configurations, such as *active regions*. At its maximum development, a large active region contains a group of sunspots and faculae. The faculae are arranged in *plages* and in an irregular network, called the *enhanced network*. The term plage indicates a tightly knit, coherent distribution of faculae; the term is inspired by the appearance in filtergrams recorded in one of the line cores of the Ca II resonance lines (see Figs. 1.1*b* and 1.3*a*). Enhanced network stands out in

---

* Some of the drawings in Father Schreiner's (1630) book show faculae near disk center.

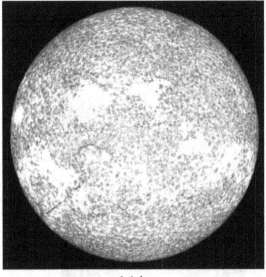

1.1 *b*

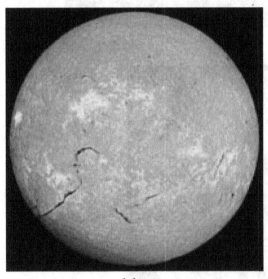

1.1 *c*

Figs. 1.2*b* and 1.2*c*. All active regions, except the smallest, contain (a group of) sunspots or pores during the first part of their evolution.

Active regions with sunspots are exclusively found in the *sunspot belts* on either side of the solar equator, up to latitudes of $\sim 35°$; the panels in Fig. 1.1 show several large active regions. In many young active regions, the two magnetic polarities are found in a nearly E–W bipolar arrangement, as indicated by the magnetogram of Fig. 1.1*e*, and better in the orientations of the sunspot groups in Fig. 1.1*a*. Note that on the northern solar hemishere in Fig. 1.1*e* the western parts of the active regions tend to be of negative

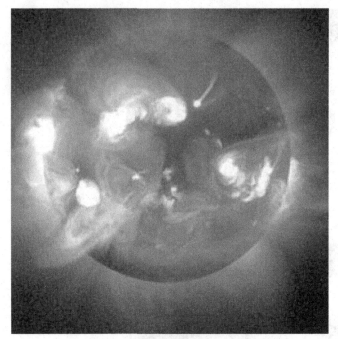

1.1 *d*

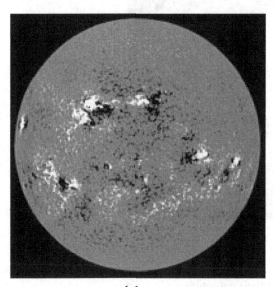

1.1 *e*

polarity, whereas on the southern hemisphere the western parts are of positive polarity. This polarity rule, discovered by G. E. Hale, is discussed in Section 6.1.

Since many active regions emerge close to or even within existing active regions or their remnants, the polarities may get distributed in a more irregular pattern than a simple bipolar arrangement. Such a region is called a complex active region, or an *activity complex*. Figure 1.2 portrays a mildly complex active region.

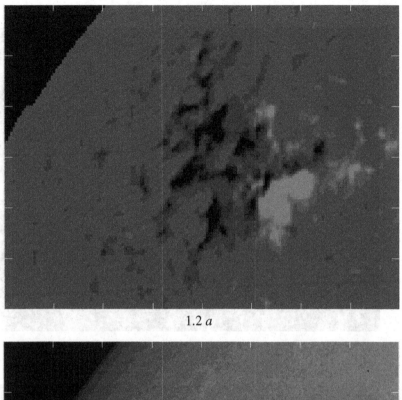

1.2 *a*

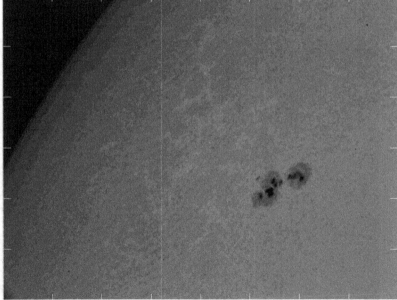

1.2 *b*

Fig. 1.1. Complex active region AR 8,227 observed on 28 May 1998 around 12 UT in various spectral windows. Panels: *a*, magnetogram (NSO-Kitt Peak); *b*, in white light (*TRACE*); *c*, in a 100-Å band centered at ∼1,550 Å, showing the continuum emission from the high photosphere and C IV transition-region emision (*TRACE*); *d*, at 171 Å, dominated by spectral lines of Fe IX and Fe X, with a peak sensitivity at $T \approx 1$ MK (*TRACE*).

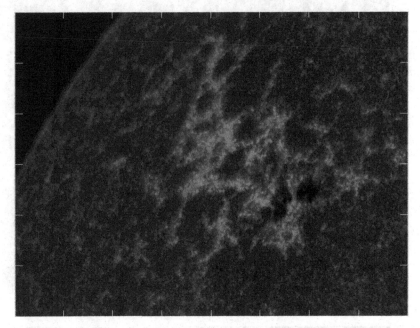

1.2 c

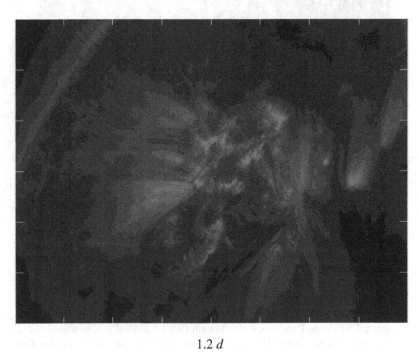

1.2 d

When a large active region decays, usually first the sunspots disappear, and then the plages crumble away to form enhanced network. One or two stretches of enhanced network may survive the active region as a readily recognizable bipolar configuration. Stretches of enhanced network originating from several active regions may combine into one large strip consisting of patches of largely one dominant polarity, a so-called unipolar region. On the southern hemisphere of Fig. 1.1*e*, one such strip of enhanced network of positive (white) polarity stands out. Enhanced network is a conspicuous configuration on the solar disk when activity is high during the sunspot cycle.

Outside active regions and enhanced network, we find a *quiet network* that is best visible as a loose network of small, bright mottles in Ca II K filtergrams and in the UV continuum. Surrounding areas of enhanced network and plage in the active complex, the quiet network is indicated by tiny, bright mottles; see Fig. 1.2*c*. Quiet network is also visible on high-resolution magnetograms as irregular distributions of tiny patches of magnetic flux of mixed polarities. This mixed-polarity quiet network is the configuration that covers the solar disk everywhere outside active regions and their enhanced-network remnants; during years of minimum solar activity most of the solar disk is dusted with it. The areas between the network patches are virtually free of strong magnetic field in the photosphere; these areas are often referred to as internetwork cells. Note that in large parts of the quiet network, the patches are so widely scattered that a system of cells cannot be drawn unambiguously.

The distinctions between plages, enhanced network, and quiet network are not sharp. Sometimes the term plagette is used to indicate a relatively large network patch or a cluster of faculae that is too small to be called plage.

Bright chromospheric mottles in the quiet network are usually smaller than faculae in active regions and mottles in enhanced network, but otherwise they appear similar. Historically, the term facula has been reserved for bright mottles within active regions; we call the bright mottles outside active regions *network patches*. (We prefer the term patch over point or element, because at the highest angular resolution these patches and faculae show a fine structure.)

The comparison between the magnetograms and the photospheric and chromospheric images in Figs. 1.1 and 1.1 shows that near the center of the solar disk there is an unequivocal relation between sites of strong, vertical magnetic field and sunspots, faculae, and network patches. As a consequence, the adjectives magnetic and chromospheric are used interchangeably in combination with faculae, plages, and network.

In most of the magnetic features, the magnetic field is nearly vertical at the photospheric level, which is one of the reasons for the sharp drop in the line-of-sight magnetic signal in plages and network toward the solar limb in Fig. 1.1*e*. Markedly inclined photospheric fields are found within tight bipoles and in sunspot penumbrae.

Filtergrams obtained in the core of Hα are much more complex than those in the Ca II H and K lines (see Fig. 1.3, and Zirin's 1988 book, which is full of them). In addition to plages and plagettes consisting of bright mottles, they show a profusion of elongated dark fibrils. These fibrils appear to be directed along inclined magnetic field lines in the upper chromosphere (Section 8.1); they are rooted in the edges of plages and in the network patches. The fibrils stand out particularly well in filtergrams obtained at $\sim$0.5 Å from the line core (see Fig. 1.3*b*).

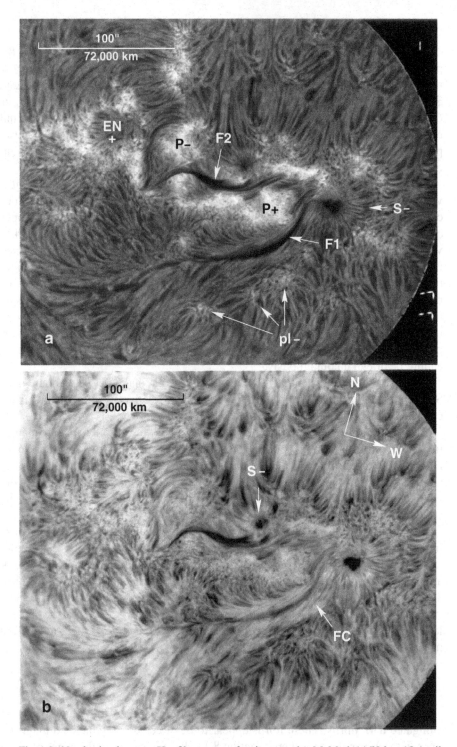

Fig. 1.3. Nearly simultaneous Hα filtergrams of active complex McMath 14,726 on 18 April 1977, observed in the line core (panel *a*) and at Δλ = +0.65 Å in the red wing (panel *b*). The letter symbols indicate the following: S, sunspot; P, plage; pl, plagette; F, filament; FC, filament channel; EN, enhanced network cell. Signs are appended to indicate the magnetic polarities. Fibrils are prominent in both panels. Exceptionally long and well-ordered fibrils are found in the northwestern quadrants. Several features are discussed in Sections 8.1 and 8.2. The chirality of filament F1 is sinistral (figure from the archive of the Ottawa River Solar Observatory, National Research Council of Canada, courtesy of V. Gaizauskas.)

The longest dark structures visible in the core of the H α line are the *filaments* (Figs. 1.1*c* and 1.3). Many filaments are found at borders of active regions and within active complexes, but there are also filaments outside the activity belts, at higher latitudes. Most filaments differ from fibrils by their length and often also by their detailed structure. Small filaments can be distinguished from fibrils by their reduced contrast at distances $|\Delta\lambda| \gtrsim 0.5$ Å from the line core. Large filaments are visible outside the solar limb as *prominences* that are bright against a dark background.

The *corona* is the outermost part of the Sun, which is seen during a total eclipse as a pearly white, finely structured halo, locally extending to several solar radii beyond the photospheric limb; see Figs. 8.4 and 8.11, Fig. 1.2 in Golub and Pasachoff (1997), or Fig. 9-10 in Foukal (1990). The coronal plasma is extremely hot ($T \sim 1 \times 10^6 - 5 \times 10^6$ K) and tenuous. The radiation of the white-light corona consists of photospheric light, scattered by electrons in the corona and by interplanetary dust particles; the brightness of the inner corona is only $\sim 10^{-6}$ of the photospheric brightness. The thermal radiation of the corona is observed in soft X-rays, in spectral lines in the ultraviolet and optical spectrum, and in radio waves. The corona is optically thin throughout the electromagnetic spectrum, except in radio waves and a few resonance lines in the extreme ultraviolet and in soft X-rays.

The coronal structure in front of the photospheric disk can be observed from satellites in the EUV and in X-rays; see Figs. 1.1*d* and 1.2*d*. In these wavelength bands, the coronal plasma, however optically thin, outshines the much cooler underlying photosphere. The features depend on the magnetic field in the underlying photosphere. The corona is particularly bright in "coronal condensations" immediately above all active regions in the photosphere and chromosphere. Coronal loops trace magnetic field lines connecting opposite polarities in the photosphere. Note that in Fig. 1.1*d* there are also long, somewhat fainter, loops that connect magnetic poles in different active regions. The finest coronal structure is displayed in Fig. 1.2*d*, where the passband reveals radiation from bottom parts of loops with $T \lesssim 1 \times 10^6$ K, without contamination by radiation from hotter loops with $T \gtrsim 2 \times 10^6$ K.

Coronal holes stand out as regions that emit very little radiation; these have been identified as regions where the magnetic field is open to interstellar space. Usually large coronal holes are found over the polar caps; occasionally smaller coronal holes are observed at low latitudes.

2

Stellar structure

This chapter deals with the aspects of stellar structure and evolution that are thought to be independent of the presence of magnetic fields. In this classical approach to global stellar structure, the effects of stellar rotation are also ignored. Rather than summarize the theory of stellar structure, we concentrate on features that turn out to be important in understanding atmospheric structure and magnetic activity in Sun-like stars, that is, stars with convective envelopes. For more comprehensive introductions to stellar structure we refer to Chapter 4 in Unsöld and Baschek (1991), and to Böhm-Vitense (1989a, 1989b, 1989c).

We present a brief synopsis of the transfer of electromagnetic radiation in order to indicate its role in the structuring of stellar atmospheres and to sketch the possibilities and limitations of spectroscopic diagnostics, including Zeeman diagnostics of magnetic fields.

In addition, in this chapter we summarize the convective and purely hydrodynamic wave processes in stellar envelopes and atmospheres. In this framework, we also discuss the basal energy deposition in outer atmospheres that is independent of the strong magnetic fields.

2.1 Global stellar structure

2.1.1 Stellar time scales

Stars are held together by gravity, which is balanced by gas pressure. Their quasi-steady state follows from the comparison of some characteristic time scales.

The *time scale of free fall* $\hat{t}_{ff}$ is the time scale for stellar collapse if there were no pressure gradients opposing gravity. Then the only acceleration is by gravity: $d^2r/dt^2 = -GM/r^2$, where r is the radial distance to the stellar center, G is the gravitational constant, and M and R are the stellar mass and radius, respectively. This leads to the order-of-magnitude estimate:

$$\hat{t}_{ff} \approx \left(\frac{R^3}{GM} \right)^{1/2} = 1{,}600 \left(\frac{M}{M_\odot} \right)^{-1/2} \left(\frac{R}{R_\odot} \right)^{3/2} \ (s), \tag{2.1}$$

where $M_\odot$ and $R_\odot$ are the solar mass and radius, respectively.

For a star virtually in hydrostatic equilibrium, local departures from equilibrium are restored at the speed of sound:

$$c_s = [(\gamma p)/\rho]^{1/2}, \tag{2.2}$$

where ρ is the mass density, p is the gas pressure, and $\gamma \equiv c_p/c_V$ is the ratio of the specific heats at constant pressure and constant volume. Using the order-of-magnitude estimate from Eq. (2.7) for hydrostatic equilibrium, $\bar{p}/R \approx GM\bar{\rho}/R^2$, we find the *hydrodynamic time scale* $\hat{t}_{hy}$:

$$\hat{t}_{hy} \equiv \frac{R}{\bar{c}_s} = \left[\frac{R^2\bar{\rho}}{\gamma\bar{p}}\right]^{1/2} \approx \left[\frac{R^3}{\gamma GM}\right]^{1/2} = \gamma^{-1/2}\hat{t}_{ff}, \qquad (2.3)$$

which is of the same order of magnitude as the free-fall time scale $\hat{t}_{ff}$.

The *Kelvin–Helmholtz time scale* $\hat{t}_{KH}$ estimates how long a star could radiate if there were no nuclear reactions but the star would emit all of its present total potential gravitational energy E_g at its present luminosity L:

$$\hat{t}_{KH} \equiv \frac{|E_g|}{L} \approx \frac{GM^2}{RL} \approx 3 \times 10^7 \left(\frac{M}{M_\odot}\right)^2 \left(\frac{R}{R_\odot}\right)^{-1} \left(\frac{L}{L_\odot}\right)^{-1} \quad \text{(yr)}. \qquad (2.4)$$

From the virial theorem (see Sections 2.6.4 and 4.12.4 in Unsöld and Baschek, 1991 or Section 2.3 in Böhm-Vitense, 1989c) applied to a star in hydrostatic equilibrium, it follows that the internal (thermal) energy E_i is half $|E_g|$. Hence the Kelvin–Helmholtz time scale is of the order of the *thermal time scale*, which a star would need to radiate all its internal energy at the rate of its given luminosity L.

The *nuclear time scale* $\hat{t}_{nu}$, the time that a star can radiate by a specific nuclear fusion process, is estimated from stellar evolution calculations. The time scale for hydrogen fusion is found to be

$$\hat{t}_{nu} \approx 1 \times 10^{10} \left(\frac{M}{M_\odot}\right) \left(\frac{L}{L_\odot}\right)^{-1} \quad \text{(yr)}. \qquad (2.5)$$

The comparison of the stellar time scales shows

$$\hat{t}_{ff} \approx \hat{t}_{hy} \ll \hat{t}_{KH} \ll \hat{t}_{nu}. \qquad (2.6)$$

Consequently, a star is in both mechanical (that is, hydrostatic) and thermal equilibrium during nearly all of its evolutionary phases.

2.1.2 Shell model for Sunlike stars

In classical theory, the stellar structure is approximated by a set of spherical shells. The *stellar interior* of the Sun and Sunlike stars consists of the central part, the radiative interior, and the convective envelope.

The central part is the section where nuclear fusion generates the energy flux that eventually leaves the stellar atmosphere. In the Sun and other main-sequence stars, hydrogen is fused into helium in the spherical *core*, on the time scale $\hat{t}_{nu}$ [Eq. (2.5)]. In evolved stars, the central part consists of a core, in which the hydrogen supply is exhausted, which is surrounded by one or more shells, which may be "dead" (and hence in a state of gravitational contraction), or which may be in a process of nuclear fusion.

In the Sun and in all main-sequence stars, except the coolest, the core is surrounded by the *radiative interior*, which transmits the energy flux generated in the core as electromagnetic radiation.

The *convective envelope* (often called the convection zone) is the shell in which the opacity is so high that the energy flux is not transmitted as electromagnetic radiation; there virtually the entire energy flux is carried by convection.

The atmosphere is defined as the part of the star from which photons can escape directly into interstellar space. In the Sun and other cool stars, the atmosphere is situated immediately on top of the convective envelope. It consists of the following domains, which are often conveniently, but incorrectly, pictured as a succession of spherical shells:

1. The *photosphere* is the layer from which the bulk of the stellar electromagnetic radiation leaves the star. This layer has an optical thickness $\tau_\nu \lesssim 1$ in the near-ultraviolet, visible, and near-infrared spectral continua, but it is optically thick (Section 2.3.1) in all but the weakest spectral lines.
2. The *chromosphere* is optically thin in the near-ultraviolet, visible, and near-infrared continua, but it is optically thick in strong spectral lines. The chromosphere can be glimpsed during the first and the last few seconds of a total solar eclipse as a crescent with the purplish-red color of the Balmer spectrum.
3. The *corona* is optically very thin over the entire electromagnetic spectrum except for the radio waves and a few spectral lines (see Section 8.6). Stellar and solar coronae can be observed only if heated by some nonradiative means, because they are transparent to photospheric radiation. The chromosphere and the corona differ enormously both in density and in temperature; the existence of an intermediate domain called the *transition region*, is indicated by emissions in specific spectral lines in the (extreme) ultraviolet.

In the present Sun, with its photospheric radius $R_\odot$ of 700 Mm, the core has a radius of $\sim$100 Mm. The radiative interior extends to 500 Mm from the solar center, and the convective envelope extends from 500 to 700 Mm. The photosphere has a thickness of no more than a few hundred kilometers; the chromosphere extends over nearly 10 Mm. The appearance of the corona is of variable extent and complex shape; see Figs. 1.1*d* and 1.2*d*. The corona merges with the interplanetary medium.

2.1.3   *Stellar interiors: basic equations and models*

The models for the quasi-static stellar interiors are determined by four first-order differential equations.

The force balance is described by *hydrostatic equilibrium*:

$$\frac{\mathrm{d}p(r)}{\mathrm{d}r} = -g(r)\,\rho(r) = -\frac{G\,m(r)\,\rho(r)}{r^2}, \tag{2.7}$$

where $p(r)$ is the gas pressure, $\rho(r)$ is the mass density, and $g(r)$ is the acceleration by gravity, all at a radial distance r from the stellar center; $m(r)$ is the mass contained in a sphere of radius r. The contribution of the radiation pressure $p_{\mathrm{rad}} = 4\sigma T^4/c$ to the total pressure balance is negligible in Sun-like stars.

In the conditions covered in this book, the perfect gas law is applicable:

$$\rho = \frac{\mu p}{\mathcal{R}T}, \tag{2.8}$$

where μ is the average molecular weight per particle and $\mathcal{R}$ is the gas constant. From hydrostatic equilibrium follows the local *pressure scale height*:

$$H_p \equiv \frac{\mathcal{R}T}{\mu g}, \tag{2.9}$$

which is the radial distance over which the pressure p drops by a factor e.

The mass distribution $m(r)$ is related to the density distribution $\rho(r)$ through

$$\frac{\mathrm{d}m(r)}{\mathrm{d}r} = 4\pi r^2 \rho(r). \tag{2.10}$$

In a steady state, the total net outward flux of thermal energy $L(r)$ through a sphere of radius r equals the total energy production through nuclear processes within it; hence

$$\frac{\mathrm{d}L(r)}{\mathrm{d}r} = 4\pi r^2 \rho(r)[\epsilon(r) - \epsilon_\nu(r)], \tag{2.11}$$

where $\epsilon(r)$ is the rate of nuclear energy production per unit mass and $\epsilon_\nu(r)$ is the energy production lost with the neutrinos. Outside the stellar central part with its nuclear processes, $L(r)$ is constant and equal to the stellar luminosity $L(R)$, the total radiative flux from the stellar surface.

In stellar interiors, the heat flow is transported either by electromagnetic radiation alone, or by convection in some combination with electromagnetic radiation. Thermal conduction is negligible in stellar interiors except in very dense cores.

If in a layer the energy flux L is carried by electromagnetic radiation alone, that layer is said to be in *radiative equilibrium* (RE). The corresponding temperature gradient is

$$\left(\frac{\mathrm{d}T}{\mathrm{d}r}\right)_{\mathrm{RE}} = -\frac{3\kappa_R \rho}{16\sigma T^3} \frac{L}{4\pi r^2}, \tag{2.12}$$

where T is the temperature, κ_R is the Rosseland mean opacity (discussed in Section 2.3.1), and σ is the Stefan–Boltzmann constant.

Often it is more convenient to use double-logarithmic temperature gradients defined by the following, and rewritten by means of Eqs. (2.8) and (2.9):

$$\nabla \equiv \frac{\mathrm{d}\ln T}{\mathrm{d}\ln p} = -\frac{\mathcal{R}}{\mu g} \frac{\mathrm{d}T}{\mathrm{d}r} = -\frac{H_p}{T} \frac{\mathrm{d}T}{\mathrm{d}r}. \tag{2.13}$$

(This ∇ symbol is not to be confused with the abbreviation ∇ for vector operators!) The corresponding RE gradient is

$$\nabla_{\mathrm{RE}} = +\frac{3\kappa_R \rho H_p}{16\sigma T^4} \frac{L}{4\pi r^2}. \tag{2.14}$$

The assumption of radiative equilibrium is consistent if the gradient ∇_{RE} is smaller than the adiabatic gradient ∇_{ad}; this Schwarzschild criterion for convective stability is discussed in Section 2.2. The adiabatic gradient is given by

$$\nabla_{\mathrm{ad}} = \frac{\gamma - 1}{\gamma}, \quad \text{with} \quad \gamma \equiv \frac{c_p}{c_V}, \tag{2.15}$$

where c_p is the specific heat at constant pressure and c_V is the specific heat at constant volume. Section 2.2 discusses an approximate procedure to determine the mean temperature gradient ∇ in convective envelopes. There it is shown that for the largest part of such a zone the adiabatic gradient ∇_{ad} is a good approximation, with the exception of the very top layer immediately below the photosphere.

In order to compute $p(r)$, $T(r)$, $L(r)$, $m(r)$ and the other r-dependent parameters that determine the time-dependent chemical composition, the dependences of ρ, μ, ϵ, ϵ_ν, κ_R, and γ on p, T and the chemical composition are needed.

The set of equations are to be completed by boundary conditions. There are two conditions at the stellar center:

$$L(0) = 0 \quad \text{and} \quad m(0) = 0. \tag{2.16}$$

The two boundary conditions at the outer edge $r = R$ concern T and p. For the temperature we have

$$T(R) = T_{eff} \equiv \left[\frac{L(R)}{4\pi R^2 \sigma} \right]^{1/4}, \tag{2.17}$$

where T_{eff} is the stellar effective temperature, sometimes called the "surface temperature." The corresponding gas pressure $p(R)$, at the same photospheric plane where $T = T_{eff}$, is best obtained from a model atmosphere for the star of interest. The location of this stellar "surface" is discussed in Section 2.3.1.

Although the parameters of the stellar interiors are well defined by the set of equations, their solution requires sophisticated numerical methods, which are discussed in specialized texts; see, for example, in Chapter 11 of Kippenhahn and Weigert (1990).

The models indicate a strong concentration of mass toward the stellar center. In the present Sun, nearly half of the mass is contained within $r = 0.25 R_\odot$. The convective envelope contains more than 60% of the solar volume but less than 2% of the solar mass.

2.2   Convective envelopes: classical concepts

2.2.1   *Schwarzschild's criterion for convective instability*

When a blob of gas is lifted from its original position, it may be heavier or lighter than the gas in its new environment. In the first case, the blob will move back to its original location, and the medium is called convectively stable. In the second case, it continues to rise, and the gas is said to be *convectively unstable*. The criterion for this (in)stability was derived by K. Schwarzschild in 1906.

Consider a blob lifted by some disturbance over a height δr. Let us assume that initially the thermodynamic conditions within the blob, indexed "in," were equal to those outside, for the mass density: $\rho_{in}(r) = \rho(r)$. It is further assumed that the rise is sufficiently fast, so that the blob behaves adiabatically (without heat exchange), yet so slow that the internal pressure p_{in} adjusts to balance the ambient pressure p during the rise. These assumptions are plausible in stellar envelopes because of the high opacity and the short travel time for sound waves across the blob, respectively. The blob continues to rise if the internal density remains smaller than the external density, so the condition for instability is

$$\rho_{in}(r + \delta r) - \rho(r + \delta r) = \delta r \left[\left(\frac{d\rho}{dr} \right)_{ad} - \frac{d\rho}{dr} \right] < 0, \tag{2.18}$$

where $(d\rho/dr)_{ad}$ is the density gradient under adiabatic conditions. With substitution of the perfect gas law [Eq. (2.8)] and use of $p_{in}(r) = p(r)$, the instability condition in Eq. (2.18) becomes

$$\left(\frac{dT}{dr}\right)_{ad} - \frac{T}{\mu}\left(\frac{d\mu}{dr}\right)_{ad} > \frac{dT}{dr} - \frac{T}{\mu}\frac{d\mu}{dr}. \tag{2.19}$$

The mean molecular weight μ is a function of p and T because it depends on the degree of ionization of the abundant elements hydrogen and helium. If the adjustment of the ionization equilibrium in the rising blob is instantaneous, the function $\mu(p, T)$ is the same for the plasma inside and outside the blob. Hence, the gradients $d\mu/dr$ and $(d\mu/dr)_{ad}$ in Eq. (2.19),

$$d\mu/dr = (\partial\mu/\partial p)_T \cdot dp/dr + (\partial\mu/\partial T)_p \cdot dT/dr \quad \text{and}$$
$$(d\mu/dr)_{ad} = (\partial\mu/\partial p)_T \cdot (dp/dr)_{ad} + (\partial\mu/\partial T)_p \cdot (dT/dr)_{ad},$$

differ only in the second term containing the gradient dT/dr, because from pressure equilibrium it follows that $(dp/dr)_{ad} = dp/dr$. Hence the *Schwarzschild criterion for convective instability* is:

$$\left|\frac{dT}{dr}\right| > \left|\left(\frac{dT}{dr}\right)_{ad}\right|. \tag{2.20}$$

In the case $|dT/dr| < |(dT/dr)_{ad}|$, the medium is convectively stable. Then a rising blob is heavier than its new surroundings, so it returns toward its initial position where it may overshoot and start to oscillate about its equilibrium position. Such oscillations are called *gravity waves* because gravity is the restoring force (see Section 2.6).

Another formulation of the Schwarzschild criterion is with double-logarithmic gradients [Eq. (2.13)]:

$$\nabla > \nabla_{ad}. \tag{2.21}$$

Stellar models computed under the assumption of radiative equilibrium (indicated by the subscript "RE") are consistent in layers where the computed temperature gradient ∇_{RE} turns out to be smaller than the adiabatic gradient ∇_{ad}, and there we have $\nabla = \nabla_{RE} < \nabla_{ad}$. If stability does not apply, we have

$$\nabla_{ad} \leq \nabla_{in} < \nabla < \nabla_{RE}, \tag{2.22}$$

where the index "in" refers to the interior of a moving gas blob and quantities without an index refer to a horizontal average over the ambient medium. The first inequality allows for a departure from adiabatic conditions in the moving blob by radiative exchange with its surroundings; this difference is very small except in the very top layers of the convective envelope. The second inequality accounts for the driving of the motion of the blob relative to its environment by the convective instability. The third inequality stands for a relatively large difference; see Fig. 2.1.

The adiabatic temperature gradient ∇_{ad} [Eq. (2.15)] depends on the degree of ionization in the plasma. In a strictly monoatomic gas that is either completely neutral or completely ionized, $\gamma = 5/3$, and hence $\nabla_{ad} = 2/5$. It is easy to see that in a partly ionized gas $\nabla_{ad} < 2/5$: during adiabatic expansion, the cooling leads to recombination, which releases latent energy so that the temperature decreases less rapidly than in the case of a gas that

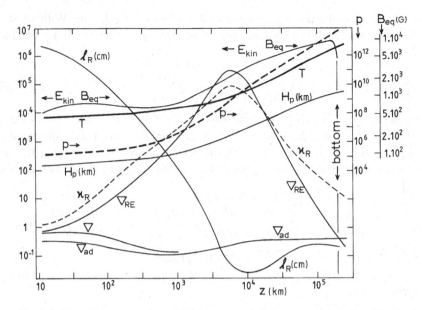

Fig. 2.1. Run of parameters through the solar convection zone, as a function of depth z below the photosphere (from tables in Spruit, 1977b).

is not ionized at all or that remains completely ionized. The quantities γ and ∇_{ad} depend particularly on the ionization equilibria in the most abundant elements, H and He; for the formulas, see Chapter 14 in Kippenhahn and Weigert (1990).

In the solar envelope, ∇_{ad} is found to drop to nearly one quarter of the monoatomic value of 2/5 (see Fig. 2.1), but this dip is restricted to a shallow layer immediately below the photosphere. For the convective instability, the extremely large value of the radiative-equilibrium gradient ∇_{RE} is much more important: in the solar convective envelope, ∇_{RE} exceeds the comparison value of 2/5 by several orders of magnitude; see Fig. 2.1. The large value of ∇_{RE} is caused primarily by the large opacity κ_R, which is explained in Section 2.3.1.

Note that stellar atmospheres, in which the characteristic optical depth (based on the mean Rosseland opacity) is smaller that unity, are stable against convection because from there photons can escape readily. Indeed, classical models for stellar atmospheres show that $\nabla = \nabla_{RE} < \nabla_{ad}$.

In O- and B-type stellar envelopes there is no convective instability. There the ionization of H and He is so complete that κ_R remains small so that everywhere $\nabla = \nabla_{RE} < \nabla_{ad} = 2/5$. Nor is convective instability important in the envelopes of A-type main-sequence stars.

Cooler stars of spectral types F, G, K, and M are convectively unstable in the envelopes below their photospheres. The term *cool stars* is often used to indicate stars that have convective envelopes.

2.2.2   The mixing-length approximation

In layers where the Schwarzschild criterion Eq. (2.21) indicates instability, convective motions develop. There are two commonly used approaches to the modeling of convection extending over many pressure scale heights. In hydrodynamic modeling, numerical simulations are used, as discussed in Section 2.5 – this ab initio method is limited by computational resources. The alternative is the classical mixing-length formalism which attempts to derive some properties of convective envelopes from a strongly simplified description of the convective processes. We first discuss some results obtained from mixing-length modeling because nearly all quantitative models of stellar convective envelopes are based on this approach. Moreover, many current ideas about processes in convective envelopes are still dressed in terms of the mixing-length theory or some similar theory of turbulent convection.

The mixing-length (ML) concept was developed by physicists, including L. Prandtl, between 1915 and 1930. The *mixing length* ℓ_{ML} is introduced as the distance over which a convecting blob can travel before it disintegrates into smaller blobs and so exchanges its excess heat, which can be positive or negative, depending on whether the blob is moving up or down. In some applications, this ℓ_{ML} is taken to equal the distance to the nearest boundary of the convective layer, but generally some smaller characteristic length scale of the medium is used. After 1930, the concept was introduced in astrophysics; many applications have followed the formalism put forward by Vitense (1953) – see also Böhm-Vitense (1958). In such applications, the mixing length is assumed to be a local quantity related to the pressure scale height [Eq. (2.9)]

$$\ell_{ML} \equiv \alpha_{ML} H_p, \tag{2.23}$$

where α_{ML} is introduced as an adjustable parameter. From the extremely small viscosity of the gas in stellar envelopes it was inferred that the convective flow is very turbulent; hence α_{ML} was expected to be small, of the order of unity.

Only the mean value of the upward and downward components in the velocity field are considered. Other quantities are also represented by their mean values that vary only with distance r from the stellar center. The chief aim is to determine the mean temperature gradient $\nabla(r)$ in order to complete the model of the internal thermodynamic structure of the star.

A basic equation for the stellar envelope is the continuity equation for the energy flux:

$$\frac{L}{4\pi r^2} \equiv \mathcal{F}(r) = \mathcal{F}_R + \mathcal{F}_C, \tag{2.24}$$

where L is the stellar luminosity. For the radiative energy flux density $\mathcal{F}_R(r)$ we have

$$\mathcal{F}_R = -\frac{16\sigma T^3}{3\kappa_R \rho}\frac{dT}{dr} = \frac{16\sigma T^4}{3\kappa_R \rho H_p}\nabla \tag{2.25}$$

[see Eq. (2.14)]. The mean gradient ∇ adjusts itself so that whatever part of the total flux $\mathcal{F}$ cannot be carried as radiative energy flux $\mathcal{F}_R$ is carried by convection. The convective energy flux $\mathcal{F}_C$ is estimated from assumed mean properties of rising and sinking blobs, depending on the local conditions; for reviews, we refer the reader to Section 6.2 in Stix (1989) or to Chapter 6 in Böhm-Vitense (1989c). Eventually a relation is found that expresses the mean gradient ∇ in terms of local quantities. This relation can be numerically solved in the framework of the equations determining stellar structure given in Section 2.1.

The principal adjustable parameter is the mixing length; in the classical, strictly local theory, values of $1 \lesssim \alpha_{ML} \lesssim 2$ have been preferred, because in solar models these parameter values are found to correctly reproduce the measured solar radius. This assumption of a strictly local mixing length ℓ_{ML} is not consistent close to the boundaries of the convection zone; hence models invoking the distance z to the nearest boundary have been constructed, for instance, with $\ell_{ML} = \min(z + z_0, \alpha_{ML} H_p)$, where the depth z_0 allows for convective overshoot (explained below).

Mixing-length models provide a first-order estimate for the *superadiabaticy* $\nabla - \nabla_{ad}$, which for the Sun turns out to be extremely small for all depths except in the thin, truly superadiabatic top layer at depths $z < 1,000$ km. The extremely small values of $\nabla - \nabla_{ad}$ in the bulk of the convective envelope follow from the relatively high mass density ρ: near the bottom, the convective heat flux can be transported at a very small temperature contrast $\delta T / T \simeq 10^{-6}$. Even though the computed $\nabla - \nabla_{ad}$ is not accurate, the mean $\nabla \simeq \nabla_{ad}$ is fairly well defined; hence so is the temperature run $T(r)$. Because of hydrostatic equilibrium, the values of the other local parameters that depend on thermodynamic quantities are also reasonably well established.

Figure 2.1 shows the profiles of several parameters as computed for the convective envelope of the Sun. This figure shows that the extent of the convection zone is determined by the radiative-equilibrium gradient ∇_{RE}. The depth dependence of $\nabla_{RE}(r)$ reflects that of the Rosseland mean opacity $\kappa_R(r)$. The mean-free path of photons ℓ_R drops from approximately 10 km near the surface (where $\tau_R \approx 1$) to values $\lesssim 1$ mm for $z \gtrsim 10,000$ km.

Dynamic quantities, such as the mean convective velocity v_{ML}, depend critically on the crude assumptions of ML theory. Nonetheless, the mean convective velocity v_{ML} has been used to estimate the kinetic energy density in the convection by

$$E_{kin} \equiv \frac{1}{2} \rho v_{ML}^2 \tag{2.26}$$

(see Fig. 2.1). Despite the uncertainty in v_{ML}, there is little doubt that even in the top of the convection zone the mean kinetic energy in the convective flow E_{kin} is smaller than the internal thermal energy density E_{th}:

$$E_{th} = \frac{3}{2} p + [n_{H^+} \cdot \chi_H + n_{He^+} \cdot \chi_{He} + n_{He^{++}} \cdot (\chi_{He} + \chi_{He^+})], \tag{2.27}$$

where n_i and χ_i represent the number density and the ionization energy of particles of species i. (Note that the ionization energy contributes appreciably to the total thermal energy E_{th} in the layers where H and He are partially ionized.) In the deep layers of the convection zone, E_{kin} is many orders of magnitude smaller than E_{th}. The statement $E_{kin} \ll E_{th}$ is equivalent to $v_{ML} \ll c_s$, where c_s is the sound velocity [Eq. (2.2)].

Within the ML approximation, there is an estimate for the characteristic velocity $\hat{v}$ of the turbulent convection in terms of the stellar energy flux density. From ML formulas given in Chapter 6 in Böhm-Vitense (1989c), one finds that the convective flux density can be estimated by

$$\mathcal{F}_c \simeq \frac{5}{\alpha_{ML}} \rho(r)\, \hat{v}^3(r). \tag{2.28}$$

Throughout the convective envelope, convection carries the entire energy flux; hence

$$\frac{5}{\alpha_{ML}} \rho(r)\, \hat{v}^3(r) \simeq \sigma T_{eff}^4 \frac{R^2}{r^2}. \tag{2.29}$$

Application to the top of the solar convection zone, assuming $\alpha_{ML} = 1.6$, leads to $\hat{v} \simeq 3.9$ km/s, which is a substantial fraction of the local sound velocity $c_s = 8.2$ km/s. Near the bottom of the convection zone, the characteristic velocity is approximately 60 m/s: there the energy flux is carried by small convective velocities (and very small temperature contrasts) in large-scale flows.

In discussions of large-scale flows in convective envelopes, often a convective turnover time $\hat{t}_c$ of convective eddies near the bottom of the convection zone is introduced. In turbulent convection, this time scale is defined as the typical length scale of the convective eddies, divided by the typical velocity in the eddies; hence in the ML approximation,

$$\hat{t}_c \equiv \frac{\ell_{ML}}{v_{ML}}. \tag{2.30}$$

Near the bottom of the solar convection zone, one finds values for $\hat{t}_c$ between approximately one week and one month.

Although the Schwarzschild criterion predicts sharp boundaries between the convective zone and the adjacent stably stratified layers, convective overshoot is to be expected because of the inertia of the convecting matter. Overshooting blobs suddenly find themselves in a subadiabatic domain, hence the force acting on the blob is reversed: the blob is decelerated. Section 2.5 discusses observational data on the convective overshoot into the atmosphere in connection with the convective dynamics and the radiative exchange.

Theoretical studies of the overshoot layer at the base of the convection zone that consider the nonlocal effects of convective eddies on the boundary layer (see Van Ballegooijen, 1982a; Skaley and Stix, 1991) indicate that this layer is shallow. In the solar case, it is only $\sim 10^4$ km thick, that is, no more than 20% of the local pressure scale height. The temperature gradient is found to be only slightly subadiabatic: $\nabla - \nabla_{ad} \simeq -10^{-6}$.

Basic assumptions in the ML approach are not confirmed by the observed solar granulation and numerical simulations of the structure of the top layers of convective envelopes. In Section 2.5 we discuss the resulting change in the picture of the patterns in stellar convective envelopes.

2.3   Radiative transfer and diagnostics

2.3.1   *Radiative transfer and atmospheric structure*

For the description of a radiation field, the *monochromatic specific intensity* $I_\nu(x, y, z, \theta, \phi, t)$ is the fundamental parameter. It is defined as the proportionality factor that quantifies the energy dE_ν flowing during dt through an area dA within a solid angle $d\omega$ about the direction $\mathbf{l}(\theta, \phi)$ within a frequency interval $d\nu$ around ν (see Fig. 2.2):

$$dE_\nu = I_\nu(x, y, z, \theta, \phi, t) \, dA \cos\theta \, d\omega \, d\nu \, dt. \tag{2.31}$$

I_ν is measured in units [erg s^{-1} cm^{-2} Hz^{-1} ster^{-1}]. Often I_ν is just called the monochromatic intensity, without the qualifier "specific."

The intensity $I_\nu(s)$ along a pencil of light remains constant unless there is emission or extinction along the path s:

$$dI_\nu(s) = I_\nu(s + ds) - I_\nu(s) = j_\nu(s) \, ds - \kappa_\nu(s)\rho(s)I_\nu(s) \, ds, \tag{2.32}$$

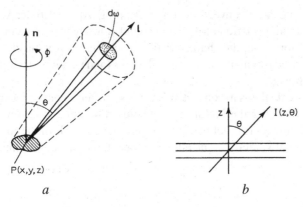

Fig. 2.2. Panel *a*: the definition of the specific intensity and related quantities. Panel *b*: the concept of a plane–parallel atmosphere.

where $j_\nu(s)$ [erg s^{-1} cm^{-3} Hz^{-1} ster^{-1}] is the monochromatic emission coefficient, $\kappa_\nu(s)$ [cm^2 g^{-1}] is the monochromatic extinction coefficient, and $\rho(s)$ [g cm^{-3}] is the mass density.

The *optical thickness* of a layer of geometrical thickness D follows from the definition of the dimensionless *monochromatic optical path length* $d\tilde{\tau}_\nu$ across a layer of thickness ds, $d\tilde{\tau}_\nu \equiv \kappa_\nu(s)\rho(s)\,ds$:

$$\tilde{\tau}_\nu(D) = \int_0^D \kappa_\nu(s)\rho(s)\,ds, \tag{2.33}$$

where the tilde indicates that the optical thickness is measured along the propagation direction of the beam.

If there is no emission within the layer, it transmits a fraction $\exp[-\tilde{\tau}_\nu(D)]$ of the incident intensity. The layer is called *optically thick* when $\tilde{\tau}_\nu(D) \gg 1$ and *optically thin* when $\tilde{\tau}_\nu(D) \ll 1$. The *local mean-free path* ℓ_ν for photons is

$$\ell_\nu(s) = \frac{1}{\kappa_\nu(s)\rho(s)}. \tag{2.34}$$

With the use of optical path length $d\tilde{\tau}_\nu$ and the introduction of the *source function* S_ν, defined as the emission per optical path length is

$$S_\nu(s) \equiv \frac{j_\nu(s)}{\kappa_\nu(s)\rho(s)}, \tag{2.35}$$

and the transfer equation (2.32) takes the form

$$\frac{dI_\nu}{d\tilde{\tau}_\nu} = S_\nu - I_\nu. \tag{2.36}$$

The source function has the same dimension as the specific intensity. If local thermodynamic equilibrium (LTE; discussed in the following paragraphs) applies, the source function is given by the local Planck function $B_\nu(T)$. In the case of pure scattering that is both coherent and isotropic, the source function equals the *angle-averaged monochromatic intensity*:

$$J_\nu \equiv \oint I_\nu \, d\omega/(4\pi). \tag{2.37}$$

The Planck function is

$$B_\nu(T) = \frac{2h\nu^3}{c^2} \frac{1}{e^{h\nu/kT} - 1}, \tag{2.38}$$

where h is the Planck constant. For $h\nu/kT \ll 1$, the Planck function simplifies to the Rayleigh–Jeans function,

$$B_\nu(T) \simeq \frac{2\nu^2 kT}{c^2}, \tag{2.39}$$

which describes the source function in the far-infrared and radio domain of the spectrum. In radio astronomy, intensity is usually expressed as (brightness) temperature, which is a linear measure for radiative emission.

The Stefan–Boltzmann function is the total LTE source function, integrated over the entire spectrum $B(T) \equiv \int_0^\infty B_\nu(T) \, d\nu$:

$$B(T) = \frac{\sigma}{\pi} T^4, \tag{2.40}$$

where σ is the Stefan–Boltzmann constant.

Consider a beam of intensity $I_\nu(0)$ passing through a homogeneous layer characterized by an optical thickness $\tilde{\tau}_\nu(D)$ and a source function S_ν. The emergent intensity follows from Eq. (2.36):

$$I_\nu(D) = I_\nu(0)e^{-\tilde{\tau}_\nu(D)} + \int_0^{\tilde{\tau}_\nu(D)} S_\nu e^{-[\tilde{\tau}_\nu(D) - t]} \, dt$$

$$= I_\nu(0)e^{-\tilde{\tau}_\nu(D)} + S_\nu \left[1 - e^{-\tilde{\tau}_\nu(D)} \right]. \tag{2.41}$$

In the optically thin case, the result is

$$I_\nu(D) = I_\nu(0) + [S_\nu - I_\nu(0)] \, \tilde{\tau}_\nu(D) : \tag{2.42}$$

the emergent intensity $I_\nu(D)$ is larger or smaller than the incident $I_\nu(0)$, depending on whether S_ν is larger or smaller than $I_\nu(0)$. The emission or extinction of an optically thin layer is proportional to its optical thickness.

When the layer is optically thick, Eq. (2.41) leads to

$$I_\nu(D) = S_\nu. \tag{2.43}$$

The mathematical formulation of radiative transfer in optically thick media such as stellar atmospheres is greatly simplified if the structure of the atmosphere is approximated by a plane–parallel horizontal stratification in which the thermodynamic quantities vary only with the height z perpendicular to the layers (Fig. 2.2). In this geometry, the *monochromatic optical depth* is defined by $d\tau_\nu(z) \equiv -\kappa_\nu(z)\rho(z)\,dz$, measured against the height z and the direction of the observed radiation, from the observer at infinity into the atmosphere:

$$\tau_\nu(z_0) \equiv \int_\infty^{z_0} \kappa_\nu(z)\rho(z)\,dz. \tag{2.44}$$

The symbol $\mu \equiv \cos\theta$ is then used for the perspectivity factor, so that the radiative transfer equation, Eq. (2.36) becomes in plane–parallel geometry

$$\mu\frac{dI_\nu(\tau_\nu, \mu)}{d\tau_\nu} = I_\nu(\tau_\nu, \mu) - S_\nu(\tau_\nu). \tag{2.45}$$

The radiative energy flow is described by the *monochromatic flux density*[*] $\mathcal{F}_\nu(\tau_\nu)$:

$$\mathcal{F}_\nu(\tau_\nu) \equiv \oint I_\nu(\tau_\nu, \mu)\,\mu\,d\omega = 2\pi \int_{-1}^{+1} I_\nu(\tau_\nu, \mu)\,\mu\,d\mu, \tag{2.46}$$

measured in [erg s^{-1} cm^{-2} Hz^{-1}]. Note that $\mathcal{F}$ is the net outwardly directed flux density: the downward radiation is counted as negative because of the perspectivity factor μ.[†]

The formal solution of Eq. (2.45) for the intensity that emerges from the atmosphere is

$$I_\nu(\tau_\nu = 0, \mu) = \int_0^\infty S_\nu(\tau_\nu)e^{-\tau_\nu/\mu}\,d\tau_\nu/\mu \tag{2.47}$$

and yields the *Eddington–Barbier approximation*,

$$I_\nu(\tau_\nu = 0, \mu) \approx S_\nu(\tau_\nu = \mu), \tag{2.48}$$

which is exact if S_ν is a linear function of τ_ν. It implies that solar limb darkening (seen in Fig. 1.1a) is a consequence of the decrease of the source function with height in the photosphere.

The corresponding approximation for the radiative flux density emerging from a stellar atmosphere is

$$\mathcal{F}_\nu(\tau_\nu = 0) \approx \pi S_\nu(\tau_\nu = 2/3), \tag{2.49}$$

which is also exact if the source function varies linearly with optical depth.

Spectroscopic studies of the center-to-limb variation of the intensity emerging from the solar disk and application of the Schwarzschild criterion (Section 2.2) have shown that stellar photospheres are close to radiative equilibrium, that is, the outward energy flow is transmitted almost exclusively as electromagnetic radiation. The condition for radiative equilibrium in an optically thick medium is that through the chain of extinction

---

[*] $\mathcal{F}$ is often called just "flux."

[†] In many texts on radiative transfer and stellar atmospheres, the net flux density $\mathcal{F}$ is written as πF. The so-called astrophysical flux density F corresponds to the mean intensity as averaged over the stellar disk, as seen from infinity.

and emission processes the total radiative flux density $\mathcal{F}$ is transmitted unchanged; hence at all depths z,

$$\mathcal{F}(z) \equiv \int_0^\infty \mathcal{F}_\nu(z)\, d\nu \equiv \sigma T_{\mathrm{eff}}^4 = L/(4\pi R^2), \qquad (2.50)$$

where L is the luminosity of the star, R is its radius, and T_{eff} is its *effective temperature*.

An alternative formulation of the condition of radiative equilibrium is

$$\int_0^\infty \kappa_\nu \rho (S_\nu - J_\nu)\, d\nu = 0, \qquad (2.51)$$

which is a continuity equation: a volume element emits as much radiative energy as it absorbs.

Equation (2.50) completely determines the variation of the source function with depth z, and hence the temperature stratification, but in a very implicit manner. The determination of an accurate model atmosphere satisfying the RE condition of Eq. (2.50) for a specific T_{eff} requires a sophisticated numerical technique. The main characteristics of such a model in RE can be illustrated by approximations, however.

The simplest approximation is that of a "gray atmosphere," that is, the assumption that the extinction is independent of frequency: $\kappa_\nu(z) \equiv \kappa(z)$. A single optical depth scale $\tau(z) = \int_\infty^z \kappa(z)\rho(z)\, dz$ then applies to all frequencies, and all monochromatic quantities Q_ν may be replaced by the corresponding integral over the spectrum: $Q \equiv \int_0^\infty Q_\nu d\nu$. The integrated source function $S(\tau)$ in a gray atmosphere in RE is found to be a nearly linear function of τ (the *Milne–Eddington approximation*):

$$S(\tau) \simeq \frac{3\mathcal{F}}{4\pi}(\tau + 2/3), \qquad (2.52)$$

where $\mathcal{F} = \sigma T_{\mathrm{eff}}^4$ is the depth-independent radiative flux density. In the case of LTE, the spectrum-integrated source function equals the Stefan–Boltzmann function, Eq. (2.40), so that in a gray atmosphere in RE and LTE the temperature stratification is given by

$$T(\tau) \simeq T_{\mathrm{eff}}\left(\frac{3}{4}\tau + \frac{1}{2}\right)^{1/4}. \qquad (2.53)$$

In actual stellar atmospheres, however, the extinction is far from gray; see Fig. 2.3. Numerical modeling shows that the nongrayness of the extinction steepens the temperature dependence on optical depth. In the deepest layers, both the temperature and the temperature gradient must be somewhat higher than in the gray case in order to push the radiative flux through the atmosphere, using the restricted spectral windows of low opacity. The outermost layers radiate more efficiently in the opaque parts of the spectrum than in the gray case, so that the temperature settles at a lower level. The detailed temperature stratification depends on the variation of the extinction through the spectrum, the wavelengths of the extinction peaks and valleys, and the measure of the departure from LTE in the outer part of the atmosphere.

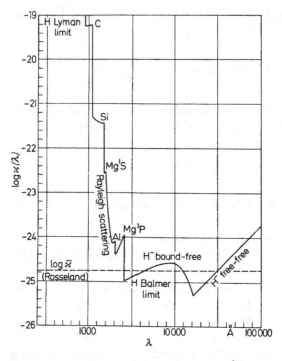

Fig. 2.3. Continuum extinction coefficient κ [cm^2] per heavy particle in the solar photosphere at the optical depth of $\tau_5 = 0.1$ at 5,000 Å, where $T = 5,040\,K$ and $\log p_e = 0.5$, plotted against wavelength λ. The spectral lines (not shown) add to the nongrayness in the form of numerous spikes on top of the continuous extinction curve. The dashed line specifies the Rosseland mean opacity (κ_R in the text) defined by Eq. (2.54), for continuum extinction only (from Böhm-Vitense, 1989b).

In the subsurface layers, below the depth $\tau_\nu = 1$ at the extinction minimum, radiative transfer can be handled simply by Eq. (2.25), using the *Rosseland mean opacity* κ_R:

$$\frac{1}{\kappa_R} \equiv \int_0^\infty \frac{1}{\kappa_\nu} \left(\frac{dB_\nu/dT}{dB/dT} \right) d\nu. \qquad (2.54)$$

Note that Eq. (2.25) describes radiative transfer as a process of photon diffusion, with the *photon mean-free path*

$$\ell_R \equiv (\kappa_R \rho)^{-1}, \qquad (2.55)$$

which is representative for the entire ensemble of photons. This diffusion approximation for radiative transport holds when the bulk of the photons are locked up within the local environment. In that case LTE applies as well, so that the source function S_ν equals the Planck function $B_\nu(T)$.

The Rosseland opacity is a harmonic mean, thus favoring the lower values of κ_ν; Fig. 2.3 shows a solar example. The weighting function between the parentheses in Eq. (2.54) selects the flux window fitting the local temperature; its shape resembles that of the Planck function for the same temperature, but its peak is shifted to higher frequency ν.

The Rosseland optical depth τ_R is based on the Rosseland opacity as the extinction coefficient:

$$\tau_R(z_0) \equiv \int_\infty^{z_0} \kappa_R \rho \, dz. \tag{2.56}$$

As another reference optical depth scale in solar studies, the monochromatic optical depth at $\lambda = 5,000\,\text{Å}$ is often used. In this book, it is indicated by τ_5. The level $\tau_R = 1$ corresponds approximately to $\tau_5 = 1$.

In Section 2.2, it is mentioned that the Rosseland opacity is particularly high for conditions in which the most abundant elements, hydrogen and helium, are partly ionized. This may be understood as follows. At temperatures less than $10^4\,\text{K}$, the elements H and He are not ionized and can only absorb efficiently in their ultraviolet ground-state continua (for H: the Lyman continuum). At such temperatures, these continua contribute very little to the mean opacity κ_R, because they fall outside the flux window, which is in the near-ultraviolet, visible, and infrared. The main contribution to κ_R then comes from the H^- ions, whose concentration is very low (typically less than $10^{-6}\,n_H$) because free electrons, provided by the low-ionization metals, are scarce. At higher temperatures, as long as H is only partly ionized, the energy levels above the ground state (which are close to the ionization limit) become populated, so that the resulting increased extinction in the Balmer, Paschen, and other bound-free continua boosts the mean opacity κ_R. At yet higher temperatures, the bound-free continua of He and He^+ add extinction. The mean opacity remains high until H is virtually completely ionized and He is doubly ionized. Then κ_R drops because the remaining extinctions (Thomson scattering by free electrons and free-free absorption) are very inefficient.

The approximation of radiative transport by photon diffusion with the Rosseland opacity as a pseudogray extinction coefficient holds to very high precision in the stellar interior, including the convective envelope. It holds approximately in the deepest part of the photosphere, for $\tau_R \gtrsim 1$. Hence, Eq. (2.53) is a reasonable approximation of the temperature stratification in the deep photosphere, provided that τ is replaced by τ_R. The location of the stellar surface, defined as the layer where T equals the effective temperature, is then at

$$T(\tau_R \simeq 2/3) = T_{\text{eff}}; \tag{2.57}$$

see Eq. (2.49).

Higher up, for optical depths $\tau_R < 1$, this approximation of radiative transport by diffusion with some mean opacity breaks down. Numerical modeling of radiative equilibrium, in which the strong variation of the extinction coefficient with wavelength is taken into account, fairly closely reproduces the empirical thermal stratification throughout the solar photosphere; small departures from this stratification caused by overshooting convection are discussed in Section 2.5. In other words, the thermal structure of the solar photosphere is largely controlled by the radiative flux emanating from the interior.

Radiative-equilibrium models for stellar atmospheres predict that the temperature decreases steadily with height until it flattens out where the medium becomes optically thin to the bulk of the passing radiation. Yet spectra of the solar chromosphere and corona indicate temperatures above and far above the temperature $T_{\text{min}} \approx 4,200\,\text{K}$ found at the top of the photosphere. Hence some nonthermal energy flux must heat the outer

Table 2.1. *Mean, integrated radiative losses from various domains in the solar atmosphere*

Domain	F (erg cm^{-2} s^{-1})	Ref.[a]
Photosphere	6.4×10^{10}	1
Chromosphere	$2\text{--}6 \times 10^6$	1
Balmer series	5×10^5	1
H$^-$	4×10^5	1
Ly α	3×10^5	1
metal lines (Mg II, Ca II)	34×10^5	1
Transition region	$4\text{--}6 \times 10^5$	1
Corona, quiet Sun	$\sim 6 \times 10^5$	2
Coronal hole	$\sim 10^4$	2

[a] References: 1, Schatzman and Praderie (1993); 2, Section 8.6.

solar atmosphere. The radiative losses from its various parts (Table 2.1) indicate that the nonthermal heating fluxes required to balance the losses are many orders of magnitude smaller than the flux of electromagnetic radiation leaving the photosphere. The nature of these nonthermal heating fluxes is discussed at some length in subsequent sections and chapters; here, we confine ourselves to the observation that undoubtedly some sort of transmission of kinetic energy contained in the subsurface convection is involved.

Even though electromagnetic radiation is not involved in the heating and thus the creation of coronal structure, radiative losses are essential as one of the mechanisms for cooling the corona and also for coronal diagnostics. The corona has temperatures $T \gtrsim 1 \times 10^6$ K in order to make the emission coefficient per unit volume j_ν equal to the local heating rate, which is small per unit volume but extremely high per particle.

In the extreme ultraviolet and soft X-rays, the coronal emission consists of the superposition of free-bound continua and spectral lines from highly ionized elements (mainly Fe, Mg, and Si). The radiative transfer is simple for the many transitions in which the corona is optically thin, although the conditions are very far from LTE. The excitations and ionizations are induced exclusively by collisions, predominantly by electrons. The photospheric radiation field at $T \simeq 6{,}000$ K is much too weak in the extreme ultraviolet and soft X-rays to contribute to the excitations and ionizations. Each collisional excitation or ionization is immediately followed by a radiative de-excitation or recombination. If the corona is optically thin for the newly created photon, then that photon escapes (or is absorbed by the photosphere). Hence, in any spectral line or continuum, the number of emitted photons per unit volume is proportional to $n_e n_y$, where n_e is the electron density and n_y is the number density of the emitting ion. The ion density n_y is connected with the proton density n_p through the abundance of the element relative to hydrogen and the temperature-dependent population fraction of the element in the appropriate ionization stage.

The total power $P_{i,j}$ emitted per unit volume in a spectral line $\lambda_{i,j}$ of an element X may be written as

$$P_{i,j} \equiv A_X G_X(T, \lambda_{i,j}) \frac{hc}{\lambda_{i,j}} n_e n_H, \tag{2.58}$$

where A_X is the abundance of element X relative to H, and temperature-dependent factors are assembled in the contribution function $G_X(T, \lambda_{i,j})$. A similar expression applies to the power emitted per unit wavelength in a continuum.

Coronal spectral emissivities $\mathcal{P}(\lambda, T)$ are defined such that $\mathcal{P}(\lambda, T)n_e n_H$ is the power emitted per unit volume per unit wavelength at λ. This emissivity is obtained by adding the contributions of all relevant spectral lines and continua; see Fig. 3.11 in Golub and Pasachoff (1997) for such theoretical spectra.

The total *radiative loss function* $\mathcal{P}(T) \equiv \int_0^\infty \mathcal{P}(\lambda, T) \, d\lambda$ determines the total *radiative loss rate*:

$$E_{\mathrm{rad}} \equiv \mathcal{P}(T)n_e n_H. \tag{2.59}$$

Since the proton density approximately equals the electron density, in many applications of Eqs. (2.58) and (2.59) the factor $n_e n_H$ is replaced by n_e^2.

Figure 2.4 shows $\mathcal{P}(T)$ for one spectral code. The bumps in the curve are caused by contributions from the strongest spectral lines emitted by a small number of elements. For $T \gtrsim 2 \times 10^7$ K the emission in spectral lines drops below the level of the continuum. Recall that the function $\mathcal{P}(T)$ is valid only for optically thin plasmas! For a more detailed discussion of radiation from hot plasmas, see Section 3.3 in Golub and Pasachoff (1997) and references given there.

At the other end of the electromagnetic spectrum, radio waves present another important diagnostic on coronal physics. Throughout the corona, the propagation of radio

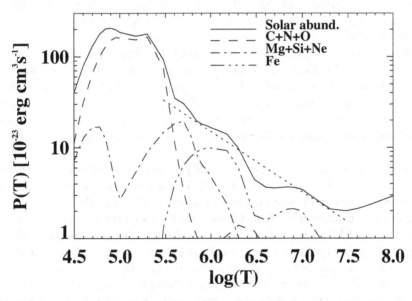

Fig. 2.4. The radiative loss function $\mathcal{P}(T)$ for an optically thin plasma [Eq. (2.59)], for the passband from 0.1 keV up to 100 keV, using the spectral code by Mewe, Kaastra, and colleagues (private communication). The curve is computed for solar photospheric abundances; the contributions by some of the major elements are also shown. The dotted straight line is an approximation to the radiative loss function between 0.3 MK and 30 MK: $\mathcal{P}(T) \approx 1.5 \times 10^{-18} T^{-2/3}$ erg cm^{3}s^{-1}.

waves is significantly affected by refraction for all frequencies smaller than $\sim 300\,\text{MHz}$ ($\lambda \gtrsim 1\,\text{m}$). The index of refraction n at frequency ν is given by

$$n^2 = 1 - \left(\frac{\nu_p}{\nu}\right)^2, \tag{2.60}$$

where

$$\nu_p = \left(\frac{n_e e^2}{\pi m_e}\right)^{1/2} = 9 \times 10^3 \, n_e^{1/2} \ (\text{Hz}) \tag{2.61}$$

is the *plasma frequency* with which electrons can oscillate about the relatively stationary ions. Consequently, only waves with frequency $\nu > \nu_p$ can propagate – the plasma frequency determines a low-frequency cutoff. With the electron density n_e, the plasma frequency decreases with height. The consequence is that meter waves, with $\nu \lesssim 300\,\text{MHz}$, can escape from the corona but not from deeper layers, decimeter waves ($\nu \lesssim 3\,\text{GHz}$) can reach us from the transition region, and microwaves ($\nu \lesssim 30\,\text{GHz}$) can leave the chromosphere.

Ray paths generally are curved because the refractive index varies throughout the corona. A ray is bent away from a higher-density region when its frequency is close to the local plasma frequency. Hence, in a hypothetically spherically symmetric, hydrostatic corona, an inward propagating ray is "reflected" upward, with the turning point somewhat above the level where its frequency equals the local plasma frequency. In the actual corona, density fluctuations affect the ray path. More complications in the wave propagation are caused by the magnetic field. For more comprehensive introductions to solar radio emission, we refer to Sections 3.4, 7.3, and 8.5 in Zirin (1988); Dulk (1985) provides a broad introduction to solar and stellar radio emission.

The emergent intensity equals the integral along the ray path $\int B_\nu \exp(-\tilde{\tau}_\nu)\,d\tilde{\tau}_\nu$, where $d\tilde{\tau}_\nu = \kappa_\nu \rho \, ds$ is an optical path length. The optical depth of the corona reaches unity in the meter range; it increases with wavelength because the extinction coefficient κ_ν is determined by free-free absorption, which is proportional to ν^{-2}. Most of the observed emission originates from just above the level where the frequency corresponds to the local plasma frequency, because there the free-free opacity, proportional to n_e^2 is highest.

Radiative transfer in the chromosphere and photosphere presents a complicated problem because the optical thickness of the order of unity in some continuum windows and many spectral lines necessitate a detailed treatment, with proper allowance for the strong variation of the extinction coefficient throughout the spectrum. In detailed spectroscopic diagnostics, the LTE approximation of the Saha–Boltzmann population partitioning must be replaced by the equations for statistical equilibrium that relate the populations of the stages of ionization and of the discrete energy levels in the atoms and ions to the local temperature and to the radiation fields of nonlocal origin that may influence the transition of interest. If the latter is a bound-bound transition, its spectral-line source function is given by

$$S_\nu = \frac{b_u}{b_l} B_\nu \left(1 - \frac{b_l}{b_u} e^{-h\nu/kT}\right), \tag{2.62}$$

where b_l and b_u measure the population departure $b \equiv n/n_{\mathrm{LTE}}$ from LTE for the lower level l and the upper level u, respectively. The statistical equilibrium equations specify the interactions between the radiation field and the local plasma by including all pertinent collisional and radiative processes that contribute to populating and depopulating the energy levels of interest. The general non-LTE problem is complex because conditions in distant source regions interlock with local conditions in the radiative transitions. Hence, the equations of statistical equilibrium must be solved simultaneously for all depths in the atmosphere, together with the equation of transfer in all the relevant radiative transitions and at all angles. For a practical introduction to the more advanced theory of radiative transfer and spectroscopic diagnostics, we refer to R. J. Rutten's lecture notes at URL http://www.astro.uu.nl/~rutten.

The interaction between the local plasma and the radiation field is complicated where the radiative exchange with the environment is a major factor in determining the local thermal structure, namely in the photosphere and chromosphere. These are precisely the layers that provide the spectroscopic diagnostics that measure the heat flows, velocity fields, and magnetic fields that are enforced upon the atmosphere from the interior.

In the Sun and Sunlike main-sequence stars, the vertical extent of the photosphere is very small in comparison with the stellar radius. The sharp solar limb is readily explained by the relative smallness of the pressure scale height $H_p \simeq 150\,\mathrm{km}$ in the photosphere. The "bottom" of the photosphere is even thinner because the continuous opacity increases very steeply in the deep photosphere. The H^- opacity increases proportionally with the electron pressure p_e, which increases more steeply with depth than the gas pressure because the degree of ionization increases very steeply with depth. In addition, with the increase of the temperature, the contribution from the bound-free continua of atomic hydrogen to the continuous opacity increases with depth.

For a specific frequency ν in the spectrum, the mean location of the layer from which the intensity or the flux density I_ν originates can be estimated by Eq. (2.48) or (2.49). In the interpretation of a feature in a line profile, such as caused by Doppler shifts, it is not the origin of the residual intensity I_ν or flux $\mathcal{F}_\nu$ that counts, but more likely the region that contributes most to the intensity depression $(I_c - I_\nu)/I_c$, where I_ν is the intensity at some position ν in the line profile, and I_c is the interpolated continuum intensity at the position of the spectral line. In general, the sensitivity of a line to a quantity requires the evaluation of a response function including that quantity. For further discussions of spectral-line response functions, see Magain (1986) for solar intensity spectra and Achmad *et al.* (1991) for stellar flux spectra.

In general, it is not practical to attempt to derive the spatial dependence of some parameter through direct inversion of the spectral profile because of the intricacy of the processes involved in the formation of the spectral features under study. In order to select a model that best fits the observational data, one must usually rely on comparing the observed spectral features with the features that are numerically synthesized, using sufficiently detailed models. Such fits are rarely unique and often questionable.

2.3.2 *Magnetic field measurements: possibilities and limitations*

We now briefly discuss techniques for measuring magnetic parameters using the Zeeman splitting of spectral lines. In the treatment of radiative transfer in a spectral line formed in a magnetized plasma, the monochromatic intensity I_λ describing an unpolarized radiation field must be replaced by four parameters characterizing a polarized radiation field; for this, the four Stokes parameters $\{I_\lambda, Q_\lambda, U_\lambda, V_\lambda\}$ are frequently used. Below, we suppress the subscripts λ or ν in the indications for the monochromatic Stokes parameters. The parameter I is the total intensity, as defined in Eq. (2.31), the parameter V [Eq. (2.64)] specifies the magnitude of the circular polarization, and the parameters Q and U together specify the magnitude and the azimuth of the linear polarization.

The single transfer equation for I must be replaced by a set of four differential equations that relate the Stokes parameters. The corresponding matrix equation is

$$\frac{d\mathbf{I}}{ds} = -\mathcal{K}(\mathbf{I} - \mathbf{S}), \qquad (2.63)$$

where $\mathbf{I} \equiv (I, Q, U, V)^{\mathrm{T}}$ is the Stokes vector and $\mathcal{K}$ is the propagation matrix containing the extinction coefficients for the various Stokes parameters and the coefficients for the magneto-optical effects producing cross talk between the Stokes parameters. The vector $\mathbf{S} \equiv (S_I, S_Q, S_U, S_V)^{\mathrm{T}}$ stands for the source function vector in the Stokes parameters and s measures the geometrical length along the chosen line of sight. The superscript T denotes the transpose of the matrix. We do not discuss the theory here, but we refer to Rees (1987) for a fine introduction to the treatment of polarized radiation and the Zeeman effect. Other references are Semel (1985), Stenflo (1985), and Solanki (1993).

In this brief introduction we consider possibilities and limitations of some standard procedures of magnetic-field measurements that take into account only the Stokes parameters V and I. We consider a simple two-component model: magnetic flux tubes of strength B with an inclination angle γ with respect to the line of sight, which fill a fraction f of the photospheric surface, with the rest of the atmosphere being nonmagnetic. Except for the magnetic field, the other plasma conditions inside and outside the magnetic structures are considered to be identical, as far as the formation of the spectral line is concerned. Only the spectra $I_{\mathrm{L}}(\lambda)$ and $I_{\mathrm{R}}(\lambda)$ in the left-handed and the right-handed directions of circular polarization are used. From these spectra, the total intensity $I(\lambda)$ and Stokes parameter $V(\lambda)$ follow:

$$I(\lambda) = I_{\mathrm{L}}(\lambda) + I_{\mathrm{R}}(\lambda) \quad \text{and} \quad V(\lambda) = I_{\mathrm{L}}(\lambda) - I_{\mathrm{R}}(\lambda). \qquad (2.64)$$

Consider a spectral line of the simplest magnetic splitting (Zeeman triplet), yielding one central, undisplaced π component and two σ components equidistant from the π component (Fig. 2.5). The wavelength displacement of the σ components is proportional to $Bg\lambda^2$, where g is the Landé factor, which depends on the atomic transition that produces the line.

If the Zeeman splitting is so large that the π and σ components do not overlap significantly (Fig. 2.5a), then the wavelength separation $\Delta\lambda_V$ between the extrema in the V profile simply equals the separation between the σ components, which is proportional to $Bg\lambda^2$. In this case, the amplitude A_V of the V profile scales with $f \cos\gamma$ and with a factor depending on the formation of the spectral line; it does not depend on B. In

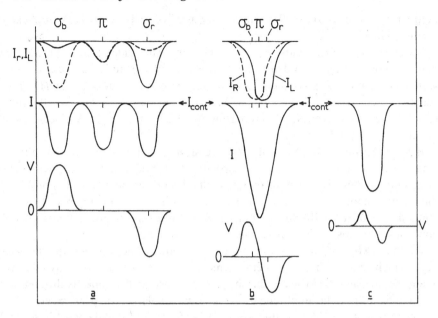

Fig. 2.5. Sketches of spectral-line profiles in the intensities I_L and I_R in left-handed (solid curves) and right-handed (dashed curves) polarization, in total intensity I, and in the Stokes parameter V for simple Zeeman splitting in a magnetic field that is inclined with respect to the line of sight. The filling factor is $f = 1$. Panels: a, large magnetic splitting; b, overlapping π and σ components; c, extremely small magnetic splitting.

principle, the inclination γ of the field with respect to the line of sight can be determined from the comparison of I and V profiles, but this deduction is model dependent because it depends on details of the line-formation process.

If the σ components partly overlap (Fig. 2.5b), the separation of the extrema in the V profile is not a linear measure for $Bg\lambda^2$. The deduction of B, γ, and f requires a detailed model for the line formation.

If the σ components almost completely overlap (Fig. 2.5c), then the expression for the Stokes parameter V in Eq. (2.64) can be approximated by

$$V(\lambda) \propto Bg\lambda^2 \cdot f \cos\gamma \cdot \frac{\partial I(\lambda)}{\partial \lambda}, \tag{2.65}$$

where $I(\lambda)$ is the line profile for $B = 0$. In this case, the wavelength separation $\Delta\lambda_V$ between the extrema in the V profile contains no independent information on the field strength B, but simply measures the distance between the inflection points in the $I(\lambda)$ profile. All that can be derived from a V profile in the case of small Zeeman splitting is a (model-dependent) measure of the magnetic flux density along the line of sight. In this book, we use the symbol φ for this quantity $Bf \cos\gamma$.

Magnetically sensitive spectral lines in the infrared present a better opportunity to determine intrinsic field strengths B than do lines in the visible, because the Zeeman splitting increases proportional to λ^2, while the line width, determined by Doppler broadening, increases linearly with λ. In the case of the Sun, in the spectral region around $\lambda = 1.6\,\mu m$

field strengths down to $\sim$400 G can be measured (see Harvey, 1977); applications are discussed in Section 4.6.

For strongly inclined fields, with $\gamma \to \pi/2$, the amplitude of the V profile is smaller than the noise in the recorded spectra. However, if the magnetic flux density Bf is sufficiently large, the presence of such a field is apparent in the total $I(\lambda)$ profile by "ears" caused by the σ components. Moreover, in case of a sufficiently strong line, its total strength increases with increasing inclination $\gamma \to \pi/2$. Applications of these effects are given in Brants (1985b).

If the polarities in the field of view are strongly mixed, the net V profile may be smaller than the noise in the recorded spectra. If such fields are strong, with sufficiently large filling factors, however, the magnetic field is detectable in the $I(\lambda)$ profile through the (nearly unpolarized) σ components. This fact is used in the improved Unno (1956) method developed by Robinson *et al.* (1980) to measure magnetic fields in cool stars; applications of this method are discussed in Section 9.2.

A simple two-component model, containing static magnetic and nonmagnetic components which do not radiatively interact, predicts V profiles that are strictly antisymmetric about the line center. However, the shape of many solar V profiles is different. In most Fe I and Fe II lines, the amplitude of the blue-wing polarization peak exceeds that of the red-wing peak by $\sim$10% (Stenflo *et al.*, 1984; Sánchez Almeida and Lites, 1992). For comprehensive reviews of the observations and the interpretation of these departures of antisymmetry we refer to Solanki (1993, 1997).

In many practical applications, the degree of circular polarization is measured in the two flanks of a suitable line, which through a calibration procedure is then converted into the mean Doppler shift and the mean flux density in the line of sight $\varphi = Bf \cos \gamma$ over the resolution element. A scan across the solar disk yields a *longitudinal magnetogram*, such as shown in Figs. 1.1*e* and 1.2*a*.

For sufficiently strong fields, the intrinsic field strength B can be estimated separately from the factor $f \cos \gamma$ by using two or more spectral lines, or at least two window pairs in one spectral line. If one such line, or window pair, is chosen such that the magnetograph signal does *not* depend linearly on B, then the ratio of the two signals is a measure for B. This line-ratio method was introduced by Howard and Stenflo (1972); see Stenflo (1985) for a review. Applications are discussed in Section 4.2.3.

Filter magnetographs, equipped with a narrow-band birefringent filter as the spectroscopic analyzer and with a two-dimensional detector, trade spectral resolution for spatial extent. Filter magnetographs are used for detecting low magnetic flux densities and for following rapidly evolving features in the magnetic field – in these applications, pioneering work has been accomplished by means of the video magnetograph at Big Bear Solar Observatory. The magnetic calibration of filter magnetographs is less accurate than that of scanning spectrographic magnetographs.

The determination of the *vector* magnetic field **B** is much more difficult than the measurement of the line-of-sight component $B \cos \gamma$, because of the high polarimetric precision required in the Stokes parameters Q, U, and V: better than $10^{-4}I$ to achieve an accuracy of 10 G and 6° (Harvey, 1985). Moreover, the interpretation of the linear polarization in terms of the transverse component of **B** depends sensitively on the strength B and on the filling factor f. We do not discuss the interpretation of the full Stokes vector $(I, Q, U, V)^{\mathrm{T}}$ here but we refer to Landi Degl'Innocenti (1976) and to Stenflo (1985);

for developments in vector field measurements, see November (1991). For an application of vector spectropolarimetry, see Skumanich *et al.* (1994).

The determination of the part of the atmosphere that contributes to the magnetic signal is more involved than the determination of the contribution functions in the case of spectral line formation outside magnetic fields (Section 2.3.1) – the result depends sensitively on the magnetic fine structure in the atmosphere. Consequently, quantitative determinations of height-dependent effects in the magnetic structure require caution. We refer to Van Ballegooijen (1985) for a convenient method to treat line formation in a magnetic field including the determination of contribution functions for the individual Stokes parameters.

2.3.3 Stellar brightness, color, and size

This section is a synopsis of stellar terminology and two-dimensional classification. For general introductions to stellar terminology, data, and classification, we refer to Chapter 4 of Unsöld and Baschek (1991), to Böhm-Vitense (1989a), and to Chapters 1–3 in Böhm-Vitense (1989b).

Traditionally, stellar brightnesses are expressed in logarithmic scales of stellar *magnitudes* m_Q, which decrease with increasing stellar brightness,

$$m_Q \equiv -2.5 \log \left(\frac{R^2}{d^2} \int_0^\infty \mathcal{F}_\lambda \, Q_\lambda \, d\lambda \right) + C_\lambda, \tag{2.66}$$

where Q refers to a specific photometric filter combination (which includes the Earth atmosphere) with transmission function Q_λ, R is the stellar radius, d is the stellar distance, and $\mathcal{F}_\lambda$ is the radiative flux density at the stellar surface. The normalization constant C_λ is settled by a convention.

The *absolute magnitude* M_Q is the magnitude of the star if it were placed at a distance of 10 pc; hence:

$$m_Q - M_Q = 5 \log \frac{d}{10}, \tag{2.67}$$

where d is the distance in parsecs (1 pc equals 3.09×10^{18} cm, or 3.26 light years).

Bolometric magnitudes m_{bol} are set by the total stellar luminosity L, that is, taking $Q_\lambda \equiv 1$. For the absolute bolometric magnitude we have

$$M_{bol} = -2.5 \log \frac{L}{L_\odot} + 4.72, \tag{2.68}$$

where $L_\odot$ is the solar luminosity.

The *UBV photometric system* is often used for stellar brightnesses and colors, with U for ultraviolet, B for blue, and V for visual. The corresponding magnitudes are usually indicated as $U = m_U$, $B = m_B$ and $V = m_V$, and the absolute magnitudes are indicated by M_U, M_B and M_V.

The *color index B–V* describes the color of the star. It is a measure of the spectral energy distribution, and hence of the effective temperature T_{eff}; B–V also depends, though to a lesser extent, on gravity g and chemical abundances. Note that stellar magnitudes and color indices are affected by interstellar extinction and reddening, respectively.

The *bolometric correction, BC*, is defined by

$$BC \equiv m_V - m_{bol}. \tag{2.69}$$

By visual inspection of their low-resolution spectra, the majority of stars are readily arranged in a one-dimensional spectral sequence, which consists of the *spectral types* (Sp): O–B–A–F–G–K–M (with a decimal subdivision). In this order, these spectral types form a temperature sequence, the O-type stars being the hottest and the M-type stars being the coolest. As a historical artifact, the hot stars are often described as "early," and the cool stars as "late"; spectral type A is "earlier" than spectral type G. Somewhat more appropriately, the stars are sometimes indicated by their color impression: A- and early F-types are called "white," G-types "yellow," K-types "orange," and M-types "red."

In addition to temperature characteristics, stellar spectra also contain more subtle information bearing on the pressure in the atmosphere, and thus on the stellar radius and luminosity. Such characteristics are used in the Morgan–Keenan (MK) two-dimensional spectral classification, which labels a star with a spectral type O through M and a *luminosity class* (LC) which is indicated by Roman numerals I–VI and the following terms: I, supergiant; II, bright giant; III, giant; IV, subgiant; V, dwarf or main-sequence star; VI, subdwarf. For instance, the MK classification of the Sun is G2 V.

Occasionally, a one-letter code is added to the spectral type. The appendix p, for peculiar, indicates that the spectrum differs in some aspects from the standard spectrum of the corresponding type, usually because the stellar chemical composition differs from that of the majority of the stars in the solar neighborhood. The addition e, for emission, refers to emission seen in some of the strongest lines in the spectrum.

For some stars that have not yet been classified in the MK system, an earlier classification is indicated by a letter preceding the spectral type, with d for dwarf, g for giant, and sg for supergiant. For instance, dM4e indicates a dwarf M4-type star, with emission lines in its spectrum.

2.3.4   *Radiative diagnostics of outer atmospheres*

The nonradiative heating of the outer atmospheres of the Sun and other cool stars leads to emission in a multitude of spectral lines and continua. Table 2.2 summarizes the diagnostics that are used in this book. Throughout this book, we define chromosphere, transition-region, and corona in terms of the associated temperatures as follows:

$$\text{chromosphere}: \qquad T \lesssim 20,000 \, \text{K},$$
$$\text{transition region}: \quad 50,000 \, \text{K} \lesssim T \lesssim 5 \times 10^5 \, \text{K},$$
$$\text{corona}: \qquad T \gtrsim 10^6 \, \text{K}$$

(see 14 for the radiative properties of these domains).

The Ca II H and K lines are the strongest and broadest lines in the visible solar spectrum. The strength of the lines and the double ionization of calcium limits the lines' formation to the chromosphere. In solar observations, the line cores of this doublet locally display an emission reversal, with a central absorption minimum; the naming convention for these features is shown in Fig. 2.6. Clearly, the source function S_ν [Eq. (2.35)] corresponding to such a line shape cannot be a monotonically decreasing function of height in the atmosphere, indicative of (intermittent) deposition of nonradiative energy.

Table 2.2. *Frequently used radiative diagnostics of atmospheric activity, together with the characteristic temperatures of maximum contribution*[a]

Diagnostic	λ (Å)	Characteristic Temperature (K)	Formed in
TiO	*bands*	<4,000	Spots
UV cont.	1,600	4,500	Temp. minimum
CO	*bands*	<4,400	Temp. minimum
Ca II H and K	$\begin{cases} 3,967 \\ 3,933 \end{cases}$	$(4\text{–}7) \times 10^3$	Chromosphere
H α (core)	6,563	$(\sim 5\text{–}10) \times 10^3$	Chromosphere
Mg II h and k	$\begin{cases} 2,803 \\ 2,796 \end{cases}$	$(4\text{–}20) \times 10^3$	Chromosphere
Ly α	1,215	...	High Chrom.
C II	1,335	1.5×10^4	low TR
Si II	$\begin{cases} 1,808 \\ 1,871 \end{cases}$	2×10^4	low TR
C III	977	4×10^4	TR
C IV	$\begin{cases} 1,548 \\ 1,551 \end{cases}$	10^5	TR
Si IV	$\begin{cases} 1,394 \\ 1,403 \end{cases}$	10^5	TR
O IV	554	2×10^5	TR
O VI	1,032	3×10^5	TR
Mg X	625	1.5×10^6	Corona
X-rays	10–200	$>10^6$	Corona

[a] Some diagnostics, particularly the hydrogen Lyman α line and the soft X-ray emission, are formed over a broad range in temperature. The temperature domains (see Vernazza *et al.*, 1981, for the chromospheric diagnostics) are crude indications only. "TR" indicates the transition region.

The detailed formation of the H and K lines is complicated, and its modeling requires the inclusion of resonant scattering, that is of partial redistribution (e.g., Uitenbroek, 1989), which couples different wavelengths within the line profile in the radiative transfer. Moreover, these resonance lines are linked to the Ca II infrared triplet (at 8,498, 8,542, and 8,662 Å), with which they share a common upper level. Their complex formation notwithstanding, the strong doublet of the H and K lines is one of the standard measures of solar and stellar magnetic activity, because the lines are strong and easily accessible to ground-based observatories.

The largest database of stellar Ca II H+K measurements is derived from the observations obtained at the Mount Wilson observatory. The HK photometer measures the signal in two narrow passbands centered on the emission cores relative to the signal in two 20-Å "continuum" bands on either side of the doublet (Vaughan *et al.*, 1978b). For dwarf stars the central passbands are generally triangular with a FWHM of 1 Å (Fig. 2.6), whereas for giant stars broader, flat-topped profiles are often used to accommodate their broader line cores (Section 9.3). The ratio of these signals is converted to an absolute flux

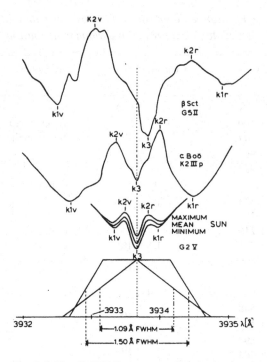

Fig. 2.6. The Ca II K line-core profiles for luminosity class II (bright giant) and III (giant) stars, and for the Sun during maximum and minimum activity. The passbands used in the Mount Wilson Ca II HK photometer are shown near the bottom. The central reversal in the line core is generally referred to as the K_3 minimum. The two peaks on the red and blue sides are the K_{2R} and K_{2V} peaks, respectively; the locations of lowest intensity on the edges of the emission reversal are referred to as the K_{1R} and K_{1V} minima. A similar nomenclature is used for other resonance lines of the same shape, most notably the Ca II H and Mg II h and k lines (figure from Rutten, 1984a).

density at the stellar surface following a procedure developed by Middelkoop (1982b). The Ca II H+K flux density is frequently given in relative units, also referred to as arbitrary units; see Fig. 9.4. The transformation constant from relative to absolute units within the triangular 1-Å Mt. Wilson bandpass is 1.3×10^6 erg cm^{-2} s^{-1} (Rutten and Uitenbroek, 1991; see Schrijver *et al.*, 1989a, for a discussion of the effects of the instrumental profile).

The chromospheric Mg II resonance lines in the ultraviolet, named the h and k lines in analogy to their Ca II counterparts, can only be observed from space. The cosmic abundance of magnesium is approximately 14 times larger than that of calcium, so the Mg II lines reach an optical depth that is at least one order of magnitude larger at a given geometric depth. The greater strength of the Mg II h and k lines makes them more sensitive to weak chromospheres, but also to the circumstellar and interstellar matter through which this radiation reaches us.

The hydrogen Balmer α line is frequently used in studies of solar magnetic activity. The interpretation of the total line emission is complicated because the line is formed over a geometric range spanning the photosphere in the line wing up to the middle and high chromosphere in the core. The line formation is particularly complicated because it is a high-excitation line. The use of the Hα line for both solar and stellar work is also complicated by its ambiguous response to magnetic activity: both absorption and emission indicate a heated atmosphere (see Section 11.5). These complications have limited the use of the Hα line largely to geometrical studies of the solar chromosphere (Chapter 1, Sections 8.1 and 8.2), where it is of tremendous value because scanning through the line profile with a filtergraph with a narrow transmission band reveals a wealth of magnetically dominated structure throughout the chromosphere (Fig. 1.3).

The hydrogen Lyman α line is formed over a particularly broad range of temperatures, from the low chromosphere several ångstrom away from line center, through the steep onset into the transition region within the peak itself, into the transition region at several tens of thousands of degrees in the line center. The observed stellar Lyα profiles are generally strongly affected both by solar Lyα light that scatters in the geocorona and in the region at the limits of the solar system where the solar wind interacts with the interstellar matter, as well as by interstellar extinction. The Lyα line shows no clear limb darkening for quiet or active regions (Withbroe and Mariska, 1975, 1976; Schrijver, 1988).

The C II, C III, and C IV lines measure the very upper part of the chromosphere and the lower transition region between chromosphere and corona. They are essentially optically thin (with effects of optical thickness only affecting the line formation beyond a distance of more than 80% from disk center in dense active-region environments; Schrijver, 1988).

The O IV and O VI lines suffer from complications: the O IV line lies in the Lyman continuum and may be weakened by continuum extinction, whereas the O VI line has a high-temperature tail in the contribution function, which places part of its formation in the corona (e.g., Jordan, 1969). These lines represent the very high transition region. The C and O ions are the main contributors to the radiative loss curve at transition-region temperatures (Fig. 2.4; e.g., Cook *et al.*, 1989).

The most direct indicator of a corona is its soft X-ray emission. This emission is virtually optically thin, so that all sources contribute to the total observed emission. The large emission scale height of the corona lets coronal features over both active and quiet regions that lie behind the limb contribute to the total X-ray emission, which in some stars may amount to a considerable contribution. The generally negligible optical thickness of the corona results in a pronounced brightness increase across the edge of the solar disk on X-ray images; the increase of the total path length above the limb causes a near-doubling of the brightness (see Section 8.8). The soft X-ray passband is dominated by emission lines, predominantly from iron, and to a lesser extent from other elements such as O, Si, Ne, and Mg (Fig. 2.4).

The calibration of soft X-ray fluxes observed from different stars and with different instruments is complicated by the sensitivity of this transformation to interstellar extinction and to the details of the temperature structure and chemical composition of the corona. Within the soft X-ray passband of 0.2 keV to 2 keV (6 Å to 60 Å), often explored by stellar instruments, interstellar hydrogen column densities $N_H \lesssim 10^{19}$ cm^{-2} have little impact, regardless of the coronal temperature. For larger values of N_H, the interstellar extinction increases rapidly. The extinction is stronger at longer wavelengths, and thus affects spectra from cooler plasma stronger than from warmer plasma. This increases the apparent color temperature – or hardness ratio, as it is commonly called – of the source. For many of the measurable main-sequence stars, on one hand, interstellar extinction is negligible; for the generally more distant giant stars, on the other hand, corrections often have to be applied.

In addition to the diagnostics listed in Table 2.2, there are others, such as the Ca II infrared lines, microwave and radio passbands, and so on, but these have received less attention in the comparison of solar to stellar activity.

In order to compare the emissions from different stars, all stellar flux densities observed at Earth have to be converted to some unit that is independent of the distance of the star from the Earth. Three measures are most commonly used: the stellar luminosity L, the flux density $\mathcal{F}$ at the stellar surface, and the ratio $\mathcal{R}$ of the luminosity (or flux density) in the activity diagnostic to the corresponding bolometric value. In this book we use the surface flux density $\mathcal{F}$ most frequently; the rationale is discussed in Section 11.2. The transformation of flux densities f at the Earth to flux densities at the stellar surface is given by

$$\log\left(\frac{\mathcal{F}}{f}\right) = 2\,\log\left(\frac{d}{R_*}\right) = 0.33 + \frac{V + BC}{2.5} + 4\,\log\left(T_{\text{eff}}\right), \qquad (2.70)$$

where d is the distance to the star, R_* is the stellar radius, BC is the bolometric correction, V is the apparent visual magnitude, and T_{eff} the stellar effective temperature. In most diagrams and relationships discussed in this book, the last expression is used, but in some cases the transformation is made using the ratio of distance to radius (note that accurate distances from the *Hipparcos* mission became available after most studies referred to in this book were published). For that last expression, the size and distance of the star need not be known; for this calibration, the transformation from f to $\mathcal{F}$ relies on the use of the bolometric correction and the effective temperature, which are both a function of stellar color indices such as $B-V$ and the stellar surface gravity (see Oranje *et al.*, 1982b).

In the case of the Sun, intensities I are measured. To compare these to stellar data, one must convert them to flux densities. To do this accurately, one should know the angular distribution of the intensity throughout the outward solid angle. To this end, the intensity variation of a given type of atmospheric structure should be observed as it rotates across the solar disk to determine the appropriate transformation from I to $\mathcal{F}$ using Eq. (2.46).

2.4   Stellar classification and evolution

2.4.1   *Hertzsprung–Russell diagram and other diagrams*

In 1913, H. N. Russell published a diagram with the absolute visual magnitude M_V plotted against spectral type Sp, showing that the data points are distributed very unevenly across the diagram: there is a high concentration in a strip called the *main sequence* (see Fig. 2.7a). In 1905, E. Hertzsprung pointed out that among the yellow, orange and red stars there are *dwarfs* (which include the Sun) and *giants*. Hence an M_V–Sp plot (Fig. 2.7a) is called a *Hertzsprung–Russell diagram* (H–R diagram). Similar diagrams are obtained when the ordinate M_V is replaced by $\log L$ or some other absolute magnitude M_Q, and the abscissa Sp by $\log T_{\text{eff}}$ or some color index such as, for instance, $B-V$ (see Fig. 2.8); the term Hertzsprung–Russell diagram is often also used for

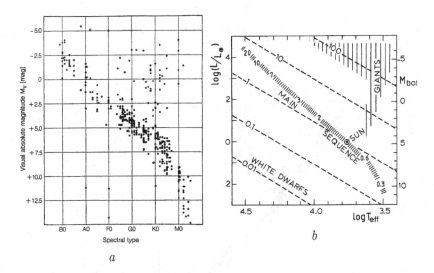

Fig. 2.7. Hertzsprung–Russell diagrams. Panel *a:* the traditional H–R diagram, with absolute visual magnitude M_V plotted against spectral type. The Sun is a G2 star, with $M_V = 4.8$ mag (figure from Unsöld and Baschek, 1991). Panel *b* is the equivalent physical H–R diagram, with the logarithm of the luminosity L (in units of the solar luminosity $L_\odot$) plotted against the logarithm of effective temperature T_{eff} (increasing from right to left). The dashed lines are loci of constant radius, labeled by $R/R_\odot$.

such equivalent diagrams. Comparing different diagrams, one sees distortions because the relationships between the various parameters describing the stellar properties are not linear, although they generally are monotonic.

For theoretical studies, the temperature-luminosity diagram $\log L$ versus $\log T_{eff}$ (Fig. 2.7*b*) is convenient; with the aid of the definition of T_{eff} in Eq. (2.17), lines of equal stellar radius R are readily included. Note that in such diagrams T_{eff} decreases toward the right to conform with the original M_V–Sp diagram.

The accurate distances determined with the Hipparcos satellite have greatly improved the quality of empirical color–magnitude diagrams. In Fig. 2.8*a* the scatter is largely intrinsic, caused by differences in age and in chemical composition and by the inclusion of unresolved binaries.

Figure 2.8*b* shows the MK classification grid superimposed on a M_V–$(B-V)$ diagram. Note that the lines of equal spectral type are curved. Lines of equal T_{eff} (not shown here) are nearly vertical but also slightly curved.

An empirical *mass–luminosity relation* has been derived from observations of double stars (for the determination of stellar masses, see Section 9.6 in Böhm-Vitense, 1989a; or Section 4.6 in Unsöld and Baschek, 1991). For main-sequence stars in binaries with accurately measured orbital elements, the relationship is remarkably tight; see Fig. 2.9.

2.4.2 Stellar evolution

The density of data points across the H–R diagram reflects evolutionary patterns. Indeed the most crowded strip, the main sequence, is occupied by stars that shift very little in position during the longest phase of their evolution, that is, while converting

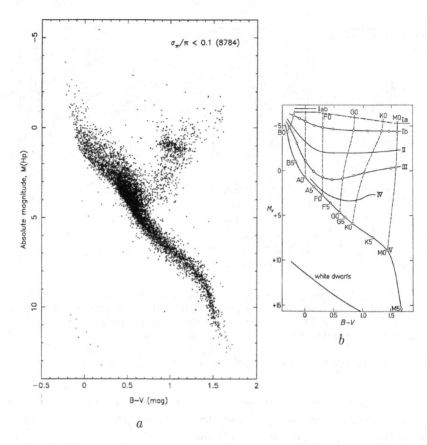

Fig. 2.8. Color–magnitude diagrams. Panel *a*: magnitude M(Hp) against $B-V$ for the 8,784 stars for which *Hipparcos* determined the distances with an accuracy better than 10%. The magnitude M(Hp) measured by *Hipparcos* is close to the visual magnitude M_V; the color $B-V$ is from ground-based observatories (figure from Perryman *et al.*, 1995). Panel *b*: the grid of the two-dimensional MK classification (spectral type vs. luminosity class) superimposed on an (M_V, $B-V$) diagram; see text for explanation (from Unsöld and Baschek, 1991).

hydrogen into helium in their cores. In the sections that follow, we briefly sketch how main-sequence stars are formed and how they develop from there. We concentrate on stars with masses less than approximately $10M_\odot$, because observations of magnetic activity in more massive stars are very rare. We do not consider the final stages of evolution.

2.4.2.1 Protostars, pre-main-sequence contraction, and ZAMS

Star formation starts by gravitational collapse in interstellar clouds. In these clouds, fragments with relatively dense cores are formed, which heat up because the increasing optical thickness impedes the loss by radiation of the thermal kinetic energy gained from gravitational collapse. Within such a core, gas pressure builds up; a *protostar* is formed when hydrostatic equilibrium is established. The envelope continues falling into that protostar, usually through an accretion disk, until a slowly contracting star remains. In this phase, magnetic activity starts. The release of gravitational energy heats up the star and maintains its luminosity. In the H–R diagram, these young, contracting

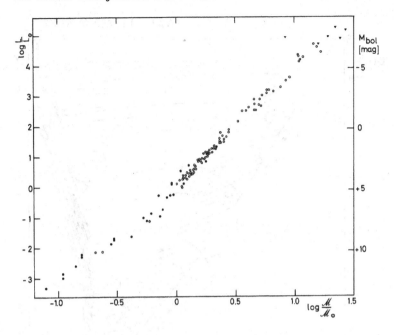

Fig. 2.9. Empirical mass–luminosity relation for main-sequence stars (from Unsöld and Baschek, 1991, after data from Popper, 1980).

stars are found above the main sequence: the *T Tauri stars*, which are surrounded by an accretion disk (Section 11.6), and the more massive *Herbig Ae* and *Be stars*.

The contraction continues until the central temperature has increased sufficiently for the nuclear reactions converting hydrogen into helium ("hydrogen burning") to begin. The star has then reached the *zero-age main sequence* (ZAMS). The duration of the slow contraction phase [estimated by the Kelvin–Helmholtz time scale $\hat{t}_{KH}$ in Eq. (2.4)] is much shorter than the time the star spends subsequently on the main sequence.

2.4.2.2 Main-sequence phase and beyond

Upon arrival on the ZAMS, the star is chemically homogeneous because during the contraction toward the main sequence the matter has been thoroughly mixed and nuclear fusion has not yet signifcantly affected the chemical composition. Model calculations for stars of different masses according to the stellar-structure equations in Section 2.1 reproduce the ZAMS and the mass–luminosity relation.

Evolution away from the ZAMS is caused by the gradual, depth-dependent changes in the chemical composition caused by nuclear fusion. The observable changes are pictured by evolutionary tracks in the H–R diagram in Fig. 2.10. During the hydrogen fusion in the stellar core, the luminosity increases, the star swells, and, if it is somewhat heavier than the Sun, its effective temperature drops slightly. Once the hydrogen concentration in the core has dropped below a critical limit, the fusion rate decreases rapidly, and the core begins to contract again. In stars like the Sun or in lighter stars, which have radiative cores, the hydrogen depletion and core contraction start in the very center and then gradually expands to include more distant parts of the core. In the convective cores of stars heavier than the Sun, the hydrogen concentration

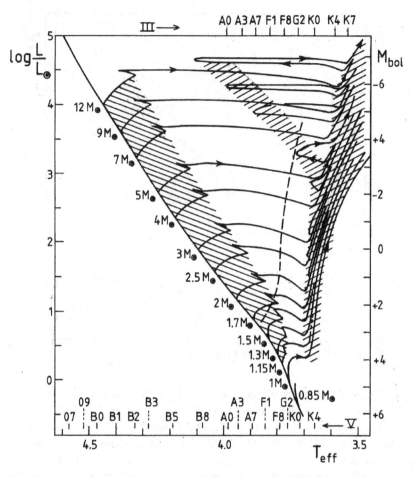

Fig. 2.10. Evolutionary tracks for stars with masses between 1 and 12 $M_\odot$, and a solar chemical composition. The tracks are shown from the zero-age main sequence (the solid curve to the left), and they continue up to the He flash (for $M < 2.0M_\odot$) and throughout the He burning up to the ignition of ^{12}C burning for $M \geq 2.0M_\odot$. Phases of slow evolution are indicated by hatched areas, with the main sequence to the left and the giant branch to the right. The tracks are from Maeder and Meynet (1989). The dashed curve is the granulation boundary (from Gray and Nagel, 1989): stars that show indications of a convective envelope in their spectra (see Section 2.5) are located to the right of it.

drops uniformly, so that the hydrogen fusion dwindles simultaneously over the entire core, which makes it contract. This core contraction produces short hooks to the left in the tracks for stars with $M > 1.2\ M_\odot$.

Because of the contraction and release of gravitational energy, both the density and the temperature increase in the core and in the adjacent layer, with the result that the conversion of hydrogen to helium continues in a shell around the "dead" helium core. This shell with hydrogen fusion works its way outward, adding more helium to the dead core, which thus grows in mass while contracting and heating up. Initially the star swells only gently during hydrogen fusion in the shell; the effective temperature decreases.

This phase is called the subgiant stage (although the corresponding region in the H–R diagram does not precisely overlap with that of stars of LC IV). During this phase, the depth of the convective envelope increases rapidly. Once the convection zone reaches the shell of H fusion, the star enters the giant branch. Thereupon the star moves steeply upward in the L–T_{eff} diagram, increasing considerably in radius and in luminosity. The star must expand in order to connect the increasingly hot core with the cool atmosphere by a virtually adiabatic temperature gradient.

Once the He core has reached a sufficiently high temperature and density, helium fusion starts, converting ^{4}He to ^{12}C and, at still higher temperatures, to ^{16}O. The course of events depends on the stellar mass. For light stars, with $M \lesssim 2.0\,M_\odot$, the electron gas in the He core is degenerated, which makes the helium fusion start in a flash. During this *helium flash* the core temperature increases, the degeneracy is removed, and thereupon the helium fusion proceeds at a more quiet pace. During these events, the star drops a little in luminosity, and then it dwells for a while in a small domain of the H–R diagram which is a part of the "helium main sequence." For stars with $M \gtrsim 2\,M_\odot$, the He core is not degenerated, so for these the He fusion starts without a flash.

During the He fusion in the core, H fusion continues in a shell. After the helium in the core is depleted, a He-fusing shell starts moving outward. After a short while, the star expands rapidly again, moving along the *asymptotic giant branch*, which runs at close distance parallel to the giant branch and extends higher. In the cores of stars with $M < 2M_\odot$ the temperatures and densities never reach the levels required to start fusion processes after the helium fusion. Such stars start to pulsate when they approach the top of the asymptotic giant branch at a luminosity somewhat in excess of $1{,}000L_\odot$. Such a star loses much of its mass outside the shells where fusion occurs; while it maintains its luminosity, it shrinks and its photosphere becomes much hotter. Consequently, the star moves to the left in the H–R diagram; near the end of that trek, part of the ejected gas is visible as a bright planetary nebula. When the fusion processes in the shells stop, the luminosity drops and the star contracts. Eventually the electron gas in the core degenerates, and the star has become a compact white dwarf.

For a more massive star, during He fusion in the core the track in the H–R diagram loops to the left over a distance depending on stellar mass and chemical composition. For stars with solar composition, these loops are increasingly conspicuous for $M > 5M_\odot$ (see Fig. 2.10).

In cores of stars with masses larger than $\sim 5M_\odot$, the core densities and temperatures become high enough for carbon fusion to follow, first in the core and then in the surrounding shell. For higher masses, more fusion stages follow but stars with masses larger about $\sim 10M_\odot$ fall outside the scope of this book.

Table 2.3 shows that the longest of the phases with H fusion is that of the H-core fusion; this defines the main sequence in the H–R diagram. The H-shell fusion phases are relatively short, particularly for stars with $M \gtrsim 2.0\,M_\odot$. The first phase of H-shell fusion takes an appreciable fraction of the star's lifetime only for stars with $M \lesssim 1.0\,M_\odot$; most of the subgiants in the H–R diagram belong to this category.

The time spent during He fusion in the core is the second-longest phase in stellar evolution; it amounts to approximately 10% of that of the phase of H fusion in the core for stars with $M \gtrsim 7\,M_\odot$, up to approximately 24% for $M < 3\,M_\odot$. During the He-fusion stages, stars with $M < 5\,M_\odot$ remain quite close to the giant branch, along which the star

Table 2.3. *Duration of the phases of stellar evolution from the zero-age main sequence, during phases of H fusion, in 10^6 yr*[a]

			H Fusion in Shell	
$M(M_\odot)$	Main Sequence	Contraction	Subgiant phase	Giant branch
1.0	(7,422	1,998)	2,259	590
1.5	2,633	58	66	154
2.0	1,094	22	13	19
3.0	346	6.3	2.5	2.3
5.0	92.9	1.5	0.5	0.3

[a] From Schaller *et al.* (1992).

moved during the final part of its H-shell fusion stage. For $M \gtrsim 5\,M_\odot$, the evolutionary tracks loop to the left, covering only a fraction of the distance between the giant branch and the main sequence. As a consequence, stars during the last phase of H-shell fusion and the He-fusion phases determine a concentration of data points in the upper right of the H–R diagram; see Figs. 2.7, 2.8, and 2.10. Without further indications, one cannot establish the precise evolutionary status of such a giant: it may be in H-shell fusion, or in one of the He-fusion stages, or (most likely) in the He-core fusion stage. Particularly after the main-sequence stage, the precise shapes of the evolutionary tracks depend on the chemical composition and on various assumptions in the modeling, such as the internal mixing at the lower boundary of the convective zone.

The evolutionary models explain the *Hertzsprung gap* in the H–R diagram between the top of the main sequence and the giant domain. The scarcity of data points in that gap corresponding to stars with $M \gtrsim 2\,M_\odot$ is caused by the short duration of the first phase of H-shell fusion, and the rapidity of protostar contraction toward the main sequence.

2.4.3 Stars with convective envelopes

The domain of stars with convective envelopes lies to the right of the dashed line in Fig. 2.10; this domain is delineated by application of Schwarzschild's criterion, Eq. (2.21), and by asymmetries in the spectral-line shapes that are characteristic for convective overshoot producing granulation (Section 2.5).

Stars with masses $M < 1.3\,M_\odot$ (spectral types F and later) possess convective envelopes from the ZAMS onward. More massive stars (spectral types O, B, and A) have radiative envelopes during their main-sequence phase; such stars develop convective envelopes during their evolution toward the giant branch.

On the main sequence between spectral types F0 V and F2 V, the convective envelope has a depth barely exceeding a few pressure scale heights. With decreasing effective temperature, the convective envelope encompasses more and more of the stellar volume; in the coolest M-type stars the entire star is convective (Section 11.5). Figure 2.11 shows how the depth $\tilde{d}_{CE}$ of the convective envelope and other stellar parameters vary with T_{eff} for stars along the main sequence.

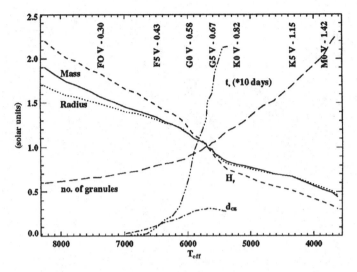

Fig. 2.11. The change of some stellar parameters along the main sequence. Shown as a function of effective temperature are the stellar mass M and radius R, the depth d_{CE} of the convective envelope, the photospheric pressure scale height H_p, and the convective turnover time $\hat{t}_c$ at the bottom of the convective envelope for a mixing-length model with $\alpha_{ML} = 2.0$ (data from Gilman, 1980). All quantities are scaled to solar units, except $\hat{t}_c$, which is in days. The number of granules on the surface is estimated from the scaling factor given in Eq. (2.72). The spectral type and color index $(B-V)$ are indicated along the top.

The high degree of correlation between these parameters is one of the complicating factors in finding the fundamental parameters that rule stellar magnetic activity, as discussed in later chapters: scalings of activity with any one of these parameters can be transformed into equally functional relations with other parameters.

2.5   Convection in stellar envelopes

The atmospheric magnetic phenomena discussed in this book are driven by convective flows. In the Sun, the convective envelope extends from a depth of $0.287 \pm 0.003 \, R_\odot$ (Christensen–Dalsgaard *et al.*, 1991) below the surface up to the photosphere, yet it contains only 1.5% of the total solar mass. The density spans a factor of approximately 10^7 from the top to the bottom of the envelope. As the temperature decreases outward, the pressure scale height (Eq. 2.9) rapidly decreases: the corresponding density contrast in the uppermost 6,000 km, for instance, is a factor of slightly over 1,000.

Above the convection zone lies the photosphere, which is convectively stable. The sudden transition from a domain where convective energy transport dominates to a domain where energy radiates away efficiently causes the cellular patterns in the stellar photosphere to be very different from the flow geometry below the visible layers, as we discuss in Section 2.5.3.

Shortly after their emergence into the photosphere, small concentrations of intrinsically strong magnetic field are moved about by plasma flows through drag coupling (Section 4.1.2). Understanding the field dynamics therefore requires insight into the flows. This section discusses these flows for environments in which the action of the magnetic

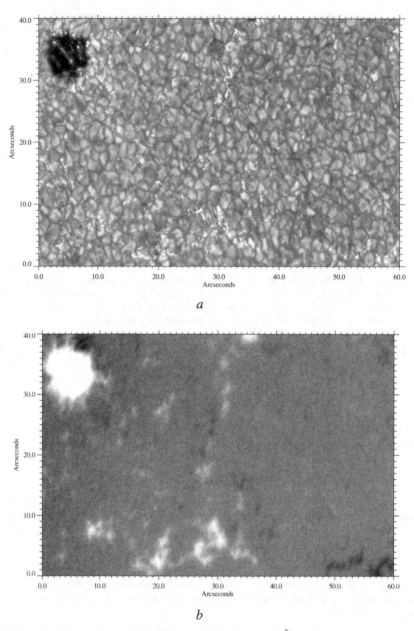

Fig. 2.12. High-resolution G-band (CH band head, in a 12-Å bandpass centered on 4,305 Å) image (*a*) and the corresponding magnetogram (*b*) of a moderately active area including quiet Sun. Note the fine structure within the large pore, and the filigree up to its otherwise sharp edge (courtesy of T. Berger and M. Löfdahl).

field on the flows can be largely ignored, as is the case in the quiet photosphere. Convection in magnetic plages and around spots is discussed in Sections 4.3 and 5.4, respectively.

In the solar photosphere, the two most conspicuous flow patterns are the granulation (Fig. 2.12), with a characteristic cell diameter of the order of 1,000 km, and the

23 Feb. 1996,16:44 to 21:03 UT

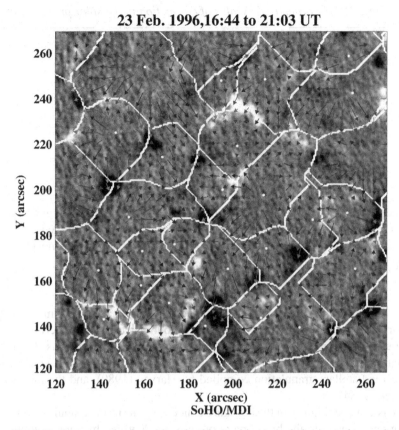

Fig. 2.13. Supergranular flow determined by tracking granulation, superimposed on a magnetogram, both observed by *SOHO/MDI* in the high-resolution mode with a 1.2-arcsec (or 870-km) angular resolution. The vectors represent averages of the flow over a 3-h interval. The horizontal divergence [Eq. (2.71)] shows a pattern of cells of central upflows (with local extrema indicated by white dots) surrounded by downflow. The cells – based on the horizontal divergence pattern – are shown here by their outlines; note that flows associated with the curl component of the pattern can cross boundaries of divergence cells. The magnetic field (shown in black and white for opposite polarities) clusters predominantly within the downflow lanes; where this appears not to be the case, the offsets are likely caused by flow evolution within the 3-h averaging interval or by the emergence of new flux.

supergranulation (Fig. 2.13), with cell diameters ranging from approximately 10,000 km to 40,000 km. Table 2.4 summarizes the time and length scales. Granulation stands out most clearly in intensity, while supergranulation is most easily visible in the large-scale flow patterns or indirectly in the chromospheric network. Mesogranulation, on a scale intermediate to granulation and supergranulation, is best seen in the horizontal divergence [Eq. (2.71)] of the flow field.

The flows on the granular scale are particularly important for the geometry of the canopy field within the chromosphere, and likely important as well for the nonradiative heating of the outer atmosphere (Chapter 10). The supergranulation is one of the drivers of the field dispersal that determine the large-scale magnetic field patterns of the Sun (Section 6.3).

Table 2.4. *Characteristic scales of convection in the solar photosphere*[a]

Name	Size Scale (km)	Time Scale (s)	Max. Horiz. Velocity (m/s)	Refs.[b]
Granulation	600–1,300	5×10^2	400–800	1,2
Mesogranulation	5,000–10,000	$>7 \times 10^3$	300–500	3
Supergranulation	15,000–30,000	$1–4 \times 10^5$	200–400	4,5

[a] The ranges result from different measurement techniques.
[b] References: 1, Bray *et al.* (1984); 2, Spruit *et al.* (1990); 3, November *et al.* (1981); 4, Simon *et al.* (1995); 5, Hagenaar *et al.* (1997).

As a consequence of the compressibility of the plasma, convection generates sound waves. These waves lead to the standing wave patterns (Section 2.6) studied in helioseismology, as well as to traveling waves that deposit energy in the outer atmosphere as they develop into shock waves (Section 2.7).

2.5.1   *Observed properties of the granulation*

At angular resolutions of close to one second of arc ($\approx$725 km) or better, the solar surface appears as a foamlike pattern with bubbles the size of Germany or Texas present everywhere except in sunspots and magnetic plage regions (Figs. 1.2 and 2.12). Granulation has been scrutinized for more than a century; see the 38-page bibliography on solar (and stellar) granulation compiled by Harvey (1989), and the monograph by Bray *et al.* (1984).

There is some ambiguity in the literature concerning the term "granule": some authors define granules as only the bright parts of cells as identified by some suitably chosen threshold, whereas others include in their definition both the bright upflow and the associated dark intergranular lanes containing the downflows. To avoid confusion, we use granule for the former and granular cell for the latter. Granulation refers to the entire surface pattern.

On unprocessed white-light movies of the quiet photosphere, the study of the granulation is hampered by a fluctuating intensity pattern with an apparent length scale of some 10,000 km and a time scale of 5 min. This pattern is caused by sound waves (Section 2.6). When this acoustic signal is suppressed by an appropriately chosen Fourier filter (Section 2.6.2), the residual intensity image reveals a smoothly evolving pattern with a typical scale $d_g \equiv \sqrt{\langle A \rangle}$, with $\langle A \rangle$ the average area, of approximately $1,300$ km. The granulation pattern is characterized by a network of connected dark lanes. These lanes are associated with relatively cool, downflowing matter. They surround irregularly shaped, but often roughly polygonal, bright regions that are associated with upflows. In the quiet photosphere, the intensity contrast between dark lanes and bright upflows reaches up to 14% in high-quality images taken near 5,000 Å, but the intrinsic contrast may be as high as $\sim$20% (see discussion in Spruit *et al.*, 1990).

The difference in the definition of granule and granular cell leads to very different, but compatible, frequency distribution functions for their sizes. The histogram of granular areas reveals no typical size scale: the number density of granules continues to increase

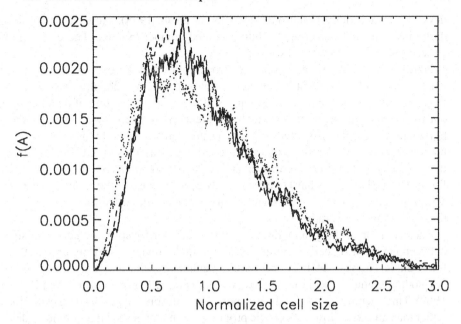

Fig. 2.14. Histograms of areas of the (supergranular) cells in the Ca II K quiet-Sun network (solid curve), the granular cells in white-light images of the granulation (dotted curve), and of a realization of a Voronoi tesselation of a two-dimensional random pattern of generator points (dashed curve; see Fig. 2.16 and Section 2.5.4). The histograms have been scaled to the average cell size of 1.33 Mm^2 for the granulation and 530 Mm^2 for the supergranulation (figure from Schrijver *et al.*, 1997a).

toward smaller scales (Roudier and Muller, 1986). The histogram $h_g(A)$ of areas of granular cells, in contrast, peaks at $\sqrt{\langle A \rangle}$ of some 1,300 km, with a distribution width that is comparable to the mean value (Schrijver *et al.*, 1997a; Fig. 2.14). These two distributions are compatible because of the dark ring around granules with a characteristic width of ≈ 0.4 arcsec: the presence of this ring transforms the monotonic distribution of granular areas into a peaked distribution for granular cells (Schrijver *et al.*, 1997a). The visual impression of a well-defined typical length scale for the granular cells of $\sim 1,000$ km is the result of differential area coverage: $A\, h_g(A)$ shows a peak that is even more pronounced than that in $h_g(A)$ itself.

Within regions with magnetic flux densities exceeding approximately 50 G on scales of some tens of thousands of kilometers, the granulation changes markedly in character; see Sections 5.3 and 5.4.

The flow profile of the plasma in the granulation is hard to determine, primarily because there is no unambiguous tracer for this flow. The characteristic magnitude of the flow is readily estimated from Doppler measurements. Canfield (1976) and Durrant *et al.* (1979) derive rms vertical velocities for the granulation of 1–1.3 km/s in the middle of the photosphere. Beckers and Morrison (1970) find the same magnitude for the horizontal component.

Intensity features of an as yet unknown nature are seen to move within granular cells. They reveal a flow with a rapid initial concentration into the intergranular lanes, often deviating from strictly radial outflows (Berger *et al.*, 1998). Flows directed along the dark intergranular lanes converge into a limited number of sinks. These lane flows are slower than the outflow by a factor of 2 to 3.

Nordlund *et al.* (1997) argue that the broad upflows in the granules have a relatively small ratio of turbulent to bulk velocities; that is to say, they are almost laminar. This is largely the consequence of the smoothness of the gentle upwelling of material throughout the convective envelope, in which turbulence is markedly weakened by the rapid horizontal expansion associated with the low pressure scale height in the photosphere. In the downflows, however, one expects more turbulence as a result of the strong gradients, if not shocks, in the flows, further enhanced by the compression of the material in the flows. This turbulence is hard to observe, however, because of the small scales involved. The close agreement of simulated and observed line profiles (Section 2.5.3) demonstrates that this turbulence is rather weak.

The evolution of the granulation is complex: cells grow, split, merge, or are squeezed between neighboring cells (see the discussion in the review by Spruit *et al.*, 1990, and, for example, Noever, 1994). A peculiarity of granular evolution is the phenomenon of the "exploding granule" (see, for instance, Callier *et al.*, 1968; Namba and van Rijsbergen, 1969). This phenomenon, which corresponds to a subsonically expanding evolution front rather than an actual flow, reflects the property that many of the larger granules develop a dark interior, from which an intensity front moves outward at 1–2 km/s. Within this front, new granular cells begin to form. Some of these disappear after a short time, whereas others develop into new granular cells. The exploding granule is attributed to radiative cooling of the central upflow, and the subsequent slumping back of the associated matter (see review by Spruit *et al.*, 1990).

The characteristic time scale for the intensity cross correlation to decrease to $1/e$ of its original value is ~ 8 min (see Spruit *et al.*, 1990). Part of that decrease is associated with the displacement of cells as a result of the larger-scale flows in which they are embedded: cells often travel a substantial fraction of their size during their lifetime. The eye can pick out cells that drift away from their original location, subject to fragmentation and merging, and yet remain sufficiently distinct to suggest that the central upflow is a long-lived but dynamically changing and drifting structure. "Lifetimes" for such features range from minutes to tens of minutes.

2.5.2 *Properties of supergranulation and mesogranulation*

On a scale much larger than the granulation, and associated with little if any intensity contrast, we find the supergranulation. Hart (1954, 1956) first reported the existence of this system of large-scale horizontal flows in the photosphere, but the spectroheliographic techniques developed by Leighton *et al.* (1962) established the ubiquity of this phenomenon. At first, the only reliable way to study the flows directly was to measure the Doppler shifts associated with the horizontal motions, visible only if observed away from disk center. Local correlation tracking using granules as tracers of the large-scale flows provides another way to identify supergranulation cells, assuming that granular displacements do indeed measure the larger-scale plasma flow. The interaction of the flows with the magnetic field, discussed in Section 4.1.2, enables the study of supergranulation even near disk center through the chromospheric network (Chapter 1 and Section 8.4.2); this chromospheric pattern suggested the existence of the supergranulation well before its actual discovery as a flow field.

The pattern in the upflows and downflows can be inferred from the horizontal displacement field $\mathbf{v}(x, y)$ by the anelastic continuity approximation (assuming negligible horizontal gradients in density ρ):

$$\nabla \cdot \rho \mathbf{v}(x, y) \equiv 0 \leftrightarrow h\, div\, \mathbf{v} \equiv \frac{\partial v_x}{\partial x} + \frac{\partial v_y}{\partial y} \approx -v_z \frac{d \ln \rho}{dz} - \frac{\partial v_z}{\partial z}. \qquad (2.71)$$

The right-hand approximation expresses the horizontal divergence in terms of vertical motions (see November, 1989; and Strous, 1994); where the horizontal divergence is positive, matter rises (and generally decelerates in the process; see Section 2.5.3), and matter flows down where it is negative (generally accelerating along the way).

Various combinations of direct and indirect techniques have resulted in quite different scales for the supergranulation: values of a "mean" diameter have been reported that range from approximately 15,000 km up to 35,000 km, with values for individual cells spanning an even larger range. Hagenaar *et al.* (1997) demonstrate that these differences largely result from the use of different methods. Autocorrelation measures, for instance, are preferentially weighted toward large cells, because of the area integration in the correlation function. The average length scale of the supergranular flow pattern (defined as the square root of the average area) determined from a frequency distribution (Fig. 2.14) of areas of chromospheric network cells lies between 15,000 km and 18,000 km.

The horizontal flow velocities associated with the supergranulation reach up to 0.2–0.4 km/s (Simon *et al.*, 1995). As in the case of granulation, the supergranular flows are not purely radial, signifying that not only the nearby, central upflows are important in determining the pattern of the flow. A complementary component of the flow, associated with the horizontal curl, is also present, corresponding to flows of 0.1–0.3 km/s directed predominantly along and sometimes crossing the downflow lanes (Schrijver *et al.*, 1996a). The rms vertical velocity associated with supergranulation is measured to be approximately 40 m/s (November *et al.*, 1981).

Because of the smooth, continuous evolution of the supergranulation, the lifetime of supergranules is not a well-defined quantity. Cross-correlation measurements using the network or the Doppler velocity pattern give lifetimes ranging from 20 to 30 h, but values of up to 50 h are quoted for individual cells (Table 2.4).

The evolution of the supergranulation has remained obscure for a long time, in large part because the typical time scale is close to one day. Ground-based studies are consequently severely hampered by the day–night cycle. Results from the *SOHO* satellite, however, have shed light on this topic. Our studies of the pattern of the horizontal divergence show that, as is the case for granules, the supergranules evolve predominantly by fragmentation and merging. When cells fragment, a downflow lane forms between the old extremum in the positive divergence and a newly formed secondary maximum at some distance. The maxima then drift apart, pushing aside surrounding lanes. The merging of supergranular cells is much like a reversed splitting. Relatively few cells appear as upflows between existing cells, and few supergranules disappear by being crushed by expanding surrounding cells. The phenomenon of the exploding granule has no counterpart in the supergranular evolution.

November *et al.* (1981) have reported on a flow pattern with an intermediate scale based on time-averaged Doppler measurements. This *mesogranulation* shows up as a pattern in divergence maps. The size scale is approximately 5,000 km to 10,000 km, and the rms vertical velocity is 60 m/s. There does not appear to be a chromospheric counterpart to the mesogranulation; apparently the persistence of the larger-scale motions dominates the positioning of the magnetic flux. Mesogranules persist for some 2 h as recognizable features in the divergence pattern and drift in the supergranular flow to the downflow network. The details of their origin and topology remain to be studied.

2.5.3   *Numerical models of convection in stellar envelopes*

Understanding stellar convection requires numerical simulations, because the stellar environment is vastly different from what can be achieved in laboratory conditions, both because of the scale of the flow and of the importance of radiative transport. The Sun provides the means to validate numerical experiments, but this is limited to the thin top layer of the convection zone that is, as we argue in the following paragraphs, atypical of most of the convective envelope. Those who perform these numerical studies find themselves in the remarkable position of falling far short of covering the required range of scales, yet being able to validate experiments for the layer of the envelope that appears to be the most important in the driving and shaping of convection.

The complexity of the convective flows is quantified by the Reynolds number $\mathcal{R}_e$ (see Table 2.5), which compares the inertial and viscous (diffusive) forces. For the solar granulation, $\mathcal{R}_e$ is estimated to be of the order of 10^{11} (see, for example, Bray *et al.*, 1984) at the base of the photosphere. Such a value would allow a vast range of flow scales to coexist. Another characteristic number is the Prandtl number $\mathcal{P}_\nu$, which compares the viscous and thermal diffusivity. The very low value of $\mathcal{P}_\nu$ in the low photosphere (and below) implies that vortices can be stretched and folded before being dissipated (Brummell *et al.*, 1995), resulting in a very complex, tangled velocity pattern.

A drastic simplification is the mixing-length approximation, described in Section 2.2, in which the convective motion is assumed to consist of eddies of a characteristic size, the mixing length $\hat{\ell}_{\mathrm{ML}}$. Near the photosphere, however, the mixing-length concept fails to describe the stratification of the superadiabatic layer, as well as the run of convective velocities with depth (see Nordlund and Dravins, 1990). Hence, detailed numerical modeling of convection, including radiative transfer, is required to understand not only the dynamics but also the structure of the layers near the surface of cool stars.

The numerical modeling of stellar envelope convection is subject to rather severe restrictions. It is not possible, for example, to approach realistic values of the controlling dimensionless numbers; sometimes the values used in simulations are a few orders of magnitude off, but often – particularly where the Reynolds number $\mathcal{R}_e$ is concerned – by a factor of a million or more. Yet, results from increasingly large computational domains (for 2^n points in cubic rasters, with n growing from 15 to 18 over the years) indicate that the best current models approach an accurate description of the observations, and that they provide their explanation in terms of the dominant processes.

The simplifications include a restriction of the size of the simulated volume, generally to a few granules horizontally with periodic boundary conditions, and to a few pressure scale heights in depth. Radiative transfer has been modeled to satisfactory accuracy by computing it along a limited number of rays in different directions, whereas opacities are

Table 2.5. *Some frequently used dimensionless numbers[a]*

Name	Symbol[b]	Char. Phot. Value[c]	Time Scales Involved
		General	
Mach	$\mathcal{M} = \dfrac{\upsilon}{c_s}$	$\lesssim 1$	sound to flow
Prandtl	$\mathcal{P}_\nu = \dfrac{\nu}{\kappa}$	(1) $\sim 10^{-9}$	thermal conduction to viscous
Rayleigh	$\mathcal{R}_a = \dfrac{\alpha g \hat{\ell}^3 \Delta T}{\kappa \nu}$	(1) $\sim 10^{-11}$	geom. mean of conductive and viscous to buoyancy
Reynolds	$\mathcal{R}_e = \dfrac{\hat{\upsilon} \hat{\ell}}{\nu}$	(1) $\gtrsim 10^{11}$	viscous to advection
		Rotation	
Rossby	$\mathcal{R}_o \sim \dfrac{P}{\hat{t}_c}$		rotation to conv. turnover
Taylor	$\mathcal{T}_a = \dfrac{4\Omega^2 \hat{\ell}^4 \Delta T}{\kappa \nu}$		conductive-viscous to rotation
		Electromagnetic	
plasma β	$\beta = \dfrac{8\pi p}{B^2}$		
resistive Lundquist	$\mathcal{N}_L = \dfrac{\hat{\upsilon}_A \hat{\ell}}{\eta}$		resistive diff. to Alfvén wave
magnetic Prandtl[d]	$\mathcal{P}_\eta = \dfrac{\eta}{\kappa}$		thermal cond. to resistive diff.
	$\mathcal{P}_\eta^* = \dfrac{\nu}{\eta}$		resistive diff. to viscous
magnetic Reynolds	$\mathcal{R}_m = \dfrac{\hat{\upsilon} \hat{\ell}}{\eta}$	(2) $\gtrsim 10^{12}$	field diffusion to advection

[a] Some estimated values are given for the base of the solar photosphere, and for convective scales $\hat{\ell}$ from granulation up to supergranulation.

[b] Symbols: $\hat{\upsilon}$, $\hat{\ell}$ characteristic velocity and length scale; $\hat{\upsilon}_A$, characteristic Alfvén velocity; ν, kinematic viscosity; η, magnetic diffusivity; κ, thermal diffusivity; Ω, angular velocity of rotation; P, rotation period; $\hat{t}_c$, characteristic time scale of convective turnover; α, coefficient of expansion.

[c] References: (1) Bray *et al.* (1984); (2) Brummell *et al.* (1995).

[d] Note the two different definitions of the magnetic Prandtl number.

treated by grouping spectral lines into a few ensembles for which averages are computed, including an average continuum.

Despite these simplifications, the simulations require formidable computational efforts. Some general trends for solar and stellar convection emerging from these efforts inspire confidence through their striking agreements with observables (see Stein and

Table 2.6. *Properties of photospheres of selected cool stars*[a]

Star	T_{eff}[b] (K)	$\log(g)$[b] (cm/s^2)	H_p (km)	$\log(p_g)$ (dyn/cm^2)	$\hat{v}_t$	F_a (erg cm^{-2}s^{-1})
Procyon	6,600	4.14	320	4.6	7.4	3×10^9
Sun	5,800[c]	4.44	140	5.1	3.9	6×10^8
α Cen A	5,800	4.14	280	4.8	5.0	8×10^8
β Hyi	5,800[d]	3.84[e]	560	4.7	5.5	1×10^9
α Cen B	5,200	4.44	130	5.1		

[a] These parameters lie near those of the stars identified in the first column. Also listed are the photospheric pressure scale height, H_p [Eq. (2.9)], photospheric gas pressure, p_g, and characteristic turbulent velocity, $\hat{v}_t$ [estimated from the mixing-length energy transport; see Section 2.2, Eq. (2.29)]. The table also lists the acoustic flux density, F_a, estimated from Eq. (2.76). Values for $\hat{v}_t$ and F_a for α Cen B are not given; see Section 2.5.3.
[b] Effective temperatures T_{eff} and surface gravities g for which Nordlund and Dravins (1990) performed granulation simulations.
In her study of solar analogs, Cayrel de Strobel (1996) gives the following values:
[c] 5777 K; [d] 5780; [e] 4.08.

Nordlund, 1998, for an overview of the state of understanding of solar near-surface convection). Nordlund and Dravins (1990) and Dravins and Nordlund (1990a) performed model computations for near-surface convection in stars with effective temperatures and surface gravities that differ somewhat from those of the Sun and other cool stars (see Table 2.6). The topology of the granulation in all simulated stars is qualitatively the same: at the photospheric level, plasma moves from isolated broad upflows to a connected network of narrow downflows. The intensity pattern correlates quite well with that of the vertical velocity: bright over upflows and dark over downflows. These models show no characteristic wave number in the Fourier transform of either temperature or flow velocity (Nordlund and Dravins, 1990); apparently there is no dominant convective scale, despite the appearance of the surface.

Ascending plasma reaches the photosphere from which radiation readily escapes and where the pressure scale height decreases rapidly with the decreasing temperature. As a result, the flow overturns in a thin layer. This is achieved through a pressure front that builds up over the upflow, which slows the flow, and deflects it into a nearly horizontal expansion. Where flows of adjacent cells meet, other regions of excess pressure form, which slow the horizontal flow and deflect it downward (aided by the increase in density as the plasma cools), but also with a component along the intergranular lanes toward the points of strongest downflow at the shared vertices of more than two cells. Near the surface, the work done by the convection in downflows is, on average, about twice as large as the work done by the upflows. This implies that convection is predominantly driven by the downflows that form because of the radiative cooling near the top (Stein and Nordlund, 1998).

The simulations suggest that most of the upflowing plasma overturns well below the surface because of the strong density stratification and associated expansion of ascending flows. This same density stratification results in a convergence within the downflows,

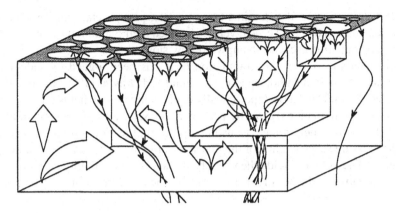

Fig. 2.15. Illustration showing the proposed successive merging of convective downdrafts on larger and larger scales (from Spruit *et al.*, 1990).

which probably extend to large depths. Numerical models suggest that the topology of the flow below the surface is fundamentally different from that at the surface (see the discussion in Spruit *et al.*, 1990, and Fig. 2.15): the narrow, elongated downflow structures that form the network at the surface converge into whirling downdrafts. The numerical models suggest that these downflows form a treelike structure, with thin downdraft twigs converging in branches and eventually in trunks. At every height, the horizontal scale of the flow appears to be at least an order of magnitude larger than the vertical scale, which is given by the local pressure scale height (Nordlund, 1986).

The downflows are expected to collect more and more matter with increasing depth, because in the shear with the ambient hotter plasma at their perimeters, turbulent mixing occurs; this captures some of the surrounding plasma by entrainment. Hence, locally the increase in lateral temperature contrast as matter sinks is weakened, the speed of the downflow is reduced, and the mass involved is increased.

Upward-moving plasma overshoots the stable photospheric layer. This matter does not cool as rapidly as one might expect: adiabatic cooling and radiative losses are largely compensated by radiative heating from below. Despite this damping, the overshooting upflows are associated with a fundamental modification of the temperature profiles: the plasma high over the upflows is relatively cooler than the downflowing matter at the same height. The temperature contrast decreases with height. The result is that the average temperature stratification in the photosphere differs only slightly from the theoretical radiative-equilibrium stratification (Section 2.3.1).

The simulations by Nordlund and Dravins (1990) show that deep in the photosphere, some 50% of the energy is carried by convection and 50% by radiation for the stars in Table 2.6, with the exception of the coolest star, α Cen B. For this star the 50% level lies some 80 km below the photosphere (which they define as the layer where the mean temperature equals T_{eff}), or more than half a pressure scale height, whereas at the photospheric height already 95% of the energy is being transported by radiation. Nordlund and Dravins argue that this is a general property of cooler stars, in which radiative diffusion can carry a substantial fraction of the stellar flux in the layers immediately below the surface, because both the opacity and the temperature sensitivity of the opacity decrease with temperature.

The deep transition in the dominant energy transport mechanism for cool main-sequence stars means that the vertical velocities reach their maximum values well below the surface: as radiation escapes, the upward flow slows down.

The characteristic size of the granulation cells has been argued to scale with the pressure scale height in the photosphere. Nordlund and Dravins (1990) assume that two stars, labeled "a" and "b," have topologically equivalent velocity fields (consistent with the results of their simulations) given by $v_a(x, y, z)$ and $v_b(sx, sy, sz)$, respectively, where s is a scale factor to be determined. Densities $\rho_{a,b}(x, y, z) = f(x, y)\exp(-z/H_{a,b})$, with $H_{a,b}$ as the pressure scale heights for the two stars, crudely approximate the atmospheric stratification. The velocity fields then satisfy the anelastic continuity equation (with a zero time derivative of the density in a Eulerian frame of reference suppressing acoustic waves),

$$\nabla \cdot (\bar{\rho}_a v_a) = \nabla \cdot (\bar{\rho}_b v_b) = 0, \qquad (2.72)$$

only if the scaling factor equals $s = H_a/H_b$. The typical granular diameter should therefore scale with the pressure scale height (listed for a few stars in Table 2.6). If this argument is correct, the number of granules on the stellar surface scales with $(M_*/T_{\text{eff}}R_*)^2$; see Fig. 2.11.

Although we can now determine the relative size of stellar granulation, we are left with the question why solar granular upflows have the specific mean separation of $\approx 1,000$ km. Some arguments have been made based on radiative transfer. Nelson and Musman (1978) argue that the theoretically expected maximum horizontal size d_{max} of the granulation and the horizontal and vertical velocities are related through a comparison of time scales, so that

$$d_{\text{max}} = 2\pi H_p \frac{v_{\text{hor}}}{v_{\text{vert}}}. \qquad (2.73)$$

The velocity v_{vert} at which plasma ascends in order to be able to replenish energy sufficiently quickly to compensate for the surface radiative losses (Nordlund, 1982) is at least 2 km/s just below the photosphere, so that with $v_{\text{hor}} \leq c_s$ (c_s is the photospheric sound speed), $d_{\text{max}} \leq 4,000$ km. At the other extreme of the size spectrum, very small-scale granules are not expected to occur, because at scales below approximately 200 km, radiation exchange tends to damp out temperature differences (Nelson and Musman, 1978). The allowed range of 200 km to 4,000 km agrees well with observations (Section 2.5.1).

Simulations of stellar granulation can be validated using spectral line profiles: the mean line shift and width, and the shape of the line bisector depend on the details of the convective flows. Convection causes photospheric line profiles with line bisectors with a peculiar C-shaped asymmetry (the observed onset of that asymmetry in the H–R diagram is shown in Fig. 2.10). This asymmetry is the result of the mixture of bright upflows, dark downflows, and all intermediate regions with appropriate filling factors. Dravins and Nordlund (1990b) computed line profiles based on their granulation simulations, and they show that their C-shaped, disk-integrated model line profiles agree well with observed stellar line profiles, as do the line widths. In the past, photospheric line profiles have been fitted by using an adjustable parameter called microturbulence, which was interpreted as evidence for the presence of small-scale turbulence. However, the successful fit of models to observations makes "microturbulent line broadening" largely superfluous.

Granulation introduces a net blueshift of typically a few hundred meters per second to the mean wavelength of photospheric lines (partly compensating the gravitational redshift), because the relatively bright upflows outshine the darker downflows (Schröter, 1957; see also Bray *et al.*, 1984). This leads to a bias in the determination of the stellar frame of reference when comparing, for example, Doppler shifts of chromospheric lines or of umbral lines relative to the photospheric "rest" frame.

Although simulations of convection do well on scales of several granules or so, larger scales still defy modeling. Recent efforts using time–distance helioseismology (Section 2.6.2) to image subsurface flows suggest that the downdrafts are not as deep as suggested by an analogy with numerical experiments for granulation simply scaled up to larger scales: the inversion results suggest that the supergranulation flows overturn at a depth of a few thousand kilometers (Duvall *et al.*, 1997). The methods used to derive these subsurface flows are still being tested, however, and they are likely to miss any narrow downdrafts, so that we will have to wait for the validation of these results.

Let us briefly look at convection deep in the envelope. The simulations for the layers near the top of the envelope suggest that downflows are highly intermittent both in space and in time. It is inferred that the highly asymmetric flow pattern continues down to the bottom of the convective envelope, because the convective instability keeps driving the cool downdrafts of heavy plasma. Many of the very local downflows have substantially higher speeds than the "typical" convective velocities computed in the mixing-length approximation. Hence these flows probably penetrate deeper in the overshoot layer below the convection zone, yet not deeper than a fraction of the pressure scale height because these flows are subsonic. In the boundary layer, the downflows are deflected, their relatively cool plasma mixes with the ambient hotter plasma, and the radiative flux from the interior is absorbed. The intricate processes near the bottom of the convection zone are beyond the present possibilities for numerical modeling.

We have yet to discuss the impact of stellar rotation on convection. Brummell *et al.* (1996), for example, simulated how the surface pattern on the scale of supergranulation is distorted by the influence of the Coriolis force, provided that the Taylor number $\mathcal{T}_a$ is large enough and the Rossby number $\mathcal{R}_o$ small enough; a rotation rate of only a few days is required for a solar-type star to produce an effect on supergranular scales. Under the influence of rotation, the intersupergranular lanes becomes more "curvaceous," and the vertices between three or more cells often contain vortices that have a "curly, eyelike or hooked" appearance. These intersections of the downflowing lanes tend to destruct themselves by the dynamical buoyancy of vertical vorticity: fast-spinning vortices evacuate themselves, thus creating an area of low pressure and low density, which can trigger a destructive upflow in their cores by which they are ultimately destroyed.

Ab initio numerical modeling of hydrodynamic convection throughout entire convective envelopes is not within reach of present means, because of the large range in physical parameters that has to be included in such models. Not only is there an enormous range of densities, but also in time scales: the thermal (Kelvin–Helmholtz) time scale of the solar convection zone is 10^5 yr, whereas typical flow time scales are close to a month near the bottom of the convection zone, and some 10 min in the top layers. For quantitative models of horizontally averaged parameters we must consequently manage with mixing-length models, using the adjustable parameter $\alpha_{\mathrm{ML}} = \hat{\ell}_{\mathrm{ML}}/H_p$ [Eq. (2.23)] to arrive at the best possible solution for the problem at hand. Nordlund and Dravins (1990) found from their sample of stars that mixing-length models with $\alpha_{ML} \simeq 1.5$ best matched the run of the mean temperature against depth found in the hydrodynamic models. Hence such mixing-length models are expected to yield the proper relationship between stellar radii and effective temperatures in the range of stellar types under consideration. Note that Schaller *et al.* (1992) find $\alpha_{ML} = 1.6 \pm 0.1$ from the best fit to the red-giant branches of more than 75 stellar clusters, obtained by their stellar-model computations.

2.5.4 *Mimicking the photospheric flow patterns*

The networklike pattern of downdrafts in both the granulation and the super-granulation is determined by a competition for the available space in the photosphere between the outflows of neighboring cells. Schrijver *et al.* (1997a) investigate the formation of the observed pattern through an analogy with Voronoi tesselations. We describe these patterns in some detail, because they are useful in the discussion of the dynamics of the network field (Section 5.3) and in the discussion of radiative losses from the chromosphere (Section 9.5.1).

Two-dimensional Voronoi tesselations are segmentations of a plane into disjoint regions containing all points closer to what is referred to as a generator point than to any other such generator point in the plane (see Okabe *et al.*, 1992, for a description and applications). The formal representation of the Voronoi regions $V(\mathbf{p}_i)$, or cells, can be written as

$$V(\mathbf{p}_i) = \left\{ \mathbf{r} \mid \frac{w_i}{f(\|\mathbf{r} - \mathbf{p}_i\|)} \geq \frac{w_j}{f(\|\mathbf{r} - \mathbf{p}_j\|)}, \text{ for } j \neq i, j \in [1, \ldots, n] \right\}, \quad (2.74)$$

where $P = \{\mathbf{p}_1, \ldots, \mathbf{p}_n\}$ is the set of generator points in the Euclidean plane that is the basis of the segmentation. In a classical Voronoi segmentation, for which $w_i \equiv 1$ and $f(r) \equiv r$, the cell perimeters are convex polygons with boundaries halfway between neighboring generator points, as shown approximately by the example in the right-hand panel of Fig. 2.16.

Because the pattern of convective surface flows is the result of a balance of outflows, the classical Voronoi pattern cannot represent this pattern because it would imply that all cells had the same intrinsic strength and that all forces were central, directed radially away from a set of point sources. If we allow only central forces but allow a range of source strengths w_i as well as a more general dependence of the strength of the outflow on distance to the upflow center than a simple proportionality $f(r) \equiv r$, then the pattern is called a "multiplicatively weighted Voronoi tesselation" if $f(r \downarrow 0) \downarrow 0$. Note that the pattern of such a Voronoi tesselation as defined by Eq. (2.74) is insensitive to a multiplicative scaling applied to all strengths, or to any transformation that does not affect where the balance between neighboring generator point occurs, such as raising the balancing terms (weights and weighting function alike) to some power. Hence, different combinations of strengths and functions $f(r)$ lead to the same tesselation; the pattern does not allow a unique determination of w_i and $f(r)$.

Schrijver *et al.* (1997a) study the case in which $f(r) \propto r^\alpha$ for $\alpha > 0$. For this tesselation, the bisectors between neighboring points are curves that are shifted away from the midpoints toward the generator point with the lowest strength. If the strengths differ enough, the cell of a weak generator point may be embedded entirely within the cell associated with a strong generator point, a situation that is not observed in solar convection (see Figs. 2.12 and 2.13).

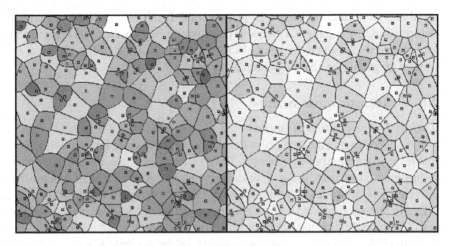

Fig. 2.16. Multiplicatively weighted Voronoi tesselations of a Poisson distribution of generator points (small squares), as described by Eq. (2.74) for $f(r) = r$. The source strengths w_i are chosen from a Gaussian distribution with a standard deviation $\sigma_S/\langle w \rangle$ of 0.16 for the left-hand panel (for which best agreement is found with the observed patterns), and 0.05 for the right-hand panel. The latter case is very similar to the classical Voronoi patterns for which $\sigma_S \equiv 0$. The gray scale indicates the source strength (from Schrijver *et al.*, 1997a).

Figure 2.16 shows examples of generalized Voronoi patterns for $\alpha = 1$, and for a two-dimensional random, or Poisson, distribution for the positions of the generator points. The strengths w_i were drawn from essentially Gaussian distributions with unit average and standard deviations σ_S. Schrijver *et al.* (1997a) use the properties of the histogram of resulting cell sizes to obtain a best-fit value of $\sigma_S/\alpha \approx 0.16$ for granulation and supergranulation alike (corresponding to the left-hand panel in Fig. 2.16); note the scaling with α, which holds for sufficiently small values of σ_S. This range in strengths is remarkably small, requiring some (self-) limiting process for the upflows or outflows. This may reflect the fact that the granulation is driven primarily through the radiative cooling of plasma within the photosphere, where sinking matter is replaced by smoothly upflowing material containing only relatively weak fluctuations (Section 2.5.3).

From further analyses of area histograms of both granulation and supergranulation, Schrijver *et al.* (1997a) concluded that the positioning of the upflow centers is very nearly random. There appears to be a slight clustering in upflow positions for the granulation, consistent with the phenomenon of exploding granules in granular evolution.

If the Voronoi analogy for the formation of the surface pattern of the (super-) granulation is correct, this simple model can be used to mimic the convective pattern. It also explains why simulations of compressible convection in a strongly stratified environment all lead to a similar pattern of upflows and downflows at the top: the pattern is largely insensitive to the details of the physical processes that are included in the model, as long as horizontal outflows dominate, which is the case as long as the pressure scale height is sufficiently small near the photosphere.

The similarity of the observed convective patterns and the Voronoi model requires that the evolution of cells during their initial formation and eventual demise is rather rapid relative to their existence as mature cells. If this were not the case, a significant deformation of the pattern's characteristics would be expected reflecting the phases of growth and decay when the strengths (or "weights") of the cells are not yet within the limited range allowed by the pattern analysis.

2.6   Acoustic waves in stars

2.6.1   The generation of sound

The flows in the convective envelopes of cool stars generate pressure fluctuations that result in traveling acoustic waves. The recent progress in numerical simulations holds promise for an accurate estimate of the wave power as a function of fundamental stellar parameters. Unfortunately, no broad study has yet been performed for the full range of effective temperatures and surface gravities covered by cool stars, so that we necessarily resort to the old estimates based on the mixing-length theory.

Stein (1967) gives the following equation for the acoustic power, F_a, generated in a stellar convective envelope:

$$F_a = \frac{\hat{\rho}\hat{v}_t^3}{\hat{\ell}}(\alpha_q\mathcal{M}^5 + \alpha_d\mathcal{M}^3 + \alpha_m\mathcal{M}), \tag{2.75}$$

where α_q, α_d, and α_m represent the emissivity coefficients of quadrupole, dipole, and monopole sources, respectively; the Mach number $\mathcal{M} \equiv \hat{v}_t/c_s$ is the ratio of the characteristic plasma velocity $\hat{v}_t$ and the sound speed, $\hat{\ell}$ is a characteristic length scale of the convection, and $\hat{\rho}$ is the horizontally averaged density. In a turbulent environment, the acoustic power is generally dominated by the quadrupole contribution [as argued initially by Lighthill (1952) and consistent with computations by Stein (1967, 1968) – which according to Musielak *et al.* (1994), overestimate the fluxes – and by Bohn (1981, 1984)]. However, numerical simulations of near-surface convection suggest that the excitation is dominated by the monopolar generation of sound in the sudden cooling of matter near the surface (Nordlund, 1986), which finds some observational support in time–distance helioseismology (Goode *et al.*, 1998).

Within the convective envelope, the velocities of the convective flows must increase with increasing height so that the flows can carry the convective flux as the density decreases; see Eq. (2.28). This simple scaling argument no longer holds, however, as the main energy transport mechanism shifts from convective to radiative just below the photosphere.

Numerical experiments (e.g., Nordlund and Dravins, 1990) indicate that the rms velocities continue to increase with height throughout the photosphere: the rms value of vertical velocities reaches a maximum some tens of kilometers below the photosphere, but the rms power in the horizontal velocities continues to increase up to a few hundred kilometers above that, depending on the stellar spectral type. The rapid decrease of the density with height and the functional dependence of F_a on $\mathcal{M}$ limit the predominant source region for acoustic waves to a relatively thin layer just below the stellar surface (*e.g.*, Musielak *et al.*, 1994). Observations confirm this: Kumar and Lu (1991) argue that the wave source region for the Sun lies at $\sim$300 km (or 1.5 pressure scale heights) below the photospheric level (at $\tau_5 = 1$), with a depth range of only some 50 km. This is 150 to 200 km deeper than predicted by the standard mixing-length theory, reflecting the inadequacy of the mixing-length approximation.

Analytical approximations show that in all cool stars the associated acoustic spectrum extends from the acoustic cutoff period, P_A, beyond which acoustic waves cannot travel [discussed in Section 2.6.2, and given by Eq. (2.79)] to roughly a factor of 10 lower, with a pronounced peak between these extremes at $\sim P_A/5$ or below (see, e.g., Ulmschneider, 1990, 1991).

The acoustic fluxes are usually computed by using the mixing-length concept to estimate the convective velocities $\hat{v}_t$. Unfortunately, the mixing-length concept fails to describe convection, particularly near the stellar surface. This makes the choice of "characteristic" mixing-length parameters ambiguous, yet a deviation from the proper characteristic value of α_{ML} by only a factor of 2 is amplified to an uncertainty of approximately an order of magnitude in the acoustic flux. Estimates by Bohn (1984), based on Eq. (2.75) and the (apparently erroneous) assumption that quadrupolar emission dominates over monopolar emission, can be approximated by

$$F_s \approx 6 \times 10^8 \left(\frac{T_{eff}}{T_{eff,\odot}} \right)^{9.75} \left(\frac{g_\odot}{g} \right)^{0.5} \left(\frac{\alpha_{ML}}{1.5} \right)^{2.8} \text{ (erg cm}^{-2}\text{ s}^{-1}\text{)}. \qquad (2.76)$$

The steep increase of F_s with T_{eff} reflects the rough proportionality of $\hat{\rho}\hat{v}_t^3$ to σT_{eff}^4 [Eq. (2.29)]. In a crude approximation, therefore, $F_s \propto \hat{v}_t^8 \propto T_{eff}^{10.7}/\hat{\rho}^{2.7}$, modified to Eq. (2.76) through the correlation of $\hat{\rho}$ with T_{eff} and g. Table 2.6 gives acoustic flux densities of several stars as predicted by Eq. (2.76); these fluxes for stars not too different from the Sun are of the order of one percent of the total luminosity. The fit of Eq. (2.76) can be used up to $T_{eff} \approx 8,000$ K; the decreasing efficiency of convection in even warmer stars results in an abrupt decrease of F_s with increasing T_{eff}.

The decrease in the generation of sound with decreasing effective temperature is readily explained by the decrease of the characteristic velocity $\hat{v}$ with decreasing T_{eff} predicted by the scaling rule in Eq. (2.29). Moreover, the mass density ρ in the top of the convection zone increases with decreasing T_{eff} because the opacity κ_c decreases sharply with decreasing temperature. As a result of virtually hydrostatic equilibrium, the gas pressure p_g near and below $\tau_R = 1$ increases with decreasing T_{eff}. The estimated characteristic turbulent velocities $\hat{v}_t$ that are expected given the mean photospheric density are listed in Table 2.6 for a few Sun-like stars.

An estimate by Musielak *et al.* (1994), limited to the solar case, but using a more advanced model for the energy spectrum of the plasma flows, yielded

$$F_s \approx 5 \times 10^7 \left(\frac{\alpha_{ML}}{1.5} \right)^{3.8} \text{ (erg cm}^{-2}\text{ s}^{-1}\text{)}. \qquad (2.77)$$

For the Sun, and with $\alpha_{ML} = 1.5$, this flux is approximately an order of magnitude lower than Bohn's value. Note that Eq. (2.77) shows a dependence on α_{ML} that is even stronger than that in Eq. (2.76).

The wave flux given by Eq. (2.77) agrees fairly well with the lower bound for the observed wave flux of $F_s(h) = 2 \times 10^7$ erg cm^{-2} s^{-1} (Deubner, 1988) in the low and middle photosphere at a height of $h \approx 300$ km over the level at which $\tau_5 = 1$. This rough agreement is remarkable, given the weakness of the mixing-length approach, the differences between actual convection and true turbulence, and the fact that the model computation does not incorporate radiative damping (Section 2.7): such damping, which has yet to be properly modeled quantitatively, substantially lowers the acoustic fluxes above the photosphere. Proper estimates for acoustic fluxes will have to await numerical results. Fortunately, the consequences of these substantial uncertainties for atmospheric heating appear to be limited, as we discuss in Section 2.7.

2.6.2   Acoustic waves, resonance, and asteroseismology

The acoustic waves generated by stellar convective motions propagate in all directions. This allows the study of the solar interior through the use of helioseismological tools (and of stellar interiors with asteroseismology).

The peak in the power spectrum of the solar acoustic (or p) modes occurs at $\sim$5 min, whereas the granulation evolves on a time scale that is only two to three times longer (see Section 2.5). In studies of granulation, the intensity fluctuations associated with the acoustic waves can be filtered out quite efficiently, however, by applying Fourier filters based on the dispersion relation and resonance conditions. Similarly, the rms Doppler signal of approximately 400 m/s to 600 m/s (Keil and Canfield, 1978) that is associated with the p modes can be filtered out to facilitate the study of, for instance, the supergranular Doppler pattern.

In the simplest description of horizontally traveling waves, the dispersion relationship reads

$$k_h = \frac{\omega}{c_s}, \tag{2.78}$$

for horizontal wave number k_h. This equation defines a cone in $(\mathbf{k}, \omega)$ space centered on the origin. Such waves can be filtered out of photospheric observations by setting all Fourier components in $(\mathbf{k}, \omega)$ space that lie on this cone to zero. The "subsonic" filter that was developed by the Lockheed group (e.g., Title *et al.*, 1989) zeroes all Fourier components that lie at $\omega > vk$ for a value of v that is set well below the sound speed of 7 km/s, usually as low as 3–5 km/s, depending on the observer and the planned use of the data set.

The most important waves in the solar photosphere travel nearly radially. These waves traverse the highly stratified interior. The interference patterns of these waves have resulted in the field of helioseismology. We discuss the wave properties here only briefly; interested readers are referred to the monograph on stellar oscillations by Unno *et al.* (1989) or the reviews by Deubner and Gough (1984) and Gough and Toomre (1991).

Outward-propagating waves experience a decrease in temperature, with an associated decrease in the phase velocity. This, together with the virtually hydrostatic equilibrium of the atmosphere, leads to a reflection of these waves for periods with a sufficiently low frequency. The limiting wave frequency is given by the *acoustic cutoff frequency*

$$\omega_A \equiv \frac{2\pi}{P_A} = \frac{c_s}{2H_\rho}\left(1 - 2\frac{dH_\rho}{dr}\right)^{1/2} \tag{2.79}$$

(with c_s as the sound speed and H_ρ as the density scale height). The cause of this reflection is that waves with periods exceeding P_A cannot propagate because such disturbances lift an entire pressure scale height without inducing pressure fluctuations. Because every pressure scale height contains roughly as much mass as all overlying layers together (this is exactly true in the case of an isothermal atmosphere), the bulk of the atmosphere is raised or lowered in response to the photospheric motions without inducing restoring forces (e.g., Schrijver *et al.*, 1991), leaving only an evanescent wave in the atmosphere. If the outer atmospheric temperature would decrease to zero, all waves – regardless of frequency – would be reflected at some finite height. In an atmosphere with a temperature inversion above the photosphere, which all magnetically active stars have over at least part of their surface, some fraction of the energy in waves with periods above the cutoff period will "tunnel" through the temperature minimum region as near-evanescent waves. Should an atmosphere contain a temperature inversion covering the entire surface, an upper limit is set to ω_A; the minimum resonant *p*-mode frequency is a measure for the minimum temperature reached in the atmosphere.

When all acoustic waves of a given frequency are reflected at some frequency-dependent height, the constructive interference of waves results in standing wave patterns for appropriate wave numbers. Leighton *et al.* (1962) were the first to detect the oscillatory pattern associated with this wave interference. Ulrich (1970) suggested that these oscillations were related to the solar resonant cavity; this was confirmed observationally by Deubner (1975). The millions of eigenmodes on the Sun, each with a characteristic velocity amplitude of the order of 10 cm/s, result in undulations of the photospheric surface by only some 15 km. These are small compared to the corrugation produced by the convective granulation.

The *p* modes are well represented by the wave equation (Deubner and Gough, 1984)

$$\frac{\partial^2 \Psi}{\partial r^2} + k^2(r)\Psi = 0 \tag{2.80}$$

for wavelengths much shorter than the stellar radius, and for positions not too close to the stellar core. Here the wave function $\Psi = \sqrt{\rho}c_s^2 \nabla \cdot \delta \mathbf{R}$, and ρ and c_s are the local density and sound speed, and $\delta \mathbf{R}$ is the fluid displacement vector.

The eigenfunctions for the resonant modes can be separated into a radial function, $f_{n\ell}(r)$, and a spherical harmonic, $Y_{\ell m}(\theta, \phi)$ $(-\ell \leq m \leq \ell)$:

$$E = f_{n\ell}(r)Y_{\ell m}(\theta, \phi)e^{i\omega_{n\ell m}t} \tag{2.81}$$

$$= f_{n\ell}(r)\left[\text{sgn}(m)^m \sqrt{\frac{2\ell+1}{4\pi}\frac{(\ell-m)!}{(\ell+m)!}} P_\ell^m(\cos\theta)e^{im\phi}\right] e^{i\omega_{n\ell m}t},$$

where P_ℓ^m are associated Legendre polynomials. The *radial order n* is the number of radial nodes. The *azimuthal order m* is the number of nodes around the equator. The *spherical harmonic degree ℓ* is a number not easily recognized in the wave pattern; ℓ is the number of nodes along a great circle that intersects the equator at an angle of $\cos^{-1}\{(m/[\ell(\ell+1)])\}$. The degree of standing waves is related to the horizontal wave number at radius r by

$$k_h r = \sqrt{\ell(\ell+1)}. \tag{2.82}$$

Deubner and Gough (1984) give the radial component of the local wave number as

$$k_r^2(r) = \frac{\omega^2 - \omega_A^2}{c_s^2} + \frac{\ell(\ell+1)}{r^2}\left(\frac{N^2}{\omega^2} - 1\right), \tag{2.83}$$

with the acoustic cutoff frequency ω_A given by Eq. (2.79), and where N is the Brunt-Väisälä frequency,

$$N^2 = g \left[\frac{d \ln \rho}{dr} - \frac{1}{\left(\frac{\partial \ln p}{\partial \ln \rho} \right)_s} \frac{d \ln p}{dr} \right], \tag{2.84}$$

at local gravity g and pressure p. The Brunt-Väisälä frequency is the frequency of an oscillating fluid parcel that remains in pressure balance with its surroundings. Note that waves only propagate where $k^2 > 0$, requiring that at least $\omega > \omega_A$ and $\omega > \ell(\ell + 1)c_s^2/r^2$; see Deubner and Gough (1984) for the detailed conditions.

Constructive interference occurs if an integer number of halfwaves fits between the turning points. This requirement has to be modified if phase changes occur at the inner and outer turning points. At the upper boundary, for instance, a phase change of nearly $\pi/2$ is expected, because of the reflection against an open boundary; radiative transfer introduces nonlocal coupling that modifies this. In general, the resonance condition for the radial wave number k_r reads

$$\int_{r_i}^{r_o} k_r dr = (n + \epsilon)\pi, \tag{2.85}$$

where $r_{i,o}$ are the inner and outer turning points of the wave and ϵ is a function of the properties of the layers near the surface of the star. Equation (2.85) is an integral equation over the entire interior of the star through which the wave can propagate. The use of this relationship for a set of waves traveling to different depths allows the determination of the run of the sound speed in the interior of the star.

For values of ℓ that are very large compared to n, Eq. (2.85) can be approximated by

$$\omega \approx \left[\frac{2\gamma(n + \epsilon)}{(\mu + 1)} \right]^{1/2} \sqrt{g\,k}, \tag{2.86}$$

for a polytrope of index μ and assuming that the adiabatic exponent γ is constant. These high-ℓ waves penetrate only a shallow surface layer, so that these approximations have to be valid only there. Note that only truly global waves, i.e., those that travel around the entire Sun, demonstrate the full resolution into spikes in the Fourier-power domain. Locally resonant modes lead to unresolvable ridges in the **k**–ω diagrams.

The surface gravity (f) modes are not described by Eq. (2.80) but rather satisfy $\Psi = 0$. Their dispersion relationship is independent of the stratification of the atmosphere (Deubner and Gough, 1984):

$$\omega \approx \sqrt{g\,k}. \tag{2.87}$$

Both Eqs. (2.86) and (2.87) describe parabolic ridges in an ω–**k** diagram. The existence of the lower-bound described by Eq. (2.87) leads to the so-called subfundamental filter, which is a more conservative filter to remove oscillatory fluctuations in sequences of photospheric observations than the subsonic filter. Once these oscillations have been removed, it is much easier to study granular evolution and, for instance, to interpret the temporal cross correlations (e.g., Spruit *et al.*, 1990).

Duvall, Kosovichev, and colleagues (e.g., Duvall *et al.*, 1997) are spearheading a new development in helioseismology, referred to as time–distance helioseismology. They use cross correlation measurements to determine the direction-dependent travel times of signals from one location on the Sun to another. An inversion of the ray paths then provides information not only on the subsurface temperature and density, but also on the flow patterns and, at least in principle, on the magnetic field. The results are tantalizing but need further confirmation at the time of this writing. The potential of this method is, however, enormous.

2.7 Basal radiative losses

Diagrams plotting stellar atmospheric radiative losses in chromospheric lines versus the effective temperature show a well-defined lower limit (Figs. 2.17 and 2.18) below which very few stars are found (such stars have peculiar abundances). Studies of relationships between radiative losses from different temperature intervals in solar and stellar atmospheres among themselves (Section 9.4), or with rotation rate (Section 11.3) have uncovered the existence of a steady background, the so-called *basal emission*. This emission exists for all cool stars; its magnitude depends primarily on the effective temperature and is rather insensitive to surface gravity or abundances.

Very near this lower-limit flux, the global stellar atmospheric variability vanishes (Schrijver *et al.*, 1989b). This suggests that dynamo cycles, active regions, or even active network do not exist on the least active stars. Stars near the observed lower-limit flux rotate very slowly, so that it is likely that a large-scale dynamo is operating at most very weakly in stars near the lower limit (Section 11.3). The minimum observed stellar flux densities (Table 2.7) are close to what is observed over the centers of supergranules on the Sun, where magnetic flux densities are low. Apparently, the basal energy deposition is not related to the intrinsically strong magnetic fields involved in cyclic activity.

The current best estimates of the basal fluxes in a number of spectral lines for stars of different effective temperatures are listed in Table 2.7 (see Fig. 2.18 for Mg II h and k). All basal fluxes decrease toward lower effective temperatures. The structure of the atmosphere depends on the fundamental stellar parameters, as demonstrated by the lower half of Table 2.7: the relative strengths of the line emissions change nonmonotonically with T_{eff}. For instance, the Mg II h+k basal emission in F-type stars is only an order of magnitude larger than the upper chromospheric C II basal emission, but this difference increases to nearly 3 orders of magnitude for mid-K-type stars, and it decreases again after that.

Rutten *et al.* (1991) have argued that basal emission in coronal soft X-rays is negligible for all mid-F- through late-G- or early-K-type stars. Schmitt (1997) finds a minimum flux of 10^4 erg cm^{-2} s^{-1} from *ROSAT* observations of a complete sample of A-, F-, and G-type stars in the solar vicinity. Whether this emission, which lies well below the values that Rutten *et al.* (1991) could determine as basal, reflects a basal coronal activity is as yet unclear.

The "basal" radiative losses are believed to be (largely) caused by the dissipation of acoustic waves in stellar atmospheres (see Schrijver, 1995, and references therein). When acoustic waves generated by the convection in a stellar convective envelope (Section 2.6) propagate outward into the stellar atmosphere, their amplitude increases rapidly because of the precipitous decrease of the density. Starting with Biermann (1946) and Schwarzschild (1948), theorists have advanced from the basic idea that these waves develop into shock waves that dissipate as they propagate. This dissipation puts the time-averaged radiative losses of even the least active parts of the solar chromosphere (Sections 2.8 and 8.1) significantly above what is expected from (hypothetical)

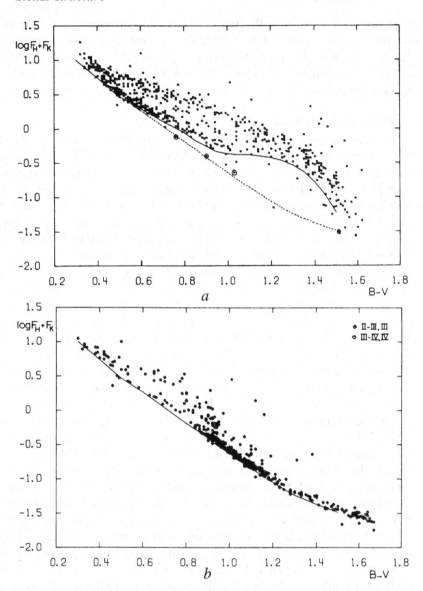

Fig. 2.17. The Ca II H+K surface flux density (in "arbitrary units" – Section 2.3.4 – of 1.3×10^6 erg cm^{-2} s^{-1}/unit) as a function of $B-V$ color for a main-sequence stars (LC V and IV–V) and b (sub)giants (LC II–III to IV). The solid curves are the minimal fluxes adopted in that study. The dashed curve in panel a repeats the minimal flux for giants from the bottom panel. Rutten (1984a; the source of the figure) finds that the encircled stars in the top panel are likely misclassified evolved stars.

radiative-equilibrium atmospheres. The same is true for the disk-integrated emissions from the least active stars.

The computation of the propagation of acoustic waves for the upper photosphere and the chromosphere is complicated by the dependence of the thermodynamic and radiative properties on the time-dependent radiation field that links nearby regions that are not

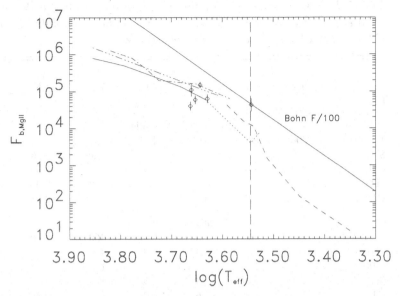

Fig. 2.18. Diagram showing the minimum or basal flux in Mg II h+k as a function of stellar effective temperature. The solid curve shows the results from Rutten *et al.* (1991; see Table 2.7), the triple-dotted/dashed curve shows the earlier results from Schrijver (1987a). The dotted curve represents the upper limits derived for very cool dwarf stars. The dashed curve shows the fluxes of the least active evolved stars from Fig. 3 in Judge and Stencel (1991). The dashed-dotted line shows the results from theoretical computations including partial redistribution in Buchholz and Ulmschneider (1994). To the right of the vertical dashed line at $T_{eff} = 3,500\,K$, most stars in the observed sample are asymptotic giant-branch stars, and the lower-limit flux is poorly defined. The diamonds show the detected fluxes in metal-deficient field giants from Dupree *et al.* (1990). The upper solid line shows the expected acoustic flux as computed with Eq. (2.76) for solar surface gravity (figure from Schrijver, 1995).

in the same phase of wave evolution. Nevertheless, substantial progress has been made over the past decade. The primary result is that the expected radiative losses resulting from the dissipation of acoustic waves are rather insensitive to a variety of details in the chain of computations. For one thing, the generated acoustic flux is relatively insensitive to the details of the turbulent energy spectrum or to the abundances.

The acoustic flux generated in the top layers of convective envelopes depends on the surface gravity g of the star [as in Eq. (2.76)]. It turns out that photospheric radiative damping, strongly sensitive to surface gravity, reduces this dependence. Ulmschneider (1988, 1989) compares the wave period P to the photospheric radiative relaxation time, $\hat{t}_{rad}$, which is approximated by

$$\hat{t}_{rad} \approx 3 \left(\frac{g_\odot}{g}\right)^{0.43} \left(\frac{T_{eff,\odot}}{T_{eff}}\right)^{5.88} \quad (s) \qquad (2.88)$$

(after Ulmschneider, 1988, but normalized to solar units). A wave with period $P \gg \hat{t}_{rad}$ suffers from strong radiative damping, whereas a wave with $P \ll \hat{t}_{rad}$ does not. Waves that are to heat the chromosphere must propagate through a layer of strong radiative

Table 2.7. *Best estimates of basal fluxes (top) as a function of B–V (in units of $10^3 erg\ cm^{-2}\ s^{-1} = W/m^2$) and relative fluxes (bottom) normalized to the C II basal flux level at each spectral color $(B-V)^a$*

	B–V						
Line	0.3	0.5	0.6	1.0	1.2	1.5	1.6
	Absolute flux densities[b]						
Mg II h+k	800	500	380[1]	130	60	<4[2]	<8[3]
Si II	100	27[1]	14[1]	1	0.6[1]	0.26[1]	0.2
C II	76[4]	5.6[4]	3.1[4]	0.16	0.13[1]	0.09[1]	0.08
Si IV	80	11[1]	4	0.4	0.19[1]	0.06[1]	0.04
C IV	160	22[1]	8	0.16	0.08[1]	0.03[1]	0.02
	Relative flux densities						
Mg II h+k	10.	90.	120.	810.	460.	<40	<100
Si II	1.3	4.8	4.5	6.3	4.6	2.9	2.5
C II	1	1	1	1	1	1	1
Si IV	1.0	2.0	1.3	2.5	1.5	0.7	0.5
C IV	2.1	3.9	2.6	1.0	0.6	0.3	0.3

[a] The flux densities in the top half of the table are from Rutten *et al.* (1991) except where noted otherwise. The Si IV and C IV flux densities are relatively uncertain.

[b] These flux densities are derived from the following: 1, logarithmic-linear interpolation from Rutten *et al.* (1991); 2, from Doyle *et al.* (1994) for Gl 813; converting $T_{eff} = 3,500$ to $B-V \approx 1.5$; 3, from Byrne (1993) for Gl 105B, the lowest detection at $B-V = 1.6$; and 4, from Schrijver (1993a).

damping. A crude estimate of the relative magnitude of radiative damping in different stars can be made by comparing the period at which the acoustic power spectrum peaks to the time scale $\hat{t}_{rad}$ (Ulmschneider, 1988). This comparison suggests that the acoustic flux reaching chromospheric heights in (super-)giants is much more strongly reduced than in main-sequence stars, in warmer stars more strongly than in cooler ones. But it is difficult to estimate how strongly the gravity dependence of wave generation is compensated by radiative damping in the photosphere; radiation damping is poorly known even for the solar atmosphere (see, for instance, Gomez *et al.*, 1987).

Interestingly, models suggest that radiation damping does not affect the total radiative output of an acoustically perturbed atmosphere, as long as sufficient power remains to form shocks low enough in the atmosphere. For the Sun, the distance between the photosphere and the mean transition-zone height corresponds to a density decrease by a factor of the order of one million. Under these circumstances, acoustic waves quickly reach a shock strength that remains approximately constant with height and that is largely independent of the initial amplitude. At this *limiting shock strength*, the shock balances dissipation against the increase in amplitude as the density decreases. The shock strength determines the wave energy flux density, F_s, passing through a layer. For a monochromatic wave in an isothermal, plane–parallel atmosphere in which transient ionization is ignored,

$$F_s = \frac{1}{12} \frac{\gamma^3 g^2}{(\gamma + 1)^2 c_s} P^2 p \approx \left(\frac{3}{\delta^2}\right) c_s p, \tag{2.89}$$

where p is the gas pressure and $\gamma = 5/3$ is the ratio of specific heats (Ulmschneider, 1991). The approximation in the right-hand term (Ulmschneider, 1990) is made by choosing a period P corresponding to the peak in the acoustic power spectrum typically at P_A/δ with the factor δ in the range of 5 to 10, and with $P_A = 4\pi c_s/\gamma g$, so that F_a is $2 \times 10^4 p$ to $9 \times 10^4 p$ erg cm^{-2} s^{-1} for the solar atmosphere.

The limiting-shock-strength behavior is also observed in nonisothermal, ionizing atmospheres, but there Eq. (2.89) is only valid as an approximation. Surprisingly, at the limiting shock strength, F_s does not depend on the wave flux generated by the convective envelope: raising or lowering the photospheric wave flux mainly affects the initial height of shock formation, lowering it, if necessary, down to levels where continuum emission serves as a major coolant.

In reality, the wave spectrum is polychromatic. This complicates matters, because acoustic waves that propagate in stellar atmospheres are dispersive (see, for example, Carlsson and Stein, 1992). Long-period waves with periods above the acoustic cutoff period produce wakes of standing waves behind them, oscillating at the acoustic cutoff period. Dispersion causes short-period waves to form shocks at intervals of the acoustic cutoff period (see also Rammacher and Ulmschneider, 1992), provided that their amplitude is large enough to shock at low heights. Fleck and Schmitz (1993) show that in a one-dimensional, vertically stratified atmosphere, relatively strong shocks run quickly and catch up with slower, weaker shocks, so that the high-frequency shock waves eventually merge to form strong long-period shocks. The merging of overtaking shocks shifts the power down to lower frequencies, into a spectrum that is apparently insensitive to the details of the input spectrum. The process of shock overtaking is limited by the finite thickness of the atmosphere. If the power in the input spectrum is increased, shocks form at a lower height, giving the shock waves an increased distance (or time) to overtake, so that the average period increases with increasing power until finally the chromospheric acoustic cutoff period is reached (see the discussion in Carlsson and Stein, 1992). Simulations by Sutmann and Ulmschneider (1995a, 1995b) for the simplified case of adiabatic waves suggest that a limiting shock strength is reached regardless of the initial acoustic spectrum.

Dynamic models by Carlsson and Stein (1994, 1997) show that efficient radiative cooling makes the intervals of shock heating so short lived that there is no time-averaged chromospheric temperature inversion in acoustically heated domains of the outer atmosphere. Carlsson and Stein use the observed photospheric motions as a piston to drive their atmosphere to model a particular feature in the nonmagnetic solar chromosphere, namely the Ca II K$_{2V}$ bright points (see Section 2.8). These bright points typically appear a few times in a row as features with asymmetric Ca II H and K line profiles, in which the violet peak in the profile is particularly brightened during certain phases (see Rutten and Uitenbroek, 1991, for a review). Carlsson and Stein show that although shock dissipation (through viscous dissipation and external work) produces short-lived heated intervals, the time-averaged temperature is a smoothly decreasing function with height, decreasing from the photospheric temperature to below 4,000 K. The radiative losses from the intermittent periods of strong shock dissipation weigh strongly in the total, time-averaged radiative losses, however. This probably explains why one finds a chromospheric temperature rise if the time-averaged radiative losses in the ultraviolet continuum are used to derive the stratification in a mean, static atmosphere; the numerical simulation shows this rise to be spurious, however. Consequently, acoustic heating is not a properly chosen term; we therefore refer to the process as acoustic energy deposition.

Although the hydrodynamic simulations by Carlsson and Stein (1994) reproduce the properties of Ca II K_{2V} bright points quite well, the relatively low area coverage by these bright points makes it unlikely that the Ca II K_{2V} grains emit the bulk of the basal emission for Ca II K. Another, more indirect, argument is that the observed distribution function of Ca II K intensities at low magnetic flux densities is nearly symmetric (see Schrijver *et al.*, 1989a), whereas occasional strong brightenings of the bright points would produce a rather skewed distribution. The Ca II K_{2V} bright points probably represent the most extreme cases in a multitude of less-obvious brightenings that add up to the basal radiative losses.

The net result of the entire chain of computations is that the predicted radiative losses from an atmosphere pervaded by acoustic waves are largely independent of the acoustic energy spectrum, of surface gravity, and even of elemental abundances: numerical experiments for acoustically heated chromospheres of moderately cool giant stars (Cuntz *et al.*, 1994; see also Cuntz and Ulmschneider, 1994), show that the Mg II line-core emission *increases* by only some 25% as the metallicity *decreases* from solar values by over 2 orders of magnitude (Fig. 2.18). This leaves the effective temperature as the major controlling parameter.

2.8   Atmospheric structure not affected by magnetic fields

The solar outer atmosphere is highly dynamic and inhomogeneous, even in regions where the magnetic fields can be ignored. Acoustic waves, both traveling and evanescent, in a range of frequencies and wavelengths result in a complicated and rapidly evolving pattern of beats in which the compression waves are modified by three-dimensional radiative transfer that has yet to be modeled in detail. In addition to these waves, there are the more gradual undulations associated with adjustments of the mean upper photospheric structure in response to granulation and other scales of convective overshoot discussed in Section 2.5. In this section we first concentrate on the dynamics of the nonmagnetic atmosphere, and then we discuss its temperature structure. The term nonmagnetic does not imply that there are regions in the outer atmosphere that are entirely free of field (see Section 4.6), but rather that in these regions the magnetic field is weak and does not inhibit the plasma motions in flows or waves. And although the focus here is on the nonmagnetic outer atmosphere, we discuss its dynamics in comparison to the magnetically dominated regions.

At photospheric heights, the oscillatory properties of the centers of supergranules and of the magnetic network appear to be rather similar: both are characterized by the broad *p*-mode spectrum with periods near 5 min (e.g., Kulaczewski, 1992). Higher up in the atmosphere, however, a clear difference is observed. The power spectrum of intensity fluctuations in chromospheric lines observed over cell centers are dominated by 3-min oscillations (Von Üxküll *et al.*, 1989, Deubner and Fleck, 1990, and Lites *et al.*, 1993b), which lose their coherence with increasing height, until only occasional coherent wave trains are observed at transition-region temperatures above $\sim$50,000 K. Deubner and Fleck (1990) argue that the 3-min periodicity reflects at least in part a trapping of modes between the temperature minimum and the transition-region temperature rise, while small-scale inhomogeneities in the structure and vertical extent of the cavity cause a broad distribution in the power spectrum. It is not necessary to have a resonating cavity to see these waves, however, because any disturbance will excite waves at the local cutoff frequency.

The evolving wave patterns result in a patchy undulation of the intensity in the chromospheric domain, similar to what is seen in the photosphere. Apart from these relatively large patches, much more localized phenomena are also observed. Rutten and Uitenbroek (1991), for instance, review the properties of the Ca II K_{2V} grains or bright points. They point out that narrow-band spectroheliograms taken in the H and K line cores show, apart from the many clusters of small bright elements in the magnetic network, the so-called internetwork bright points, cell flashes, or cell grains (referred to in Section 2.7) that are observed within supergranular interiors. These grains are 1,500 km or smaller in size, and persist for $\sim$100 s each time they brighten. They tend to recur several times with 2- to 5-min intervals between successive brightenings. Rutten and Uitenbroek (1991) proposed that these features are related to the inner-cell bright points seen in the ultraviolet continuum near 1,600 Å (which originates near the classical temperature minimum, which appears, as we pointed out in Section 2.7, not to exist as such in nonmagnetic environments) and they argued that these grains are a hydrodynamical phenomenon of oscillatory nature. They speculate that these grains are a direct result of the acoustic processes in the nonmagnetic chromosphere, which is now supported by simulations (Section 2.7).

An even lower coherence of waves at still higher formation temperatures is found in the early work of Vernazza *et al.* (1975). Using raster spectroheliograms obtained with the Apollo Telescope Mount on board Skylab, taken 1 min apart with a resolution of 5 arcsec, they found substantial power in the short-term fluctuations that occur synchronously in a few spectral lines. The average lifetime of the brightenings is 70 s, and their mean occurrence is spaced by 5.5 min, but no strong periodic oscillations are seen. The brightenings occur in the central regions of the network cells as well as in the boundaries.

Athay and White (1979), in contrast, found that the C IV λ 1,548 line intensities, observed by *OSO*-8 with an effective aperture of $2'' \times 20''$, show relatively frequent periodic oscillations in the 3- to 5-min range. The periodic oscillations have a short coherence length and a tendency to be mixed with prominent aperiodic fluctuations. Athay and White propose that this mixing prevented other studies from uncovering these periodic variations. They interpret the much more frequent low-amplitude aperiodic intensity fluctuations as sound waves whose periodicity and coherence are destroyed by the variable transit time through an irregularly structured chromosphere (see the review by Deubner, 1994).

Over the magnetic network, the oscillations are much more stochastic in nature. The power spectra for the strong network are dominated by low-frequency oscillations with periods of 5 up to 20 min. Interestingly, Deubner and Fleck (1990) find that the spectra of areas of atmospheric emission intermediate to the cell interior and the strong network are intermediate to the abovementioned spectra, displaying a clear 3-min component; this suggests that magnetic and acoustic energy dissipating processes coexist within the same resolution element.

During the past two decades, evidence has been found – mainly from carbon-monoxide infrared spectra – that there are also very cool parts in solar and stellar outer atmospheres, with temperatures down to below 4,400 K, which have been interpreted as being caused by a thermal bifurcation of the chromosphere (*e.g.*, Heasly *et al.*, 1978, for α Boo; Ayres, 1981; Ayres *et al.*, 1986; and Solanki *et al.*, 1994). Initial numerical modeling suggested that the strong CO-line cooling could cause such a thermal bifurcation (*e.g.*, Kneer, 1983; Muchmore and Ulmschneider, 1985), but more recent simulations by Anderson (1989) and Mauas *et al.* (1990), which incorporate many CO lines and the CO dissociation equilibrium in time-independent models, yield no evidence for an instability.

The CO lines go into emission some 300 to 1,000 km above the solar limb (Solanki *et al.*, 1994; Uitenbroek *et al.*, 1994), so that the cool material is truly superphotospheric. The formation height of the CO lines apparently samples a substantial height interval because of the presence of power near 3 min – characteristic of chromospheric oscillations – as well as 5 min – characteristic of photospheric oscillations – in CO spectra (Uitenbroek *et al.*, 1994). Ayres (1991a) proposed that a distribution of cold columnar regions with a collective surface filling factor of less than some 20% (causing increasing shadowing of cool areas in front of others as one looks progressively nearer to the solar limb) would be consistent with the observational constraints (see the discussions by Athay and Dere, 1990, and Ayres *et al.*, 1986). The response of the atmosphere to acoustic waves makes these warm and cool regions transient, resulting in evolving patterns across solar and stellar disks.

3

Solar differential rotation and meridional flow

In this chapter, we summarize data on solar rotation, meridional flow, and other large-scale flows, that is, flows at length scales of $\ell \gtrsim 100,000$ km, substantially larger than that of the supergranulation. There is not yet a quantitative theory for large-scale convection, differential rotation, and meridional flow; hence this chapter is restricted to observational data. The solar rotation $\Omega(\theta, r)$ is found to be differential; that is, dependent on both the heliocentric latitude θ as well as the radial distance r to the solar center; it may vary in time. This chapter is a summary; for comprehensive reviews we refer to Schröter (1985) and Snodgrass (1992); Bogart (1987) presented a review with a broad historical scope.

A consistent description of rotation and meridional flow requires a coordinate system based on rotational data. The heliocentric coordinate system depends on the orientation of the solar rotation axis with respect to the ecliptic plane, usually described by the inclination i of this axis with respect to the normal of the ecliptic plane and the longitude of the ascending node $\hat{\Omega}$ of the solar equator on the ecliptic. Carrington (1863) determined i and $\hat{\Omega}$ by minimizing meridional displacements of sunspots, and his numerical results are still used; see Schröter (1985) for a discussion of the errors in these values and their effects on the flow determination. Carrington also determined an average synodic (i.e., as seen from the Earth) rotation period for sunspots of 27.275 days, which corresponds to the true sidereal period of 25.38 days, or a Carrington rotation rate of 14.18°/day (this rate turned out to correspond to the rotation rate of long-lived sunspots at a latitude of 16°). Solar rotations are numbered, starting with Carrington rotation No. 1 on 9 November 1853. The Carrington rotation period defines the grid of Carrington longitudes that underlies the synoptic (or Carrington) maps, in which any feature on the Sun is given the Carrington longitude corresponding to the central meridian at the moment that it crosses the central meridian.

There are three methods for measuring the velocity field in the solar atmosphere. The first is the measurement of displacements, using solar features as tracers, for instance by a correlation analysis of series of images. The second is the measurement of Doppler shifts in a spectral line over (a large part of) the solar disk. Most of such data have been collected with the efficient and sensitive measurements obtained with the Doppler compensators in magnetographs, which measure the shift of the spectral line. A large data library has been collected with the full-disk magnetograph at Mt. Wilson Observatory; the set-up and data analysis have been described by Howard and Harvey (1970). The third method is the application of helioseismology, discussed in Section 3.3.

Although the first two methods are in principle straightforward, the accuracy is limited by random and systematic errors (discussed by Schröter, 1985, and Snodgrass, 1992). Below we touch upon a few of these errors and uncertainties.

As for the tracer method, an inaccuracy in the determination of the orientation of the solar axis affects the deduced velocity field: an error of 0.1° in i introduces an error in v_{rot} of 4 m/s near the equator, and it distorts the meridional flow pattern. Clearly, the interpretation of the velocity field indicated by a specific tracer depends on the depth where those tracers are effectively anchored, and this depth is not accurately known.

The method using Doppler shifts is affected by several systematic errors. One is stray light of instrumental and Earth-atmospheric origins that veils the solar image and that has a spectrum that is little affected by the local solar velocities at the observed positions. In addition, measurements have to be corrected for the limb shift, which is observed as a redshift of medium-strong spectral lines with respect to their wavelengths measured at the center of the disk. This line-dependent shift is not yet completely understood. A major contribution comes from the blueshift of the reference wavelengths, because spectral line are preferentially blueshifted near the center of the disk by the rising component of granular convection (Section 2.5.3). Moreover, the asymmetry of the C-shaped line profile, also caused by the granulation, varies across the disk.

3.1   Surface rotation and torsional patterns

Usually the latitude-dependent velocity of the solar surface features is represented by

$$\Omega(\theta) = A + B \sin^2 \theta + C \sin^4 \theta, \qquad (3.1)$$

where Ω stands for the sidereal angular velocity and θ for heliocentric latitude; A, B, and C are adjustable coefficients. In Table 3.1 we have selected a few of the $\Omega(\theta)$ determinations that are derived by various methods applied to large bodies of observational data; most of these $\Omega(\theta)$-relations are plotted in Fig. 3.1.

The rotation rates determined from Doppler shifts and various tracers observed at the photospheric level differ by less than 5%. At the 1% level, however, differences between results derived by various methods and various groups of authors appear. For the appreciation of the differences, note that a rotation rate of 14.4°/day corresponds to an equatorial velocity of 2.025 km/s; a relative accuracy of 1%, corresponding to 0.14°/day or 20 m/s, can be reached within one method. In the case of velocities determined from Doppler shifts, however, the systematic errors may be larger than 1%. For measurements using tracers, a relative accuracy of ~5 m/s may be reached.

As for the tracer measurements, the function $\Omega(\theta)$ derived from recurrent spots (i.e., spots crossing the solar disk for at least a second time) by Newton and Nunn (1951) is virtually identical to that determined for long-lived spots by Balthasar *et al.* (1986); hence we did not plot the latter relation in Fig. 3.1.

The rotation rates are significantly larger when small and short-lived sunspots are included as tracers; see Howard (1984) and Table II in Schröter (1985). Presumably this is caused by the rapid forward motion of the dominant leading sunspots during the emergence and growth of the active regions (Section 5.1.1.1); indeed in multi-spot

Table 3.1. *Solar differential rotation as measured by various methods*[a]

A	B	C	Remarks	Refs.[b]
Photospheric rotation, from Doppler shifts				
14.049	−1.492	−2.605	1967–1984, Mt. Wilson	(1)
14.07	−1.78	−2.68	1979–1983, Kitt Peak	(2)
From long-lived and recurrent sunspots				
14.368	−2.69		1878–1944, recurrent spots	(3)
±0.004	±0.04			
14.37	−2.86		1874–1939, long-lived spots	(4)
±0.01	±0.12			
From cross-correlation analysis of Kitt Peak magnetograms				
14.38	−1.95	−2.17	1975–1991, all magnetic features	(5)
±0.01	±0.09	±0.10		
14.42	−2.00	−2.09	1975–1991, small features only	(5)
±0.02	±0.13	±0.15		
From filaments				
14.42	−1.40	−1.33	1919–1930, Meudon	(6)

[a] The coefficients A, B, and C in Eq. (3.1) give the sidereal rotation rate expressed in deg/day.
[b] References: (1) Snodgrass (1984); (2) Pierce and LoPresto (1984); (3) Newton and Nunn (1951); (4) Balthasar *et al.* (1986); (5) Komm *et al.* (1993a); and (6) d'Azambuja and d'Azambuja (1948).

groups, the leading spots indicate a larger rotation rate than the following spots (see Fig. 6 in Howard, 1996). We argue that the rotation rate of single, long-lived sunspots represents a standard that is related to a large-scale rotation in the part of the convection zone where these spots are anchored. The rotation rates indicated by small and short-lived spots are affected by the local and transient proper motions within growing active regions, and they are not a feature of the convective envelope as a whole. Consequently, we did not enter rotation data for short-lived spots in Table 3.1 and Fig. 3.1.

The differential rotation rates deduced from cross-correlation analyses of pairs of Kitt Peak magnetograms observed on consecutive days by Komm *et al.* (1993a) for small magnetic features only differ only very slightly from the results for complete magnetograms (see Table 3.1). In Fig. 3.1 only the $\Omega(\theta)$ curve for small-scale features is shown. This curve differs little from the curve for recurrent sunspots.

The differential rotation relation determined by Snodgrass (1984) from Mt. Wilson Dopplergrams measured in the line $\lambda 5,250$ Å differs little from that determined by Pierce and LoPresto (1984). The latter authors used a set of 13 spectral lines of different strengths that cover the photosphere and the low chromosphere; they found no significant variation of $\Omega(\theta)$ with formation height. Schröter (1985) also reviewed other attempts

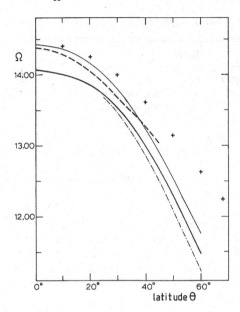

Fig. 3.1. Solar sidereal rotation rate (deg/day) against heliocentric latitude, using various diagnostics. From photospheric Doppler shifts: heavy solid curve (Snodgrass, 1984) and dotted–dashed curve (Pierce and LoPresto, 1984); from recurrent sunspots: heavy dashed curve (Newton and Nunn, 1951); from cross correlation of small features in magnetograms: thin line (Komm *et al.*, 1993a); from filaments: crosses (d'Azambuja and d'Azambuja, 1948).

to measure a height dependence of $\Omega(\theta)$; he concluded that these studies either show an insignificant height effect, or disagreements.

Figure 3.1 suggests that sunspots and cross correlations of relatively high-resolution magnetograms indicate a significantly higher rotation rate than that of the photospheric plasma derived from Doppler shifts, and this is the conclusion of several authors, including Snodgrass (1992). Schröter (1985) wondered, however, whether a more precise treatment of the correction for stray light could raise the photospheric rotation rate, possibly to the value derived from long-lived sunspots. But even if the difference suggested by Fig. 3.1 is completely real, it is small, less than 2%, which indicates that the layers in which the magnetic structure, including the long-lived sunspots, are anchored rotate quite nearly as fast as the photospheric plasma. Since sunspots are believed to be anchored deep in the convection zone (see Chapters 4–6), it follows that the angular rotation rate varies little with depth; this conclusion is confirmed in Section 3.3.

We mention in passing that a variety of rotation curves $\Omega(\theta)$ have been found from chromospheric and coronal tracers; see the review by Schröter (1985). Small features are found to rotate at the same rate as the underlying photosphere; that is, these features rotate with their magnetic feet in the photosphere. Some large-scale magnetic structures and patterns rotate at rates that differ markedly from the rotation rate of their magnetic feet in the photosphere. For later reference (Section 8.2) we include in Fig. 3.1 the rotation indicated by filaments which move relative to the photospheric magnetic field at their roots, particularly at higher latitudes. The case of the coronal holes is explained in Section 8.7, and the decay patterns of active regions are discussed in Section 6.3.2.2.

We do not discuss rotation curves for plages, because plages are deformed during their decay. For the study of the structure and the evolution of plages, their rotation data are of interest; for a review, see Howard (1996).

Scrutinizing the Mt. Wilson Dopplergram data, Howard and LaBonte (1980) discovered a latitude- and time-dependent pattern, which is often labeled with the misleading term "torsional oscillation". It is a slight undulation with an amplitude of ~5 m/s about the mean, smoothed differential rotation curve. This undulation becomes visible only after averaging over several solar rotations. It is approximately symmetrical about the equator, and it travels from near the poles to the equator in ~17 years. A better term for this pattern is *traveling torsional wave*, and as such has been found in the rotation of magnetic features in Mt. Wilson magnetograms (Snodgrass, 1991) and in Kitt Peak magnetograms (Komm *et al.*, 1993a).

The wave in the rotation curve derived from the cross correlation of magnetographic data is similar to the wave in the rotation curve from Mt. Wilson Dopplergrams. The wave in the magnetic features is displaced toward the equator by some 10° (Snodgrass, 1992). In other words, the wave in the magnetic features precedes the wave in the Doppler shifts by ~2 years. In addition, the wave amplitude in the magnetic data is larger by a factor of about 1.5, and the wave shape, in terms of the temporal variation of the normalized coefficients B/A and C/A in Eq. (3.1), differs somewhat (Komm *et al.*, 1993a). The remarkable connection between the torsional wave pattern and the magnetic activity cycle is discussed in Section 6.1.

3.2 Meridional and other large-scale flows

Meridional circulation in the solar convection zone is a key element in the physics of convective envelopes and of the magnetic dynamo. Consequently, it has been the subject of many studies.

The determination of the weak meridional flow in the photosphere from Doppler shifts is hard because of the uncertainties in the orientation of the solar axis and in the limb shift, as mentioned in the introduction to this chapter. Despite some controversial results, there is a growing consensus that there is such a flow, which amounts to ~10 m/s at midlatitudes.

The two-dimensional cross-correlation analysis of pairs of Kitt Peak magnetograms for two consecutive days by Komm *et al.* (1993b) has been very successful in bringing out the meridional flow $M(\theta)$; see Fig. 3.2 for the average flow over 12 y. It is well represented by

$$M(\theta) = (12.9 \pm 0.6) \sin 2\theta + (1.4 \pm 0.6) \sin 4\theta \text{ (m/s).} \tag{3.2}$$

Figure 3.2 includes determinations of the meridional flow from Mt. Wilson Dopplergrams that are in good agreement with the flow derived from consecutive Kitt Peak magnetograms.

Komm *et al.* (1993b) found that the amplitude of the meridional flow varies with the solar cycle: it is smaller than average during the cycle maximum and larger than average during cycle minimum. The shape of the latitude dependence of the flow, i.e., the ratio of the coefficients in Eq. (3.2), does not change during the cycle, however.

Cross-correlation analyses of Mt. Wilson magnetograms by Latushko (1994) and Snodgrass and Dailey (1996) yielded long-term meridional flow patterns that are quite

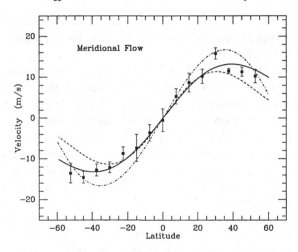

Fig. 3.2. The meridional flow derived from 514 magnetogram pairs during the period 1978–1990; the solid curve corresponds to the fit in Eq. (3.2). Positive (negative) velocities mean that the flow is to the North (South). Two fits derived from Mt. Wilson Dopplergrams are shown by the dashed curve (1982–1984; Snodgrass, 1984) and the dotted–dashed curve (Ulrich, 1993; figure from Komm *et al.*, 1993b).

similar to those found by Komm *et al.* (1993b) in the active-region belts between 10° and 40° latitude, but that differ in zones at higher latitudes. In addition, Snodgrass and Dailey (1996) found a markedly different flow pattern near the equator, as well as rapid changes in the flow during the course of the activity cycle. We point out that the analyses of the Mt. Wilson data are quite sensitive to slowly changing, large-scale activity patterns because the active regions are not masked, and the cross correlations include pairs of magnetograms that are separated by many days. Displacements of large-scale magnetic patterns (see Sect. 6.2.1) may differ from those of the small magnetic features between two consecutive days that were analyzed by Komm *et al.* (1993b). We suggest that the flow deduced from the Kitt Peak magnetogram pairs probably reflects the meridional flow in the photosphere more closely than the results derived from the Mt. Wilson magnetograms.

Meridional displacements of sunspots have an amplitude of only a few meters per second, and several authors agree that at low latitudes the displacements tend to be toward the equator, and at higher latitudes poleward; see Howard (1996) for references. From the Mt. Wilson sunspot data, Howard (1991a) deduced somewhat higher amplitudes of ∼5.5 m/s, and a marked separation at ∼15° latitude, with spots moving poleward at the higher latitudes and toward the equator at lower latitudes.

The meridional displacements of sunspot nests (Section 6.2.1) are also very small; they amount to no more than 3 m/s. The meridional flow data on spots and nests are compatible with notions that sunspots and sunspot nests are anchored deep below the photosphere, deeper than the magnetic network structure analyzed by Komm *et al.* (1993b). At that greater depth, the meridional flow is expected to be slower, and there the flow may differ from that in the photosphere.

The magnitude and the depth-dependence of the meridional flow have been investigated by Braun and Fan (1998) from helioseismic measurements by means of the

MDI on board *SOHO* and the ground-based Global Oscillations Network Group. The significant frequency difference between poleward- and equatorward-traveling waves measured over solar latitudes ranging from 20° to 60° is found to be consistent with a poleward meridional flow of the order of 10 m/s. From the variation of the frequency shifts of the *p*-modes with degree, Braun and Fan (1998) infer the speed of the meridional flow, averaged over the indicated latitudes, for the top half of the convection zone. They find no evidence for a significant equatorward return flow within that depth range.

The determination of large-scale flows by means of Doppler shifts and tracers automatically entails a search for possible "giant cells" that have been expected as the fundamental convective mode on the basis of conventional convection theory. The size of such cells would be of the order of the depth of the convection zone; various cell shapes have been considered, such as "banana" rolls that are strongly elongated in latitude. Considerable efforts have been invested in searches for a signature of such giant cells in the solar surface layer, but as yet no evidence has been found (see Snodgrass, 1992, for references). In fact, during one such search, Howard and LaBonte (1980) discovered the torsional wave phenomenon.

3.3   Rotation with depth

In a nonrotating star, the eigenfrequencies of the acoustic *p*-modes of given spherical harmonic degree ℓ but different azimuthal order m (see Section 2.6) are identical. In a rotating star, however, these modes are split into a multiplet. The simplest of these splittings, the doublet, results from the summation of prograde and retrograde traveling waves. These waves travel at different speeds relative to the observer because of the stellar rotation: in the stellar reference frame, they form a standing wave, but in the observer's frame two frequencies are observed.

These splittings provide information on the internal rotation rate. The magnitude by which waves are affected by the internal rotation depends on the ray paths (the expression for the splitting involves the rotation rate at all depths, as a function of latitude, e.g., Hansen *et al.*, 1977): high-ℓ modes probe shallow layers of the Sun, whereas low-ℓ modes travel through the deep domains of the solar radiative interior. Hence, the information in all of the observed modes can in principle be used to derive the rotation rate of the complete interior of the Sun. As the modes travel through the Sun at the local sound speed, they spend most of their time in the outer layers. Consequently, the *p*-mode spectra contain relatively little information on the domains of the deepest interior. The domain near the rotation axis is particularly hard to probe, because at a given angular velocity, the rotation velocity introducing the frequency shifts decreases toward the axis of rotation.

As with any inversion process, a regularization criterion is used to remove high-frequency excursions from a mean, smooth trend: the solutions are based on the assumption that the variation of rotational velocity with depth and latitude is smooth and slow. Moreover, the rotation profiles are assumed to be symmetric about the equatorial plane.

The helioseismic inversions (for example, Goode *et al.*, 1991; Kosovichev *et al.*, 1997) show that the differential rotation observed at the solar surface persists with only relatively small differences throughout the convective envelope (Fig. 3.3): at low latitudes, the rotation rate is nearly independent of depth, whereas at higher latitudes there appears to be a gradual change in the rotation rate with depth. Just below the bottom of the convection zone, the rotation curves converge, and the radiative interior appears to rotate

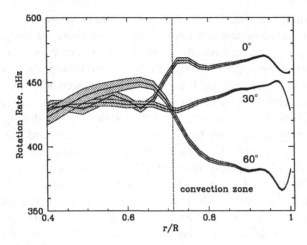

Fig. 3.3. Internal sidereal solar rotation $1/P(\theta)$ as inferred from helioseismological measurements. Three curves show the rotation profile for $0°$, $30°$, and $60°$ in latitude θ (from Kosovichev *et al.*, 1997).

rigidly, at least down to $r/R_\odot \sim 0.4$. Within the solar convective envelope, the rotation profile resembles that of a set of nested cones, with their tops near the center of the Sun and axes aligned with the rotation axis. This is in marked contrast to early models of solar rotation that predicted near-rigid rotation on cylinders parallel to the rotation axis. That expectation was based on the conservation of angular momentum, which restricts convective motions to paths at a fixed distance from the rotation axis. The meridional flow and other transport routes of angular momentum appear to be much more effective than previously thought.

The transition from the rigidly rotating radiative interior to the differentially rotating envelope appears to occur in a thin layer that is located immediately below the convection zone proper. Spiegel and Zahn (1992) named this zone the "tachocline." This layer is probably turbulent as a result of stresses exerted by the convection zone overhead; Spiegel and Zahn draw a parallel with the wind stress of the Earth's ocean surfaces. The thickness of the tachocline has been observationally limited to less than $0.09R_\odot$, perhaps as small as $0.02R_\odot$ (see Elliot and Gough, 1999, and references therein).

With the possible exception of the inner regions within $\sim 0.3R_\odot$ of the center, the radiative interior rotates with an angular velocity that lies between the rotation rates of equator and poles near the surface. Because of the magnetic braking associated with the solar magnetized wind (Section 8.9), this synchronization suggests a rather efficient coupling between the two parts of the interior that counteracts the shear induced by the magnetic brake acting on the envelope (Section 13.3).

Near the surface, there appears to be a shallow layer with a strong gradient in angular velocity: within 35,000 km to 70,000 km below the surface, depending on latitude, the angular velocity *increases* with depth by $\sim 5\%$ at low latitudes, but *decreases* by a comparable amount at latitudes near $60°$. The nature of this strong shear is not yet understood. Its magnitude of up to 100 m/s is a substantial fraction of the surface velocity associated with supergranulation. Despite this substantial magnitude, there appears to be no significant anisotropy in the patterns of the supergranulation (Section 2.5.2). The sense of the shear at active latitudes agrees with the notion that sunspots are anchored relatively deep compared to smaller-scale concentrations, resulting in a difference in rotation velocities, but that connection requires further study.

The differential rotation of the convective envelope, where convective flows move matter throughout the volume, and the nearly rigid rotation of the interior, with a shear layer in between, are maintained by the transport of angular momentum. Several mechanisms have been proposed for the process that redistributes angular momentum: meridional circulation, rotation-induced turbulent diffusion (Endal and Sofia, 1978b, Pinsonneault *et al.*, 1989), gravity waves (Zahn *et al.*, 1997), wind-driven circulation (Zahn, 1992), or a torque applied by a magnetic field (Charbonneau and MacGregor, 1993). Among the few studies that provide parameterizations for other than solar conditions – to which we shall return in Section 12.7 – we find that by Kitchatinov and Ruediger (1995). Their model reproduces the internal differential rotation profile quite well within the convective envelope, but the transition from differential to rigid rotation in the tachocline is much thicker in their model than that allowed by the helioseismic results. They use a mixing-length model to describe the turbulent transport of heat, which they argue is anisotropic because of the influence of rotation. This would result in a 4-K temperature difference between pole and equator for a solar model that acts to oppose the meridional flow that is driven by the centrifugal force. Consequently, the meridional circulation is countered much more efficiently than could otherwise be achieved by viscous drag alone. This flow prevents the star from settling into a cylindrically symmetric rotation pattern, which would occur if the flows were dominated by the conservation of angular momentum. The magnitude of the invoked pole–equator temperature difference is, however, substantially larger than observations allow: Kuhn *et al.* (1998) find no significant systematic temperature difference from *SOHO/MDI* roll data, although these data were accurate enough to provide evidence for a 2-K ripple of the temperature along the limb on smaller scales.

4

Solar magnetic structure

The magnetic field that corresponds to conspicuous features in the solar atmosphere is found to be confined to relatively small magnetic concentrations of high field strength in the photosphere. Between such concentrations, the magnetic field is very much weaker. The strong-field concentrations are found at the edges of convective cells: convective flows and magnetic field tend to exclude each other. These time-dependent patterns indicate that the magnetic structure observed in the solar atmosphere is shaped by the interplay between magnetic field, convection, and large-scale flows.

The construction of a comprehensive model for the main phenomena of solar magnetic activity from basic physical principles is beyond our reach. The complex, nonlinear interaction between turbulent convection and magnetic field calls for a numerical analysis, but to bring out the main observed features would require simulations of the entire convection zone and its boundary layers. The intricacy of convective and magnetic features indicates that extremely fine grids in space and in time would be required. Such an ambitious program is far beyond the power of present supercomputers. Hence, we must gain insight into the solar magnetic structure and activity first by studying the observational features, and subsequently by trying to interpret and model these with the theoretical and numerical means at hand.

Our approach is to map the domain of solar magnetic activity by a mosaic of models. Some of these models are well contained; others are based on ad hoc assumptions. Some models fit together nicely, as pieces in a jigsaw puzzle, but others are like mosaic tiles cemented by broad strips of guesswork.

Section 4.1.1 sketches the theory of magnetic fields in the convective envelopes of the Sun. In subsequent sections, the phenomena of magnetic structure and its dynamics in the solar atmosphere are summarized and complemented by theoretical ideas and models.

The solar phenomena discussed in this chapter are restricted to those observed in the solar photosphere having length scales up to those of large active regions ($\ell \lesssim 250$ Mm) and time scales up to a few weeks. Phenomena on progressively larger length and time scales are addressed in Chapters 5, 6, and 7. Chapter 8 is dedicated to the outer atmosphere of the Sun.

4.1 Magnetohydrodynamics in convective envelopes

4.1.1 Basic concepts of magnetohydrodynamics

In this section, we summarize basic concepts and formulas needed to discuss solar magnetic phenomena and their interpretation. For a comprehensive discussion, we refer to Chapter 4 in Parker's (1979) book, where Gaussian cgs units are used; this is a convention we also follow. Mestel (1999) provides a broad introduction to stellar magnetism from a theorist's point of view.

For application in the convection zone and lower solar atmosphere, we use the standard magnetohydrodynamics (MHD) approximation, which assumes that the typical plasma velocities v are low ($v \ll c$), so second-order terms in v/c are negligible (c is the speed of light). Moreover, the effects of local electric charge densities and displacement currents, of viscosity, and of radiation pressure are neglected.

The force-balance equation is:

$$\rho \frac{d\mathbf{v}}{dt} = -\nabla p + \rho \mathbf{g} + \frac{1}{c} \mathbf{J} \times \mathbf{B}, \tag{4.1}$$

where ρ is the gas density, p is the gas pressure, $\mathbf{g}$ is the gravitational acceleration, $\mathbf{J}$ is the electric current density, and $\mathbf{B}$ is the magnetic field strength; $\mathbf{J} \times \mathbf{B}/c$ is the Lorentz force.

The vectors $\mathbf{J}$ and $\mathbf{B}$ are related by Maxwell's equations, which under the above-mentioned conditions are approximated by Ampère's law:

$$\nabla \times \mathbf{B} = \frac{4\pi}{c} \mathbf{J}, \tag{4.2}$$

and by

$$\nabla \cdot \mathbf{B} = 0 \tag{4.3}$$

(magnetic field lines close on themselves; there are no magnetic monopoles).

The electric field $\mathbf{E}$ is tied in by the Faraday and approximated Gauss laws

$$\nabla \times \mathbf{E} = -\frac{1}{c} \frac{\partial \mathbf{B}}{\partial t}, \tag{4.4}$$

$$\nabla \cdot \mathbf{E} = 0. \tag{4.5}$$

Ohm's law is frequently used with a scalar electric conductivity σ:

$$\mathbf{J} = \sigma \left(\mathbf{E} + \frac{1}{c} \mathbf{v} \times \mathbf{B} \right). \tag{4.6}$$

By combining Eqs. (4.2) through (4.6), we can write Eq. (4.4) in the form of the *MHD induction equation*:

$$\frac{\partial \mathbf{B}}{\partial t} = \nabla \times (\mathbf{v} \times \mathbf{B}) + \eta \nabla^2 \mathbf{B}$$

$$\mathcal{O}\left(\frac{B}{\hat{t}}\right) \sim \mathcal{O}\left(\frac{vB}{\ell}\right) + \mathcal{O}\left(\frac{\eta B}{\ell^2}\right) \tag{4.7}$$

local change by advection and diffusion

where the ohmic *magnetic diffusivity* η (assumed constant in the above derivation) is given by

$$\eta = \frac{c^2}{4\pi\sigma}. \tag{4.8}$$

In the dimensional equation in Eq. (4.8), $\hat{t}$ is a characteristic response time, while v, B and ℓ are characteristic values for velocity, magnetic field strength, and length, respectively. The ratio of the advection term to the diffusion term on the right-hand side of the induction equation is known as the *magnetic Reynolds number*:

$$\mathcal{R}_m \equiv \frac{v\ell}{\eta}. \tag{4.9}$$

If $\mathcal{R}_m \ll 1$, the temporal evolution of the field is dominated by the diffusion term $\eta\nabla^2 \mathbf{B}$, and the plasma motions are not important. Magnetic flux leaks out of any concentration so that the gradients are reduced. The corresponding characteristic time scale for such ohmic diffusion is

$$\hat{t}_d \approx \frac{\ell^2}{\eta} = \frac{\ell}{v}\mathcal{R}_m. \tag{4.10}$$

In the laboratory $\hat{t}_d$ is small, being less than 10 s even for such a good conductor as a copper sphere of radius 1 m. In most astrophysical conditions, however, the typical length scale ℓ is so large that $\mathcal{R}_m$ and $\hat{t}_d$ are very large, even though the magnetic diffusivity η is several times that of copper. Cowling (1953) showed that the ohmic diffusion time of a sunspot is $\sim$300 yr, and for a global solar magnetic field $\hat{t}_d$ is of the order of 10^{10} yr. Even for a thin flux tube (radius $\ell = 100$ km) located in the top layers of the solar convection zone ($\eta = 2 \times 10^6$ cm^2/s), we find $\hat{t}_d = 1.6$ yr.

If $\mathcal{R}_m \gg 1$, the first term on the right-hand side of Eq. (4.8) dominates. In this case, the magnetic flux is constant through any closed contour moving with the plasma, and the field is said to be frozen into the plasma (see Section 4.2 of Parker, 1979): there is virtually no relative motion between the magnetic field and the plasma in the direction perpendicular to the field. The plasma can, however, flow *along* the field without changing it.

In most astrophysical conditions, such as in stellar atmospheres and convective envelopes, the magnetic Reynolds number $\mathcal{R}_m$ is large except for extremely small length scales, which even for the Sun lie well below observable scales. Hence, *observable* changes in the photospheric magnetic field *must involve transport of field by plasma motions* in the photosphere and convective envelope.

The long diffusion times in astrophysical plasmas make the relation between magnetic field and electric current different from everyday experience with electromagnetism. In domestic and most laboratory conditions, the effects of electric resistivity are predominant: the electromotive force of a generator or a battery is needed to maintain electric currents and magnetic fields. In such conditions, changes of the magnetic field are caused by changes in the electromotive force or the resistance in the current circuit. In many astrophysical conditions, however, the effects of resistivity are minor: once magnetic fields and electric currents are there at sufficiently large scales, they last for a very long time: no electromotive force is needed to maintain them. Electric current and magnetic field then are two complementary vector quantities connected by Ampère's law [Eq. (4.2)]. Local changes in the magnetic field are caused by plasma motions through

$\partial \mathbf{B}/\partial \mathbf{t} = \nabla \times (\mathbf{v} \times \mathbf{B})$. In such conditions, the electromagnetic state of the medium may be described by the magnetic field **B** alone. The magnetic field lines and flux tubes may be pictured as strings and rubber tubes that are transported, and may be deformed, by plasma flows.

This picture of strings and tubes of long-lasting identity being carried by flows breaks down at small $\mathcal{R}_m$, that is, at extremely small length scales ℓ in Eq. (4.9). In the photosphere, $\eta \approx 10^8$ cm^2/s, so that in flows as slow as $v = 10$ m/s, $\mathcal{R}_m$ drops below unity only for scales as small as $\ell < 1$ km. Only in such extreme conditions does ohmic dissipation become important, causing the magnetic field to diffuse and fields of opposite polarities to reconnect. In cases of ohmic diffusion that do not involve reconnection, the picture of strings can be maintained in the sense that the "field lines slip across the plasma." Such an example is shown in Fig. 4.1*b*: fluid flow is crossing a steady pattern of field lines; see the stream lines in Fig. 4.1*a*. Field reconnection may also occur for $\mathcal{R}_m \lesssim 1$ if adjacent field lines are in different directions; antiparallel fields may cancel altogether (these processes are illustrated in Fig. 4.1*e*–4.1*j*). Such field reconnections change the field pattern.

4.1.2 *Interaction between flows and magnetic field*

The structure and evolution of the magnetic field in a stellar atmosphere are primarily determined by the interaction with flows. Some aspects of this complex interaction have been studied in theoretical and numerical studies of simplified configurations. Early studies by Parker (1963a) and Weiss (1964) investigated the influence of systems of persistent two-dimensional velocity rolls on initially weak and homogeneous magnetic fields. These studies have shown that, for prescribed flow patterns at large magnetic Reynolds number $\mathcal{R}_m$, the magnetic field ends up in the regions between the convective rolls; see Fig. 4.1. Even within one turnover time of the rolls, an expulsion or concentration is seen (Fig. 4.1*c* and 4.1*d*). After several turnover times, the magnetic field in the roll interiors is strongly sheared. Hence there the diffusive term $\eta \nabla^2 \mathbf{B}$ becomes dominant in Eq. (4.8), so reconnection and field annihilation occur (Fig. 4.1*e*–4.1*j*). The net result is that the field disappears from the roll interiors and becomes concentrated at the boundaries. Ultimately a steady state is reached when the diffusion of the field from between the roll boundaries balances the advection of the field.

These classical studies indicate that persistent flows concentrate the magnetic field into regions between the cells, until the Lorentz force $(\nabla \times \mathbf{B}) \times \mathbf{B}/(4\pi)$ is sufficiently intensified to hamper the flow locally. Field strengths locally produced by this process are estimated by equating the magnetic energy density $B^2/(8\pi)$ to the kinetic energy density in the flow $\rho v^2/2$ to yield the *equipartition field strength*:

$$B_{\mathrm{eq}} \equiv (4\pi \rho v^2)^{1/2}. \tag{4.11}$$

The actual state of convection in stars is much more complex than in such simulations of persistent, well-ordered rolls. The magnetic field in a zone of turbulent convection must be very inhomogeneous; such a magnetic field is generally called *intermittent*. At high magnetic Reynolds number $\mathcal{R}_m$, the magnetic field concentrations outlive the individual turbulent eddies by a large factor because the field cannot slip back. Consequently, the local concentration by convective flows is expected to form and maintain a field configuration

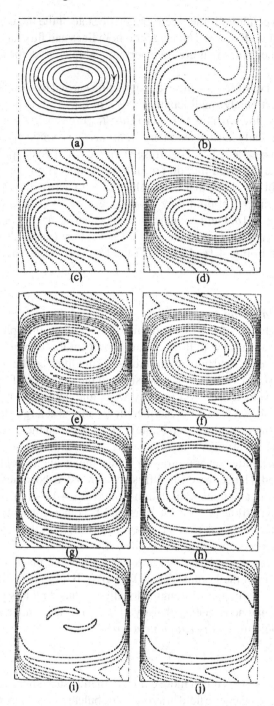

Fig. 4.1. The effect of a persistent velocity cell (a two-dimensional roll) on an initially weak, vertical, and homogeneous field. The streamlines of the flow are plotted in panel *a*. Panel *b* shows the final equilibrium state of the magnetic field lines for a modest Reynolds number $\mathcal{R}_m = 40$. For $\mathcal{R}_m = 10^3$ (averaged over the system), the development of the magnetic field lines is displayed in panels *c*–*j* at intervals $\Delta t = 0.5\hat{t}_c$, where $\hat{t}_c$ is the turnover time of the roll, beginning at $t = 0.5\hat{t}_c$ in *c* and ending in *j* with $t = 4\hat{t}_c$ when the asymptotic state is nearly reached (from Weiss, 1966).

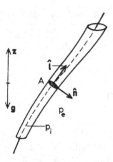

Fig. 4.2. The geometry of a magnetic flux tube with an internal field $\mathbf{B}(l)$, cross-section A, internal gas pressure p_i, and external gas pressure p_e.

consisting of tortuous strands of magnetic flux, which change shape and position while interacting with the flows. Within such strands, the field strength is expected to be of the order of the equipartition value, and the gross topology of the system should evolve on a time scale much longer than the characteristic lifetime of the convective eddies. In other words, *magnetic field and convective flows* are expected to *coexist side by side: most of the field is concentrated in strands without appreciable internal turbulence, while the medium between the magnetic strands is vigorously convective, and without appreciable magnetic field.* This qualitative description is consistent with sophisticated numerical simulations (see Nordlund *et al.*, 1994) and with the observed intermittence of the magnetic field in the solar photosphere (see examples in Section 1.1 and later in this chapter).

The partition between convection and magnetic field suggests, as a useful approximation for magnetic fields in convective zones, a two-component model consisting of distinct flux tubes separated by field-free plasma. Parker considered the physics of such flux tubes (Fig. 4.2) in a convective envelope (see Parker, 1979, especially his Chapter 8). The mechanical equilibrium [Eq. (4.1)] is described by magnetohydrostatic equilibrium, assuming that the flows within and outside the flux tube are subsonic:

$$-\nabla p + \rho \mathbf{g} + \frac{1}{4\pi}(\nabla \times \mathbf{B}) \times \mathbf{B} =$$

$$-\nabla p + \rho \mathbf{g} - \frac{1}{8\pi}\nabla B^2 + \frac{1}{4\pi}(\mathbf{B} \cdot \nabla)\mathbf{B} = 0. \tag{4.12}$$

In the second line of this equation, the Lorentz force has been separated into two terms: the first, $-\nabla B^2/(8\pi)$, is the gradient of the *isotropic magnetic pressure* $B^2/(8\pi)$, and the second is the force arising from the *magnetic tension* $B^2/(4\pi)$ *along the field lines*. If the magnetic field lines have a local radius of curvature R, their tension $B^2/(4\pi)$ exerts a transverse force $B^2/(4\pi R)$ per unit volume (hence this force is zero for straight field lines). A net force is exerted on the plasma insofar as this curvature stress is not balanced by the gradient of the magnetic pressure $B^2/(8\pi)$.

The magnitudes of the pressure terms p and $B^2/(8\pi)$ in Eq. (4.12) are compared in the *plasma-β* parameter, which is defined as

$$\beta \equiv \frac{8\pi p}{B^2}. \tag{4.13}$$

If $\beta \gg 1$, the pressure balance is virtually completely determined by the gas pressure distribution, and the magnetic field has a negligible influence on the local medium. For $\beta \ll 1$, in contrast, the gas pressure has virtually no influence. In that case, the (slowly evolving) magnetic field is essentially *force free*:

$$\frac{1}{c}\mathbf{J} \times \mathbf{B} = \frac{1}{4\pi}(\nabla \times \mathbf{B}) \times \mathbf{B} = 0. \tag{4.14}$$

In a force-free field, the current $\mathbf{J}$ runs parallel to $\mathbf{B}$. A *current-free field* is a special case of a force-free field. A current-free field is a *potential field*, because $\nabla \times \mathbf{B} = 0$.

In a potential field, the gradient of the magnetic pressure $-\nabla B^2/(8\pi)$ is exactly balanced by the curvature force $(\mathbf{B} \cdot \nabla)\mathbf{B}/(4\pi)$; hence there hydrostatic equilibrium $\nabla p = \rho \mathbf{g}$ applies [see Eq. (4.12)] wherever the plasma velocities are appreciably less than the sound speed.

In stellar convective envelopes, photospheres, and chromospheres, the gas pressure drops off approximately exponentially with height (Section 2.1.3), whereas the magnetic field strength drops off with some modest power of height. Hence β drops with height within nearly vertical field structures. In deep layers where $\beta \gg 1$, the pressure balance is dominated by the gas pressure distribution to which the magnetic field adjusts, whereas in outer atmospheres where $\beta \ll 1$ the magnetic fields dominate in the total pressure balance and hence there they are nearly force free.

Many basic properties of the magnetic field can be deduced by considering a *slender* (or *thin*) flux tube, i.e., a tube whose diameter d is smaller than both the radius of curvature of the field lines and the pressure scale height. Since the Lorentz force has no component along a line of force, the component of Eq. (4.12) in the direction of the unit vector tangent to the field $\hat{\mathbf{l}}$ (see Fig. 4.2) is a restatement of hydrostatic equilibrium,

$$-\frac{\partial p}{\partial l} + \rho(\mathbf{g} \cdot \hat{\mathbf{l}}) = 0, \tag{4.15}$$

whereas in a direction $\hat{\mathbf{n}}$ perpendicular to the tube we have

$$\frac{\partial}{\partial n}\left(p + \frac{B^2}{8\pi}\right) = 0. \tag{4.16}$$

If the magnetic field inside the tube is a potential field, the pressure p_i is constant across the tube radius, and there is a gas pressure jump $p_e - p_i$ between the internal gas pressure p_i and the external pressure p_e. This is compensated by the Lorentz force in a thin current sheet bounding the flux tube, so that

$$p_i + \frac{B^2}{8\pi} = p_e. \tag{4.17}$$

This bounding current sheet is called the *magnetopause*. Equations (4.15)–(4.17) imply that the variation of the field strength and cross section along the tube can be evaluated if the temperature distributions inside and outside the tube are known, together with the field strength, tube cross section, and either the internal or the external gas pressure at one position. The field strength at any position z is given by

$$B(z) = \{8\pi[p_e(z) - p_i(z)]\}^{1/2}. \tag{4.18}$$

The cross-sectional area $A(z)$ is given by flux conservation: $B(z)A(z) \equiv \Phi$ is constant. The distributions $B(z)$ and $p_i(z)$ depend on the height z, but not on the path (i.e., the shape and the inclination of the axis) of the flux tube through the medium. The path of the flux tube in the convection zone is determined by the velocity shear in the ambient medium that couples to the tube through drag, the buoyancy of the flux tube, and the magnetic tension along the tube. Spruit (1981a, 1981b) treated the physics of slender flux tubes moving through a field-free ambient convection zone.

If a flux tube is stretched by a sheared flow, the field strength is enhanced because of mass and flux conservation. Consider some volume of a flux tube of strength B, cross section A, length l, and density ρ. Conservation of flux implies that BA remains constant; ρlA is constant because the plasma cannot flow across the field. Hence $B \sim \rho l$; if the mass density ρ remains constant, we have $B \sim l$.

A flux tube and its supposedly nonmagnetic surrounding medium are coupled through hydrodynamic *drag*. For a tube with a circular cross section with radius a, the drag force F_d per unit length of the tube is

$$F_d = \frac{1}{2}\rho v_\perp^2\, a\, C_d, \tag{4.19}$$

where $v_\perp$ is the relative velocity perpendicular to the flux tube. The drag coefficient C_d depends on the (hydrodynamic) Reynolds number and is of the order of unity in stellar convective envelopes (Section 8.7 of Parker, 1979).

A flux tube that is (nearly) in thermal equilibrium with its environment $[T_i(z) = T_e(z)]$ is buoyant since the necessarily reduced internal gas pressure implies a lower mass density ρ. For a horizontal tube, the buoyancy force F_b per unit length is

$$F_b = \pi a^2 g\,(\rho_e - \rho_i). \tag{4.20}$$

By balancing buoyancy against drag, Parker (1979, his Section 8.7) determined the rate of rise $v_\uparrow$ for flux in thermal equilibrium with its surroundings:

$$v_\uparrow = v_A\,(\pi/C_d)^{1/2}\,(a/H_p)^{1/2}, \tag{4.21}$$

where v_A is the *Alfvén speed* at which magnetic disturbances travel along the field:

$$v_A \equiv \frac{B}{(4\pi\rho)^{1/2}}. \tag{4.22}$$

With this velocity and an inferred field strength at the base of the convection zone of 10^4–10^5 G (see Section 4.5), and flux tube radii $a \simeq 0.1H_p$, the time scale for them to drift up through the convection zone ranges from days to months. Since this is much shorter than the time scale of the solar-cycle period on which they are replenished, some mechanism is needed to keep the flux tubes within in the convection zone; we return to this issue in Section 4.5.

For flux tubes that are in thermal balance with their surroundings $[T_i(z) = T_e(z)]$, the pressure scale height is the same inside and outside the tube; hence Eqs. (4.15)–(4.18) imply that the field strength decreases exponentially with height, with a scale height of twice the pressure scale height H_p (Parker, 1979, his Section 8.2):

$$B(z) = B(z_0)\exp\left[\frac{-(z - z_0)}{2H_p}\right]. \tag{4.23}$$

The increase of the field strength with depth according to Eq. (4.23) is drastic: for instance, a flux tube in thermal equilibrium with its surroundings with a field strength $B = 1$ kG at the top of the solar convection zone that extends down to the bottom of the convection zone would reach a field strength B in excess of 10^7 G near the bottom, and it would be impossible to keep it there for even a small fraction of the time scales observed in magnetic activity. Consequently, flux tubes residing in the convection zone for more than, say, a month must be cooler than their environment. The indication that flux tubes have to be cooler than their environment raises the problem of their thermal time scales. Since flux tubes of field strengths larger than approximately the equipartition field strength exclude turbulence, convective heat exchange between the plasma inside the tube and the ambient plasma is severely hampered. Hence heat exchange is through radiative diffusion, which is also rather inefficient in convective envelopes because the mean-free path for photons and the radiative diffusivity η_{rad} are very small (see Section 2.2 and Fig. 2.1). Hence the radiative diffusion time scale for a tube of radius a,

$$\hat{t}_{rad} = a^2/\eta_{rad}, \tag{4.24}$$

is very large (Fig. 4.3): for a tube with $a = 100$ km radius we find $\hat{t}_{rad} > 3$ months for all depths larger than 1,500 km; near a depth of 8,000 km, $\hat{t}_{rad} \simeq 600$ yr; see Zwaan and Cram (1989).

We conclude that in the convective envelope, flux tubes at strengths $B \gtrsim B_{eq}$ are thermally well insulated. Consequently, plasma within moving flux tubes or streaming within a flux tube adjusts virtually adiabatically to the pressure balance along the tube. Once the flux tube is out of thermal balance with its environment, the thermal imbalance ($T_i \neq T_e$) may last for months.

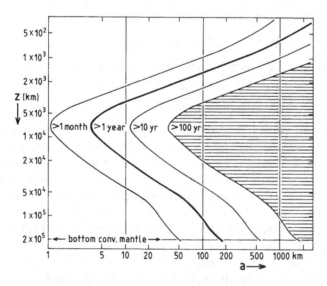

Fig. 4.3. Radiative relaxation time scale $\hat{t}_{rad}$ (Eq. 4.24) of flux tubes as a function of tube radius a and depth z in the solar convection zone. The radiative diffusion coefficient η_{rad} was borrowed from Spruit (1977b; his Chapter 2).

4.1.3 Magnetic reconnection

The term reconnection shows up in many places in this book and in the literature in general. Let us briefly explore the meaning of the term, by exploring the cause of the phenomenon. The concept of reconnection is intimately tied to the concept of the field line, which is defined as follows: at any point in space, there is a path everywhere tangent to the local direction of the magnetic field, that connects ultimately onto inself (because there are no magnetic monopoles), but that can be thought of as connecting two "sources" of opposite polarity (for instance formed by the photospheric concentrations of magnetic flux in the case of outer-atmospheric field lines), sometimes through a magnetic null point ($\mathbf{B} \equiv \mathbf{0}$) or through infinity. For a perfectly conducting plasma, the induction equation Eq. (4.8) contains only the advection term, and – as already stated in Section 4.1.1 – the field and the plasma move together in such a way that flows occur only along the field. In other words, two plasma elements located anywhere along a field line always remain on the same field line: reconnection does not occur in the hypothetical case of perfect conduction. More realistically, reconnection occurs if the length scales as measured by the magnetic Reynolds number [Eq. (4.9)] are sufficiently small.

Strong changes in the magnitude and direction of the field are not sufficient for reconnection to occur. Imagine, for example, a square with two positive sources placed on two corners diagonally across from each other, and two negative sources on the remaining two, all of equal absolute strength. The field in the center of the square forms a saddle point, around which field lines resemble hyperbolic arcs with the two diagonals as asymptotes. If there are no currents within this volume other than those maintaining the sources, then the field outside the sources is a potential field for which $\nabla^2 \mathbf{B} \equiv 0$. The induction equation Eq. (4.8) then reduces to the case of infinite conductivity, and the field lines are frozen in (and in fact static if $\mathbf{v} = \mathbf{0}$). In the case of a constant-α force-free field, the diffusion term in Eq. (4.8) equals $-\eta\alpha^2\mathbf{B}$, reflecting the exponential decay of the currents and their associated magnetic field. But because the direction of $\mathbf{B}$ is not affected by the temporal evolution, no reconnection occurs in this case either.

We introduce one more qualification: imagine an axially symmetric field of parallel field lines, with a radial gradient in the field strength, associated with some cylindrical current system (like the flux tubes discussed in Section 4.1.2). In this nonforce-free case, in which gas pressure is required to balance the magnetic pressure (in this case equal to the total Lorentz force, because the field lines are straight), these currents – if not somehow maintained – decay on a resistive time scale. The result is that the field weakens, and the continued balancing of forces requires that plasma slips across field lines, even though in this case no reconnection takes place. Hence, reconnection also generally involves an imbalance in forces, in which material is being accelerated by the Lorentz forces.

In view of this, we conclude that reconnection requires the dissipation of electric currents (either those maintaining the "sources" of the field, or those induced by motions of those sources), in a situation in which the field is not force free. Only in that case does the diffusion term cause a local redirection of the magnetic field that occurs other than by advection; in other words, this redirection of the field is associated with plasma no longer being tied to the field lines.

The dissipation of currents leading to the reconnection process is fundamentally the way in which the outer-atmospheric system irreversibly responds to the evolution of the photospheric magnetic field. As the volume of space pervaded by the field tries to resist changes in fluxes through any one of an unlimited number of closed-loop integrals, these motions induce currents with their own corresponding magnetic fields that initially counter the changes. As these currents decay, the field configuration is forced to adjust to the new situation, effectively "forgetting" its previous state, in the process accelerating plasma along as well as across the magnetic field: reconnection is nothing but the failure of the conceptual isomorphism of field lines and elastic bands, through which the field simplifies a tangled connectivity.

4.2   Concentrations of strong magnetic field

In Chapter 1 we pointed out that magnetic features that are conspicuous in the photosphere or in the chromosphere are rooted in a relatively strong magnetic field – markedly stronger than the equipartition field strength [Eq. (4.11)] $B_{eq} \approx 500$ G in the top of the convection zone. Fields stronger than 1 kG we call *strong fields*. These strong fields are found in patches with a wide range of fluxes and varying degrees of compactness ("filling factor"). Because of their intricate fine structure, down to the highest spatial resolution, we avoid terms suggesting a monolithic structure, such as "element;" we refer to all these patches of intrinsically strong field as *concentrations*.

These concentrations of strong magnetic field form a *hierarchy*: their properties depend on the total magnetic flux Φ they contain (Zwaan, 1978), with bifurcations that are related to the evolution of the features (Section 5.1). The properties are summarized in Table 4.1 and in Section 4.2.4.

4.2.1   Sunspots

Sunspots are the largest compact magnetic concentrations, with fluxes in the range $5 \times 10^{20} \lesssim \Phi \lesssim 3 \times 10^{22}$ Mx. A complete sunspot consists of a dark central *umbra*, which is surrounded, at least partly, by a *penumbra* (Fig. 4.4). Thomas and Weiss (1992a) provide an excellent survey of sunspot physics; Bray and Loughhead (1964) wrote a classical monograph on sunspot morphology. In this section, the discussion is limited to properties that appear most relevant in the present context of characterizing the flux concentrations.

Photospheric umbrae show a pattern of fine ($\lesssim 1''$) *umbral dots* that are brighter than the umbral background (Danielson, 1964; Bray and Loughhead, 1964). The estimates for the lifetimes of dots range from 30 min to more than 1 h. In addition, many umbrae contain extended bright features, such as light bridges, which are composed of small roundish or elongated elements. Usually the pattern of umbral dots across the umbra is irregular, both in the number density and the brightness of the elements (Zwaan, 1968); the pattern differs strongly from spot to spot. Many umbrae, and certainly the stable ones, show one or two dark *umbral cores*, where dots are very faint or seemingly absent. We refer to Muller (1992) for a review of umbral structure.

Maltby (1994) briefly reviews the problems in the deduction of semiempirical models for the thermodynamic stratification in umbrae. Such models are based mainly on the umbral intensity in its dependence on wavelength and the spot's position on the disk. Surprisingly, the umbral intensity varies slightly but significantly during the solar activity cycle: it is lowest at the beginning of the cycle and then it increases monotonically

Table 4.1. *The hierarchy of magnetic concentrations* Φ *is the magnetic flux,* R *is the radius of a sunspot,* R_u *is the radius of a sunspot umbra or of a smaller magnetic concentration, and* B *is the magnetic field strength at its center.*

Property	Sunspot with Penumbra		Pore	Magnetic Knot (micropore)	Faculae, Network Clusters	Filigree Grain
	large	small				
Φ (10^{18} Mx $=10^{10}$ Wb)	3×10^4	500	250 − 25	≈ 10	$\lesssim 20$	≈ 0.5
R (Mm)	28	4	–	–	–	–
R_u (Mm)	11.5	2.0	1.8 – 0.7	≈ 0.5	–	≈ 0.1
B (in G $= 10^{-4}$ T)	2900 ± 400	2400 ± 200	2200 ± 200	≈ 1500–2000	–	≈ 1500
Overall contrast in continuum:	dark			–	bright	
Cohesion:	single, compact structure				$\Rightarrow$ cluster of	
Behavior in time:	remain sharp while shrinking during decay			?	= modulated by granulation	
Occurrence:	exclusively in active regions				both inside and outside active regions	

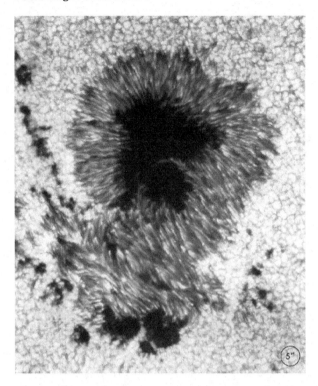

Fig. 4.4. Sunspot and pores. The large sunspot consists of a dark umbra surrounded by the penumbra. The penumbra shows relatively bright, radially aligned, and elongated elements. This print does not show the umbral structure consisting of tiny, bright umbral dots and the relatively very dark umbral core(s). Several pores are visible at the bottom and on the left-hand side of the frame. Penumbral patches connecting umbras and pores, displayed near the bottom of the frame, are typical for compact sunspot groups of complex magnetic structure (Observatoire Pic du Midi, courtesy of R. Muller).

through the cycle (Albregtsen and Maltby, 1978). Consequently, Maltby *et al.* (1986) derived three models for umbral cores, representative of the beginning, middle, and end of the cycle. We return to the dependence on the activity cycle in Section 6.1.

The umbral temperature stratification is markedly out of radiative equilibrium. For one thing, radiative equilibrium requires a steep temperature gradient in the deepest umbral atmosphere that is not compatible with the well-established wavelength dependence of the umbral intensities in the infrared. Hence there must be a substantial nonradiative energy transport in the deeper parts, even in the darkest umbral cores.

The thermodynamic structure of the umbral cores hardly depends on the size of the spot, except perhaps for the smallest spots (Zwaan, 1965; see also Maltby, 1994). The magnetic field strength in umbrae depends only weakly on spot size: the values found from measurements in the visible range from $2,900 \pm 400$ G for the largest spots down to $2,400 \pm 200$ G for the smallest spots with penumbrae (Brants and Zwaan, 1982). These findings suggest that, besides the nonmagnetic convection zone and photosphere, a second equilibrium state exists in the solar mantle (Zwaan, 1968), which is characterized in the photosphere by a field strength of $B = 2,900 \pm 400\,\text{G}^*$ and an effective temperature of $T_{\text{eff}} = 3,950 \pm 150\,\text{K}$ (quiet photosphere: $T_{\text{eff}} = 5,780 \pm 15\,\text{K}$), which corresponds to a total radiative flux density of $22 \pm 3\%$ of the photospheric flux density.

Near the center of a sunspot, the magnetic field is vertical within $\sim 10°$. With increasing radial distance, the inclination of the field increases, becoming nearly horizontal in the penumbra. In other words, the magnetic field of a sunspot resembles the potential field at the end of a long solenoid. This configuration was already discovered by Hale and his collaborators (see Hale and Nicholson, 1938) and has been confirmed by measurements with vector magnetographs (see Skumanich *et al.*, 1994).

Sunspots of all sizes show an intrinsic spread in continuum intensities as well as in field strengths. These spreads can be attributed to different amounts of bright fine structure in the resolution element and to an apparent variation in the umbral structure during the solar activity cycle (Section 6.1.2). In the near-infrared around $\lambda = 1.6\,\mu\text{m}$, where the intensity measurements are less sensitive to stray light and the Zeeman splittings are large (Section 2.3.2), Kopp and Rabin (1994) found that the relation between magnetic field strength and continuum intensity (hence temperature) across sunspots has a characteristic, nonlinear shape, despite intrinsic differences between different (parts of) spots. This finding holds a clue for the mechanism of energy transport in spots.

Sunspots are saucer-shaped depressions of the photosphere, as is indicated by the *Wilson effect*, the asymmetric distortion of spots observed near the solar limb (illustrated in Plate 3.23 of Bray and Loughhead, 1964). The corresponding Wilson depressions in the surfaces of constant optical depth τ for $0.1 \lesssim \tau \lesssim 1.0$ amount to approximately 400–600 km (see Gokhale and Zwaan, 1972).

The *penumbra* differs markedly from the umbra, and the transition between these parts is quite sharp. Whereas in the umbra the field is nearly vertical, the penumbral field is strongly inclined. Penumbrae transmit some 75% of the average solar energy flux density. Usually there is a conspicuously radial striation in the penumbral brightness structure. In contrast with the umbra, the penumbra is in a complex dynamic state, as indicated by proper motions of bright and dark features, and by the *Evershed flow*, a radial outflow with velocities up to several kilometers per second as measured by Doppler shifts of photospheric spectral lines and by feature tracking. For observational descriptions, see Muller (1992) and Shine *et al.* (1994).

There is a *radial similarity* in the variations of the parameters across quasi-stable spots with $\Phi \gtrsim 2 \times 10^{21}\,\text{Mx}$. This similarity is most readily described for nearly circularly symmetric spots: the parameters vary with the relative radial distance $r/R(t)$ to the spot center, where $R(t)$ is the time-dependent radius of the spot (Gokhale and Zwaan, 1972). For instance, the umbral area is a constant fraction (0.17 ± 0.03) of the total area of the spot. Indeed, decaying spots shrink without noticeable changes in the normalized radial variations of the physical parameters (Martínez Pillet *et al.*, 1993). This self-similarity is confirmed by vector-magnetographic measurements (Skumanich *et al.*, 1994).

---

* Occasionally field strengths larger than 3,300 G have been measured (e.g., Von Klüber, 1947). In well-documented cases, such high values are found to occur as a transient phenomenon in a small fraction of a complex sunspot in a δ configuration (Section 6.2.1); these strong fields are nearly horizontal (see Livingston, 1974). We conclude that the range $2,900 \pm 400\,\text{G}$ is representative for mature umbrae.

4.2.2   Pores

Magnetic elements with fluxes $2.5 \times 10^{19} \lesssim \Phi \lesssim 2.5 \times 10^{20}$ Mx are seen as dark *pores*: small umbrae without penumbrae (Figs. 4.4 and 2.12). Near 4×10^{20} Mx, either pores or small spots with (often incomplete) penumbrae are found; abnormally large pores are observed as a transient phase in the evolution of sunspots (see Section 5.1.1.3).

In pores and in small umbrae, the magnetic field strengths have often been underestimated because of stray light. Field strengths between 2.0 and 2.5 kG are measured during excellent seeing conditions, or from magnetically sensitive lines that are weak in the photospheric spectrum but enhanced in spectra of umbrae and pores (Brants and Zwaan, 1982). Measurements in the near-infrared confirm field strengths over 2 kG (Muglach *et al.*, 1994). The magnetic field appears to be close to vertical near the centers of pores; in most cases, the transition from the strong, nearly vertical field within the pore's interior to the weak field outside is as sharp as seeing permits (Fig. 2.12).

4.2.3   *Faculae and network patches*

Outside sunspots and pores, localized concentrations of strong magnetic flux are observed with intrinsic field strengths near 1.5 kG that are bright relative to the upper photosphere and chromosphere. These features are best seen in the cores of strong spectral lines, such as the Ca II resonance lines (Fig. 1.1*b*) and in Hα (Fig. 1.3*a*). In active regions, the bright clumps called *faculae* are arranged in tightly knit, coherent distributions called *plages* and more widely in the *enhanced network* (Figs. 1.1, 1.2 and 1.3). Outside active regions, the bright *network patches* form the so-called *quiet network* coinciding with the supergranular downflows (Section 2.5).

Observations with high spatial resolution ($\lesssim 400$ km) in continuum, weak lines, and line wings reveal an intricate bright structure in the faculae and network. Its collective appearance has been called *filigree* by Dunn and Zirker (1973), who describe it as consisting of roundish dots and somewhat elongated segments, with smallest dimensions of $\lesssim 300$ km (Fig. 2.12); in this book these elements are referred to as *filigree grains*. These grains lie between granules; they change in shape and sharpness over a time comparable with the lifetime of the granules (Dunn *et al.*, 1974; Mehltretter, 1974; Title and Berger, 1996). Apparently the filigree inside and outside the active regions only differs in the spatial density of the filigree grains (Mehltretter, 1974).

From the linear relation between the number density of filigree grains and the magnetic fluxes measured in magnetograms observed at the National Solar Observatory at Kitt Peak, Mehltretter (1974) estimated an average flux per filigree grain of approximately 5×10^{17} Mx. This flux is consistent with a magnetic flux tube of a few hundred kilometers in diameter at a field strength between 1 and 2 kG. Presumably many of the filigree elements in Mehltretter's data are not resolved; hence the estimated mean flux per element presents an upper estimate. We have taken Mehltretter's average value as a typical flux for a filigree grain in Table 4.1. In Section 5.3, we discuss the transient nature of filigree grains in the vivid interaction between the field and convective flow at scales smaller than 1 arcsec.

The fine structure in plages and network is not resolved in magnetograph measurements. Consequently, magnetographs measure flux densities φ averaged over resolution elements, and not the intrinsic magnetic field strength B in the truly magnetic fine structure. Intrinsic field strengths can be determined by means of the line-ratio technique using sets of spectral lines (Section 2.3.2). More than 90% of the measured magnetic flux is contained in strong fields, with intrinsic strengths between ~ 1 and $2\,kG$ (see Stenflo, 1985). More recently, detailed magnetic-field measurements have been obtained in the near-infrared near $\lambda = 1.6\,\mu m$, where in spectral lines of high Zeeman sensitivity the Zeeman components are completely separated for field strengths down to less than $\sim 500\,G$ (Rabin, 1992; Rüedi *et al.*, 1992). In plages and in enhanced network, intrinsic field strengths lie in the range 1.2–$1.7\,kG$; the scatter is largely intrinsic, but Rabin noticed that the field strength tends to increase with increasing mean flux density (or filling factor). In plages, filling factors up to 50% are found locally. In regions of lower mean flux densities, some patches of field are found that are weaker than $1\,kG$; we discuss these findings in a broader context in Section 4.6.

The line-of-sight magnetic signal in magnetographs drops sharply to zero near the solar limb (Fig. 1.1*e*), which suggests that the photospheric magnetic field in plages and network is close to vertical. This is confirmed by vector-magnetograph measurements (Lites *et al.*, 1993a), except close to a polarity inversion or close to sunspots. From measurements in bright grains in the quiet Ca II network, Sánchez Almeida and Martínez Pillet (1994) infer that the inclination of the magnetic field with respect to the vertical is less than $10°$(see also Section 6.2.2).

The expansion of the magnetic field with height can be seen by comparing magnetograms obtained in different spectral lines: although in lines formed deep in the photosphere the fields are strong and sharply defined, the magnetic features become coarser and more diffuse as the height of line formation increases (Giovanelli and Jones, 1982). This fanning is also seen in strong Fraunhofer lines: the bright grains are extremely fine as seen in the far line wings, but they become increasingly coarse closer to the line center (Section 8.1).

Within the magnetic elements in plages and network, the plasma is practically static: recent attempts to measure flows along the magnetic field have yielded only upper limits. Muglach and Solanki (1992), for example, set an upper limit of $\pm 0.3\,km/s$ for vertical flow in the magnetic structure in a quiet network region.[†] Transient flows during flux emergence are discussed in Section 4.4.

In the hierarchy of magnetic structure, *magnetic knots* take position between the dark and the bright features. At moderate resolution, they are not conspicuous in continuum or line wings (Beckers and Schröter, 1968; see also Spruit and Zwaan, 1981). At high resolution, these knots presumably correspond to the "micropores" observed by Dunn and Zirker (1973). The photospheric appearance of a small flux concentration depends strongly on the spatial resolution in the observation (Title and Berger, 1996). At a resolution of 140 km FWHM achieved with the technique of speckle interferometry, Keller (1992) found that magnetic concentrations larger than $\sim 300\,km$ appear with a dark core. In the chromospheric Ca II H and K line cores, the magnetic knots appear as bright faculae.

4.2.4   Hierarchic pattern

In summary, the hierarchy of the strong-field concentrations reveals the following pattern (Table 4.1):

[†] Steady downdrafts, which had been reported earlier, have been refuted by measurements of high spectral resolution; see Stenflo and Harvey (1985).

Despite the large range of total fluxes, spanning more than 4 orders of magnitude, the intrinsic magnetic field strengths in the concentrations cover a range of no more than a factor of 3.

Near the axis of the concentration, the magnetic field as recorded in the photosphere is nearly vertical, with the exception of magnetic grains close to a sunspot or in some polarity inversion zones. In the photosphere, persistent, strongly inclined fields are observed only in penumbrae, that is, in the fringes or in the immediate vicinity of umbrae.

The magnetic field lines in a concentration fan out with height: the magnetic field resembles that near the end of a long solenoid as if it stood vertically in the convection zone, with its top somewhere in the photosphere.

Apart from oscillations, the plasma within the strong magnetic field in sunspot umbrae, pores, faculae, and network patches is static: no vertical flows have been detected.

In the photosphere, the largest magnetic concentrations, namely the sunspots and the pores, are dark. The facular and network grains are bright, if observed at a sufficiently high spatial resolution.

Sunspot umbrae and pores are compact magnetic structures, with magnetic filling factors of 100%; presumably this compactness applies to magnetic knots as well. During the decay process, sunspots and pores maintain their properties for quite a while: they shrink while keeping their compactness and sharp outlines. Note that this temporal behavior is contrary to what is to be expected in the case of simple ohmic diffusion. In contrast, faculae and network patches are clusters composed of small magnetic grains – nearly everywhere the magnetic filling factor is less than 25%, except very locally in peaks up to 50%.

Sunspots, pores, and plages are exclusively found in active regions; probably the same applies to magnetic knots. There is no qualitative difference between faculae and network patches. The difference is that the number density of magnetic grains constituting the faculae tends to be larger than that in network patches outside active regions.

4.3   Magnetohydrostatic models

4.3.1   *Flux-tube models*

In order to model the magnetic concentrations in the solar atmosphere, the concept of the *magnetohydrostatic flux tube* has been developed based on assumptions suggested by the properties listed in the previous section. The notion of magnetostatic models was developed for sunspots in a correspondence between Biermann and Cowling; see Cowling (1985) and Biermann (1986). The darkness of sunspots suggests a strongly reduced nonradiative energy transport, which was explained as a consequence of the suppression of freely overturning convection by the strong magnetic field (Biermann, 1941). Cowling (1946) argued that the forces caused by the lateral inequalities of the gas pressure inside and outside the sunspot mold the magnetic field into such a configuration that the magnetic forces balance the lateral gas pressure gradients. We begin the development of magnetohydrostatic models with the simple case of a single, relatively thin, flux tube.

Consider an axially symmetric, untwisted[‡] ($B_\phi = 0$) flux tube, surrounded by a non-magnetic medium (Fig. 4.5). We assume that mass flows both inside and outside the flux

---

[‡] This is no limitation, because the effects of twists on the structure of flux tubes in the solar convection zone and atmosphere are very small (Parker, 1979, his Chapter 9; see also Spruit, 1981c). If the flux

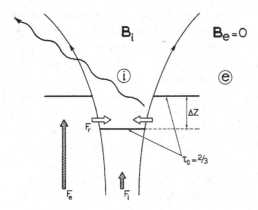

Fig. 4.5. Concept of the magnetohydrostatic flux-tube model. One level of constant optical depth in the continuum, $\tau_0 = 2/3$, is shown, with the Wilson depression Δz. The hatched arrows F_i and F_e stand for the flux densities in the (nonradiative) energy flows inside and outside the flux tubes, respectively. The horizontal arrows indicate the influx of radiation into the transparent top part of the tube. The resulting bright walls are best seen in observations toward the solar limb (as seen along the oblique wavy arrow; figure adapted from Zwaan and Cram, 1989).

tube are so small that magnetohydrostatic (often abbreviated to magnetostatic) equilibrium, as described by Eqs. (4.15) and (4.16), determines the pressure balance throughout the flux tube and its surroundings. In the top part of the convection zone, just below the photosphere, the field strength in the tube is assumed to be sufficiently large to hamper energy transfer by convection (see Section 4.1.2). This assumption implies that (a) the energy flux density F_i inside the tube is much smaller than the energy flux density F_e in the exterior convection zone; and (b) the only lateral energy transfer is due to radiative processes, which is important only in the transparent top part of the tube. The deeper part of the tube is thermally insulated from the surrounding convection zone.

Because $F_i < F_e$, the plasma inside the tube is cooler than the plasma outside at the same height: $T_i(z) < T_e(z)$. This temperature difference is largest in the deepest atmospheric layers within the flux tube, where the radiative cooling is most effective and where the interior of the tube is well shielded against the lateral influx of radiation by the relatively high opacity.

Because of magnetostatic equilibrium, the gas pressure in the tube is lower than the exterior pressure; see Eq. (4.17). In the deep part, where $T_i(z) < T_e(z)$, the pressure scale height $H_{p,i}$ (Eq. 2.9) within the tube is smaller than the external scale height $H_{p,e}$. With the assumption that for large depth $p_i(z) \uparrow p_e(z)$, the isobars within the top of the tube are depressed relative to those outside: there is a positive pressure difference $\Delta p(z) = p_e - p_i > 0$, which is balanced by the Lorentz force in the bounding current sheet. The depressed isobars are associated with depressed density isopleths, which leads to lower opacities and thus to the *Wilson depression* Δz; see Fig. 4.5.

tube were twisted, the twists would propagate toward the most expanded part of the tubes, i.e., to the upper atmosphere. Small twists left in the top of the convection zone would not affect the field near the axis at all and the field in the outer parts only slightly.

In order to study the properties of magnetic elements with diameters up to $\sim 1\,$Mm, Spruit (1976, 1977a, 1977b) constructed numerical models of flux tubes, assuming a potential field within the tube. Such models possess an infinitely thin current sheet at the interface (the *magnetopause*) between the field and the field-free external plasma. Spruit computed the temperature distribution in the flux tube, using an energy equation based on a depth-dependent heat diffusion coefficient, which is anisotropic for the nonradiative component: reduced transport along the field and no energy transport other than radiation perpendicular to the field. Clearly, a diffusion equation provides but a crude model of energy transport by radiation and convection, particularly in the part of the tube that is not optically thick.

Spruit's flux-tube models contain three adjustable parameters: (1) the ratio $q < 1$ of the nonradiative energy flux density inside the tube relative to that outside, (2) the Wilson depression Δz, and (3) the radius a_0 at some specified depth. Spruit found an array of models mimicking the observed magnetic elements by adopting a ratio $q \approx 0.2$ independent of the radius a of the flux tube. The observed magnetic field strengths between 1 and 2 kG are reproduced if a Wilson depression of approximately one photospheric pressure scale height ($100 \le \Delta z \le 200\,$km) is adopted.

The lateral influx of radiation causes the appearance of the flux tube to depend markedly on its radius a_0. Its effect is small in the (dark) pores, but it increases with decreasing radius until sufficiently thin tubes show up as bright faculae. In the thinnest flux fibers, the lateral radiation field produces a *warm bottom*, such that at the same *optical* depth the interior of the tube is warmer than the photosphere well outside the tube (see Grossmann-Doerth *et al.*, 1989). Remember that at the same geometrical depth the tube interior is cooler than the convection zone outside. In white light, the brightest part of the flux tube is its cylindrical *bright wall*, which is best exposed when the element is observed obliquely, that is, toward the solar limb (see Fig. 4.5). That explains why at moderate angular resolution the brightness contrast of faculae and network patches relative to the photosphere increases toward the solar limb (Spruit, 1976). At a high angular resolution, the contributions of the wall and the bottom can be recorded separately. In such conditions, near the center of the solar disk, micropores (observed by Dunn and Zirker, 1973) may be distinguished, as features between truly dark pores and bright faculae, exposing a relatively dark bottom surrounded by a bright wall.

Spruit's simplifications of an infinitesimally thin current sheet, radiative transport as a diffusion process, and his treatment of convection at the outer boundary of the tube have been relaxed in more recent investigations of magnetic flux concentrations. Numerical simulations treating the full time-dependent, three-dimensional problem with sophisticated radiative transfer in the atmospheric layers are still too demanding for present computational means. Consequently, the published investigations depend on some combination of simplifications, such as stationary models, gray opacity in the radiative transport, and either a coarse computational grid or a reduction to a two-dimensional problem with flux slabs instead of flux tubes (see Grossmann-Doerth *et al.*, 1989; Schüssler, 1990; Steiner *et al.*, 1998). These models resulted in a better description of the warm bottom around $\tau \sim 1$. Moreover, both steady and dynamic models indicate a narrow downflow region immediately outside the magnetopause, with velocities up to several kilometers per second. This convective downflow is triggered by the increased radiative cooling through the flux tube, which is more transparent than its environment. This leads to a symbiosis between flux tubes and the thin convective down-drafts (Section 2.5), which is fundamental in the interplay between convection and the magnetic field (Section 5.2).

4.3.2 Sunspot models

Apart from some fine structure, the sharply bounded umbra appears to be characterized by a rather uniform brightness and nearly vertical magnetic field lines, moderately inclined at the edges but nearly uniform in strength. These data suggest that the observed umbra is the top of a almost current-free column (Gokhale and Zwaan, 1972). The surrounding penumbra is identified as the top of an extended penumbral current system, which tapers down to a thin current sheet wrapping the deep part of the umbral column; see Fig. 4.6.

Jahn (1989) demonstrated that a realistic global spot model may be constructed with a thin current sheet in the magnetopause wrapping the entire sunspot tube, plus an extended current system occupying the penumbral fringe of the structure. The shape of the current sheet and the currents in the sheet and the penumbral current system were adjusted such that the observed magnetic-field strength distribution in the photospheric layer across the spot is approximated. From the current distribution and the gas pressure distribution outside the flux tube, the pressure distribution within the sunspot is derived, and from that the thermal distribution. The resulting thermal stratification indicates that most of the energy is transported by some other means than radiation.

From the existence of umbral dots, the processes of formation and decay of sunspots (see Sections 4.4 and 5.1.1.1), and from considerations of energy transport, several authors arrived at a *cluster model*: sunspots and their umbrae were suggested to consist of a bundle of flux tubes. From the brightness and high color temperatures of some dots (Beckers and Schröter, 1968), it had been concluded that umbral dots would be the tops of virtually field-free columns of hot gas between the cool flux tubes, which, if interpreted

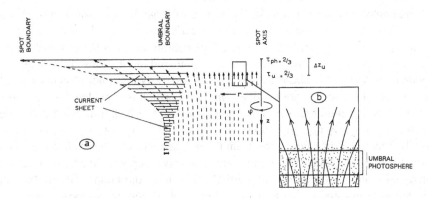

Fig. 4.6. Panel *a*: a schematic diagram of the sunspot magnetic field, with its current-free umbral column surrounded by a broad penumbral current system near its top and a relatively thin sheet with azimuthal currents in deeper layers. The field lines are indicated by dashed curves; the arrow heads indicate the level of optical depth $\tau = 2/3$. The current system is shown by the horizontal hatching (figure adapted from Gokhale and Zwaan, 1972). Panel *b*: the umbral segment framed by the rectangle in panel *a* is enlarged to show a possible configuration for some umbral dots.

as a mode of convection, would show appreciable upward velocities during part of their lifetimes (see Choudhuri, 1992). These expectations are not borne out by observations, however: the mean field strength in umbral regions with dots is only slightly lower than it is in the dark cores (Buurman, 1973). In spectrograms of sufficient angular resolution, the individual bright structures show field strengths that are slightly, if at all, reduced with respect to the ambient umbral values; there are no indications of markedly deviant Doppler shifts either (Zwaan *et al.*, 1985, Lites *et al.*, 1991).

Degenhardt and Lites (1993a, 1993b) suggested a way out of this quandary by proposing that umbral dots are the thin tops of tapering columns of gas, which, below the umbral photosphere, are much hotter and which possess a much weaker magnetic field than the ambient umbra (see Fig. 4.6*b*). Solutions of the steady-state MHD equations in the thin-flux-tube approximation, allowing for a steady internal upward flow and lateral exchange of radiation, reproduce observed features of umbral dots for a range of input parameters. This gently dynamic model is only a first step toward the solution of the problem of umbral dots and the nonradiative energy transport in umbrae and pores: the vertical transport of radiation still has to be included, and the assumed conditions just below the umbral atmosphere are to be connected with some magnetoconvective process(es) deeper down.

The model proposed by Degenhardt and Lites differs from the cluster model in that the hotter columns are permeated by the magnetic field. Moreover, isolated umbral dots would correspond to hot convective upwellings of short duration. Yet their sunspot structure cannot be called monolithic either, because the magnetic field is much weaker in the hot columns than it is in the ambient umbra.

The tapering of the hot columns makes the umbral atmosphere nearly uniform (Fig. 4.6*b*), which fits in with the weakness of fine structure in spectrograms in the visible. However, in the spectral region near 1.6 μm, where the continuum extinction is lowest, a clear signature of hot columns is found in spectral Si I lines of very high excitation potential (Van Ballegooijen, 1984a).

Penumbrae are considered to be the fringes of large and compact magnetic concentrations, resulting from the adjustment of a nearly magnetostatic sunspot system to the ambient gas pressure distribution. In itself the penumbra is not static, however, because of the complex system of flows associated with mixed inclinations of the magnetic field (penumbral anemones). These are probably caused by dynamic adjustments to local pressure imbalances, and the *interchange convection*, which is considered to be responsible for the detailed penumbral structure and dynamics. The heat impressed on the penumbra from the underlying convection zone causes relatively hot, nearly horizontally and radially oriented, flux tubes to rise from the magnetopause, whereas relatively cool flux tubes descend. The hotter rising tubes are relatively steeply inclined with respect to the horizontal plane – presumably along these flux tubes the clumps of gas descend that are seen as bright grains that flow toward the umbral border and locally move into the umbra (Ewell, 1992; Sobotka *et al.*, 1995). The descending cooler tubes are more nearly horizontal – along these tubes most of the Evershed flow moves toward the outer border (Jahn, 1992; see also Spruit, 1981c).

For detailed theoretical studies of penumbral structure and dynamics, we refer to Thomas and Weiss (1992b) and to Thomas (1994), and references given there.

4.3.3 Discussion of magnetostatic models

The lateral pressure balance in magnetostatic equilibrium [Eq. (4.12)] yields magnetic-field strengths in the observed range if a value of one to a few pressure scale heights is adopted for the Wilson depression. The observed trend from the largest field strength in sunspot umbrae down to the smallest in fine filigree grains is compatible with the expected dependence of the lateral influx of radiation in the top of the flux tube bundle on its radius. Also, the transition from the darkness of sunspots and pores to the brightness of facular and network grains is readily explained by the dependence of the lateral influx of radiation on the diameter of the (composite) tube.

In the photospheric tops of magnetostatic flux tubes, the mass density ρ is much lower than it is in the ambient top of the convection zone. Hence the flux tubes are buoyant, which is consistent with the assumption of verticality of the tubes. Large-scale flows transport thin flux tubes by drag; because of their stiffness and buoyancy, however, the tubes resist bending during their transport.

The sharp drop-off toward the limb of the longitudinal magnetic field signal in plages and network (Fig. 1.1e) results from three properties of concentrations of strong magnetic field: (1) the magnetic field is nearly vertical, (2) the flux tubes fill only a fraction of the photosphere, and (3) in each tube there is a Wilson depression of the order of one pressure scale height or more. Because of the Wilson depression, the layer contributing to the formation of the diagnostic spectral line is hidden when the flux tube is observed very obliquely.

The exponential decrease of the gas pressure with height, with a scale height of ~ 150 km in and near the photosphere, implies that also the absolute difference between the external and the internal gas pressure $\Delta p(z) = p_e - p_i$ drops rapidly with height. Hence the Lorentz force drops with height, and so do the azimuthal electric currents in the current sheet. Consequently, the field becomes an unbounded, force-free field within a few pressure scale heights above the photosphere; presumably that field differs little from a potential field. In other words, in the atmosphere the field should resemble the field at the end of a long solenoid, as is observed.

The degree of divergence of the field lines depends on the total magnetic flux in the structure: for increasing flux, the divergence starts at a progressively lower level in the atmosphere. For small magnetic features, with photospheric diameters no more than a few times the pressure scale height, flux tubes expand several times their diameter between the photosphere and a few scale heights higher in the low chromosphere, and in this way they cope with the rapid decrease of the gas pressure. For sunspots, with diameters of many times the pressure scale height, such an adaptation is not possible: a strong horizontal component in the magnetic field is required on their periphery at levels as deep as the photosphere. This horizontal component is responsible for penumbrae in sunspots.

Flux-tube models predict that a tight cluster of flux tubes, presenting itself as one entity in observations of insufficient resolution, displays an overall field configuration nearly identical to that of one single flux tube containing the same amount of flux. That is why it is not easy to tell whether an observed single filigree grain is a single flux tube or an unresolved bunch of thin flux fibers. In any case, measurements of intrinsic field strengths indicate that such a grain consists of some number (one, a few, or many) of magnetostatic flux tube(s) of *strong* field.

Spruit (1992) summarizes theoretical studies concerning the effects on the Sun of the heat flux blocked by sunspots. Such effects turn out to be small. One reason is the enormous heat capacity of the convection zone, giving it a large thermal time scale $\hat{t}_{th}$ of some 10^5 years. Hence on time scales shorter than $\hat{t}_{th}$, almost all changes in the heat flux are absorbed by the convection zone. In Spruit's original paper, the absence of conspicuous bright rings observed around sunspots is explained by the efficient heat transport by turbulent diffusion in the convection zone; for the explanation in the picture of convection in stellar envelopes that emerges from ab initio hydrodynamic modeling (Section 2.5), see Spruit (1997).

Faculae and network grains provide heat leaks to the top of the convection zone because of the Wilson depression in their flux tubes; as a consequence, they are surrounded by narrow dark rings. These dark rings compensate for only a fraction of the excess emission by faculae and network grains. Hence the solar irradiance variations of observational interest reflect the *instantaneous* coverage by spots, faculae and network (see Section 6.1).

In principle, sunspots and pores could be destroyed by the *interchange* or *fluting instability*, which occurs if the total energy of the structure can be decreased by a nonaxisymmetric change of the shape of the tube, for instance, by shredding the tube lengthwise into a bunch of thinner tubes (Parker, 1975). This instability is prompted by the curvature force where the magnetopause is concave toward the external medium, which applies to magnetostatic flux tubes because the field lines diverge with height. This instability is not observed, however, suggesting that some stabilizing process occurs. Flux tubes may be stabilized by buoyancy, because the gas in the diverging part of the flux tube is less dense than the gas below the magnetopause (Meyer *et al.*, 1977; Jahn, 1992). The dynamic partitioning between patches of strong field and convective flows suggests that splitting by the fluting instability is countered by convective flow converging on the magnetic patches (Section 5.3).

We now turn to two urgent problems remaining at the basis of the magnetohydrostatic modeling of magnetic concentrations. The theoretical and numerical studies of flux-tube models to date do not explicitly treat the physics of the nonradiative energy transport within the tubes; usually that transport is simply parameterized by a reduced efficiency factor relative to that in the free convection zone for energy transport along the field, and by a complete inhibition of nonradiative energy transport perpendicular to the field. Semiempirical thermodynamic models for sunspot umbrae, and models for flux tubes mimicking pores and facular and network grains, indicate that there is a substantial nonradiative energy flux density along the magnetic field of some 20% of the energy flux density in the normal convection zone (Section 4.3.1).

Note that the Wilson depression Δz provides an indication of the thermal stratification of the deeper layers, and thus on the energy transport. In order to explain the Wilson depressions, ranging from approximately one pressure scale height in thin flux tubes to several scale heights in sunspot umbrae, the internal temperature deficit must extend over many pressure scale heights.

In the present magnetostatic models, the processes in the current sheet are not considered – in many models, it is even approximated by an infinitesimally thin sheet bounding the flux tube. The actual current wall is the transition between the supposedly current-free interior and the virtually field-free convection zone outside. Within the current wall, the thermodynamic quantities vary radially and there the radial component of the gas pressure gradient is balanced by the Lorentz force. The lateral influx of radiation from the hotter ambient convection zone into the tube is absorbed in the current wall, and there is additional heating by ohmic dissipation. The heat deposited in that wall is transferred somehow, presumably resulting in a net upward flux along the field (Gokhale and Zwaan, 1972).

4.4   Emergence of magnetic field and convective collapse

The strong-field concentrations that determine the face of the Sun surface in *active regions*, which usually show a clearly bipolar organization; see Figs. 1.1*e* and 1.2. The birth phenomena of such an emerging active region (Fig. 4.7) are readily pictured by the emergence of the top of an Ω-shaped flux loop (Fig. 4.8); see Zwaan (1985). Some of the chromospheric phenomena have been known for quite some time: the first manifestation is the appearance of a small, compact and very bright (bipolar) plage (Fox, 1908; Waldmeier, 1937; Sheeley, 1969) in the Ca II H and K line cores and in high-resolution magnetograms. Nearly simultaneously the *arch filament system* (AFS) becomes visible in the H α line core: this is a set of roughly parallel dark fibrils connecting faculae of opposite polarity (Bruzek, 1967, 1969). Each fibril is a loop whose top ascends with a speed of several kilometers per second, up to 10 km/s. In both legs, matter flows down at speeds of some tens of kilometers per second, locally reaching up to 50 km/s, as measured in H α.

The development of an *emerging flux region* (EFR; term coined by Zirin, 1972) follows distinct patterns. At first, the intruding bipolar feature does not disturb the ambient "old" magnetic structure in the photosphere and chromosphere. The faculae of opposite polarity move apart: during the first half hour of emergence the rate of separation of the footpoints may exceed 2 km/s; then it drops to a value between 1.3 and 0.7 km/s during the next 6 hours (Harvey and Martin, 1973). The expansion of the active region continues at a gradually slower pace for up to some days. New magnetic flux may emerge somewhere between the diverging opposite polarities in the previously emerged flux, or in their immediate vicinity. The arch filament system lasts as long as flux is emerging, but the individual dark H α fibrils live only for $\sim$20 min.

Initially the emerging bipoles may be oriented in rather arbitrary directions, but they rotate and after approximately 1 day most EFRs have assumed the "correct" orientation of a mature active region, nearly parallel to the equator, as sketched in Fig. 4.9 (Weart, 1970, 1972).

At closer examination, many active regions appear to be formed by contributions from several EFRs that emerge separately but in close proximity, nearly simultaneously or in rapid succession, within a few days (Weart, 1970, 1972; Schoolman, 1973).

The photospheric granulation in the central part of an EFR is disturbed: it looks fuzzy, there are conspicuous alignments in the dark intergranular lanes, and there is no filigree as yet. In the paragraphs that follow, we summarize results presented in the theses by Brants (1985c) and Strous (1994), and in review papers by Zwaan (1985, 1992), with reference to Fig. 4.7.

The dark alignments in the granulation are thought to be caused by tops of magnetic loops passing through the photosphere (Fig. 4.7*a*). These alignments are parallel to the fibrils in the overlying arch filament system (Fig. 4.7*c*), as expected. Well-defined

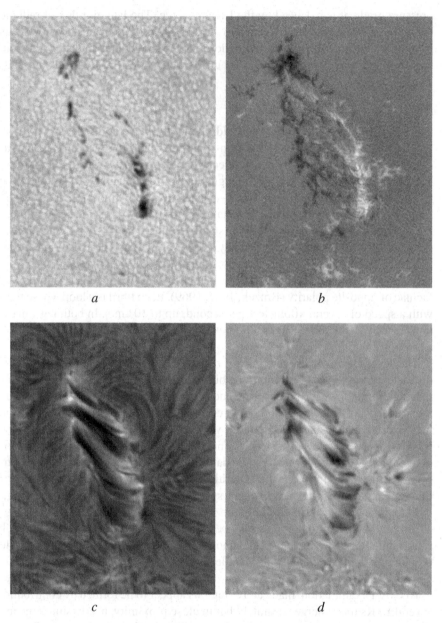

Fig. 4.7. Nearly simultaneous aspects of an emerging active region (AR 5,617 on 29 July 1989). Panel *a*: the image in continuum intensity; dark pores are located in the outer strip and the internal granulation looks fuzzy, with nearly parallel alignments in the dark intergranular lanes. Panel *b*: line-of-sight component of the magnetic field. Panel *c*: arch filament system recorded in the H α-line core. Panel *d*: Doppler shift in H α; white marks upward velocity and black downward flow. The tops of the loops are rising. At the feet there are strong downward flows, which are sharply bounded near the inner edge of the strip with pores. The combination of the various aspects shows the emerging flux region as a sharply outlined intrusion from below. The field of view is ∼65 Mm × 85 Mm; North is toward the left and West toward the top (courtesy of L. Strous).

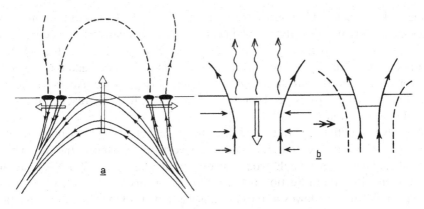

Fig. 4.8. Panel *a*: top of an Ω-shaped loop as a model for the emergence of magnetic flux, separation of the polarities, and coalescence of sunspots. Broad open arrows indicate local displacements of flux tubes. This figure is very schematic: emerging flux regions consist of many more separate flux tubes. Panel *b*: diagram of the convective collapse of a flux tube.

alignments have a width of ∼1,500 km, are visible in the photosphere for some 10 min, and have a speed of rise $v \simeq 3$ km/s – here *dark* matter is rising! If this speed equals the Alfvén speed [Eq. (4.22)], then the field strength is approximately 600 G and the flux is approximately 1×10^{19} Mx (Zwaan, 1985).

The observed properties indicate that an EFR consists of a bundle of many, fairly discrete flux loops, which arrive at the photosphere with intrinsic field strengths of at least a few hundred gauss, up to roughly 600 G, which roughly equals the equipartition field strength [Eq. (4.11)] at the top of the convection zone. For more detailed observational data and theoretical modeling of flux emergence, we refer to Section 5.1.1.1.

At the lower tips of the strong chromospheric downflows (Fig. 4.7*d*) there are localized downflows in the photosphere, with velocities up to 2 km/s, (Brants, 1985a, 1985c). In one particularly well-documented case, a small downdraft area developed from a darkish protopore, with a diameter slightly larger than 1 arcsec, into a pore.

Zwaan (1985) interpreted the photospheric downdraft process as the *convective collapse* that transforms the emergent field of intermediate strength ($B \lesssim 500$ G) into a

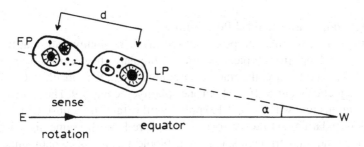

Fig. 4.9. Sketch of a large bipolar active region at maximum development. Distances *d* between the centroids of opposite polarity range from ∼100 Mm to ∼150 Mm. The inclination angle α between the bipolar axis and the solar equator ranges from 0° to ∼10° (α depends on latitude, as discussed in the text; figure from Zwaan, 1992).

strong field ($B \gtrsim 1.5$ kG), referring to Spruit's (1979) calculations of the convective collapse of slender flux tubes: the initial downdraft is enhanced by the convective instability in the top of the convection zone. As a result, the photospheric top of the tube becomes nearly evacuated relative to its environment, so it is compressed and the magnetic field is strongly enhanced until the magnetic pressure balances the excess of the external over the internal gas pressure. Then the collapsed flux tube has reached magnetostatic equilibrium (Section 4.3).

Note that an initial upward flow in the tube in and below its photosphere would work in the opposite way: the updraft would be convectively enhanced, the flux tube would expand, and the magnetic field would become weaker (Spruit, 1979). We return to such a convective blowup in Section 6.4.3. In emerging flux regions, all flows within the field point downward, however: the chromospheric parts of the tubes drain, the radiative cooling causes the photospheric plasma to sag, and the convective instability in the top of the convection zone amplifies the downdraft and carries the cooling and the partial evacuation deeper down. In Section 4.3.3 it was concluded that the Wilson depressions, ranging from one to some pressure scale heights, require that the flux-tube interiors have to be significantly cooler than the exterior down to several pressure scale heights below the photosphere. We suggest that the convective collapse is the mechanism accomplishing that feat.

In models for the convective collapse, a substantial reduction of turbulent heat flux is invoked already at field strengths somewhat smaller than the equipartition field strength. This assumption is supported by the observed darkness in the alignments in the granulation, which suggests that the field in the tops of the emerging flux loops is already strong enough to order the local granulation and to reduce the turbulent heat transport to the extent that the still rising flux-loop tops are darker than the granules. Note that the plasma within the emerging loop is expected to arrive relatively warm in the top of the convection zone, as do the granules.

Sunspot pores grow by an increase of area and by a decrease of continuum brightness. Both phenomena happen simultaneously in the immediate vicinity of downdrafts, indicating convective collapse. In fact, when sufficient magnetic flux becomes available fast enough, pores are formed in one continuous process with the convective collapse. The local downward flow stops, however, once a truly dark pore mass has formed, with a field strength larger than ~ 2.0 kG.

4.5   Omega loops and toroidal flux bundles

In this section, we infer properties of the magnetic configuration at the roots of an Ω-shaped flux loop[§] from typical features of bipolar active regions at maximum development, that is, shortly after the emergence of all the flux. We consider a large active region, with a magnetic flux $\Phi > 10^{22}$ Mx in one polarity; see Fig. 4.9. The great majority of large active regions follow Hale's (1919) polarity rule (Table 6.1): during one sunspot cycle the leading polarities on the northern hemisphere have the same sign, which is opposite to that on the southern hemisphere. The bipole axis is somewhat inclined with

---

[§] The shape of such a loop is not precisely known (see Section 5.1.1.1); the Ω shape is used as a heuristic device to describe the observations.

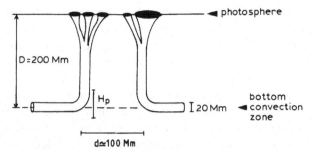

Fig. 4.10. Schematic model showing a large active region on top of an Ω-shaped flux loop emerged from a toroidal bundle located at the bottom of the convection zone. The flux in the bundle is 3×10^{22} Mx; the pressure scale height H_p at the bottom of the convective envelope is 60 Mm. The assumed field strength $B = 10$ kG in the toroidal bundle is somewhat larger than the equipartition value near the bottom of the convection zone (see Fig. 2.1; figure from Zwaan, 1992).

respect to the direction of rotation, with the leading polarity (preceding in the solar rotation) being closer to the equator (Joy's rule). The inclination angle γ varies from a few degrees near the equator to $\sim 10°$ at a latitude of 35° (Joy, 1919; Brunner, 1930; Wang and Sheeley, 1989). The distance between the centroids of opposite polarity in such large regions is between roughly 100 and 150 Mm (see Wang and Sheeley, 1989). This suggests that active regions emerge from a toroidal flux bundle below the solar surface, which has been wound into a spiral by differential rotation (see Fig. 6.10 and Section 6.2.2).

The pronounced preference for that nearly E–W orientation of active regions indicates that the magnetic field is strong enough to resist distortion by the turbulent convection, from where the Ω-shaped loop is rooted, all the way up into the atmosphere (Zwaan, 1978). In other words, the field strength $B(z)$ should everywhere be approximately equal to or exceed the local equipartition field strength $B_{eq}(z)$. The flux in an active region must have been stored in a toroidal flux bundle with a diameter smaller than the local pressure scale height H_p, to allow the top of the emerging loop to move essentially as an entity. Considering the amount of flux in a large active region, Zwaan (1978) argued that the toroidal flux system must be located close to the bottom of the convection zone, because only there is the pressure scale height large enough. For large active regions, the ratios between the scales in the resulting picture (Fig. 4.10) seem acceptable.

The buoyancy of magnetic fields presents a problem for siting magnetic fields with $B \gtrsim B_{eq}$ in the convection zone, because such fields would be expelled within a fraction of a year (Section 4.1.2). It is now believed that the toroidal flux system is located just *below* the convection zone, in the overshoot layer at the top of the radiative interior (Spiegel and Weiss, 1980; Van Ballegooijen, 1982a, 1982b; Schüssler, 1983). The direct argument is that convection and magnetic field expel each other (Section 4.1.2); hence the magnetic field is worked into the interface between the convection zone and the radiative interior (see Schüssler, 1983). An advantage is that in the subadiabatic stratification the buoyancy of flux tubes is counteracted: rising plasma becomes cooler than its surroundings and so the density contrast, $\rho_e - \rho_i$, is reduced. We discuss the flux storage problem in Section 5.1.1.2.

The substantial temperature reduction within flux tubes, brought about by the convective collapse, is limited to the very top layer of the convection zone. Over the rest of the convection zone, the interior of the tubes is expected to be quite close to thermal equilibrium with the ambient plasma: $T_i(z) = T_e(z)$. Because of hydrostatic equilibrium, the internal and external gas pressure distributions p_i and p_e are quite nearly equal, which raises the problem of how the magnetic pressure $B^2/8\pi$ is contained. Zwaan (1978) argued that "turbulent pressure" confines the magnetic field up to the equipartition field strength. Van Ballegooijen (1984b) replaced that primitive notion by a proper treatment of the departure from strict hydrostatic equilibrium caused by the turbulence in the convection zone. The upper part of the convection zone is "puffed up" by the turbulence such that the gas pressure is slightly enhanced over the gas pressure within the strictly static flux tubes (without turbulence). Van Ballegooijen found that toroidal flux bundles, with $B_{eq} \lesssim B \lesssim 10B_{eq}$, can be connected with observed active regions by vertical legs with a field strength that smoothly decreases with height but that nowhere drops below $0.6B_{eq}(z)$. Thus, it appears that nearly vertical flux tubes in an Ω loop can be contained at field strengths of the order of the local equipartition values, provided that the Ω loop is quasi-static and rooted in a sufficiently strong toroidal system. This leaves the problem of how such an Ω loop can be formed during the emergence process; this problem is discussed in Section 5.1.1.2.

4.6   Weak field and the magnetic dichotomy

A weak magnetic-field component was discovered in the solar photosphere by Livingston and Harvey (1971); see Harvey (1977). A very sensitive magnetograph and good seeing conditions are required to make the weak magnetic features visible. Nearly all the properties reported here are based on observations collected with the NSO/Kitt Peak magnetograph and the videomagnetograph at Big Bear Solar Observatory (BBSO; for technical details on the latter, see Martin, 1984, and Zirin, 1985a). Below we have collected properties of the weak field from reviews by Harvey (1977), Martin (1988), and Zwaan (1987), and from movies obtained with the BBSO videomagnetograph.

The features of the weak field differ from those of the strong field (Section 4.2) in several respects: the weak field is found all over the solar disk, also in the otherwise empty cell interiors within the network of strong field – hence the weak field is often called the *internetwork (IN) field*.[||] It consists of small fragments, which we call *specks*. The polarities are mixed at scales of a few arcseconds ("pepper and salt"), and they are independent of the dominant polarity of the surrounding magnetic network; among the weak-field specks, conspicuous pairs of opposite polarity are not common (Fig. 4.11).

Whereas in a single magnetogram it is not always possible to tell a relatively large internetwork speck apart from a small network patch or a pole of an ephemeral bipolar region, a time series brings out their distinctly different dynamic behavior. The internetwork specks move quickly, at speeds of 0.3 km/s or larger (Zirin, 1985a), whereas the network patches move much slower. Many of the internetwork specks emerge in clusters of mixed polarities from small source areas, from where they stream more or less radially

---

[||] We prefer the term "internetwork" over "intranetwork" or "innernetwork," in order to indicate that the (majority of the) features are found outside the network, within the cell interiors.

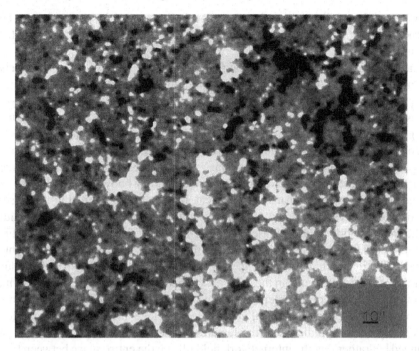

Fig. 4.11. The weak internetwork field. The conspicuous black and white patches belong to the intrinsically strong network field; the small specks are fragments of the internetwork field. The field of view is $308'' \times 237''$ near the center of the disk (from Wang *et al.*, 1995).

outward. Probably these source areas correspond to updrafts in supergranules and meso-granules (Section 2.5). When an occasional dipole is spotted in the internetwork field, its poles are seen to travel together in the same general direction, whereas the poles of an ephemeral active region always move apart, as is typical for the emergence of strong fields (Section 4.4).

Many internetwork specks disappear either by merging with an internetwork speck or a network patch of the same polarity or by canceling with a magnetic fragment of opposite polarity (see Martin, 1988). Others disappear by fading out of sight. No statistics on the lifetimes of internetwork specks have been published, but it is reported that there is no appreciable change in the field within 5 min, and that conspicuous internetwork specks may remain visible for at least several hours (see Wang *et al.*, 1995).

After separating internetwork specks from network patches in a time series of video-magnetograms on the basis of their dynamical behavior, Wang *et al.* (1995) found that the distribution of magnetic fluxes in the internetwork specks clearly differs from that in the network patches. Both flux distributions are smooth curves, each with a single peak. For the internetwork specks, the peak is at $\Phi = 6 \times 10^{16}$ Mx (in agreement with the estimate $\Phi = 5 \times 10^{16}$ Mx by Harvey, 1977), whereas the peak for the network patches is at $\Phi = 2 \times 10^{18}$ Mx. The distributions overlap in the range $4 \times 10^{17} < \Phi < 1 \times 10^{18}$ Mx. Clearly, the shapes of the distribution functions depend on seeing, instrument sensitivity, and the delicate problem of the calibration of the measurements, but the reality of two distinctly different distribution functions for the internetwork specks and the network patches stands out, because they were determined in the same field of view with the same instrumental setup.

The observational data indicate that the internetwork field differs qualitatively from the magnetic field in (ephemeral) active regions and in the network (see Zwaan, 1987). The properties point to an *intrinsically weak* internetwork field, which responds virtually passively to the local convective flows. The weak field does not impede the convective energy flow appreciably, so it does not lead to a noticeable photospheric cooling. Hence no convective collapse can occur, so there is no appreciable Wilson depression and the buoyancy in the magnetic field is very small. This implies that the weak field is brought to the photosphere in convective updrafts, rather than by buoyancy.

The lack of buoyancy means that the weak field is not set vertically: it intersects the photosphere at a variety of angles. This explains why the longitudinal magnetic signal in the internetwork field decreases only slightly toward the solar limb (see Fig. 1 in Martin, 1988), which is very different from the marked decrease toward the limb in the signal of the magnetic network (Fig. 1.1). The sensitivity and the spatial resolution of the magnetic measurements in the solar atmosphere are not yet good enough to measure the intrinsic strength of the internetwork specks accurately, but the interpretation of the internetwork field as an intrinsically weak field explains the observed properties aptly. This interpretation predicts that the features in the internetwork field are not sharply bounded but rather that the internetwork field fills up the entire space between the strong field, with a fluctuating field strength.

Apparently there is a marked *dichotomy* in the photospheric magnetic field: there are local concentrations of strong field, with intrinsic field strengths above 1 kG, surrounded by a weak, turbulent, internetwork field, with field strengths no more than a few hundred gauss; that is, smaller than or approximately equal to the equipartition field strength B_{eq} in the top of the convection zone. The origin of this dichotomy lies in the rather strict conditions for the formation of a strong field: an Ω-shaped loop rising over many pressure scale heights (in order to allow draining of sufficient mass along the legs), at a strength of approximately the equipartition value B_{eq} or larger (to provide the thermal insulation needed for the convective collapse). Under different conditions, flux tubes rising into the atmosphere do not experience a convective collapse. For instance, an initially strong, horizontal flux tube, rising nearly horizontally over several pressure scale heights, blows up because the plasma trapped in it expands to maintain pressure requilibrium. Hence the flux tube weakens while rising, then is harassed by convection, and eventually emerges in the photosphere like a sea serpent (Spruit *et al.*, 1987). The sea-serpent shape explains the pepper-and-salt appearance of the internetwork field.

Spruit *et al.* (1987) discuss various formation processes, such as the formation of so-called U loops in between two Ω loops that have emerged from the same toroidal bundle (see Figs. 6.16 and 6.17). Weak, small-scale fields are also produced by the aperiodic dynamo that is expected to operate in any region of turbulent convection (see Section 7.3).

In its response to the convective flows, the internetwork field is perpetually varying in strength and in direction. Locally the field may approach or exceed the equipartition field strength for a while, and there the internal plasma could cool a little, so the field would increase somewhat further in strength and the increased buoyancy would try to put it in a more vertical position. Locally this may lead to a partial convective collapse. This can explain the occurrence of some relatively strong local internetwork specks, which remain visible for several hours.

Magnetic specks with strengths between weak (less than a few hundred gauss) and strong (more than ~ 1 kG) are found, as expected, to be transient features. Examples are emerging flux loops during their convective collapse (Section 4.4), as well as the moving magnetic features in moats around decaying sunspots (Section 5.1.1.3), and possibly also in the decay process of other concentrations of strong field (Sections 6.4.2 and 6.4.3). Relatively long-lived internetwork specks are seen to linger close to network patches and even to merge with network patches (Martin, 1988, 1990b); conceivably these magnetic features undergo a (partial) convective collapse while being pushed by convective flow against the strong flux tubes in the network patch.

Intermediate field strengths between 400 G and 1,200 G have been measured in the near infrared ($\lambda \simeq 1.6\,\mu$m): in plages, more than 90% of the magnetic flux is present as an intrinsically strong field, with $1,200 < |B| < 1,800$ G, whereas intermediate field strengths $B < 1,200$ G were found in resolution elements of low magnetic flux (Rabin, 1992; Rüedi *et al.*, 1992). Intrinsic field strengths below 400 G could not be measured; thus the range of field strengths in the truly weak internetwork fields is not covered. Note that isolated local peaks in these measurements should not be mistaken for the typical field strength of the internetwork field.

In principle, a weak, turbulent magnetic field can be estimated from the depolarizing Hanle effect in certain photospheric resonance lines (see Stenflo, 1994, for a detailed description of this effect). Faurobert-Scholl *et al.* (1995) interpreted center-to-limb measurements of the linear polarization in Sr I λ 4, 607 Å. From their involved analysis it is concluded that the average field strength is between 20 G and 10 G at heights between 200 km and 400 km above the level $\tau_5 = 1$. This result is consistent with a ubiquitous, truly weak, turbulent magnetic field. From the high degree of polarization in the Na I D doublet, Landi Degl'Innocenti (1998) arrives at extremely low field strengths for the chromosphere: fields stronger than approximately 0.01 G are excluded. We discuss this result in Section 8.5.

5

Solar magnetic configurations

5.1 Active regions

The concentrations of strong magnetic field occur in characteristic time-dependent configurations. A prominent configuration is the bipolar *active region*, which is formed during the emergence of strong magnetic flux. Around the time of its maximum development, a large active region comprises sunspots, pores, and faculae arranged in plages and enhanced network (Fig. 1.2). In this book, the term active region is used to indicate the complete area within a single, smooth contour that just includes all its constituents. We prefer the term active region over the older term center of activity. Classical descriptions of active regions are found in Kiepenheuer (1953, his Section 4.12), De Jager (1959), and Sheeley (1981).

The present chapter is restricted to simple *bipolar* active regions. Throughout this book we use the term bipolar active region for regions consisting of only two fairly distinct areas of opposite polarity, in contrast with complex active regions in which at least one of the polarities is distributed over two or more areas. Bipolar active regions are building blocks in complex active regions and nests (Section 6.2.1).

At maximum development, directly after all magnetic flux has emerged, bipolar active regions range over some 4 orders of magnitude in magnetic flux, size, and lifetime; see Table 5.1. The division between small active regions and ephemeral active regions is based on an arbitrary historical choice: there is no qualitative difference between the two groups of bipolar features other than gradual trends.

The region's lifetime is defined rather loosely as the duration of its visibility as a bipolar system. The lifetime is roughly proportional to the magnetic flux at maximum development; the constant of proportionality for active-region decay amounts to approximately 1×10^{20} Mx/day according to Golub (1980) and to approximately 2×10^{20} Mx/day according to Chapter 12 in Harvey (1993).

In Section 5.1.1, we follow the evolution of a large bipolar active region that shows all the time-dependent features. Smaller bipolar regions share the main evolutionary pattern. In Section 5.1.2, the evolution of ephemeral regions is summarized.

5.1.1 Evolution of a bipolar active region

5.1.1.1 Emergence and growth

The emergence process (Section 4.4) takes a small fraction of the total lifetime of an active region. Even for large active regions, virtually all flux emerges well within 4 days; rise times between 4 and 5 days are exceptions (Chapter 3 in Harvey, 1993).

114

Table 5.1. *Typical magnetic fluxes and lifetimes of bipolar active regions*[a]

Region	Magnetic flux (Mx)	Lifetime
Large (with sunspots)	$5 \times 10^{21} - 3 \times 10^{22}$	weeks–months
Small (no spots, maybe pores)	$1 \times 10^{20} - 5 \times 10^{21}$	days–weeks
Ephemeral	$3 \times 10^{18} - 1 \times 10^{20}$	hours–$\sim$1 day

[a] The magnetic flux is for one polarity, at maximum development of the region.

Large active regions are formed by contributions from several emerging flux regions (EFRs) that surface separately but in close proximity, in a rapid succession, within a few days (Section 4.4). During the emergence process, the polarity distribution may be complicated because the contributing EFRs do not line up perfectly. Usually, however, the various EFRs align and merge in such a way that $\sim$1 day after all flux has emerged, one single, simply bipolar, active region results.

In a large emerging flux region (NOAA AR 5,617), Strous *et al.* (1996) followed the horizontal dynamics for 1.5 h, starting 7 h after the emergence began, by tracking individual elements in high-resolution images obtained by using several spectral windows; one set of such images is shown in Fig. 4.7. The region had an approximately elliptical shape, its long axis being 55,000 km. Most of its perimeter was outlined by pores; those of preceding polarity (black in Fig. 4.7*b*) occupied the NW boundary, those of following polarity the SW edge. The magnetic configuration inside the region was complex: small magnetic elements of both polarities were scattered across the whole region. The polarities were not distributed at random, however, because specks of the same polarity were lined up locally in threads up to 15,000 km long. (This intricacy of the magnetic field may be a feature peculiar to this region, which was unusual in having the spots of following polarity developing fastest, and eventually in decaying very rapidly. The emerging flux region studied by Brants and Steenbeek, 1985, and by Zwaan, 1985, appeared less complex.)

In contrast to the intricacy of the magnetic field, the large-scale pattern of horizontal motions in AR 5,617 was simple: faculae of opposite polarity streamed in opposite directions. All faculae of a given polarity tended to move toward the edge occupied by pores of the same polarity; even faculae at the far opposite end of the region shared this trend (Fig. 5.1). The mean (vector-averaged) separation rate between the two polarities and the mean common drift were determined for each bin on a superimposed square grid. Averaged over the active region, the separation rate of the polarities amounted to 0.84 ± 0.07 km/s; it was largest in the central part of the region. The average common drift was 0.19 ± 0.03 km/s; it was largest near the preceding edge.

The pores at the edges of the region AR 5,617 showed a similar bidirectional velocity pattern as the faculae of the same polarity, but their speeds were lower (separation: 0.73 ± 0.07 km/s; drift: 0.10 ± 0.03 km/s), and in a somewhat different direction, more along the string of pores, in the direction of the main sunspot of the corresponding polarity at the extreme ends of the region. One consequence is that in their migration toward the

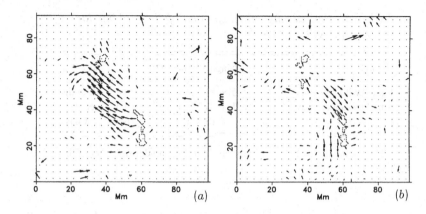

Fig. 5.1. Facular flow fields in and around an emerging flux region. Panel *a*: average velocities of faculae of leading polarity. Panel *b*: same for faculae of following polarity. The velocities are with respect to the ambient photosphere. The field of view and the orientation are nearly the same as in Fig. 4.7; the dotted contours show the position of the major sunspot pores at the middle of the observing period. The vector lengths are such that the distance between two adjacent grid points is equivalent to a velocity of 500 m/s (from Strous *et al.*, 1996).

edges of the active region, the faculae can catch up with the pores and eventually merge at the edge of the region. Furthermore, the pores are formed at the edges, and after moving along their edge, they form larger sunspots by coalescence.

Within AR 5,617, the large-scale plasma flow, as measured by the displacements of the granules, was very similar to the common drift of faculae of opposite polarities. This indicates a close relation between the plasma flow in the region and the dynamics of the emerging flux tubes. The active region appeared to be a large-scale divergence (i.e., an updraft) in the plasma flow, which extended to the edges of the region, where locally the granulation was seen to move past the pores at a speed of ∼0.1 km/s. The whole interior of the active region showed a marked clockwise vorticity. Just outside the region, counterclockwise satellite vortices were seen, which were associated with local downflows.

In general, emerging flux regions appear as sudden intrusions: there are no reports of plasma updrafts preceding flux emergence. At first, the immediate surroundings are not affected by the EFR. The phenomena suggest the following interaction between the emerging flux loop and the ambient plasma. When approaching the photosphere, a frayed, buoyant flux loop drags plasma along. That flow is enhanced by the convective instability so that flux loop and upward flow, coupled by drag, reach the photosphere at nearly the same time. The motion of the flux tubes differs markedly from the diverging plasma flow, however, because over part of the region flux tubes moved ahead of the local plasma flow, and in another part they moved against the flow. This dynamic pattern indicates that the flux tubes are not carried passively by convective flows; apparently the dynamics are predominantly determined by forces within the emerging flux loops.

Pores grow by increase of area, associated with decreasing brightness. This happens in the immediate vicinity of downdrafts, indicating convective collapse. There is some converging motion in the rapidly darkening matter, but the main impression is that of nearly *in situ* darkening, as demonstrated by Brants and Steenbeek (1985). The accumulation of faculae against pores at the edges of the region can explain the growth of pores. This crowding increases the cross section of the flux-tube bundles, so their central parts become better insulated against the lateral influx of radiation from the ambient convection zone. While the field contracts to an even larger strength, these parts cool down further, and more magnetic flux is concentrated into pores.

Larger sunspots grow by coalescence of small umbrae, pores, and faculae, converging at typical velocities near 0.25 km/s, sometimes as high as 1.0 km/s (see Vrabec, 1974). A substantial fraction of the moving magnetic patches of one polarity in the region end up in one or a few sunspots. Magnetographic movies, such as used by Vrabec (1974), indicate that the magnetic fragments ending in one sunspot travel along one or a few well-defined tracks, which may be long and markedly curved (see also figures in McIntosh, 1981). Pictures published by García de la Rosa (1987) show how some of the larger fragments in the coalescence can remain visible, or reappear, in later evolutionary stages, including the final dissolution. These observations indicate that a sunspot is not a monolithic structure. Yet some sunspots develop very dark umbral cores without a visible fine structure, which indicates that locally the constituent flux tubes fuse into a tight structure.

The growth of the leading sunspot and of the separation between the centroids of opposite polarity for the region as a whole appear to be related. The expansion of the growing active region is caused mainly by the faster motion of the leading sunspot relative to the average solar rotation characteristic of its latitude. This proper motion slows down with time, and it drops to zero when the growth of the spot stops. During the growth of the region, the main following sunspot moves backward in longitude at a much slower pace, or it stays put. These typical dynamics in growing active regions have been known for a long time (Maunder, 1919; Astronomer Royal, 1925; Section 45 in Waldmeier, 1955; summary in Section 1.7 in Kiepenheuer, 1953). The eventual separation between the centroids of the polarities in fully developed, large bipolar regions (with $\Phi > 5 \times 10^{21}$ Mx) range between 40 and 170 Mm; for such large regions, there is hardly any dependence of this separation on the total magnetic flux (see Fig. 2 in Wang and Sheeley, 1989).

From the coalescence of a substantial fraction of the flux in sunspots and the separation of the polarities, it is inferred that, somewhere below the photosphere, parts of the flux bundle are held together particularly tightly: the photospheric tops of the flux tubes drift together, like a bunch of balloons above a hand holding the strings. When the emergence and the major flux displacements ends, the flux distribution in the photosphere maps the two pillars of a supporting Ω-shaped loop.

The dynamics in emerging active regions suggest an evolution in the interplay between the magnetic field and the plasma flow. During the rise of the flux loop through the upper part of the convection zone, its flux tubes are probably somewhat weaker than the equipartition value B_{eq} [Eq. (4.11)]; hence the tubes are still pliable to the turbulence in the rising convective bubble. This may explain the initial mix of polarities within the growing active region. After the flux tubes have surfaced, they cool and collapse (Section 4.4), so their buoyancy, strength, and stiffness increase; thus they try to straighten themselves into a vertical position. The systematic separation of the opposite-polarity faculae and the ordered pattern in the H α arch filament system (Fig. 4.7) indicate the basic simplicity in the emerging flux system, after the effects of the convective turbulence in the emergence process are undone. For a description and a model of the detailed dynamic structure in emerging active regions, see Strous and Zwaan (1999).

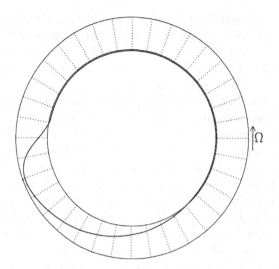

Fig. 5.2. Polar diagram for the configuration of an emerging flux tube according to a numerical simulation by Caligari *et al.* (1995), showing the tube shape projected on the equatorial plane. Note the marked asymmetry between the two legs of the emerging loop.

5.1.1.2 Storage and emergence of flux tubes

Most of the theoretical and numerical studies of the storage of flux tubes below the bottom of the convection zone and the rise of Ω loops have been based on the thin-flux-tube approximation (Section 4.1.2), as worked out by Spruit (1981c) and applied by Moreno-Insertis (1986). As to the storage, two conditions in the toroidal flux system have been considered: *temperature balance* and *mechanical balance*. In the calculations, usually a turn of the flux spiral (Fig. 6.10) is approximated by a flux ring in a plane that is parallel to the equatorial plane (Fig. 5.2).

In *thermal balance*, when $T_i = T_e$, flux tubes with magnetic field strength B in excess of the equipartition field strength B_{eq} are very buoyant. The buoyancy could be reduced by the Coriolis force associated with a plasma flow within the flux ring against the direction of solar rotation – such a counterflow could be provoked by a hypothetical slow meridional flow toward the equator (Van Ballegooijen, 1982a).

In *mechanical balance*, the sum of all forces acting on elements in the flux ring must vanish. Since no force (except those provoked by specific flows) can balance the component of the buoyancy force parallel to the rotation axis, the buoyancy must be zero, so $\rho_i = \rho_e$. Magnetostatic equilibrium [Eq. (4.17)] requires that $p_i < p_e$; hence the interior of the tube must be cooler than its surroundings. The force balance perpendicular to the rotation axis is achieved by a faster rotation of the plasma within the flux ring than the rotation of the nonmagnetic environment, such that the outwardly directed Coriolis force balances the inwardly pointing magnetic curvature force. If the flux ring is not yet in equilibrium, it seeks its proper position within the subadiabatic overshoot layer (see Caligari *et al.*, 1998). If it is buoyant, it rises and thereby cools with respect to its environment and so reduces its density contrast. An unbalanced magnetic tension force leads to a poleward slip of the ring, thus reducing its radial distance to the rotation axis;

because of conservation of angular momentum the rotational velocity of the internal plasma increases and hence the outwardly directed Coriolis force increases. Because of the large radiative relaxation time scale $\hat{t}_{\text{rad}}$ (Section 4.1.2, Fig. 4.3), the internal plasma is assumed to adjust itself adiabatically.

A flux ring in mechanical equilibrium may be disturbed by the *undulatory (or Parker) instability* (Parker, 1966): when a part of the ring is lifted by a local disturbance, plasma can flow from the crest to the troughs. This makes the crests lighter and the troughs heavier, working to amplify the perturbation.

Choudhuri and Gilman (1987) studied buoyant toroidal flux rings rising (without undulation) from the bottom of the convection zone to the top while remaining in thermal balance (this assumption is not realistic). Such rings are deflected to higher latitudes by the Coriolis force; flux rings with initial strengths $B < 10^5$ G turn out to surface outside the sunspot belts. Subsequent studies of emerging Ω loops have confirmed that initial field strengths above approximately five times the equipartition field strength $B_{\text{eq}} \simeq 10^4$ G are required for active regions to emerge in the sunspot belts close to the equator (Section 6.1.2); only then is the rise time of the loop tops shorter than the solar rotation period.

In mechanical balance, flux tubes with strengths up to $\sim 10^5$ G can be stored in the overshoot layer before they are affected by the undulatory Parker instability (see Caligari *et al.*, 1995, and references therein). This is an important confirmation that the magnetic flux tubes can stay there for a substantial fraction of the sunspot cycle period while being reinforced by differential rotation (Section 6.2.2).

Caligari *et al.* (1995) followed the rise of a flux loop bulging from a flux ring with $B \simeq 10^5$ G by numerical simulations, which were stopped when the summit was still 13,000 km below the surface because there the tube diameter and the pressure scale height became equal so that the thin-flux-tube approximation breaks down. These simulations reproduce several observational characteristics. The flux loops emerge in the sunspot belts, with the proper tilt γ of the dipole axis with respect to the direction of rotation (Section 6.2.2).

As a loop rises, it develops an asymmetry: the leg preceding in the sense of the solar rotation becomes less vertical than the following leg; see Fig. 5.2. This asymmetry is due to the conservation of angular momentum, which makes the summit of a rising loop lag behind the local rotation in the top of the zone. This asymmetry can explain why in the expansion of a growing active region the leading pole moves ahead faster than the following pole falls behind. It also clarifies why the centroid of the leading polarity tends to lie further removed from the central polarity-inversion line than the following polarity, as noticed by Van Driel-Gesztelyi and Petrovay (1990) and Cauzzi *et al.* (1996).

Caligari *et al.* (1998) discuss the influence of the initial conditions on the storage and emergence of magnetic field. They prefer mechanical balance over thermal balance because mechanical balance entails fewer free parameters and it can be made compatible with dynamo models.

Although remarkably successful in many respects, the present flux-emergence models leave some problems unsolved. The numerical simulations underlying Fig. 5.2 stop when the top of the loop is still 13,000 km below the photosphere, because the thin-flux-tube approximation does not apply to the rapidly expanding loop top and its emergence. From there on, the emergence phenomena have to be pictured qualitatively, as in the heuristic Ω-loop model in Section 4.4.

Once the top of a loop has emerged in the atmosphere, the magnetic tension is released, and the legs in the top of the convection zone are expected to become nearly vertical because of their buoyancy, while they separate [this is supported by helioseismic subsurface imaging by Kosovichev (private communication), which shows sound–speed perturbations below an active region compatible with an inclination of no more than $30°$ from the vertical in the first $\sim 5{,}000$ km below the surface]. From the configuration in Fig. 5.2 it is hard to understand why the separation of the polarities stops within a few days, once all flux has emerged. One would guess that the separation of the polarity centroids should continue for a while after flux emergence. In other words, the compactness of bipolar active regions presents a yet unresolved problem (and so does the compactness of the activity nests, discussed in Section 6.2). We find the slightly bent flux loops* as in Fig. 5.2, with the broad bases at the bottom of the convection zone and their gently upward sloping legs, hard to reconcile with the compactness of active regions.

In this book, we stick to the heuristic Ω-loop picture (Fig. 4.10), with its steep, nearly vertical legs, as an icon to evoke the deductions from observational data and numerical simulations. This model resembles the adiabatic flux-tube models constructed by Van Ballegooijen (1982a), in which the nearly vertical legs also move apart; but for a plausible range of parameters this drift speed is small, $v_d < 100$ m/s, comparable to the typical speed in the dispersal of faculae in an aging active region. Admittedly, a serious problem is that no mechanism is known that causes the initial uplift in the formation of such loops with steep legs. We ask the reader, first, to remember that the actual loops are slanted, the preceding legs being less steep than the following ones, and, second, to wait with us for the next-generation flux-emergence models.

As a step to such a next-generation model, the physics of twisted flux tubes rising in a stratified medium, without the application of the thin-flux-tube approximation is studied in papers reviewed by Moreno-Insertis (1997).

5.1.1.3 Around maximum development

After all the magnetic flux has emerged, the active region goes through a transition phase of *maximum development*. During this phase, restructuring toward a more polished bipolar appearance continues, and the first indications of decay become visible.

The coalescence of small spots into the large leading sunspot continues after flux emergence for ~ 1 day. When that process ends, usually there is one large leading sunspot, which is located very close to the leading (western) edge of the active region. There may be a conspicuous secondary spot in the following polarity, which then tends to lie close to the trailing edge of the region. Often, however, sunspot material of following polarity is found in a cluster of smaller sunspots and pores. In addition to major sunspots, usually some small sunspot and several pores of either polarity are present. Even at maximum development, no more than $\sim 50\%$ of the magnetic flux in an active region is incorporated in sunspots (Schrijver, 1987b; N. R. Sheeley, private communication).

---

* That shape of the loop, as shown in Fig. 5.2, with its small curvature, follows directly from the most unstable mode in the Parker instability of the flux ring, with the lowest azimuthal wave number $m = 1$.

Sunspot pores with diameters exceeding 5,000 km show a strong tendency to develop at least some penumbral structure (see Bray and Loughhead, 1964, their Sections 3.4 and 3.5). During flux emergence, the growing spots overshoot the mark for penumbra formation, and they form large pore masses without penumbral structure. Presumably this transient situation lasts no longer than approximately half a day.

There are only a few, brief descriptions of the formation of penumbrae. In their Section 3.5.2, Bray and Loughhead (1964) describe the formation of rudimentary and transitory penumbrae. From Bumba (1965) and Bumba and Suda (1984), Zwaan (1992) gathered that a penumbra develops in spurts, sector by sector. One such sector is completed within an hour or so, and a penumbra encompasses most of an umbral periphery within half a day. Often the first development of a penumbra is toward the nearby edge of the region. For some time, a gap toward the inside of the region remains open for more magnetic parcels to march in and to coalesce with the umbra. After the formation of a penumbra, often the sunspot assumes a more roundish overall shape.

The formation of the penumbra marks a sudden change in the magnetic sunspot structure near the photosphere, from a nearly vertical system, with little fanning, to a widely fanning sheaf. Clearly, this change affects the magnetic structure in the overlying atmosphere. In fact, the development of the penumbra appears to be preceded by some reconnection of the magnetic field. In Hα and in soft X-rays it is seen that initially closed flux loops start opening up within an hour or so after the first signs of flux emergence. The picture is that, after emergence, the expanding flux loop meets a preexisting coronal field. In most cases, the field directions do not match, so reconnections can occur. As a result, flux tubes at the leading and trailing edges of the region, which originally were connected with each other, find themselves linked with opposite-polarity flux in the surrounding photosphere.

During and shortly after their formation, many sunspots develop a *moat cell*, which is an annular zone without a stationary magnetic field that surrounds (part of) their periphery (Sheeley, 1969, 1981; Harvey and Harvey, 1973; Vrabec, 1974). Measured from the outer edge of the penumbra, these moats are 10–20 Mm wide; they persist with little change for at least several days, probably for nearly the entire lifetime of the sunspot. Moats are found around all leading spots, except the smallest. Still-growing sunspots may show partial moats (see Fig. 7 in Vrabec, 1974). Completely circular moats are only found around sunspots that are well isolated; moats show breaks where there is a substantial magnetic structure nearby. During its formation, the moat flow sweeps faculae aside into the "wreath" around the moat (see Section 2.4 in Sheeley, 1981).

Across moats, magnetic elements drift radially away from the spot with velocities from $\sim$0.2 to 2.0 km/s. These outwardly moving features differ in several aspects from the inflowing magnetic features during the coalescence of spots. All outwardly moving magnetic features are small; they are no more than $\sim$1 arcsec in diameter. They are not seen in movies in the continuum, but they are well visible in magnetograms and on Ca II H and K spectroheliograms. Contrary to the coalescing fragments, the outstreaming features are of both polarities. The pattern resembles pepper and salt, with occasional indications of opposite-polarity pairs. The estimated fluxes range from 6×10^{17} to 8×10^{19} Mx (Harvey and Harvey, 1973).

Shape, size, and horizontal flow pattern suggest that moat cells are similar to supergranulation cells. The notion is that the flow is convectively driven by the somewhat overheated layer just below the penumbra. This forcing may explain why moat cells have longer lifetimes and somewhat larger sizes and velocities than most of the normal supergranules.

The outstreaming magnetic features across moats share characteristics with the inter-network field (Section 4.6): the polarities are mixed on scales of a few arcseconds. The magnetic patches move with the photospheric plasma flow, without a tendency of oppo-site polarities to separate. There is a quantitative difference, however: the outstreaming features in the moat tend to carry much more flux per patch than do the internetwork-field specks. Harvey and Harvey (1973) verified that under the cover of mixed polarities, there is a slow, steady loss of magnetic flux from the sunspot agreeing with the polarity of the decaying spot. In Section 6.4.2 we discuss the explanation for the outstreaming magnetic features.

Even during maximum development, some 50% of the magnetic flux in the leading polarity is in bright faculae (see Fig. 1.2); this fraction is even larger in the following polarity. Part of the faculae are arranged in *plages*, whereas another part is collected in the magnetic *enhanced network*.

In Chapter 1, plages are defined as tightly knit, coherent distributions of faculae, which are best seen in the cores of strong spectral lines, such as the Ca II resonance lines (Fig. 1.1b) and in H α (Fig. 1.3a). The almost empty cells in plages show a characteristic range of diameters, from less than 10^3 km (corresponding to that of small granules) up to some 10^4 km (less than that of a supergranule); see the positive (white) polarity plage in Fig. 5.3. This suggests that the convective pattern within plages differs markedly from that in the quiet photosphere; even the granulation in plages appears to be abnormal. This and other peculiarities of the so-called plage state are discussed in Section 5.4. The term plage can be applied to coherent distributions of faculae with diameters down to roughly 2×10^4 km; smaller concentrations are better called plagettes (Fig. 1.3a). Plagettes share properties with plages; for one thing, their internal granulation is abnormal.

Some authors use the term plage (singular) for all bright material within an active region, including its enhanced network. In this book, we use the term plage either as defined in the previous paragraph or for all magnetic structures of an active region within a specific isogauss contour but outside sunspots (see Section 5.4).

During the phase of maximum development, the faculae in plages and in enhanced network exhibit the characteristic filigree pattern (Section 4.2.3). Note that the filigree becomes visible shortly after the local emergence of flux is completed; it is only absent in emerging flux (Zwaan, 1985). Apparently the filigree is a feature of strong, nearly vertical magnetic fields, after the convective collapse.

5.1.1.4 Decay phase

Even during flux emergence, some decay features are observed, but the decay phenomena become conspicuous after all available flux has surfaced. The total area of the active region, as measured by its outer contour, slowly increases by a gradual expansion. Particularly at the periphery of the region, large roundish holes of supergranular size begin to appear. The spectrum of magnetic concentrations becomes simpler: the pores and the smaller sunspots disappear. Often one leading sunspot remains for some time, slowly shrinking during a period of one or a few weeks; very rarely does the leading sunspot survive for a few months.

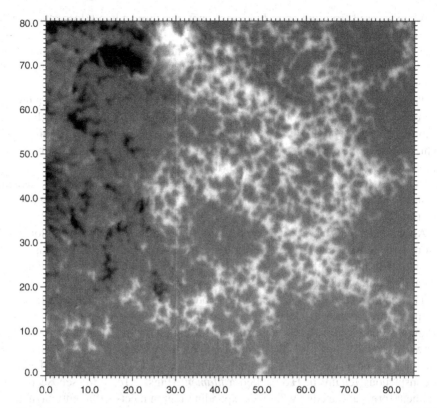

Fig. 5.3. High-resolution magnetogram of plage in AR 5,168, obtained with the Swedish Vacuum Solar Telescope on 29 September 1988 in the Fe I 6,302-Å line. The magnetogram is corrected for the telescope's modulation-transfer function to an ultimate resolution of near 0.3″. Tick marks are in arcseconds, equivalent to 725 km at disk center (figure from Tarbell *et al.*, 1990).

Most sunspots decay by fragmentation: a spot breaks up into several parts within one or a few days after maximum development. The parts disappear quickly by continued fragmentation. This fate befalls all smaller spots and most, if not all, following spots. As a result, 95% of the recurrent sunspot groups return during the next solar rotation with only their leader spot, without a trace of a follower left (Astronomer Royal, 1925). In small active regions, the leaders, too, may disappear by fragmentation (see Fig. 20 in McIntosh, 1981, for a slowly progressing case).

The sunspot fragmentation starts with the sudden appearance of very bright umbral dots (Zwaan, 1968, 1987) and thereupon the umbra crumbles away. Spots vanishing by fragmentation show one, if not all of the following properties: irregular shape, conspicuous bright structure well before the fragmentation starts, and absence of a moat cell.

Large leading sunspots go through the phase of slow gradual decay: the spot just shrinks, the initial ratio of umbral area to total area is maintained, and the umbral and penumbral brightnesses do not change substantially. In other words, the spot remains self-similar during gradual decay (Section 4.2.1). One consequence is that magnetic flux tubes initially embodied within the umbra must be transferred to the penumbra, before they are severed from the spot. For every spot, the flux and area loss rates $d\Phi/dt$ and dA/dt remain nearly constant at its particular rate throughout the gradual decay phase. The magnetic field strength in the umbra does not vary significantly either (Cowling, 1946). The gradual decay continues until the spot subsides to a fraction of its size, whereupon that remnant suddenly fragments.

There is a large range in the flux loss rates. Even though spots surviving a few months are exceptional, they present the intriguing problem of accounting for such an extreme stability. Sunspots gradually decaying for more than 1 week (a) are leading spots, (b) which possess one or two very dark umbral cores, (c) have an umbra/penumbra structure that is nearly circularly symmetric, and (d) are encircled by moats.

In many cases, the decay of sunspots poses the intriguing problem that magnetic flux vanishes without any indication of what happened. This vanishing act is discussed in Section 6.4.2.

After the sunspots have disappeared, the slow expansion of the active region continues. More large, empty cells appear in the facular material, or, in other words, plages are transformed into an enhanced magnetic network. The enhanced network keeps slowly expanding and weakening.

This picture of a gradually expanding region, with an ever-decreasing average magnetic filling factor, applies to isolated active regions. In those cases, at least a part of the originally enhanced network may become so diluted that it is lost in the common quiet network. Usually, however, the magnetic field at the edges of a bipolar active region interacts with the surrounding magnetic structure; see Chapters 6 and 8. There we find a conspicuous feature in $H\alpha$, called a *filament* (Figs. 1.1c, 1.3a; see Section 8.2). A filament forms at the edge where an expanding region meets sufficient magnetic flux of opposite polarity in its vicinity. Filaments may already form during the stage of maximum development of the region.

When there is no longer a noticeable bipolar trace in the magnetogram, the active region has found its end – this definition is as vague as is the reality. Some features of the active region survive the bipolar entity, however. Remnants of a monopolar stretch of network issuing from one of the polarities of the former active region may merge with network of the same dominant polarity and may survive for a long time (Section 6.3). A filament formed at the edge of an active region may outlive the bipolar region for many months.

5.1.2   *Evolution of ephemeral active regions*

For ephemeral active regions, the evolution sequence is very straightforward (Harvey and Martin, 1973; Chapter 6 in Harvey, 1993): the magnetic flux is seen to emerge in two patches of opposite polarity, and, if observed at sufficient resolution, the poles show fine structure. At a gradually reducing speed, the poles keep separating. Because the amount of magnetic flux is small, no sunspots are formed; a few of the largest ephemeral regions may show a short-lived pore. There is no gradual decay: during its evolution, the poles resemble patches of network from which they can only be distinguished by their history as poles of an emerging flux region. An ephemeral region comes to an end when one of its poles is lost in (or cancels against) the multitude of network patches. The majority of the (small) ephemeral regions live for a couple of hours; a few last for a few days (see Chapters 5 and 6 in Harvey, 1993).

5.2   The sequence of magnetoconvective configurations

In this brief summary section, we emphasize the temporal sequence in the appearance of magnetic configurations.

The rapid formation of an active region, reaching its maximum development within a few days, and its subsequent slow decay suggest that an active region is a prefabricated product from the deep toroidal system whence it emerged. Thereupon decay proceeds at once. Sunspots may decay very slowly, but in all cases the decay is irreversible. The notion is that the magnetic flux in an active region remains "anchored" in the deep strong-field system for no more than their lifetimes; in Sections 6.2.2 and 6.4 we discuss that deep strong-field system and its erosion.

Apparently the network configurations, consisting of small magnetic grains, with relatively thin flux tubes underneath, are a steady configuration (but the individual grains are not). A probable reason for this steadiness is that a network configuration presents no impediment to the heat flow (see Section 5.4). The nearly vertical flux tubes are rather stiff for at least some megameters, and probably for some tens of megameters, because of the cool downdraft along the flux-tube bundle and the high opacities in the top layer of the convection zone (see Fig. 4.3). They tend to stick together because of the horizontal inflow associated with the convective downflow.

The network configuration appears to be adaptable to a wide range of average flux densities. Probably the plage configuration is similar to the network configuration, the difference being that in the high magnetic flux density the large supergranules are replaced by smaller cells.

In conclusion, all magnetic configurations share radiative cooling in the atmosphere as an important mechanism. First, that cooling determines the process of the formation of the strong magnetic field during the convective collapse (Section 4.4). Second, radiative cooling is probably fundamental in the maintenance of network configurations.

5.3   Flux positioning and dynamics on small scales

It appears that all of the solar surface is covered with magnetic field. Most of the solar surface contains weak internetwork field (Section 4.6), and only a relatively small fraction of the surface is filled with intrinsically strong field with field strengths of the order of a kilogauss. The dynamic positioning of concentrations of flux on granular scales is likely to be directly related to outer-atmospheric heating (Chapter 10). The positioning of photospheric flux determines the geometry of the field in the outer atmosphere. In view of this, we address the small-scale positioning of flux here in detail, to be used later in Sections 6.3.2 and 9.5.1, and in Chapter 10.

Outside spots and plages, concentrations of strong magnetic field are found embedded within the downdraft regions of the supergranulation (Fig. 2.13); the corresponding pattern is usually called the *magnetic network*, even where it is incomplete and hard to recognize as a network. The network pattern is apparently the eventual pattern that forms by the interplay of magnetic field and heat transport in the photosphere and the top of the convection zone (Section 5.2). In plages, the field outlines cellular openings on a range of scales below the scale of the supergranulation (Fig. 5.3, Section 5.1). At the smallest scale, the magnetic grains are everywhere arranged in the filigree outlining the granules (Section 4.2).

The intermittent nature of the intrinsically strong field is shown in high-quality magnetograms, such as that shown in Fig. 5.3, or indirectly in other diagnostics, such as the G-band picture in Fig. 2.12. Magnetograms always have a somewhat fuzzy appearance. The relatively long integration times and the delay introduced to obtain image pairs of opposite polarities needed to construct a magnetogram allows atmospheric distortions to blur the images: images that have been sharpened by correcting for the telescope's modulation transfer function, and by speckle or phase-diversity methods to correct for atmospheric blurring certainly appear much sharper.

Sharper images are obtained with some indirect diagnostics: the CH molecular band head (the G band) offers a proxy of the magnetic concentrations (or at least a fraction of the strong field, since not all flux concentrations appear bright in this diagnostic), and it yields very sharp images of flux tubes, as shown in Fig. 2.12 (see Title and Berger, 1996, on the sharpness of such images). Berger and Title (1996) discuss observations of the quiet Sun made in the G band with an angular resolution of $\sim$0.3 arcsec or 220 km. In any one of the frames, easily recognizable bright structures interact with the local granular flow field. These grains move along the intergranular lanes while responding to granular flows by sideward excursions. The shapes of the individual bright features change rapidly, with frequent fragmentations and collisions as the magnetic flux reorganizes itself over the ensemble of concentrations.

On scales of up to a few supergranules, patterns in high-resolution magnetograms appear to be statistically self-similar: structures show a similar complexity, regardless of the scale at which they are observed. This suggests that the geometric pattern of the magnetic field is fractal; that is, the pattern has a "shape made of parts similar to the whole in some way" (as defined by Mandelbrot, 1986).

Fractals are commonly characterized by dimensions. Different types of fractal dimensions are in use, and a fractal pattern often requires multiple dimensions to characterize a specific pattern. The scaling dimension is frequently used: for a fractal S, the scaling dimension is defined by $\mathcal{D} \equiv \log(N)/\log(m)$, where S is geometrically similar to N replicas of itself when scaled by a factor m. This can be loosely translated into an algorithm for image processing: prepare a set of duplicates of an image, scale these by different factors m_i into a sequence of coarser and coarser images, and count the number N_i of coarse pixels that contain one or more of the original pixels that were part of the pattern under investigation, and that satisfy some criterion, for instance that a magnetogram signal exceeds some threshold. If a pattern is a true fractal, observed with sufficient angular resolution over a sufficiently large area, the relationship between $\log(N)$ and $\log(m)$ is linear, and its slope is the scaling dimension. Tarbell *et al.* (1990) use such a renormalization method to estimate the scaling dimension $\mathcal{D}_{field}$ of the magnetic pattern (shown in Fig. 5.3) in a magnetic plage and the network immediately surrounding it. They find $\mathcal{D}_{field}$ to lie between 1.45 and 1.60 for length scales from about 0.2″ up to at least 5″. The image did not have the same fractal properties over the entire field of view or scaling range, however, and therefore the $\log(N) - \log(m)$ relationship was slightly curved. In a subsequent study of the same region, Balke *et al.* (1993) used another fractal dimension that is based on the ensemble properties of clusters of field rather than on the entire image. They derived $\mathcal{D}'_{field}$ from the relationship between surface area $A(L)$ of coherent, isolated clusters of magnetic flux and the linear size L of the smallest square box that contains the cluster (these boxes can be oriented arbitrarily provided the direction is the same for all of them). They find a relationship that can be approximated by a power law $A \propto L^{\mathcal{D}'_{\text{field}}}$, with $\mathcal{D}'_{\text{field}} = 1.53 \pm 0.05$ for clusters with $L \lesssim 3''$ (with fluxes up to a few times 10^{18} Mx), while for larger clusters $A \propto L^2$. The few large clusters for which $L \geq 3''$ cover most of the core of the magnetic plage.

Schrijver *et al.* (1992b) propose that percolation theory (e.g., Stauffer, 1985) explains the scaling relationship found by Balke *et al.* (1993). Percolation theory studies patterns in incomplete, not necessarily regular, lattices. Imagine an infinite regular lattice (either a triangular, a square, a honeycomb, or some other regular lattice). Each site of the lattice may be occupied at random determined by a probability p. Clusters are defined as sets of occupied nearest-neighbor sites. A remarkable property of lattices is that there is a critical value p_c, the percolation threshold, which is the lowest value of p for which an infinitely large (percolating) cluster exists. The value of p_c is a property of the type of lattice.

The fractal dimension determined from the power-law relationship between area A and box size L for clusters in two-dimensional lattices is $\mathcal{D}_{\text{th}} = 1.56$ for $p < p_c$, *regardless of the type of lattice* (see Stauffer, 1985), provided that L is smaller than the average rms distance ξ between occupied sites in finite clusters. For $L > \xi$ the scaling $A \propto L^{\mathcal{D}_{\text{th}}}$ crosses over into $A \propto L^2$ for a lattice in two dimensions. The histogram of the number of clusters $n(A)$ of a given area A (the number of occupied sites in the cluster) varies as $n(A) \propto \exp(-A)$ for sufficiently large A and $p < p_c$ (see Stauffer, 1985).

The remarkable similarity between the geometry of the magnetogram and the geometry of randomly filled lattices, found by Balke *et al.* (1993), implies that magnetic flux is randomly positioned within the magnetic plage and the surrounding enhanced network, at least for length scales below 3 arcsec (Schrijver *et al.*, 1992b). This analogy suggests that more complicated models to explain the self-similarity of magnetograms, such as the repeated folding and stretching of strands of magnetic flux as proposed by Ruzmaikin *et al.* (1991), are not required.

This random positioning on small scales is the result of the interaction of magnetic flux with photospheric flows. The velocities with which flux concentrations travel through the intergranular lanes from and to (and sometimes even through) vertices common to three or more adjacent cells typically are some 0.1–0.5 km/s on time scales of more than a few minutes. On very short time scales velocities range from $\sim$1–5 km/s (Berger, 1996), which is substantially faster than the velocities associated with evolutionary changes in the granulation pattern (at $\sim$0.5 km/s), or the radial outflow in granules ($\sim$1 km/s). The rms displacement of these concentrations implies an equivalent diffusion coefficient [Eq. (6.6)] of approximately 50–70 km²/s, which is consistent with the rough estimate $\langle r^2 \rangle / 4t$ based on the characteristic scales of the granulation.

On scales larger than a few arcseconds, the frequency distribution for fluxes $|\Phi|$ in concentrations in the quiet photosphere shows a nearly exponential decrease with Φ (Fig. 5.4):

$$n(\Phi) \propto e^{-\beta |\Phi|}, \tag{5.1}$$

with $\langle \Phi \rangle \equiv 1/\beta \approx 3 \times 10^{18}$ Mx. This distribution was derived for fluxes in the range from $\approx 2 \times 10^{18}$ Mx up to $\approx 2 \times 10^{19}$ Mx.

The distribution in Eq. (5.1) can be understood as the consequence of the interactions to which flux concentrations are subjected (Schrijver *et al.*, 1996a, 1997c, 1997d). Where two concentrations of the same polarity meet, they merge to form a larger grain;

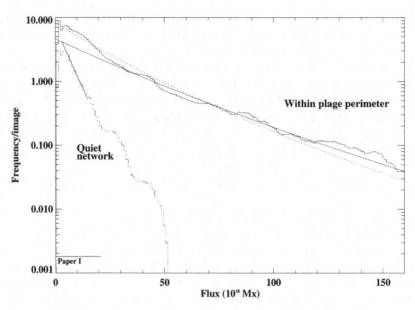

Fig. 5.4. Histograms of the number of concentrations of total magnetic flux $|\Phi|$ observed in a *SOHO/MDI* magnetogram sequence for a plage and a quiet photospheric region. The distributions are smoothed with a box width of 5×10^{18} Mx. The lower inclined solid line segment is an exponential fit to the distribution in the quiet photosphere. The upper solid curve is a model prediction based on the time-averaged magnetogram with a spatial smoothing of 1,700 km; the dotted curve is for a spatial smoothing of 8,300 km (figure from Schrijver *et al.*, 1997c).

elsewhere single grains are seen to split (see early work by Smithson, 1973a). This behavior resembles that of mercury drops in joints between floor tiles: a little push makes the drops move, whereas locally some drops coalesce and elsewhere others split. This two-dimensional picture obviously is not the full story, because these grains are but the photospheric tops of flux tubes that extend vertically over probably many pressure scale heights. A more complete picture is that of vertical sheets and sheaves in space, something like a fly curtain made up of strings. The nearly vertically arranged field responds to the dynamic pressure of surrounding convective cells: locally flux is squeezed together, and elsewhere a sheet of flux is torn apart.

As a consequence of this dynamic behavior, the concept of a lifetime of flux concentrations is not particularly useful for individual small-scale flux concentrations. Even though magnetic grains are not long-lasting entities, the concept of flux tubes remains practical for the description of plasma columns arranged along the magnetic field lines. In the collapsed, strong-field case, the flux tubes stand for the rather stiff vertical threads in the magnetic fabric. Therefore we continue to use the term flux tube for the fundamental elements in the description of the magnetic field in convection zones and photospheres.

When magnetic concentrations of opposite polarity are forced together, flux generally cancels, i.e., there is a steady loss of flux in both polarities, usually until the smallest flux grain has completely disappeared (Livi *et al.*, 1985; Martin, 1988), although sometimes opposite-polarity concentrations remain side by side until separated again. The flux-loss rate per canceling pair is estimated to be between 10^{17} and 4×10^{18} Mx/h. Concentrations of opposite polarity seem to seek each other, but that is not surprising because all grains are forced to travel exclusively along the channels between the convective updrafts. We return to the large-scale aspects of flux removal from the photosphere in subsequent sections of this chapter and Chapter 6.

Schrijver *et al.* (1996a, 1997c, 1997d) model the flux distribution over magnetic concentrations by balancing loss and gain terms for each interval in the flux histogram. The processes include the merging of concentrations upon collision, which removes concentrations from intervals. They also accommodate the effect that the splitting of concentrations creates as well as destroys concentrations of a given flux, and that cancellation of unequal flux concentrations leaves a smaller residual concentration. Three parameters, each a function of flux, control the solution: the collision frequencies for opposite and like polarities, and the fragmentation rate.

The collision frequencies are estimated by using the geometry of the supergranulation: flux displacements are almost exclusively constrained to the lanes where opposing outflows from neighboring upwellings meet and are deflected both downward and sideways. Consequently, magnetic concentrations follow paths in a pattern that is much like the Voronoi tesselation discussed in Section 2.5.4. In that geometry, the total frequency ν_c of collisions per unit area is given by

$$\nu_c = \frac{\bar{v} n_t}{\lambda_m} \approx \frac{\bar{v}}{2\sqrt{\rho_u}} n_t^2 \equiv \ell n_t^2, \tag{5.2}$$

where λ_m is the mean-free path length, n_t is the total number density of flux concentrations placed anywhere on the cell perimeters, $\bar{v}$ is their average displacement velocity, and ρ_u is the number density of upflow centers. The proportionality of ν_c and n_t^2 is expected for collision frequencies in general, whereas the collision parameter ℓ reflects the details of the geometry. For $\bar{v} \approx 0.2$ km/s and $\rho_u = (1.2-5) \times 10^{-9}$ km^{-2} (reflecting the range in reported characteristic sizes for supergranules), $\ell \approx 1,400-2,800$ km^2/s. In this geometry, the collision parameter is independent of magnetic flux, because flux concentrations that move toward each other on the intersupergranular lanes necessarily collide, so that $\nu_c(\Phi_1, \Phi_2) = \ell N(\Phi_1) N(\Phi_2)$.

Fragmentation counteracts the growth of concentrations that occurs in equal-polarity collisions. The fragmentation rate is quantified by a probability $k(\Phi_1, \Phi_2)$ for a concentration to split into two parts Φ_1 and Φ_2 per unit time. Note that the total probability for a concentration of flux Φ to fragment per unit time into any combination of fluxes is given by

$$K(\Phi) = \int_0^{\Phi} k(x, \Phi - x)\, dx. \tag{5.3}$$

Schrijver *et al.* (1997d) argued that the observed, nearly exponential distribution of fluxes contained in flux concentrations in the quiet photosphere [Eq. (5.1)] requires $K(\Phi)$ to be roughly proportional to Φ: the more flux a concentration contains, the more likely it is to fragment.

Within a hypothetical monopolar environment there is no cancellation of flux, so that stationary situations do not require a source of new flux. If $K(\Phi) = \kappa|\Phi|$, the local mean flux density $|\varphi|$, the slope β of the flux distribution in Eq. (5.1), ℓ, and κ are related through

$$\beta = \left(\frac{2\kappa}{\ell|\varphi|} \right)^{1/2}. \tag{5.4}$$

Whenever two polarities are present, the solution depends on the details of the source function for new bipoles that replace the canceled flux. Numerical simulations by Schrijver *et al.* (1997d) have shown that if the source function is not too narrow in its flux range, an exponential distribution is a good approximation over a substantial range of fluxes Φ when ℓ and κ are constants. In the case of a zero net flux density over some substantial area, that is, in perfect flux balance, the approximate e-folding value for the flux histogram equals

$$\beta' = \left(\frac{\pi^2}{3} \frac{\kappa}{\ell|\varphi|_{\pm}} \right)^{1/2}. \tag{5.5}$$

Here $|\varphi|_{\pm}$ is the average flux density observed for either polarity. This value of β' differs only slightly from the value β for the monopolar case.

The agreement between the model and observation for both plage and quiet-photospheric regions once the corresponding flux density $|\varphi|$ is specified, illustrated in Fig. 5.4, suggests that the ratio κ/ℓ of the fragmentation rate to the collision parameter is the same within the quiet and plage environments. This is rather surprising, given the apparent differences in granulation and larger-scale flows in these environments (Section 5.4). Two possible causes come to mind. First, the flows that actually move the magnetic flux about may occur at a substantial depth, where the flows are not affected significantly by the presence of magnetic flux, regardless of the changes in flows at the surface. The alternative is that the convective flows that drive the concentrations together are the same that separate the parts after fragmentation. Hence, even if the flows are affected, this may leave the ratio κ/ℓ unaltered.

Given the collision and fragmentation rates, the time scale on which flux in the mixed-polarity, quiet network is canceled by collisions between concentrations of opposite polarity can be computed. Given a distribution function as in Eq. (5.1), and provided that the collision frequencies between equal and opposite polarities are the same (i.e., in the case of flux balance), the time scale for flux cancellation is

$$\hat{t}_C = \frac{2}{\ell\beta|\varphi|_{\pm}}. \tag{5.6}$$

In the very quiet photosphere, where $|\varphi|_{\pm} \approx 2\,\mathrm{G}$, the values of ℓ and β given above lead to a time scale $\hat{t}_C$ of 1.5–3 days. Flux-emergence studies (Schrijver *et al.*, 1997d) suggest that as much flux emerges in ephemeral regions within $\sim$1.5 days as is present in the quiet network, in good agreement with the above estimate for flux cancellation. This replacement time scale is surprisingly short, which implies that most of the flux in mixed-polarity quiet-Sun regions originates quite *locally* and *not* in far-away active regions.

The above statistical model determines the relative balance of positive and negative polarities. If there is a clear excess in one polarity, then the concentrations of that polarity collide frequently to form relatively large concentrations, whereas the minority polarity is broken down to smaller concentrations and quickly cancels in collisions. That, together with the reduced survival time of the minority flux in enhanced network, causes the contrast in appearance between a very quiet photosphere and the enhanced unipolar network. This balance between polarities makes the appearance of a region depend sensitively to the relative strength of the flux imbalance, as witnessed by the clear visibility of the poleward streaks formed by decaying active regions, discussed in Section 6.3.1: as the relative flux excess in one polarity increases, the minority species is cancelled more and more rapidly, causing a rapid change from a mixed-polarity appearance to one dominated by the majority species.

5.4   The plage state

Much of the magnetic flux that resides in the solar photosphere for a significant time emerges in active regions (i.e., in the set of bipolar regions that are larger than ephemeral regions). Eventually much of this flux disperses into the quiet photosphere by a diffusive random-walk process discussed in Section 6.3.2. Before that dispersal occurs, however, magnetic plages show a remarkable coherence that is incompatible with a diffusive dispersal. This phase lasts from days to some weeks, increasing with the flux content of the region. This coherence is associated with a rather well-defined outer perimeter of the active region during this phase of its evolution (Fig. 5.5): at a resolution of a few arcseconds, a 50 G contour outlines the magnetic plages and plagettes as relatively coherent areas with only a few, relatively small openings in their interior. Note that some small areas in the surrounding quiet photosphere also exceed the perimeter threshold.

The plage outline is not particularly sensitive to the precise value of the contour level, as illustrated in the lower panel of Fig. 5.5. The sharp transition in the mean flux density across the plage boundary is inconsistent with the diffusive dispersal of flux tubes: the magnetic field in a plage appears constrained.

Within a plage perimeter, the average flux density determined for a large variety of active regions, excluding their spots, shows a limited scatter (Fig. 5.6). This average plage flux density (depending on the magnetogram calibration) lies between $100\,G$ and $150\,G$ at least for plage areas ranging from 4×10^{18} up to 5×10^{20} cm^2 (Schrijver, 1987b; Schrijver and Harvey, 1994); see Fig. 5.6. This *plage state* holds for newly emerged regions as well as for mature bipolar regions, including complexes.

The plage state leads to some scaling properties of active regions. Some of these, as determined by Schrijver (1987b) based on a small sample of regions, are listed in Table 5.2 – in which line 6 reflects the plage state. These relationships show, for instance, that the distance between the geometric centers and between the centers of gravity for the two polarities in the plage are nearly the same (line 3 in Table 5.2). The data also suggest that the projected area of the coronal condensation of relatively isolated active regions is roughly proportional to the area of the photospheric active region, but expanded by approximately a factor of 5 for a typical active region (line 5). This suggests, for one thing, that coronal magnetic loops expand with increasing height over the photosphere; this is further discussed in Sections 8.6 and 9.6. Note that the coronal Mg X emission is nearly proportional to the total absolute magnetic flux of the regions (line 7), which is further discussed in Section 8.4 and in Chapter 10. Wang and Sheeley (1989; their Fig. 2) show that the separation of centroids scales with total flux – compatible with lines 2 and 6 in the table – but the scatter about the mean relationship is very large.

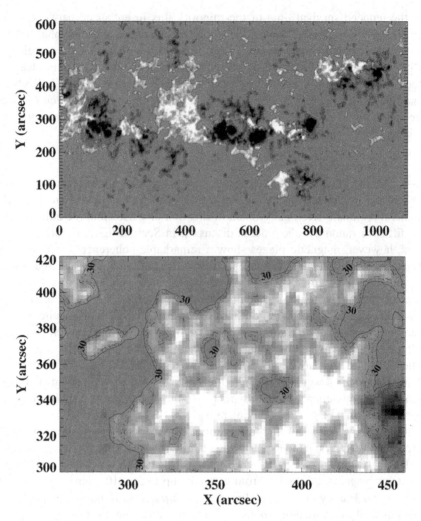

Fig. 5.5. Magnetogram obtained at the NSO/Kitt Peak Observatory on 11 September 1989, shown with 2-arcsec (1,450-km) pixels. The gray scale saturates at ±400 G. The top panel shows a ±50 G contour, while the blowup in the lower panel shows contours at ±30, ±50, and ±70 G.

Within the magnetic plage and in patches of strong network within the quiet Sun, the photospheric granulation shows a substantial change in texture and dynamics. This change occurs at a flux density of 50–75 G at a resolution of a few arcseconds, that is to say, at approximately the same value that defines the plage perimeter (see the example in Fig. 5.7; the review by Spruit *et al.*, 1990, gives more details). Instead of well-developed granular cells, the scales are smaller and the intensity pattern is more chaotic. The rms intensity fluctuations are reduced by a factor of ∼2 in the magnetic plage as compared to the quiet photosphere. Spruit *et al.* (1990) argue that this is more than expected given the presence of relatively bright magnetic flux concentrations in the dark intergranular

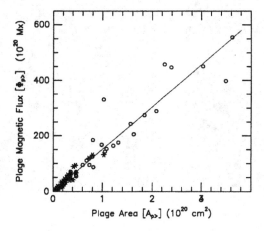

Fig. 5.6. Flux contained in magnetic complexes (excluding spots) in areas with an absolute magnetic flux density exceeding 25 G plotted against the corresponding surface area (from Schrijver and Harvey, 1994). The asterisks show 89 recently emerged active regions; the circles show 28 existing sunspot regions (five more regions substantially exceed the upper bound of the plot window). The best-fit relationship is given by $\Phi_{p>} = (-1.8 \pm 0.8) + (153 \pm 4)A_{p>}$, in which the active-region flux $\Phi_{p>}$ contained in pixels with $\varphi > 25$ G is expressed in Mx and the surface area $A_{p>}$ in cm^2.

Table 5.2. *Scaling relationships for active regions*[a]

Line	Relationship	Correlation
(1)	$f_{\Phi,s} = 5.1 \, f_{A,s}$	0.98
(2)	$A_\Phi = 1.1 \, d_\Phi^{2.0}$	0.93
(3)	$d_\Phi = 1.0 \, d_p + 2{,}600$	0.98
(4)	$A_{\text{Ca II}} = 570 \, A_\Phi^{0.70}$	0.99
(5)	$A_{\text{Mg X}} = 31 \, A_\Phi^{0.92}$	0.95
(6)	$\Phi = 2 \times 10^{11} \, A_\Phi^{1.1} + \Phi_S$	0.99
(7)	$I_{\text{Mg X}} = 0.34 \, (\Phi - \Phi_S)^{1.1}$	0.96

[a] Relationships taken from Schrijver (1987b), based on only eight regions observed on 14 different days. Symbols: $f_{\Phi,s}$ is the ratio of the magnetic flux contained in spots to the total magnetic flux in the region, $f_{A,s}$ is the ratio of the area of the spots to the total active-region area, A_Φ (km^2) is the active-region area as determined from the magnetogram, d_Φ (km) is the distance between centers of gravity of opposite-polarity flux in magnetograms, d_p (km) is the distance between geometrical centers of opposite polarity flux in magnetograms, $A_{\text{Ca II}}$ (km^2) is the area determined from Ca II K spectroheliograms, $A_{\text{Mg X}}$ (km^2) is the area determined from coronal Mg X rasters obtained with Skylab, Φ (Mx) is the total magnetic flux contained in the active region, Φ_S (Mx) is the total magnetic flux contained in the sunspots, and $I_{\text{Mg X}}$ (erg/s) is the total Mg X (625 Å) intensity of the coronal condensation.

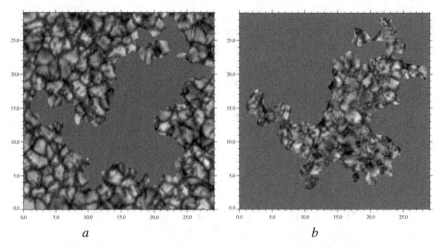

a b

Fig. 5.7. Comparison of granulation outside and inside a magnetic region. The two panels are complementary segmentations of a G-band (4,686 Å-wide passband) image based on a magnetic mask that reflects the cumulative coverage of the area by G-band bright points (compare Fig. 2.12) throughout a 70-min time series of observations of a patch of magnetic flux in the otherwise quiet photosphere. Panel a shows the region outside the magnetic region, while panel b shows its complement inside the region. Tick marks are spaced by an arcsecond, or $\sim$725 km (figure courtesy of T. E. Berger; modified after Berger *et al.*, 1998).

areas. White-light movies reveal that the evolution of the granulation in plages and in their surroundings differs markedly: instead of the gradual evolution of normal granulation, with its frequent mergings and fragmentations, we see a more slowly evolving pattern with interactions that are near the limit of resolution and hard to describe.

The network pattern of the supergranulation is absent in coherent magnetic plages. Openings of supergranular size, as seen in the quiet photospheric flows or the chromospheric images of enhanced network, are not seen within plages proper (Zwaan, 1978; see Figs. 5.3 and 5.5). Instead, there are openings on a range of scales between that of the granulation and the supergranulation. We know of no studies at present that assess the evolution of the flows in coherent, large plages directly, either using Doppler signals away from the center of the disk or correlation tracking near the disk's center.

Some measure for the large-scale convective flows in active and quiet regions can be compared indirectly by measuring the displacement velocities of concentrations of magnetic flux. On one hand, the observed velocities (Schrijver and Martin, 1990) do not differ significantly between the active region and the surrounding enhanced network for values determined from pairs of magnetograms taken up to $\sim$2 h apart. On the other hand, the rms displacements of flux concentrations that have been tracked over periods longer than a few hours and up to approximately a day appear to be lower by a factor of $\sim$2 inside the active region than in the surrounding enhanced network (compare Table 6.2; the relationship of the rms displacement to the diffusion coefficient is discussed in Section 6.3.2). Apparently, there is a fundamental difference in the geometry of the displacements between plage and quiet Sun, but the details of this difference remain unknown. We conjecture that the short-term displacements associated with (meso-)

granular flows are similar in the plage and quiet-network areas – despite the abnormal surface appearance of the granulation in plages – but that the drift and dispersal associated with supergranulation are different, leading to a reduced long-term root-mean-square dispersal (see Hagenaar *et al.*, 1999, for a discussion of flux displacement in a medium with multiple flow scales). Unfortunately, flux cannot be reliably tracked within an active region for more than $\sim$1 day, because of the frequent fragmentation and (cancellation) collisions between concentrations.

Another aspect of the plage phenomenon is that the power in surface *f*- and acoustic *p*-mode waves (discussed in Section 2.6) is reduced by approximately a factor of 2 inside plage regions (see the review by Title *et al.*, 1992). Active regions have been shown to be sites where the power in inwardly propagating *p* modes exceeds that in outward propagation waves, indicating strong wave absorption within the active regions (e.g., Braun *et al.*, 1990, Bogdan *et al.*, 1993). This absorption occurs throughout the active region, apparently even in embedded field-free regions, and is most pronounced, and perhaps saturated, within sunspots. The absorption by spots is at least 20 times that of quiet areas, whereas the active region as a whole absorbs at least twice as efficiently as the quiet regions. The absorption mechanism is not known; proposed mechanisms include resonant absorption, effects of forced turbulence, or leakage into the outer atmosphere by a modification of the magnetoacoustic cutoff frequency.

The cause of the plage state has not yet been identified. It is possible that the magnetic field is somehow held together, or is "anchored," at some depth until just before the ultimate disintegration of the corresponding active regions. It is hard to understand, however, why this would not result in more diffuse edges and a concentration of the flux in the center of gravity of each of the two polarities, unless the anchoring were shallow enough that the rigidity of the flux would prevent serious bending away from its "footpoint." Effective anchoring of the field could be the result of the repeated emergence of convective blobs at roughly the same location; this could freeze the flux in nearly stationary downdrafts of a flow pattern that evolves little (Simon *et al.*, 1995). This notion is supported by the exceptionally long-lived cells in regions of enhanced network, the so-called pukas (Livingston and Orrall, 1974) that live for many days if not an entire disk passage. Another explanation focuses on phenomena occurring near the surface: Schrijver (1989) explored the possibility that an abrupt change in the dominant scale of convection near a mean flux density of 50 G serves to contain flux. Not only would the flux dispersal within the plage be reduced because of the scale change [see Eq. (6.6)], but it would also result in a rather sharp edge of the high-flux density plage area with a fairly homogeneous distribution of flux within the perimeter: in this model, the perimeter acts as a membrane with a preferential direction of transmission, caused by the difference in scales and the associated number density of paths leading away from the membrane inside and outside the plage region.

5.5 Heat transfer and magnetic concentrations

Despite the observed difference in the convective patterns at the surface, the energy flux through the plage area is very nearly the same as that through the surrounding quiet photosphere: integrated over all outward directions, the plage is in fact marginally brighter than the quiet photosphere, primarily because of the radiation leakage that occurs through the walls of the flux tubes formed by the Wilson depression of many of the flux concentrations. Apparently, the convection below the surface adjusts to the presence of the field while transporting nearly the same amount of energy (important for very active stars that can be almost entirely covered by a magnetic field; Sections 11.3 and 12.2).

Magnetic concentrations both suppress and aid the local heat flux: convective transport is severely hampered within the tube, but in contrast the rarefied photospheric tops of the constituting flux tubes present heat leaks from the top of the convection zone (Spruit, 1976; Grossmann-Doerth *et al.*, 1989). The latter effect outweighs the former: together with the faculae, the network elements account for an important part for the solar luminosity variation in phase with the activity cycle (Foukal and Lean, 1990). In other words, the magnetic field in the network allows efficient heat transfer in the solar mantle, which helps to determine the locations of the convective patterns. Thus, the magnetic network provides a certain structural organization to the convection. Hence, immediately adjacent to and between the flux tubes in a magnetic concentration, the local convective downdraft is strengthened. These enhanced downdrafts are expected to create even longer "cold fingers" than indicated by numerical models for subphotospheric convection in the absence of a magnetic field (see Section 2.5). These enhanced downdrafts entail three stabilizing effects. First, the cool downdrafts help to keep the enclosed flux tubes at a lower temperature. Second, the downdrafts pull in a horizontal flow that converges on the bunch of flux tubes in the magnetic concentration. The aerodynamic drag of this inward flow holds the flux tubes together in the cluster. Third, the presence of downdrafts hampers updrafts from forming right below magnetic concentrations (unless there are strong vortices involved that subject themselves to a buoyancy instability because they evacuate themselves by the centrifugal forces involved; see the discussion in the penultimate paragraph of Section 2.5.3).

6

Global properties of the solar magnetic field

6.1 The solar activity cycle

6.1.1 Activity indices and synoptic magnetic maps

In 1843, H. Schwabe discovered an 11-yr cyclic variation in the number of sunspots. In 1849, R. Wolf started to tabulate daily sunspot numbers; in addition, he and others have attempted to reconstruct sunspot behavior back to the beginning of the 17th Century. The sunspot activity has been expressed in the Wolf sunspot index number (or Zürich relative sunspot number), defined as

$$R \equiv k(10g + f), \tag{6.1}$$

where f is the number of individual spots visible on the disk, g is the number of spot groups (i.e., active regions with sunspots), and k is a correction factor intended to adjust for differences between observers, telescopes, and site conditions. Usually monthly averages or smoothed mean values of the Wolf number R are used to take out the modulation by solar rotation and short-period fluctuations in solar activity.

The Wolf number R, involving spot counts by eye, is a subjective measure. Moreover, the definition in Eq. (6.1) seems to be rather arbitrary. Yet it serves in many solar activity studies, because it is the only record that covers more than two centuries with rather consistent data. The index R correlates surprisingly well with more objective, quantitative indices that are mentioned below (see Fig. 6.1).

An obvious alternative to the Wolf number R is the total area of the sunspots present on the disk, measured on broadband photoheliograms (see White and Trotter, 1977).

Two other commonly used measures of global activity are based on the plages as visible in the enhanced emission in the Ca II K line core: the total Ca II K plage area and the Ca II K plage index (based on both the area and the brightness of plages). Yet another measure is the disk-integrated Ca II K line-core flux, which measures the sum of the emissions from plages and network, plus the basal emission from the entire solar disk; this index can be compared with the Ca II H+K line-core flux from stars (see Chapter 9).

The 10.7-cm (2,800-MHz) radio flux is another chromospheric emission that measures magnetic activity (Section 8.5); it is little affected by weather and by seeing.

Some indications of activity variations before the advent of telescopes can be obtained from reports of naked-eye sunspot observations. In addition, geophysical effects caused by solar magnetic activity leave traces: aurorae prompted reports, and the carbon isotope

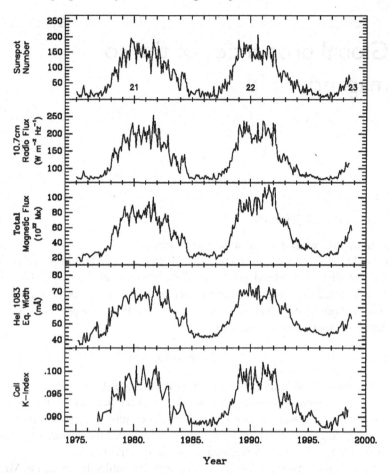

Fig. 6.1. Activity indices, averaged over a Carrington rotation, for sunspot cycles 21 and 22, from the top: relative sunspot number, the 10.7-cm radio flux (Algonquin Radio Observatory and Dom. Radio Astronomy Observatory), the total absolute magnetic flux (NSO/Kitt Peak), and the He I 10,830-Å equivalent width and Ca II K index (NSO/Sacramento Peak); note that the zero point of the Ca II Kindex is well below the range shown in the figure (figure provided by K. L. Harvey).

ratio ^{14}C/^{12}C in tree rings provides information on the activity level averaged over several decades (Eddy, 1980; Foukal, 1990, his Section 11.2.3).

A global measure for truly magnetic activity is derived from magnetograms: the total absolute magnetic flux for areas with an absolute flux density above a suitable threshold $|\varphi_0|$. The threshold isolates the magnetic flux in spots, plages, and enhanced network from the flux over the rest of the disk (see Chapter 12, Fig. 9 in Harvey, 1993). Moreover, the threshold cuts out effects of instrumental noise, and it reduces the effect of variable seeing on the absolute flux determination. For the daily NSO/Kitt Peak magnetograms, the threshold is often set at 25 G. Figure 6.14 below shows an example based on synoptic maps, with a threshold at 0 G.

Figure 6.1 demonstrates that the activity indices are remarkably similar, except for small but significant departures from monotonic relationships.

Activity indices do not bring out those essential aspects of the magnetic activity cycle that are in the (time-dependent) distribution of the phenomena across the solar disk. Nowadays studies of solar activity profit from the monthly *Synoptic Magnetic Maps*. These maps, shown in Fig. 6.12 below, are derived from the NSO/Kitt Peak daily magnetograms; see Section 6.3 for a description.

6.1.2   Characteristics of the solar activity cycle

The sunspot cycle (Fig. 6.2) varies markedly in amplitude – by at least a factor of 3 since 1860. Also the cycle period (counted from one sunspot minimum to the next) varies appreciably, between approximately 8 and 15 yr; the average period is 11.1 yr. The rise from minimum to maximum tends to be shorter than the decline from maximum to minimum, particularly so during cycles with a large amplitude.

Sunspots are confined to the *sunspot belts* which extend to ∼ 35° latitude on either side of the solar equator. The first sunspots of a cycle appear at relatively high latitudes. During the cycle, the sunspots tend to form at progressively lower latitudes. The last spots in a cycle show up close to the equator. This tendency is called Spörer's rule (compare Table 6.1), although it had been noticed by earlier observers. Maunder's *butterfly diagram* (Fig. 6.3) results when the latitudes of sunspot appearances are plotted against time.

The butterfly diagram shows that sunspot cycles overlap by 2–3 yr: the first spots of the next cycle already show up at high latitudes, whereas the last spots of the preceding cycle appear near the equator. The shape of the butterfly diagram changes when the occurrences of small Ca II plages (interpreted as small active regions without sunspots) and ephemeral regions are also included (Harvey, 1992, and Chapter 11 in Harvey, 1993). Small and ephemeral regions preceding the sunspot groups by 0.5–1.5 yr at yet higher latitudes (Fig. 6.3) extend the butterfly wings with poleward tips. In addition, the last sunspot groups seen near the equator are followed by small active regions for a period up to half a year. Hence the total overlap between two subsequent cycles is approximately 4–5 yr.

In 1914, six years after their discovery of a magnetic field in spots, Hale and his collaborators discovered the *polarity rules* (see Hale and Nicholson, 1938, for data

Table 6.1. *The rules of solar activity (compare Figs. 6.3, 6.9, and 6.12)*

Rule	Description
Hale(–Nicholson)	Regions on opposite hemispheres have opposite leading magnetic polarities, alternating between successive sunspot cycles.
Joy	The center of gravity of the leading polarity in active regions tends to lie closer to the equator than that of the following polarity.
Spörer	Spot groups tend to emerge at progressively lower latitudes as a cycle progresses.

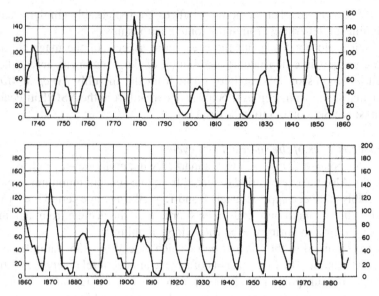

Fig. 6.2. The variation of the annual mean Wolf sunspot number $\langle R \rangle$ from 1735 to 1987 (from Foukal, 1990).

covering several decades; see the listing of this and other rules in Table 6.1). During a given cycle, the majority of the leading sunspots on the northern hemisphere are of the same polarity, whereas most of the leading sunspots on the southern hemisphere are of the opposite polarity. During the next cycle, all polarities are reversed. Thus, the period of the *magnetic activity cycle* is twice that of the sunspot cycle, that is, 22.2 yr on average. In most active regions, the bipole axes depart only slightly from the regular nearly E–W direction (Section 4.5), but some regions show an inverse orientation (Section 6.2.2).

During most of the sunspot cycle, each polar cap (defined as regions with a latitude larger than 60°) is occupied by magnetic faculae of a clearly dominant polarity, with opposite polarities on the opposite caps. The polar caps change polarity approximately 1–2 yr after the sunspot maximum, to take the sign of the following polarity of the hemisphere during the current cycle. The nearly unipolar caps reach their largest extent and largest total magnetic flux usually just before the next sunspot minimum (Howard, 1972; Sheeley, 1991; see also Chapter 11 in Harvey, 1993).

The magnetic activity cycle has a remarkable connection with the torsional wave pattern in the differential solar rotation (Section 3.1). The zones of maximum rotational shear according to the Doppler data, just equatorward of the latitude strip of relatively fast rotation, pass through the centroids of the wings of the butterfly pattern. These zones correspond to the zones of relatively slow rotation as deduced from magnetic features. The wave pattern moves with the latitude drift of the magnetic activity toward the equator – probably it starts already near the poles ∼6 yr before the first sunspot groups of a new sunspot cycle are sighted.

The solar irradiance, as monitored by satellites, displays variability caused by dark sunspots and bright faculae and network at a range of time scales (Willson and Hudson,

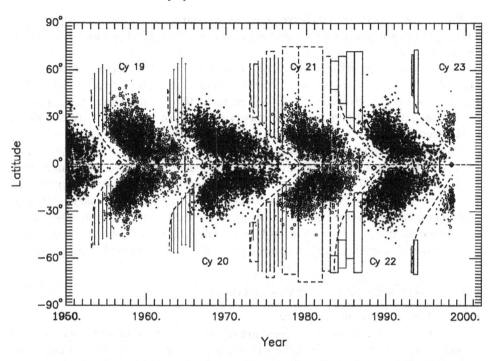

Fig. 6.3. A Maunder butterfly diagram. The sunspot cycle numbers are indicated; sunspot groups of odd cycles are shown as open circles, those of even cycles by pluses. The large/small symbols stand for sunspot areas larger/smaller than 100 millionths of the solar hemisphere. Vertical bars on the equator indicate the times of sunspot minima. The heavy dashed curves mark the boundaries between successive sunspot cycles. During the early phases of the cycles, the light solid lines indicate latitude ranges where small active regions without sunspots were seen as small Ca II K plages; dotted boxes mark the ranges of ephemeral regions (figure provided by K. L. Harvey).

1991). During the sunspot maximum, the irradiance is larger than during the sunspot minimum by nearly 0.2% (see Fröhlich, 1994) – the bright faculae and network patches overcompensate the effect of the spots. For the theoretical interpretation, see Spruit (1994). For proceedings of a conference on solar irradiance variations, see Pap *et al.* (1994).

Section 4.2.1 alludes to the cycle-dependent variation of the intensities of sunspot umbrae discovered by Albregtsen and Maltby (1978): the first sunspots of a cycle are darkest, and the umbral intensity increases linearly with time. This is a time-dependent cycle effect rather than a dependence on latitude, because the relation between umbral brightness and cycle phase shows much less scatter than that between brightness and sunspot latitude.

The irregularities in the amplitudes of the sunspot cycle have prompted many attempts to find larger periods modulating the sunspot activity cycle. The results are questionable, however, because the reliable records are too short to allow convincing conclusions. The amplitude of the solar activity variations has varied greatly. Maunder (1890) pointed out that from 1645 to 1715, practically no sunspots were reported; Eddy (1976, 1977) confirmed that during this Maunder Minimum the sunspot cycle had nearly or completely vanished. In addition, other long-period minima and maxima in the cycle amplitude can be traced from reports of naked-eye sunspots and aurora, and from the isotope ratio $^{14}C/^{12}C$ in tree rings, such as the Spörer Minimum around 1500 and the 12th Century Grand Maximum.

The cyclic variation in the activity level in the northern and the southern hemispheres may differ for an extended period of time (for an early reference, see Maunder, 1922). For instance, cycle 21 (November 1974 – September 1986) started with all major activity concentrated in the northern hemisphere; only after almost 1 yr did activity develop in the southern hemishere as well. In recent cycles, the total amounts of absolute magnetic flux surfacing in both hemispheres over an entire cycle are virtually equal (Zwaan and Harvey, 1994). A striking imbalance during the Maunder Minimum has been pointed out by Ribes and Nesme-Ribes (1993): in the records of the Observatoire de Paris, nearly all spots observed between 1672 and 1711 occurred in the southern hemisphere. All these spots appeared close to the equator, at latitudes below 20°. No spots were observed in the northern hemisphere from 1678 until 1702. Only after 1714, as the Sun came out of the Maunder Minimum, were they nearly equally distributed over both hemispheres.

6.2   Large-scale patterns in flux emergence

6.2.1   *Activity nests and complexes*

There is a marked tendency for active regions to emerge in the immediate vicinity of an existing active region, or at the site of a previous active region. This clustering tendency in the emergences, the subsequent spreading of the magnetic flux, and the merging of blankets of old flux can lead to complicated large-scale patterns, which are sometimes referred to as activity complexes (Bumba and Howard, 1965). Here we focus on the clustering tendency in the *emergence* of bipolar active regions; hence we borrow the term *nest of activity* coined by Castenmiller *et al.* (1986) to emphasize the surprising compactness of the birth site of clustering active regions. The clustering tendency is readily shown by stacking 30°-wide latitudinal strips from the activity belts cut from subsequent NSO/Kitt Peak synoptic maps of the magnetic field (Fig. 6.4); a mounting covering 100 Carrington rotations is shown by Gaizauskas (1993).

The clustering in the appearances of sunspots has been known for a long time: Cassini (in 1729; see citation in Ribes and Nesme-Ribes, 1993) and Carrington (1858) drew attention to it. A century ago, the compilers of the *Greenwich Photoheliographic Results* started to distinguish between (decaying) "recurrent groups" (rotating back onto the disk) and "revivals" (with indications of recent flux emergence). Becker (1955) published the first truly quantitative study of the clustering in sunspot appearances.

The tendency of bipolar active regions to emerge within existing active regions was confirmed by Harvey (1993; her Chapters 2 and 3) in her study of bipolar active regions. In her sample, subsequent emergences of bipoles within a nest are typically separated by 4–5 days. In Harvey's large sample, even the large, complex active regions were found to be formed by a succession of emergences of simply bipolar regions, if the formation could be followed daily.

In compilations of active-region data, as in the *Solar-Geophysical Data*, the complexes resulting from the emergence of a set of bipolar regions in rapid succession are listed as single, magnetically complex active regions. We follow this habit by calling such a structure a complex active region, or an activity complex. We reserve the term nest for

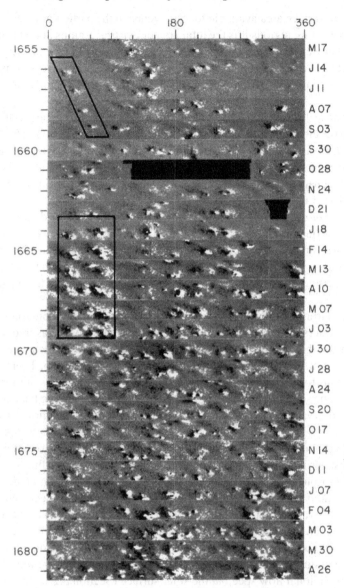

Fig. 6.4. Nests of activity in a chronological arrangement of strips taken from the northern belt of latitudes between 10° and 40° for Carrington rotations 1655–1681 (May 1977–May 1979). The horizontal scale is the Carrington longitude, increasing westward to the right. To the left of the strips are the Carrington rotation numbers, and to the right the date on which that Carrington rotation began. North is toward the top of each strip. The boxes frame complexes discussed in the text (from Gaizauskas *et al.*, 1983).

the area in which a sequence of subsequent regions emerges, as well as for that sequence of (complex or bipolar) active regions.

Harvey and Zwaan (1993) compare the *emergence rates* for emergences within existing active regions to those for emergences outside existing regions. These emergence rates are normalized to the surface area available for emergence in the particular category: for emergences within existing regions it is the total area of existing regions at that time; for the emergences outside existing regions it is the remaining area within the activity belts. Averaged over cycle 21 (1975–1986), the emergence rate within existing regions is more than 22 times larger than the rate for emergences elsewhere (see also Liggett and Zirin, 1985).

Harvey and Zwaan (1993) estimate that 44% of the regions with area A larger than 525 Mm^2 (at maximum development), and 35% of the small regions with $375 < A < 525\, Mm^2$ emerge within existing large ($A > 1,000\, Mm^2$) bipolar active regions. The actual number of regions involved in nests is larger, however, because Harvey's sampling intervals of ~ 1 month were widely separated in time. Hence the fraction of bipolar regions involved in nests is definitely larger than 50%.

The *Greenwich Photoheliographic Results* with its accurate sunspot data present a long-term basis for the quantitative analysis of the nest phenomenon. Brouwer and Zwaan (1990) found that the "single-linkage clustering method" not only yields an efficient and objective way of finding nests but also brings out properties of sunspot nests that pass unnoticed during a visual inspection of the data. Figure 6.5 shows some results of that clustering method.

Sunspot nests are surprisingly compact; Fig. 6.6 displays the spreads in longitude ϕ and in latitude θ of the centers of gravity of the subsequent sunspot groups, measured relative to the three-dimensional nest axes (in ϕ, θ, and t). The spread in longitude corresponds to less than $\sim 30,000$ km, and that in latitude is even less than $\sim 15,000$ km. There are no convincing indications for a tendency of (nearly) simultaneous occurrences of more than one nest along the same solar meridian. So the term active longitude, dating back to early solar-geophysical observations, is misleading, and hence to be avoided. Notions of simultaneous activity (nearly) $180°$ apart in longitude do not stand statistical tests either. Nests are also compact in the sense that magnetic flux emerging in a nest tends to remain confined within the nest perimeter during the nest's active lifetime; magnetic flux starts dispersing over the photospheric surface after the emergences stop (Gaizauskas *et al.*, 1983).

Sunspot nests participate in the differential rotation as determined by recurrent sunspots; the intrinsic spread about this average dependence of rotation rate on latitude is only 15 m/s (Brouwer and Zwaan, 1990). The displacements in the longitude of emergences within a nest are even less than the uncertainty of 5 m/s. The large-scale displacements in the latitude of nests during their lifetimes is extremely small; the upper limit of 3 m/s is well below the meridional flow speed at the solar surface (Section 3.2).

Lifetimes of sunspot nests range up to ~ 6 months, if no intermissions longer than two solar rotations without visible sunspots are allowed. If substantially longer intermissions are accepted, then some nests are found with lifetimes up to a few years (Van Driel-Gesztelyi *et al.*, 1992; for an early indication, see Becker, 1955).

Nests tend to cluster themselves, leading to *composite nests* (Fig. 6.7). Most of the composite nests comprise two components, but in their (small) sample Brouwer and Zwaan (1990) found one composite of four nests. The components are quite close in latitude ($\Delta\theta < 2.5°$) and in the most conspicuously nested nests, the components are within $\Delta\phi < 25°$ in longitude. The component nests of a composite overlap in time, such that a complete composite nest may last longer than a year, longer than each of

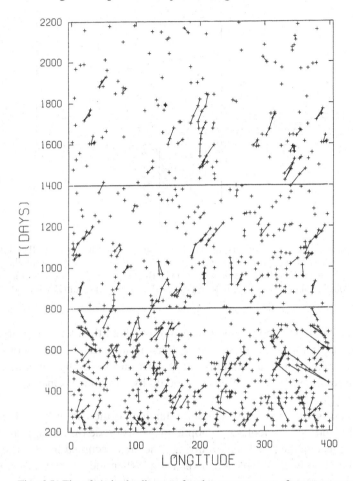

Fig. 6.5. Time-longitude diagram for the appearances of sunspot groups on the northern hemisphere between August 1959 and December 1964. Each cross represents a sunspot-group appearance. Sunspot groups that belong to one nest according to the clustering criteria are connected. Note that nearby data points in this diagram may differ substantially in latitude (from Brouwer and Zwaan, 1990).

its components. Probably most of Becker's (1955) "Sonnenfleckenherde" (sunspot foci) are composite nests in our terminology.

Day-to-day coverage of the magnetic field reveals vigorous activity in (composite) nests. In the "great complex," a composite nest indicated by the box enclosing rotations 1664–1669 and $20° < \phi < 90°$ in Fig. 6.4, Gaizauskas *et al.* (1983) counted 29 major active regions, with large sunspots, emerging during those six solar rotations. Despite the very high rate of flux emergence in that complex, however, the total absolute magnetic flux remained roughly constant at 1.3×10^{23} Mx. Since during the active life of a (composite) nest most of its magnetic flux remains within the nest's perimeter, it is inferred that on average magnetic flux emerges and disappears at equally large rates within the nest's confines.

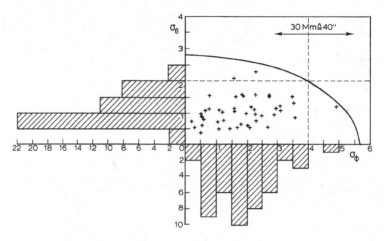

Fig. 6.6. Scatter plot of the spreads in latitude (σ_θ) and in longitude (σ_ϕ) in the clustering of sunspot groups about the (ϕ, θ, t) axis of the nest, and the histograms of the spread distributions, for the 47 sunspot nests in the data of Brouwer and Zwaan (1990) that consist of at least four members. Scales are in heliographic degrees (figure from Zwaan and Harvey, 1994).

Also in small nests, more bipolar emergences are involved than what appears from an inspection of the synoptic magnetic maps. For instance, the "simple complex," comprising four seemingly similar small bipolar regions surrounded by the oblique box in Fig. 6.4 during rotations 1656–1659, was made up by subsequent emergence of (at least) seven small bipolar regions.

The onset of a nest is sudden: within one solar rotation, it is there in its typical appearance of complexity and flux content. On synoptic maps, its crude overall appearance is maintained as long as emergences keep occurring. The area occupied by the nest parts remains bounded and without conspicuous expansion. When the emergences stop, the appearance of the nest remnant changes within one solar rotation: magnetic flux starts spreading over the solar surface, crossing the original boundary of the nest.

A complex of activity tends to be productive in outer-atmospheric ejecta and solar flares (Section 8.3); apparently the complex field configuration is unstable and relaxes through often spectacular adjustments. The most complex configurations are those containing so-called δ spots, i.e., spots having umbrae of opposite polarity enclosed by one penumbra. Transient, transverse fields with strengths exceeding 3,400 G have been measured quite locally in such spots (Livingston, 1974; Tanaka, 1991).

A truly extreme δ complex is formed with a highly stressed magnetic configuration in at least one of its major constituents, straight from its emergence; generally its polarity is inverted as compared to the Hale–Nicholson polarity rule. In order to explain the dynamical evolution of the complex magnetic fields in two extremely flare-active δ regions, Tanaka (1991) proposed the emergence of twisted flux knots.

Because the most violent flares are observed in δ configurations, such complexes receive a lot of attention from authors particularly interested in solar flares (see Section 10.4 in Zirin, 1988). In the framework of the solar activity cycle it should be emphasized, however, that only a handful of δ configurations occur during a sunspot cycle. Complex δ configurations appear to be rare accidents in the chain production of nice, bipolar active

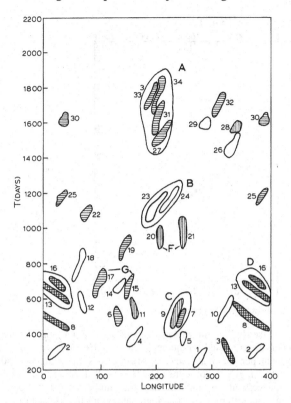

Fig. 6.7. Time-longitude diagram for the 34 nests with at least four sunspot members in the data shown in Fig. 6.5. Latitude zones are indicated by the shading within the contours: not hatched, $\theta < 10°$; horizontally hatched, $10° < \theta < 15°$; vertically hatched, $15° < \theta < 20°$; cross-hatched, $20° < \theta < 30°$. Composite nests are encircled and indicated by A, B, C and D; probably the pairs indicated by F and G are also physically related (from Brouwer and Zwaan, 1990).

regions satisfying the Hale polarity rule (see Zirin and Liggett, 1987, for ways of δ-spot development).

6.2.2 Properties of bipolar active regions

Using NSO/Kitt Peak magnetograms, Harvey (1993) investigated the properties of bipolar active regions* as functions of size, heliographic latitude, and phase of the sunspot cycle. During 29 full solar rotations distributed over slightly more than sunspot cycle 21, she recorded properties of all clearly bipolar regions larger than 375 Mm^2 whose moment of maximum development (Section 5.1) could be observed. Every bipolar region was counted only once, even if it crossed the disk more than once. Most of the properties were determined at maximum development, that is, immediately after all magnetic flux

---

* We recall that the term *bipolar active region* is used for regions consisting of only two fairly distinct areas of opposite polarity, in contrast with complex active regions, in which at least one of the polarities is distributed over two or more noncontiguous areas.

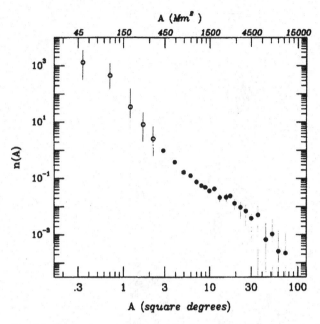

Fig. 6.8. Cycle-averaged size distribution $n(A)$ of bipolar regions measured at maximum development. The numbers are corrected for incompleteness in the counts and normalized to a number density per day and per square degree (1 square degree is approximately $150 \, \text{Mm}^2$). Results based on 978 active regions (dots) are combined with those on 9,492 ephemeral regions (circles); the solid vertical bars indicate how the numbers per day depend on the assumed (size-dependent) mean lifetime of the ephemeral regions that is used to correct for the low sampling rate. The dotted vertical bars estimate the uncertainties associated with the statistics of small samples (from Zwaan and Harvey, 1994).

had emerged. Harvey added a large sample of ephemeral regions, defined by the size range $15 < A < 375 \, \text{Mm}^2$.

At any phase of the cycle, the distribution $n(A)$ of the areas A of bipolar regions is a smooth, monotonically decreasing function of their area A (Fig. 6.8). For regions larger than $\sim 500 \, \text{Mm}^2$ (or 3.5 square degrees), the shape of $n(A)$ does not vary significantly with the phase of the cycle; for smaller bipoles the cycle amplitude decreases with decreasing size. Note that the relation between area A and total flux Φ discussed in Section 5.4 implies that $n(\Phi)$ has the same shape and properties as $n(A)$. An earlier indication of the surprising shape constancy of active-region distribution functions was found by Schrijver (1988), who discovered that the shape of the distribution function of the sizes of active regions on the solar disk is virtually constant over the solar cycle.

The (smoothed) cyclic variations are in phase for active regions in all size ranges, except for ephemeral regions for which the cycle minimum occurs ~ 1 yr earlier; this reflects the first announcement of the new cycle by high-latitude ephemeral regions mentioned in Section 6.1 (Fig. 6.3). In the cycle variation there are pulses of enhanced activity; the pulse times and amplitudes differ for active regions of different size ranges. It may take three to six solar rotations for the size distribution $n(A)$ to approach its mean shape.

Both the lifetime $\hat{t}_{\text{life}}$ (see introduction to Section 5.1) of bipolar regions and the rise time $\hat{t}_{\text{rise}}$ (from first appearance to maximum development) tend to increase with

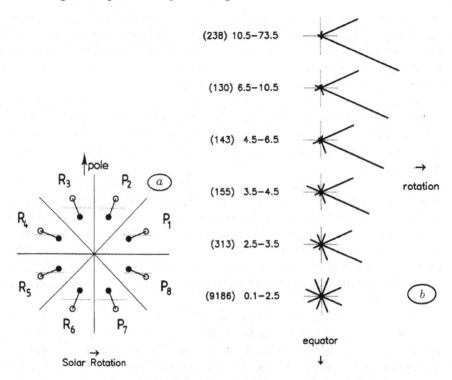

Fig. 6.9. Panel *a*: the bin division used to characterize the orientations of the bipole axes. Open circles are the leading polarity appropriate for the cycle according to Hale's polarity rule. Panel *b*: relative numbers in the orientation bins for six area ranges, which are indicated (in square degrees) on the left, preceded by the number of regions between parentheses (adapted from Harvey, 1993).

increasing area *A* (at maximum development), but there is a large intrinsic spread about these mean trends. The rise time $\hat{t}_{\text{rise}}$ is short: $\hat{t}_{\text{rise}} \leq 3$ days for most active regions, but there are some rare cases of 4 or 5 days.

Harvey binned the orientations in 45° sectors (Fig. 6.9). The results confirm Hale's polarity rule and Joy's orientation rule for sunspot groups down to small bipolar regions without sunspots: the preferred orientation is nearly E–W, with the leading polarity closest to the equator (P_8 in Fig. 6.9). The fraction of deviant bipole orientations (R_i for $i = 3, \ldots, 6$ in Fig. 6.9) increases with decreasing size. Note that the deviant orientations are not properly described by a spread around the preferred direction; there is an indication of a secondary maximum in the direction opposite to the preferred one (R_4 in Fig. 6.9). Among the ephemeral regions, Harvey (1993) found a small but significant preference (60/40) for the proper orientation; this preference is most pronounced among ephemeral regions emerging within the sunspot belts.

Quantitative data on Joy's rule are available based on measurements of orientations of active regions for much larger samples of sunspot groups and plages than used by Harvey (1993), but these compilations disregard the size and the evolutionary phase of the regions; see Section 7.1 in the review by Howard (1996). From such a large, mixed data set, Howard finds that the angle (tilt) between the mean orientation axis and the direction of solar rotation increases from 1° near the equator to 7° at 30° latitude, with the leading polarity closest to the equator (Fig. 12 in Howard, 1996). He notes that the tilt for plages is slightly more pronounced than that for spot groups. Howard also finds that changes in the tilt angles correlate with changes in the separation of the polarities; the sense of the correlation is in accordance with the Coriolis effect.

Hale's polarity rule divides the butterfly diagram in domains belonging to subsequent sunspot cycles. This determines a time-dependent *mean latitude* $\theta_0(t)$ for recently emerged regions larger than $1,000 \, \text{Mm}^2$ (6.5 square degrees). These mean latitudes $\theta_0(t)$ describe the progressive and smooth shift of the emerging regions toward the equator during the cycle (Spörer's rule).

The mean latitude $\theta_0(t)$ for large active regions is a convenient reference for the latitude distribution of various size groups of bipolar regions throughout the cycle. The main results are that (a) for all size intervals, including that of the ephemeral regions, the latitude distribution peaks precisely at the mean latitude $\theta_0(t)$ defined by the large active regions, and that (b) although for all size bins of active regions larger than $1,000 \, \text{Mm}^2$ the latitude distributions about the mean latitude are virtually identical, the distribution for smaller regions widens progressively with decreasing size, particularly so on the poleward side.

The separation of the cycles in the butterfly diagram enables the study of the cycle variation of regions in various size intervals for regions belonging to one sunspot cycle. For relatively large regions, with $A > 6.5$ square degrees, the cycle amplitude in the emergence spectrum $n(A)$ is a factor of 13.9; for small regions, with $2.5 < A < 3.5$ square degrees, the amplitude is 8.8. For ephemeral regions ($A < 2.5$) the cycle amplitude is between 2.0 and 2.5 (Chapter 5 in Harvey, 1993).

The general finding is that all properties of the bipolar regions vary *smoothly* with the size of the regions and with the phase of the cycle. In other words, there are no indications for well-defined classes of regions of distinctly different behavior. Specifically, ephemeral regions appear to be the smallest bipolar active regions, at one end of a broad spectrum of sizes: the distinction between the ephemeral region and the active region is a historical artifact. Note that the frequency distribution of active-region sizes is such that both ends of the spectrum contribute markedly to the total magnetic-flux budget of the photosphere.

Bipolar regions emerging within existing active regions possess some properties underlining the fundamental role of activity nests in the activity cycle:

1. Although the emergence *frequency* (number of emergences per day) within existing active regions varied during cycle 21 by as much as a factor of 15, the corresponding emergence *rate* (emergence frequency per unit available area) varied by a mere factor of 1.5. The frequency of emergences outside existing regions varies by a factor of 3.5 during cycle 21 (nearly the same factor applies to the emergence rate). Consequently, emergences in nests contribute most to the cycle amplitude in the total number of bipolar emergences.

2. Bipoles emerging in nests show approximately the same orientation as their parent region. Most of these are in the dominant directions: sectors P_8 and P_1 in Fig. 6.9, which is not surprising. However, all five successive emergences in the only three parent regions with a reversed orientation (R_4) were within

the same R$_4$ sector. Although these numbers are small, the result suggests that the orientation is provided by a sub-surface magnetic flux bundle from which successive bipoles emerge.

The mean East–West inclinations of the magnetic field in active regions with respect to the vertical have been determined by Howard (1991b), using Mt. Wilson magnetograms in the period 1967–1990 to compare the total magnetic fluxes for all active regions as a function of central-meridian distance. From the asymmetry of this distribution follows the inclination of the field lines in the magnetic plages with respect to the vertical, in the plane parallel to the solar equator. The sum of the absolute line-of-sight magnetic fluxes in both leading and following polarities in entire active regions indicates that the field lines are inclined toward the East, that is, trailing the rotation, by 2°. Taking the fluxes in the polarities separately, the field in the leading portions is inclined eastward ∼10°, and the field in the following portions westward by ∼6°; in other words, the fields are inclined toward each other by 16°. Note that the largest regions dominate in those averages over more than 8,000 observations, because they contribute most to the total fluxes in the longitude bands, and they are more frequently counted.

The field inclinations depend markedly on the evolutionary phase: in growing regions, the field lines are strongly inclined toward the West, by 24.5° for the entire regions, by 10° for the leading parts, and by 28° for the following parts. For regions decreasing in flux, the field in the entire regions and that in the leading polarity are inclined eastward by 5° and 13°, respectively, whereas the field in the following polarity leans westward by ∼3°. Note that in all cases the leading and following polarity fields incline toward each other by ∼16° (Howard, 1991b).

We note that the eastward inclination (i.e., trailing the rotation) that is indicated by the majority of the (decaying) active regions, is consistent with the increase of rotation rate with depth in the top of the convective envelope in the activity belt (Fig. 3.3). The reason for the strong westward inclination of growing active regions is not known.

It has been known for a long time that the distribution of sunspot sizes about the central meridian is not symmetrical (see Minnaert, 1946). The direction of the magnetic field lines is taken to be perpendicular to the sunspot surface, so that the direction of the spot field follows from the central-meridian distance where the maximum spot sizes occur. Applying this principle to the sunspot areas in the Mt. Wilson data set, Howard (1992) found that the sunspot fields are inclined toward the east by several degrees, which roughly agrees with the finding for magnetic plages. In contrast to the plages, however, the leading and the following spot fields are inclined away from each other by a degree or so.

6.2.3 Subphotospheric structure of the strong magnetic field

The preferential nearly E–W orientation of active regions indicates that the regions emerge from a nearly toroidal flux bundle below the solar surface (Fig. 6.10; see Babcock, 1961). This bundle is supposed to have been wound into a spiral by differential rotation (Chapter 3), with the rotation at low latitudes being fastest. Section 4.5 summarizes the arguments why this toroidal flux system is believed to be located just below the convection zone. For large active regions, the ratio of the separation of the poles relative to the depth of the convection zone (Fig. 4.10) seems to be of a correct

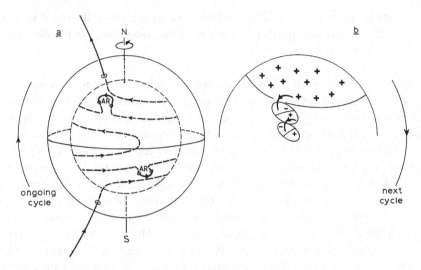

Fig. 6.10. Dynamo aspects. Panel *a*: nearly toroidal magnetic flux system. The differential rotation is invoked to stretch the field in the E–W direction. When a loop of flux tubes is released, the Ω-shaped loop may break through the photosphere, forming a bipolar active region; see Section 4.4. To the left, the direction of the poloidal field component for the depicted cycle is indicated. Panel *b* (discussed in Section 7.2): sense of the reconnecting flux loops as derived from the flux-transport model applied to an early phase of the ongoing cycle represented in panel *a* (from Zwaan, 1996).

magnitude within this picture. But for small and ephemeral regions, the pole separation is uncomfortably small with respect to the convection zone depth or to the expected size of the largest convective flows in the deep convection zone (Fig. 6.11*a*). To relieve this problem, Zwaan and Harvey (1994) suggested that smaller active regions ride piggy back on tops of subsurface magnetic flux arches (Fig. 6.11*b*). We reserve the term arch for these hypothetical subsurface, long-lived, flux loops.

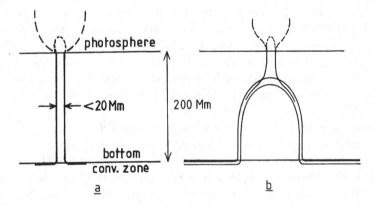

Fig. 6.11. Rooting of small active regions. Panel *a*: hypothetical flux loop supporting a small active region, if it were anchored below the bottom of the convection zone. Panel *b*: small active region supported by a flux loop emerged from a subsurface magnetic flux arch (from Zwaan and Harvey, 1994).

Computations of flux loops rooted in toroidal strands (see Van Ballegooijen, 1982a, 1982b) indicate that the eventual equilibrium position of a flux loop depends sensitively on the initial conditions: a rising flux loop may make it up to the photosphere, but if not it stops at some depth depending on initial conditions. Similarly, flux loops retracted from an active nest (discussed in Section 6.4.1) may not make it all the way back to the bottom of the convection zone.

Zwaan and Harvey (1994) suggested that the top of a flux arch provides the well-focused launching site for bipoles belonging to an active nest. The specific thickness of a given arch may explain the limited range of the sizes of the bipoles emerging in one nest. Thin arches, for example, can only support nests spawning small bipolar regions. If bipolar regions emerging in one nest spring from one arch, then their orientations are expected to correspond to the orientation of the arch. Composite nests suggest that arches may be multiple, with two or more components slightly shifted in latitude and somewhat more in longitude.

Arches supporting smaller bipoles probably tend to be thinner, or higher, or both. Thinner or higher arches would tend to be more warped by large-scale convective flows near the bottom in the convection zone, which may explain why the fraction of deviant bipole orientations increases with decreasing size (Fig. 6.9).

The lifetimes of large-scale patterns, namely, several months for nests and up to a few years for composite nests and for emergences within a narrow latitude band, indicate the local durability of the root system consisting of the toroidal flux bundle and the arches. The lifetimes are compatible with the long radiative relaxation times of flux tubes in the convection zone (Section 4.1.2 and Fig. 4.3).

Magnetic field shows a strong tendency to emerge in pulses. In other words, an initial emergence of magnetic flux appears to provoke subsequent flux emergences. The dominant features of this tendency are the following: (a) the sequence of emergences in a nest, (b) the delayed, sudden onset of major activity at the northern or southern hemisphere (Section 6.1.2), and (c) the cyclic variation of emergences being nearly in step with the total area of already existing active regions (Section 6.2.2). Apparently, the magnetic flux emerges preferentially and repeatedly from an already affected section in the toroidal rope, or from the top of a flux arch hovering below the surface.

6.2.4 Unipolar and mixed-polarity network

Once the containment of magnetic flux within an activity nest or active region comes to an end, patches of flux disperse from the site. Around the maximum of the sunspot cycle, remnants of large active regions form large areas of enhanced, unipolar magnetic network. Giovanelli (1982) defined the magnetic network as of *mixed polarity* where the ratio between the mean flux densities in the opposite polarities $|\bar{B}_{major}/\bar{B}_{minor}|$ is between 1 and 2.5, and "unipolar" if the ratio exceeds 2.5. Large areas of unipolar magnetic network, in which the dominant polarity may contribute more than 80% to the total absolute flux (Giovanelli, 1982), are built up on the poleward side of the activity belt; there is a dominance of the following polarity flux. A very small part of that flux ends up in the polar cap; see Section 6.3.2. Gradually the network is diluted into the quiet network. Giovanelli established that during five years around the sunspot minimum, mixed-polarity network dominates the latitudes up to the polar caps at 60°. Approximately 2 yr after the sunspot minimum of 1976, the mixed polarities were rapidly replaced by large areas of unipolar network over a large part of the activity belts (up to 40°); these unipolar domains originated from large active regions of the new cycle.

6.3   Distribution of surface magnetic field

6.3.1   *Distribution of photospheric flux densities*

The large-scale patterns in the photospheric magnetic field are important for the studies of stellar magnetic activity, dynamo theory (Chapters 7 and 15), and for the interpretation of radiative losses from stellar outer atmospheres (Section 9.4). We describe these patterns in this section. In Section 6.3.2 we discuss a model for their evolution.

6.3.1.1   *Large-scale patterns*

The large-scale evolution of the photospheric field can be studied by using full-disk magnetograms, strung together as a movie. An excellent example is the set of magnetograms produced by the Michelson Doppler Imager (*MDI*) on *SOHO*, which has been observing full-disk magnetograms every 96 min since shortly after its launch in December of 1995, although with some interruptions of up to a few months. Solar rotation complicates the interpretation of these sequences.

Synoptic maps provide a visualization of the global patterns of the photospheric magnetic field corrected for rotation. Such maps (see Fig. 6.12) are derived from daily magnetograms taken at the National Solar Observatory at Kitt Peak (see Gaizauskas *et al.*, 1983). Three averaging procedures are involved in the production of such maps. First, the net magnetic flux density is measured with $1''$ pixels across the entire disk, at an angular resolution typically of the order of a few arcseconds, depending on seeing conditions. These maps are subsequently binned to match the coarser resolution of approximately one heliocentric degree of the synoptic maps. Finally, a time-weighted average is determined of observations made on successive days by using a $\cos^2(\phi)$-weighting, that is, a weighting giving most weight to the period when the region to be mapped onto the synoptic summary is near the central meridian. The sample maps in Fig. 6.12 are based on the signed magnetic flux densities, which results in some numerical cancellation of flux within the pixels of the synoptic maps of opposite polarities that are present in the high-resolution data. The magnitude of this numerical cancellation appears to be some 20% to 30% of the total flux during the minimum of the cycle, whereas at cycle maximum some 10% to 15% cancels (J. W. Harvey, private communication); this difference is a consequence of the larger area coverage by mixed-polarity fields during periods of low activity in contrast to the increased importance of unipolar regions during cycle maximum.

The synoptic maps in Fig. 6.12 exhibit several of the general properties of the global distribution of magnetic flux densities, such as the Hale and Joy rules (Section 6.1.2; see also the summary in Table 6.1). Also visible are the poleward streaks formed by areas of a dominant polarity that arch in an easterly direction from low to high latitudes (see Section 6.3.2.2).

6.3.1.2   *Global time-dependent distribution of the magnetic field*

Schrijver and Harvey (1989) used a series of synoptic magnetograms to study the cycle dependence of the histogram of flux densities on the disk. They determined the time-averaged histograms $h_t(|\varphi|)$ of magnetic flux densities $|\varphi|$. Some of these are

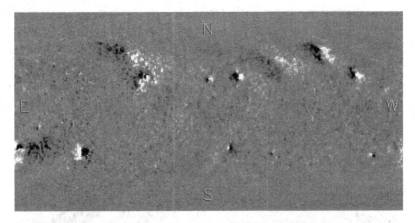

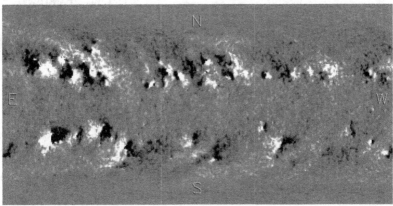

Fig. 6.12. Synoptic magnetograms of sine latitude vs. longitude of the full Sun (North is up; East is to the left), constructed from magnetograms observed at the NSO/Kitt Peak Observatory taken during the early rise phase of solar cycle 21 (Carrington rotation 1655, early 1977, shown as phase *b* in Fig. 6.14) in the top panel, and during the rise phase near cycle maximum (Carrington rotation 1669, mid-1978, shown as phase *c* in Fig. 6.14) in the bottom panel. The resolution is ≈1° in longitude and 1/90 in sine latitude (12,200 km, or 16.9 arcsec at the equator). The two polarities are shown by darker and brighter shades of gray; the gray scale saturates at ±50 G.

shown in Fig. 6.13; the phases of the cycle for these histograms are marked in Fig. 6.14, which shows the time-dependent total absolute magnetic flux density in the photosphere.

As activity increases, the tail in $h_t(|\varphi|)$ at high flux densities is raised, because of the increased surface coverage by plages and enhanced network [for $|\varphi| > 50$ G, $h_t(|\varphi|) \propto |\varphi|^{-2.05}$; Schrijver, 1990]. Counterbalancing this increase in the tail of $h_t(|\varphi|)$ is a decrease in the fraction of the solar surface that is covered by regions of low flux densities. There are associated changes everywhere between these extremes, so $h_t(|\varphi|)$ cannot be realistically approximated by, for example, two components that represent a "typical quiet photosphere" and an "enhanced network and plage."

The histograms $h_t(|\varphi|)$ show little evidence of hysteresis. For instance, the histogram for the rising flank of the cycle (interval *c*) is virtually the same as that for the decay phase of the cycle (interval *e*) when the level of the surface-averaged absolute flux

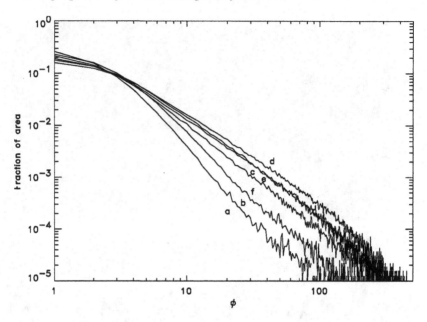

Fig. 6.13. Time-dependent distribution function h_t for the absolute magnetic flux density ϕ ($|\varphi|$ in the text) in synoptic maps (see Fig. 6.12). The bin width is 1 G. The observed magnetic flux densities are corrected for projection, assuming the field to be vertical. Latitudes in excess of 60° were excluded, because the projection effects are hard to correct for in an environment of generally low flux density. The distribution functions are averages over five (*a*) or six (*b*, *c*, *d*, *e*, *f*) consecutive rotations, from cycle minimum (labeled *a*) to maximum (*d*); compare Fig. 6.14. The rise and decline phases *c* and *e* have comparable levels of activity and flux distribution functions.

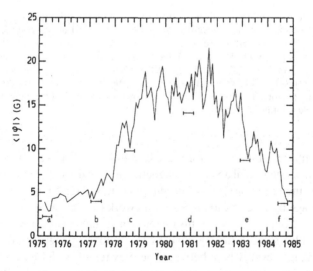

Fig. 6.14. The average magnetic flux density determined from synoptic maps. The six intervals for which the distribution functions $h_t(|\varphi|)$ are shown in Fig. 6.13 are identified. The synoptic maps used in the computation of the total fluxes for phases *b* and *c* are shown in Fig. 6.12 (figure from Schrijver and Harvey, 1989).

density $\langle|\varphi|\rangle$ is comparable (see Fig. 6.14). Consequently, for time scales exceeding no more than the temporal resolution of half a year in Fig. 6.13 (adopted to smooth out short-term fluctuations), $h_t(|\varphi|)$ appears to be primarily a function of the corresponding surface-averaged flux density $\langle|\varphi|\rangle(t)$, regardless of the phase of the cycle.

The importance of this finding lies in the comparison with stellar data: on a time scale of less than half a year, the solar photosphere "forgets" the state of activity it was in before, having processed all photospheric flux (see Section 6.3.2.4). So when temporal averages are made over intervals short compared to the solar cycle, the Sun probably mimics stars of different levels of activity, ranging over a factor of ~ 5 in the magnetic photospheric flux density for averages over a full rotation.

6.3.2 Dispersion by random walk and large-scale flows

6.3.2.1 An analytical model
Leighton (1964) suggested that for the purpose of modeling the large-scale patterns in the photospheric magnetic field outside spots and plages, the dispersal of the concentrations of magnetic flux in the photosphere can be described as a random walk, in which the flux responds passively to evolving supergranular flows (he referred to the associated diffusion coefficient D as the "mixing coefficient" to avoid confusion with the resistive diffusion associated with current dissipation; Section 4.1.1). This suggestion became a crucial argument in the development of the Babcock–Leighton model for the dynamo (see Section 7.2). Since then, this empirical model has been proven surprisingly successful, particularly through the efforts of DeVore, Sheeley, Wang, and colleagues.

On an intuitive basis, the notion of the concentration of strong field in compact flux tubes makes it easy to imagine that drag-coupling to the convective flows results in diffusive dispersal. A more formal basis for the description of the dispersal of the photospheric field as a random-walk diffusion process is found in the induction equation, Eq. (4.7). For the intrinsically strong field in the photosphere, resistive diffusion can be ignored. If we assume that the field is normal to the surface at all times (requiring that there are no vertical gradients in the horizontal flow, i.e., $\partial v_x/\partial z = \partial v_y/\partial z = 0$), then the induction equation can be rewritten in the form of a continuity equation for the scalar B_z. In other words, the field is passively advected by the horizontal flows, responding on large scales as if it were "diffusing" in the convective flows.

The assumption that there are no vertical gradients in the flow is untenable in a stratified convecting medium. But the flow and the field are not coupled rigidly at all levels: the flow can move around the intrinsically strong flux tubes, coupling through drag, while buoyancy strives to maintain the vertical direction of the field. Thus, effectively, there is some characteristic depth where the flow couples to the tubes, resulting in the impression of a random-walk diffusive dispersal. Even though we do not have an accurate description of the geometry of evolving supergranulation, the assumption of a classical diffusive dispersion involving Gaussian profiles as fundamental solutions turns out to be a good approximation to study the dispersal of flux on time scales much longer than the supergranular time scale of a day. This is because of the central-limit theorem (see Hughes, 1995, his Chapter 4, for a discussion): if the individual steps introduce no net drift, and have a variance σ^2, then the asymptotic probability density function for the position after n steps approaches a Gaussian with variance $n\sigma^2$ for sufficiently large n, regardless of the distribution function of step lengths.

Before a global description is made of the evolution of the large-scale photospheric flux, it is instructive to consider the simpler mathematical formulation of field dispersal on a Cartesian plane in the absence of large-scale flows, valid for the Sun on relatively small length scales and short time scales. In this case, the diffusion equations for the two opposite polarities, N and S, written for coordinates (x, y) become:

$$\frac{\partial \varphi_N}{\partial t} = D \left(\frac{\partial^2 \varphi_N}{\partial x^2} + \frac{\partial^2 \varphi_N}{\partial y^2} \right) + E_N - C_N, \tag{6.2}$$

$$\frac{\partial \varphi_S}{\partial t} = D \left(\frac{\partial^2 \varphi_S}{\partial x^2} + \frac{\partial^2 \varphi_S}{\partial y^2} \right) + E_S - C_S. \tag{6.3}$$

The functions E_N and E_S in these equations describe the emergence rates of magnetic flux into the photosphere. These reflect the distribution of active-region fluxes as a function of flux and position, and their orientation according to Joy's rule: the functions E_N and E_S should conserve the total flux, but reflect the offset of the two polarities within active regions at maximum development.

The functions C_N and C_S describe the cancellation rates, i.e., the disappearance of magnetic flux from the photosphere in collisions against flux of opposite polarity; C_N has to equal C_S regardless of the detailed dependence of the cancellation rate on basic physical parameters.

By subtracting Eq. (6.3) from Eq. (6.2), one obtains the equation for the net magnetic flux density φ:

$$\frac{\partial \varphi}{\partial t} = D \left(\frac{\partial^2 \varphi}{\partial x^2} + \frac{\partial^2 \varphi}{\partial y^2} \right) + \left[E_N - E_S \right]. \tag{6.4}$$

Note that this equation does not depend on the rate at which flux is canceled, which is why we can formulate flux dispersal as a diffusion process by using Eq. (6.4) (see also DeVore, 1987). This does not mean that the cancellation process is to be discarded as unimportant: flux must eventually disappear from the photosphere, and the details of the small-scale flux cancellation are likely to be linked intimately to atmospheric heating (see Chapter 10), if not to the cycle itself (Chapter 7).

The description of flux dispersal by Eq. (6.4) relies on the assumption that the diffusion coefficient does not depend on the local flux density through some magnetoconvective interaction. If D were a function of both φ_N and φ_S – for instance by a dependence on the total flux density per resolution element, $D = D(\varphi_N + \varphi_S)$, then Eq. (6.4) by itself would not suffice to model flux dispersal, because it describes only the net flux density $\varphi_N - \varphi_S$. In that case the equations for N and S polarities, reflecting the dependence of the diffusion coefficient on the local flux density, would be required, and knowledge of the details of the cancellation process would be essential. There is no evidence for such a dependence of D outside the confines of active regions (see the discussion of the plage state in Section 5.4).

The measurements required to estimate the diffusion coefficient follow from the solution of Eq. (6.4) for a point source injection of a total flux Φ. This solution is the probability distribution for a flux particle to move a distance r from its position at $t = 0$

(the time of "injection"):

$$\varphi(x, y)\,\mathrm{d}x\mathrm{d}y = \frac{\Phi}{4\pi Dt}e^{-\frac{x^2+y^2}{4Dt}}\,\mathrm{d}x\mathrm{d}y. \tag{6.5}$$

The diffusion coefficient is related to the observable rms displacement by

$$D = \frac{\langle r^2 \rangle}{4t}. \tag{6.6}$$

For a classical random walk in which each particle steps a distance $\hat{\ell}$ in a time interval $\hat{t}$, the rms displacement equals the value given by Eq. (6.6) for the order of magnitude estimate

$$D \sim \alpha\,\frac{\hat{\ell}^2}{\hat{t}} = \alpha\,\hat{v}^2\hat{t}, \tag{6.7}$$

where α is a constant of the order of unity, depending on the geometry of the displacements. Hence the diffusion coefficient is to be estimated from a characteristic step length $\hat{\ell}$ (which we take to be typically 1/6th of the supergranular circumference $2\pi R_{sg}$, i.e., $\hat{\ell} \sim R_{sg}$) and time scale $\hat{t}$. For the average radius of a supergranule ($R_{sg} \approx 8,000\,\mathrm{km}$) and a lifetime $\hat{t}$ of a day, $\hat{\ell}^2/\hat{t}$ is found to be $\approx 750\,\mathrm{km}^2/\mathrm{s}$.

Observational concerns keep us from adopting the above estimate for D unreservedly. First of all, the value of α depends on the displacement pattern (Simon *et al.*, 1995), which is not well known. In the absence of a detailed flow model, the value of D is to be estimated from observational data on rms displacements. On the smallest scales of supergranular diffusion, i.e., for distances of only a few supergranules and time scales of a few days, the dispersal of magnetic flux can in principle be studied by tracking concentrations of magnetic flux to determine their path as a function of time. A fundamental problem is that flux concentrations often fragment into smaller pieces or coalesce with neighboring concentrations (e.g., Berger and Title, 1996; Section 5.3 discusses collision and fragmentation rates), which makes tracking difficult. Relatively large flux concentrations in unipolar enhanced network can sometimes be tracked for up to several days (e.g., Schrijver and Martin, 1990), because in such an environment cancellation is less important, and the concentrations evolve to a larger average flux because of the reduction in cancellation interactions that result in small residual concentrations. The resulting apparent diffusion coefficient $\langle D \rangle$ estimated from the random displacements equals $\sim 220\,\mathrm{km}^2/\mathrm{s}$ (Table 6.2).

In estimating the diffusion coefficient from tracking results, one has to be careful in the interpretation of the value. First of all, the random-walk properties of flux concentrations depend on the flux Φ they contain (Schrijver *et al.*, 1996a). A large population of relatively small concentrations coexists with a smaller, but more conspicuous population of larger concentrations. These smaller concentrations tend to move somewhat faster than the larger ones, probably because larger concentrations tend to form in strong, relatively stable convergences in the flow. Consequently, Eq. (6.6) should not be applied directly to derive a proper flux-weighted effective diffusion coefficient from the observed displacements. Schrijver *et al.* (1996a) argue that the diffusion coefficient estimated from feature-tracking may have to be increased by a factor of 2–3 because of this bias. Potential effects of giant cells or of correlations in evolution between neighboring cells

Table 6.2. *Some published diffusion coefficients D for the dispersal of solar photospheric magnetic flux*

D (km^2/s)	Method	Reference
	Quiet-sun and active-network data:	
	Cross-correlation methods: $\langle D \rangle = 180 \pm 30$	
200	14-h K-line sequence	Mosher (1977)
150	1-day magn. sequence	Wang (1988)
120–230	Thresholded, high-res. magnetograms, assuming lifetime of 1–2 days	Komm *et al.* (1995)
	Object-tracking methods: $\langle D \rangle = 220 \pm 20$	
175	Magnetogram sequence	Smithson (1973a)
140–300	14-h K-line sequence	Mosher (1977)
175–300	Nearest-neighbor tracking on K-line sequence	Mosher (1977)
250	10-day magn. sequence	Schrijver and Martin (1990)
	Analysis of large-scale patterns: $\langle D \rangle \approx 600$	
600	Global pattern	Sheeley (1992)
	Active-region data:	
110	10-day magn. sequence	Schrijver and Martin (1990)

would not be detectable until after many days, which tracking studies cannot address because of the frequent collisions and fragmentations to which flux concentrations are subjected.

The superposition of scales in the flux dispersal has to be taken into account in the interpretation of the measurements. Hagenaar *et al.* (1999) show that on time scales shorter than approximately 4–6 h the granular dispersion dominates the rms displacements. Thereafter the slow supergranular drift begins to contribute to the dispersal, but it doesn't reach the complete randomization required for truly diffusive dispersal until at least 10–12 h have passed.

These studies are based on measurements of mean-square displacements. More detailed tests of the diffusive behavior can be made by analyzing the entire distribution function of displacements. The displacement statistics for object tracking turn out to be nearly compatible with the Gaussian distributions expected for a classical random walk (e.g., Schrijver and Martin, 1990). Lawrence (1991) pointed out, however, that there are small deviations in the distribution of displacements as a function of time that suggest that the motions were not representative of ordinary random walk in two dimensions. Lawrence and Schrijver (1993) claim that the dynamics of moving concentrations of magnetic flux are similar to a random walk of particles on clusters on random lattices, with a dimensionality near 1.56 (compare the discussion of percolation theory in Section 5.3, but now for clusters of paths, or bonds, that connect lattice sites). Probably, the relatively slow supergranular evolution imposes an underlying order on the random walk as the displacements occur primarily along the supergranular boundaries on time scales up to a few days. This would leave a signature of an underlying raster, resulting in percolation phenomena (mathematically, percolation is defined as particle dispersal on a raster with a frozen-in disorder; Hughes, 1996).

This percolation interpretation is not unique. Not only have the effects of the combination of granular and supergranular displacements yet to be explored, but coupling of the diffusion coefficient D with the local flux density is also a possible cause for apparent subdiffusive dispersal. The commonly made Fickian assumption in diffusion problems is that the rate of transfer, F, of the diffusing substance through a unit area is proportional to the gradient of the concentration, C. In one dimension that would mean $F = -D\partial C/\partial x$; compare that with the diffusion approximation of radiative transfer in Eq. (2.25). Non-Fickian or anomalous diffusion occurs when the rate of transfer also depends on the concentration itself. Schrijver (1994) derives a non-Fickian diffusion equation that has the same solution for diffusion in regular space as that for fractal diffusion on lattices, but with very specific requirements on the diffusion coefficient and the rate of flux transfer. These very specific requirements are in marked contrast to the natural explanation of the fractal analogy for diffusion on random lattices. Further study is required to tell whether anomalous diffusion does indeed occur because of the residual latticelike structure of the network for time scales of up to a few days.

6.3.2.2 The large-scale dispersal of flux

We now turn to the large-scale dispersal of flux over the entire solar surface. Let the large-scale differential rotation and the meridional flow be described by $\Omega(\theta')$ and $v(\theta')$ for colatitude θ' measured from the North Pole. On the spherical solar surface, the radial photospheric field, measured as the photospheric flux density φ, is then described by

$$
\frac{\partial \varphi}{\partial t} = -\Omega(\theta')\frac{\partial \varphi}{\partial \phi} + \frac{D}{R_\odot^2}\left[\frac{1}{\sin\theta'}\frac{\partial}{\partial\theta'}\left(\sin\theta'\frac{\partial\varphi}{\partial\theta'}\right) + \frac{1}{\sin^2\theta'}\frac{\partial^2\varphi}{\partial\phi^2}\right]
$$
$$
- \frac{1}{R_\odot \sin\theta'}\frac{\partial}{\partial\theta'}\left[v(\theta')\varphi\sin\theta'\right] + E(\phi,\theta',t) \tag{6.8}
$$

(see Sheeley *et al.*, 1983). The source term $E(\phi, \theta', t)$ allows modelers to "inject" bipolar regions, with positive and negative polarities offset in position, thus treating the photospheric magnetic flux density as an equivalent surface magnetic charge density. Such injection sidesteps the emergence and evolution of active regions, because the dynamics of the emergence of active regions into the photosphere appears to be subject to processes other than simple random walk and thus cannot be incorporated in the surface diffusion model. Matching the observed large-scale patterns requires a diffusion coefficient of $D = 600 \pm 200\,\text{km}^2/\text{s}$. The difference between this value and the results from other studies (Table 6.2) remains a puzzle, although some of the potential explanations of this paradox were mentioned in the previous section: the existence of a relatively large population of faster moving small concentrations, or of larger-scale cells or some coherence in the evolution of supergranular cells. In Section 6.4.2 we point out another potential contribution to this disparity associated with U-loop emergence that causes an apparent in situ disappearance of flux that could mimic an increased dispersal coefficient.

The model in Eq. (6.8) describes the observed patterns very well, however. Let us focus for a moment on the poleward streaks. These poleward crescents of alternating polarity (Fig. 6.12) are a peculiarity of the large-scale dispersal of magnetic flux. They are now understood to result from a balance between the poleward meridional flow and the differential rotation (DeVore, 1987; Sheeley *et al.*, 1987). Their surprising property is that whereas the streaks rotate rigidly, the flux constituting them rotates differentially. Wang has presented the following analogy to these streaks, modified here to include shear. Imagine that you are on the outside shore of a gradual bend in a stream, where the water flows more rapidly than at the opposite shore. Now release ducks that are trained to swim across the river always exactly facing the far shore, and repeat this at regular intervals. The ducks will, of course, drift downstream as they cross the water. After the first duck reaches the opposite shore, the curve that connects the swimming ducks is stationary, although each duck on that line is swimming across the flow and drifting in it.

DeVore (1987) compares the time scales that play a role in the formation of the streaks. The transport equation, Eq. (6.8), contains three distinct time scales. For them to be estimated, the differential rotation is approximated by

$$\Omega(\theta) = \Omega_{\mathrm{eq}} - \Omega_{\mathrm{d}} \sin^2 \theta, \tag{6.9}$$

and the meridional flow by

$$v(\theta) = v_0 \sin 2\theta, \tag{6.10}$$

as functions of latitude θ.

The first of the three time scales is the *shearing* time scale $\hat{t}_{\mathrm{s}}$ associated with the differential rotation (Section 3.1):

$$\hat{t}_{\mathrm{s}} = \frac{2\pi}{\Omega_{\mathrm{d}}}. \tag{6.11}$$

For a value of $\Omega_{\mathrm{d}} \approx 2°/\mathrm{day}$, one finds $\hat{t}_{\mathrm{s}} \approx 180$ days. The second time scale is that associated with the *meridional flow* (Section 3.2):

$$\hat{t}_{\mathrm{m}} = \frac{R_\odot}{v_0}, \tag{6.12}$$

which for $v_0 \approx 10\,\mathrm{m/s}$ equals $\hat{t}_{\mathrm{m}} \approx 800$ days. The time scale for *global dispersal* is given by

$$\hat{t}_{\mathrm{d}} = \frac{R_\odot^2}{D}. \tag{6.13}$$

For a diffusion coefficient of $600\,\mathrm{km^2/s}$ this yields $\hat{t}_{\mathrm{d}} \approx 9{,}500$ days.

If the meridional flow were reduced substantially, a pattern of poleward streaks would still form by the interaction of diffusion and differential rotation. The hybrid time scale important for this pattern is given by

$$\hat{t}_{\mathrm{h}} = (\hat{t}_{\mathrm{s}} \hat{t}_{\mathrm{d}})^{1/2} = \left(\frac{2\pi R_\odot^2}{D\Omega_{\mathrm{d}}}\right)^{1/2} \tag{6.14}$$

or $\hat{t}_h \approx 1,300$ days, which is of the same order of magnitude as $\hat{t}_m$. The details of the pattern formation in cases with very different meridional flows have not yet been explored for stars, but they are directly relevant to studies of coronal hole evolution (Section 8.7) and to stellar flux–flux relationships (Section 9.5.2).

The best-fit diffusion coefficient of $D \approx 600 \pm 200\,\text{km}^2/\text{s}$ for the surface diffusion equation, Eq. (6.8), allows a fraction of the flux in the leading polarities of the sources to reach the equator, where it can cancel with opposite flux originating from the other hemisphere. A substantially lower diffusion coefficient would allow the transport of comparable amounts of flux of both polarities to the poles in the streaks. The polar fields would then never grow strong enough to reach the observed values. (Wang *et al.*, 1989, and Sheeley, 1992, show the impact of varying the meridional flow and the diffusion coefficient on the global field patterns.) Note that if one hemisphere were far more active than the other at a given time, then some excess in leading polarity flux from the active hemisphere could diffuse across the equator, there to be caught in the meridional flow toward the opposite pole. That flux would have the proper polarity for that cycle. This is one process by which the two hemispheres may influence each other.

The simulations by Sheeley and colleagues (see Sheeley, 1992) demonstrated several other interesting features. First, the time scale to reach the observed patterns after injecting the bipoles as observed, but starting with a field-free photosphere, turns out to be $\sim 2\,\text{yr}$; apparently, the large-scale patterns of the field survive in the photosphere for no more than that (compare the time scales of the order of months up to half a year derived from flux histograms and flux replacement rates in Sections 6.3.1 and 6.3.2.4, respectively). Second, only the largest 300 out of 2,500 input active regions turn out to contribute substantially to the large-scale polarity patterns, because only the flux in the largest regions has a reasonable chance of surviving long enough to be part of the large-scale pattern (which also explains the insensitivity of the model results to regions emerging on the backside of the Sun: those that are large enough, live long enough to be observed as the regions rotate onto the disk and thus be recognized as sources).

The description of photospheric flux dispersal by a diffusion equation has found substantial support through its successful reproduction of observed global patterns of the field, primarily the poleward streaks. This success is, nevertheless, quite surprising: the derivation of the model bypasses the inherent three-dimensionality of the magnetic and velocity fields: the success of Eq. (6.8) implies that once the field is released from the active-region plage, it appears to be moved about completely passively by drag-coupling to the flows that are observed at the surface, from granulation to meridional circulation. Potential explanations for this include the weakening of subsurface fields, and subsurface reconnection that disconnects the surface field from the deep source regions (Schrijver *et al.*, 1999b, and references therein).

The success of the diffusion model inspires some predictions for stars. The model depends on Joy's rule – the slight preferential tilt of the active-region axes with the leading polarity closer to the equator – to make the polar flux change sign every 11 yr for the Sun. The leading-polarity flux starts out nearer to the equator, where it cancels somewhat more rapidly against the opposite leading polarity of the other hemisphere, thus leaving an excess of following polarity flux to reach the polar regions. If the differential rotation affects the inclination angle of the dipolar axes, the amount of flux reaching the poles changes accordingly. It remains to be seen how important that is to the functioning of the dynamo (Chapters 7 and 15), but classical models of the Babcock–Leighton type predict a sensitive dependence. Note that the relative magnitude of meridional flow to differential rotation is less important for the flux evolution toward the poles after emergence, because the rate of cancellation of opposite polarities by diffusion across sheared boundaries between monopolar regions does not depend on the amount of shearing of these boundaries: because the diffusion equation is linear in flux density, the solution can be seen as a summation of the evolution of point sources, which are by themselves insensitive to the differential rotation as far as the amount of flux that reaches the poles is concerned (the longitude-averaged diffusion equation can also be used to demonstrate this).

6.3.2.3 Random walk and the width of the network

A similar separation of scales as performed for supergranular and global flows above can be attempted on smaller scales, for example, between granulation and super-granulation. Consider the interaction of these two different flow scales, and in particular how granulation maintains a finite width of the supergranular magnetic network in the quiet photosphere. For a crude estimate of the width, assume a one-dimensional geom-etry and allow for only a single polarity. Imagine a flow $v(x)$ that alternates in direction on some length scale ℓ as a crude model for the outflow from neighboring supergranules toward their common downflow. In addition, assume some small-scale diffusive process, which is associated with granular motions, characterized by a diffusion coefficient D. The associated one-dimensional diffusion equation for the flux density φ,

$$\frac{\partial \varphi}{\partial t} = \frac{\partial}{\partial x}\left(D\frac{\partial \varphi}{\partial x}\right) - \frac{\partial v\varphi}{\partial x},$$

(6.15)

has as a stationary solution

$$\varphi(x) = \varphi_0 \exp\left[\int_0^x \frac{v(z)}{D(z)}dz\right],$$

(6.16)

provided that D is independent of φ. For $v(x)$ alternating between $-v$ and $+v$, the flux density profile is a decreasing exponential with a half width $\lambda = D/v$. The granular diffusion constant in the quiet photospheric network is $\sim 60\,\mathrm{km^2/s}$ (Section 5.3). For a supergranular outflow of 100 m/s, as observed near the network boundaries, the FWHM of the network is expected to be $2\lambda \approx 1,200\,\mathrm{km}$. This value is clearly sensitive to the details of the flow, but it explains why the magnetic network appears as a pattern of roughly linear structures around the supergranular upflows, despite the granular random walk. This argument leads to a standard Reynolds-number argument for the relative importance of the granular diffusion on the supergranular scale $\mathcal{R}'_e = \hat{v}\hat{\ell}/D \approx 40$ for $\hat{\ell} \approx 8,000\,\mathrm{km}$ and velocity $\hat{v} \approx 0.3\,\mathrm{km/s}$ (see Table 2.5), that is, the supergranular flow dominates. By the same argument, $\mathcal{R}'_e$ for the mesogranulation is only half that, so that the effects of granular dispersal relative to mesogranulation are expected to be somewhat stronger.

6.3.2.4 The photospheric flux budget

The lack of hysteresis in the time-dependent flux histograms $h_t(|\varphi|)$, the evolu-tion of large-scale patterns in the photospheric magnetic flux, and the rapid flux replace-ment in mixed-polarity quiet regions suggest that the time scale on which the photosphere can "forget" about its previous activity, that is, on which a new and independent realization of the field pattern can be expected, is of the order of a few days on small scales in the quietest photosphere up to ~ 1 yr for the largest-scale patterns. The question how long flux survives in the photosphere is, however, ill posed, and it should be reformulated to address time scales for pattern evolution, which depends on the size scale: diffusive processes smooth out high-frequency patterns much more quickly than low-frequency patterns. This can be seen from a time-scale argument using diffusion equations such as Eq. (6.4): $\hat{t} \propto \hat{\ell}^2 / D$.

How long does flux associated with large active regions survive in the photosphere? Schrijver and Harvey (1994) assume that the photospheric flux budget during the minimum and the maximum of the cycle is determined by an instantaneous balance between emerging and canceling flux. At cycle maximum, the flux emerging in bipolar regions larger than ephemeral regions is roughly $E_{max} \approx 6 \times 10^{21}$ Mx/day, and at minimum $E_{min} \approx 7 \times 10^{20}$ Mx/day. An upper bound for the time scale on which the active-region flux disappears again from the photosphere is then given by the ratio of flux present to the flux emergence rate: $\Phi(t)/E_e(t)$. Estimating the photospheric flux from the absolute value of synoptic magnetograms, they find flux replacement time scales of 4 months at cycle maximum to 10 months for cycle minimum. But much of the flux may never leave the active regions (in a pure diffusion model most of the flux emerging in isolated active regions should actually cancel with itself), which would result in a substantially shorter time scale. The rough estimate of less than a year is consistent with the lack of hysteresis in surface flux distributions discussed in Sections 6.3.1.2 and 6.3.2.2.

Schrijver and Harvey (1994) point out that the balancing of source and sink terms has severe repercussions for what can be said about the amplitude of stellar dynamos. The total flux Φ in the solar (or stellar) photosphere is described by

$$\frac{d\Phi}{dt} = -C(\Phi) + E(t), \tag{6.17}$$

where $C(\Phi)$ is the sink term representing canceling flux, and $E(t)$ describes the flux emergence rate into the photosphere. Because the solar cycle appears to be a series of quasi-stationary states, at any one time $C(\Phi) \approx E(t)$. The amplitude in E for 1975–1986 is 8.4, whereas that in Φ is only ~ 3.5. Hence, $C(\Phi)$ is not simply proportional to $\Phi(t)$. This isn't hard to interpret: in a thoroughly mixed-polarity environment, collision – and therefore cancellation – rates depend on the square of the number density of flux concentrations [compare Eq. (5.2)], whereas at the interface of large-scale patterns this drops to a proportionality with number density. The global cancellation rate therefore depends on the geometry of the surface patterns. These cannot be observed for stars. So if the dynamo strength proves to involve the flux emergence rate, we face a serious problem, because that quantity cannot be estimated in a straightforward way from the amount of flux present in the photosphere, but rather would involve knowledge of sources of new flux.

6.4   Removal of magnetic flux from the photosphere

Whereas the emergence of magnetic flux and its collapse to strong-field flux tubes is conspicuous (Section 4.4), magnetic flux disappears from the solar atmosphere without drum or trumpet. Yet, obviously all the magnetic flux emerging into the photosphere during one sunspot cycle must be removed at the same average rate. Locally, the rates of flux removal differ enormously: the rate per unit area is very much higher in productive nests than in the quiet network.

Because of the low electrical resistivity of the solar plasma, the magnetic field cannot free itself from the plasma and simply escape into space. In order for magnetic flux to be removed from the photosphere, it must either be destroyed by ohmic dissipation, which can only happen at extremely small length scales (Section 4.1.1), or be transported out of sight, that is, retracted below the photosphere. Note that random-walk diffusion disperses magnetic flux over a larger area but it does not by itself remove flux from the atmosphere. In the sections that follow we discuss three modes of flux removal from the photosphere; one also includes the removal of flux from the toroidal system below the convection zone.

6.4.1   *Flux cancellation and the retraction of flux loops*

In Section 5.3 we discussed the phenomenon of *flux cancellation*: magnetic grains of opposite polarity meet and there is a steady loss of flux in both polarities, usually until the smallest flux grain has completely disappeared. Figure 6.15 illustrates possible modes for the removal of magnetic flux, starting with the encounter of two magnetic grains of opposite polarity, but otherwise equal. Converging flows are needed to bring the magnetic grains together, and probably also to start the downdraft that is needed to help pull the (lowest) flux loop downward. If before the cancellation the two

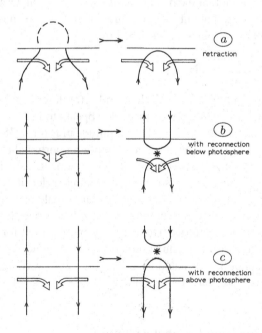

Fig. 6.15. Modes of flux removal from the solar photosphere (horizontal line). Broad open arrows symbolize converging plasma flows, and the asterisks indicate sites of field reconnection. In order to complete the flux removal, in panel *b* the upper flux loop must be lifted out of the photosphere, and in panel *c* the lower loop must submerge (from Zwaan, 1987).

poles were still connected by the original coronal field loop formed at their emergence, the flux removal is just a retraction of the original flux loop (Fig. 6.15*a*). In other cases, field reconnection is a prerequisite prior to retraction. In case of flux reconnection below the photosphere (Fig. 6.15*b*), the removal of the upper ∪-shaped flux loop is hampered by the load of gas trapped in the loop. We do not know of observations of flux cancellation suggesting this mode of flux removal.

The frequent occurrence of coronal bright points (Section 8.6) during or just preceding flux cancellation over encounters of magnetic grains suggests that flux retraction after reconnection of magnetic flux above the photosphere (Fig. 6.15*c*) is a common mode of flux removal from the photosphere. During cancellation processes in the quiet photosphere, the flux-loss rate per pair of magnetic grains is estimated to range from 10^{17} Mx/h to 4×10^{18} Mx/h (Livi *et al.*, 1985).

Bipolar active regions lose much of their flux across their outer boundaries. There flux cancellation occurs wherever magnetic patches from the active region meets flux of opposite polarity in neighboring magnetic network or in an adjacent active region. Complete retraction of flux emerged within the original bipolar region is rare. This is also true for ephemeral regions: in nearly all cases, the poles of opposite polarities keep moving apart until one of the poles is lost in the network (Martin, 1984). Two exceptional cases have been reported by Zirin (1985b) and by Martin *et al.* (1985) for a very small and a small bipolar active region, respectively. In both cases, the regions emerged in the usual fashion with the separation of their poles. Thereupon the poles moved back to their central polarity-inversion zone, as if pulled or pushed together again. Within a few days, virtually all flux canceled, and little if any trace of the active region was left.

Even in cases of large local flux-loss rates, the process involves a sequential cancellation of small magnetic grains. In addition, it appears that the magnetic flux vanishes utterly from the photosphere – whatever scenario of Fig. 6.15 applies, the (lower) flux loops are to be pulled well below the photosphere, in order to prevent a pileup of flux loops below the field overlying the photosphere.

As argued in Section 6.2.1, within a large, complex active region in a nest, magnetic flux disappears at a very high rate, which is approximately equal to the emergence rate. In active complexes, the poles of an original bipolar component within the complex are not seen to gather again, which suggests that most of the flux loss occurs through flux cancellation between patches belonging to different bipolar components of the complex. A clear example has been provided by Rabin *et al.* (1984), who documented the overnight disappearance of 4×10^{20} Mx from a relatively small area (30″, or 20,000 km), complete with a small sunspot and a plagette, at the edge of a polarity inversion within a large complex active region.

Though flux cancellation and the subsequent flux submergence occur very frequently, the mechanism of the flux retraction is not yet quantitatively modeled, but what must happen is the following. The photospheric opposite-polarity legs are brought together by a converging flow. These must be brought extremely closely together before the curvature force, $\propto (\mathbf{B} \cdot \nabla)\mathbf{B}$ (see Section 4.1.2), can exert a sufficient torque on the photospheric tubes to overcome the buoyancy. The torque bends the tubes; the higher-atmospheric field adjusts its configuration and comes down between the footpoints. We note that once the field is embedded in the photospheric downdraft, the aerodynamic drag helps to carry it down. The interior plasma is already cool, and the field is strong, so it moves

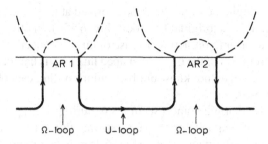

Fig. 6.16. The formation of a ∪ loop between two Ω loops forming the active bipolar regions AR 1 and AR 2 from the same toroidal flux strand. Note that one of those active regions may have decayed and been dispersed beyond recognition.

adiabatically. Consequently, the flux tube gets even cooler relative to the environment while it sinks, so, if the tube does not meet substantial magnetic flux on its way down, it may sink to the bottom of the convection zone.

6.4.2   Flux elimination

Wallenhorst and Howard (1982) and Wallenhorst and Topka (1982) found that magnetic flux may disappear from decaying active regions without any visible transport of magnetic flux. Apparently there are covert ways of flux transport. In this context, Spruit *et al.* (1987) discussed the role of ∪ *loops*, which are formed between two adjacent Ω loops emerged from the same toroidal flux strand (Fig. 6.16). The plasma inside such a ∪ loop is trapped on the field lines: whereas the top of an Ω loop can rise rapidly from the bottom of the convection zone by draining mass along its legs, the ∪ loops cannot leave the Sun until some process releases the plasma from the field lines (Parker, 1984). Spruit *et al.* (1987) argue that the bottom part of the ∪ loop comes into thermal equilibrium with its surroundings, so it is buoyant and starts to rise. During its rise, the loop expands greatly in order to remain in pressure balance and the field strength drops by some orders of magnitude. The field becomes weak and then it is at the mercy of convection. Eventually the field emerges in a sea-serpent fashion, and it may be identified with a weak internetwork field observed in the photosphere (Section 4.6). Phases of the sea-serpent process are sketched in Fig. 6.17. The emerging crests of the weak field run into the overlying chromospheric field, which is rooted in strong flux tubes in the photosphere (panel *b*). Within the contact current sheets reconnection occurs, and so a series of many, small, separate ∪ loops is formed. These loops are supposedly milled down further by convection; eventually the resulting scales are small enough for ohmic decay to occur. As a net result, flux from the toroidal system has been removed from the Sun. At the same time, seemingly unrelated elements of opposite polarity (labeled 5 in Fig. 6.17) disappear because they lost their roots. This sea-serpent emergence of ∪ loops provides an explanation for the observed *in situ* disappearance of magnetic field (see also Simon and Wilson, 1985; Topka *et al.*, 1986).

Rising ∪ loops also suggest an explanation for the mixed-polarity outflow of magnetic features across sunspot moats (Section 5.1.1.3); see Fig. 6.18. The flow in the moat cell grabs part of the rising ∪-loop system and carries it radially outward, with sea serpents and all. In this model, sunspots can decay slowly and gradually by thin flux tubes being

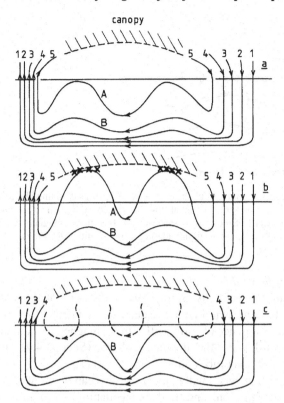

Fig. 6.17. Sketch of the elimination of toroidal flux and flux disappearance *in situ* as a result of the rising of a stretch of a toroidal flux bundle in a ∪ loop formed between two Ω loops that emerged earlier. Panel *a* suggests the sea-serpent deformation of the rising flux bundle A in the upper convective layers. In panel *b*, crosses mark the contact current sheets between expanding emerged crests and the overlying chromospheric field; there the field reconnects. Panel *c* symbolizes the resultant separate flux loops, which are subsequently shredded by convection and then destroyed by ohmic decay. The net result is that the toroidal flux tube A has been destroyed and its two photospheric ends, labeled 5, have been removed (figure adapted from Spruit *et al.*, 1987).

peeled off, from the bottom upward, by the combined action of the rising ∪ loop and the moat cell.

The effects of the rising ∪ loops and the sea-serpent process are best seen near sunspots, because these large flux concentrations contain large numbers of ascending ends of ∪ loops, and moat cells provide conspicuous flow patterns. It is to be expected, however, that these effects also work at less conspicuous sites, causing seemingly *in situ* disappearances of flux grains of opposite polarity located in distant plages or network patches.

In Section 6.3.2 it is found that the diffusion coefficient $D \approx 600\,\text{km}^2/\text{s}$, required to explain the large-scale pattern in the photospheric flux distribution, is larger than the coefficient $D \approx 200\,\text{km}^2/\text{s}$ that follows from measured flux displacements in local network and plages; see Table 6.2. We suggest that the seemingly *in situ* disappearances by flux elimination contribute to the artificially large diffusion coefficient that is required in a diffusion model for the large-scale flux transport over the solar disk.

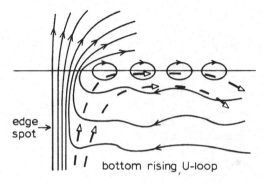

Fig. 6.18. Magnetic features streaming out across a sunspot moat as explained by a rising ∪ loop that has one ascender in the spot. The moat flow, suggested by broken stream lines and open arrow heads, grabs a part of the ∪-loop system, and thus makes magnetic patches of mixed polarities parade across the moat (figure adapted from Spruit *et al.*, 1987).

6.4.3   *Convective blowup of flux tubes*

In his treatment of flux-tube stability, Spruit (1979, 1981c) pointed out that the sense of the initial velocity disturbance within the tube determines whether that flux tube will contract or expand. In the emergence of sufficiently strong magnetic flux, the initial plasma flow is directed downward; hence the flux tubes collapse (Section 4.4). Here we consider conditions that may make flux tubes blow up.

In Section 5.3 we pointed out that very small magnetic grains can be produced by the fragmentation of magnetic concentrations and by the cancellation of flux in magnetic grains of opposite polarity but somewhat different absolute magnetic flux. In the resulting very thin flux tubes, the interior plasma may start to rise because of the temperature rise caused by an increased penetration of radiation from the hot exterior, or by being squeezed between strong granules. Such an initial updraft is expected to be convectively enhanced, so the internal temperature and gas pressure increase, and the tube expands. Hence the tube may become weak beyond repair and be torn to shreds by convection. The result is that the magnetic flux is transferred from the reservoir of the strong field to that of the weak field.

It is not surprising that this convective blowup of flux tubes has not yet been observed because these brief events are expected to occur at very small scales. Thus it is not yet known how important this effect is in the removal of strong-field flux from the photosphere, but it may be expected to play some role in quiet, mixed-polarity network where very small magnetic grains are formed by fragmentation and cancellation.

6.4.4   *Discussion of removal, transfer, and dissipation of field*

For the magnetic activity to proceed in its cyclic variation, as much strong-field magnetic flux must be removed as shows up during one sunspot cycle. This applies not only to the photosphere, but also to the solar interior, at least very nearly so. We can only picture such an ultimate removal of strong-field magnetic flux through a chain of processes transforming the strong field into a weak field, grinding the weak field by convection to smaller and smaller scales, so that it eventually can be destroyed by ohmic dissipation.

Flux cancellation plus retraction removes the strong field from the photosphere. In Chapter 7 we argue that a fraction of this retracted field may provide the seed field needed for the subsequent sunspot cycle.

The rising of the bottoms of ∪ loops is an efficient means to transfer magnetic flux from the strong-field reservoir to the weak-field reservoir, by gradually peeling off the strong-field system consisting of toroidal flux strands, subsurface magnetic arches and flux columns supporting young active regions including sunspots (Section 6.2.2). The lifetimes of long-lived sunspots and nests suggest that the destruction of some major components of the strong-field system takes a few months up to approximately half a year. Flux transfer from the strong-field to the weak-field reservoir suggests that a part of the root-mean-square strength of the weak internetwork field may depend on solar latitude and cycle phase; this is a challenging topic for observational study.

In principle, photospheric flux may also be transferred from the strong-field to the weak-field mode by the convective blowup of flux tubes (Section 6.4.3), but it is not yet known how important this process is. Particularly in the deep convection zone, tubes of strong field are expected to be eroded by the lateral influx of radiation into the flux tubes (Section 4.1.2, Fig. 4.3).

Even if the field is weakened, it still cannot be lifted out of the atmosphere easily because it is loaded with plasma (Parker, 1984). Hence the weak field is expected to be milled down by convection to smaller and smaller scales, until it can be destroyed by ohmic dissipation. Probably the top layers of the convection zone are most important in this destruction, because there the convective scales are smallest and the velocities largest.

7

The solar dynamo

In this book, the term *solar dynamo* refers to the complex of mechanisms that cause the magnetic phenomena in the solar atmosphere. Usually, however, that complex is broken down into three components: (1) the generation of strong, large-scale fields of periodically reversing polarity, (2) the rise of these fields to the photosphere, and (3) the processing in, spreading across, and removal from the photosphere of magnetic flux. Components (2) and (3) are discussed in Chapters 4–6; in this chapter, we concentrate on aspect (1). Even on this limited topic, there is a stream of papers, but, as Rüdiger (1994) remarked, "it is much easier to find an excellent ... review about the solar dynamo ... than a working model of it."

In dynamo theory, the mean, large-scale solar magnetic field is usually taken to be the axially symmetric component of the magnetic field that can be written, without loss of generality, as the sum of a toroidal (i.e., azimuthal) component $\mathbf{B}_\phi \equiv (0, B_\phi, 0)$ and a poloidal component, which is restricted to meridional planes: $\mathbf{B}_p \equiv (B_r, 0, B_{\theta'})$, where θ' is the colatitude. The poloidal component is usually pictured as if a dipole field aligned with the rotation axis were its major component, which is a severe restriction.

All solar-cycle dynamo models rely on the differential rotation $\mathbf{v}_0(r, \theta')$ to pull out the magnetic field into the toroidal direction, as sketched in Fig. 6.10a; about this mechanism there is no controversy.

The main problem in solar-cycle dynamo theory is how to generate the properly cycling poloidal field. Cowling (1934) showed that an axisymmetric magnetic field cannot be maintained by plasma motions (for a proof, see Moffatt, 1978, Section 6.5, or Parker, 1979, Section 18.1). Hence the true magnetic field of a dynamo cannot be exactly axisymmetric.

Parker (1955) indicated a solution to the dynamo problem by pointing out that rising blobs of plasma expand laterally and consequently tend to rotate because of the Coriolis effect: convective motions in a rotating body show *helicity*. Rising plasma blobs rotate clockwise in the solar northern hemisphere and counterclockwise in the southern hemisphere, and so does the magnetic field contained within them. The rising loops of field are twisted, and toroidal field B_ϕ is transformed into poloidal field B_p (and vice versa). The rate of generation of B_p is proportional to B_ϕ; Parker estimated the net effect of many convection cells by the resulting electric field

$$E_\phi = \alpha B_\phi. \tag{7.1}$$

The effect of the generation of a poloidal field by the helicity in convective motions is often called the α-*effect*, after the proportionality constant in Eq. (7.1) (although Parker used the Greek letter Γ).

The cycle dynamo cannot be simulated numerically because of the high magnetic Reynolds number $\mathcal{R}_m$ in the solar convection zone. At such high $\mathcal{R}_m$, the magnetic field is highly intermittent, with length scales down to ~ 1 km (see Section 4.1.1). Since the volume of the convection zone is $\sim 10^{18}$ km^3, full numerical coverage would require 10^{18} grid points, which is far beyond the present computational means. There are interesting attempts, however, to study specific effects in dynamo action by numerical simulations in a much smaller volume (e.g., Nordlund *et al.*, 1992). For the study of the overall dynamo action, however, the problem must be made tractable by drastic simplifications, for instance, by considering the *kinematic dynamo* in which the plasma velocity field $\mathbf{v}(r, \phi, \theta')$ is prescribed and the Lorentz force acting back on the flow is ignored. With this kinematic approximation, information on the amplitude of the dynamo action can only be included with the help of heuristic arguments. Another simplification is to retain only the largest scales by taking averages, so that the small spatiotemporal scales are washed out. It is hoped that the statistical averaging retains the essential physics at small scales adequately, so that the solution of such a severely reduced problem still bears on aspects of the actual dynamo operating in Sun and stars. As illustrations of the variety of treatments of the dynamo problem, we summarize two different approaches in the following sections. For a detailed review of theoretical work on dynamo theory, we refer the reader to Mestel (1999).

7.1   Mean-field dynamo theory

In order to remove the detailed magnetic fine structure to obtain a mathematically tractable large-scale magnetic field, each physical quantity Q is split into a large-scale component Q_0, which remains after averaging, and a fluctuating component Q_1, which is zero on average. We write the following:

$$\mathbf{v} = \mathbf{v}_0 + \mathbf{v}_1, \quad \mathbf{B} = \mathbf{B}_0 + \mathbf{B}_1, \quad \mathbf{E} = \mathbf{E}_0 + \mathbf{E}_1, \quad \mathbf{J} = \mathbf{J}_0 + \mathbf{J}_1. \tag{7.2}$$

An average is indicated by $\langle \cdot \rangle$. The averages $\langle \cdot \rangle$ should satisfy the *Reynolds rules*:

$$\langle Q + S \rangle = \langle Q \rangle + \langle S \rangle, \quad \langle Q \langle S \rangle \rangle = \langle Q \rangle \langle S \rangle, \quad \langle c \rangle = c. \tag{7.3}$$

The averaging algorithm $\langle \cdot \rangle$ should commute with ∇, $\partial/\partial t$, and $\int \ldots dt$.

Here Q and S are arbitrary functions of $\mathbf{r}$ and t; c is a constant. Since the fluctuating components are zero on average, $\langle \mathbf{B} \rangle = \mathbf{B}_0$, which is called the *mean field*, and $\langle \mathbf{v} \rangle = \mathbf{v}_0$, the *mean flow*, which usually is set equal to the differential rotation. In many applications, the average $\langle \cdot \rangle$ is taken over heliocentric longitude ϕ so that $\mathbf{B}_0$ is the axisymmetric component of $\mathbf{B}$; hence according to Cowling's theorem $\mathbf{B}_1$ cannot be axisymmetric. Note that locally the fluctuating component $\mathbf{B}_1$ may be much stronger than the mean field $\mathbf{B}_0$.

Substituting the expressions in Eq. (7.2) in Maxwell's equations, Eqs. (4.2)–(4.6), and performing the averaging $\langle \cdot \rangle$, one finds that the relations involving $\mathbf{B}_0$, $\mathbf{E}_0$, and $\mathbf{J}_0$ keep the familiar shapes of Eqs. (4.2)–(4.5), but the quadratic term $\mathbf{v} \times \mathbf{B}$ in Ohm's law, Eq. (4.6), adds an extra term in the corresponding equation when the fluctuations $\mathbf{v}_1$ and

$\mathbf{B}_1$ are statistically correlated:

$$\mathbf{J}_0 = \sigma \left(\mathbf{E}_0 + \frac{1}{c}\mathbf{v}_0 \times \mathbf{B}_0 + \frac{1}{c}\langle \mathbf{v}_1 \times \mathbf{B}_1 \rangle \right). \tag{7.4}$$

In order to handle the term with $\langle \mathbf{v}_1 \times \mathbf{B}_1 \rangle$, one uses the so-called first-order smoothing approximation to derive

$$\langle \mathbf{v}_1 \times \mathbf{B}_1 \rangle = \alpha \mathbf{B}_0 - \beta \nabla \times \mathbf{B}_0 \tag{7.5}$$

(for details, see Steenbeck and Krause, 1969). Sticking to the kinematic approximation, with $\mathbf{v}_0$ given and independent of $\mathbf{B}$, and assuming isotropic turbulence $\mathbf{v}_1$, one estimates the parameters α and β by

$$\alpha \simeq -\frac{1}{3}\langle \mathbf{v}_1 \cdot \nabla \times \mathbf{v}_1 \rangle \hat{t}_c' \, ; \quad \beta \simeq \frac{1}{3}\langle \mathbf{v}_1^2 \rangle \hat{t}_c', \tag{7.6}$$

where $\hat{t}_c'$ is the *correlation time scale* of the turbulence. The first-order smoothing approximation boils down to the assumption that the correlation time scale $\hat{t}_c'$ is much smaller than the eddy turnover time $\hat{t}_c$:

$$\hat{t}_c' \ll \hat{t}_c \equiv \lambda_c / v_1, \tag{7.7}$$

where λ_c is the correlation length of the turbulence. This assumption is not consistent with the traditional notion of convective turbulence that a convective eddy exists for approximately one turnover time, which is supported by laboratory experiments and which is fundamental in the mixing-length approximation for convection in stellar mantles (Section 2.2). Nevertheless, this smoothing approximation is commonly used.

Replacing Ohm's law, Eq. (4.6), by Eqs. (7.4) and (7.5) in the derivation of the induction equation (4.8), one obtains the *dynamo equation* for the mean field $\mathbf{B}_0$:

$$\frac{\partial \mathbf{B}_0}{\partial t} = \nabla \times [\mathbf{v}_0 \times \mathbf{B}_0 + \alpha \mathbf{B}_0 - (\eta + \beta)\nabla \times \mathbf{B}_0]. \tag{7.8}$$

An equivalent equation was first derived by Parker (1955) using heuristic arguments. Ten years later, a formal derivation was given by Steenbeck, Krause, and Rädler (see Roberts and Stix, 1971). Summaries are found in several textbooks (see Stix, 1989; Krause and Rädler, 1980; or Mestel, 1999); Hoyng's (1992) lecture contains both a broad introduction and a critical discussion of developments.

First we indicate the meaning of the terms in Eq. (7.8). The term $\nabla \times (\mathbf{v}_0 \times \mathbf{B}_0)$ describes the *advection* of mean field by the mean flow, which includes the generation of magnetic field by the velocity shear. The term $\nabla \times (\alpha \mathbf{B}_0)$ is the α *effect*, which is the formation of the poloidal field component from the toroidal field, and vice versa. The factor $\mathbf{v}_1 \cdot \nabla \times \mathbf{v}_1$ in the expression for α in Eq. (7.6) is the *kinematic helicity* caused by a coupling of convection and rotation, which induces a preferred sense of rotation in vertically moving plasma blobs. Because it depends on the Coriolis effect, α is expected to behave as

$$\alpha \propto \Omega_r \cos \theta', \tag{7.9}$$

where Ω_r is the angular rotational velocity and θ' is the colatitude.

The term $-\nabla \times [(\eta + \beta)\nabla \times \mathbf{B}_0]$ includes *turbulent diffusion*, through the *turbulent diffusivity* β, in addition to the ohmic diffusion that is associated with diffusivity η.

Note that this turbulent diffusivity applies exclusively to the mean field $\mathbf{B}_0$. If the diffusivities are constants, the diffusion term simplifies to $(\eta + \beta)\nabla^2\mathbf{B}_0$. In solar applications, $\beta \gg \eta$; hence the (turbulent) diffusion time scale [see Eq. (4.10)] of the mean field, $\hat{t}_{\mathrm{d,t}} \approx R_\odot^2/\beta \approx 10\,\mathrm{yr}$, which is of the order of the period of the sunspot cycle (which is no coincidence), is much smaller than the ohmic diffusion time for the Sun, $\hat{t}_{\mathrm{d}} \approx R_\odot^2/\eta \approx 4 \times 10^9\,\mathrm{yr}$ (both computed for the photosphere). The diffusive action of the small-scale turbulence on the mean field $\mathbf{B}_0$ may be understood as follows: imagine an initially toroidal bundle of field lines. If the kinematic approximation applies, then the turbulent convection tears the field lines apart, which thus become entangled in the turbulent flow. After averaging over longitude, it appears that the mean field has spread over a much larger volume; in other words, the mean field is diffused.

The dynamo equation, Eq. (7.8), has been used to model the global magnetic fields of the Sun, stars, and planets. The type of the dynamo described by the dynamo equation depends on α, β, and the systematic velocity pattern $\mathbf{v}_0(r, \theta')$ (primarily the differential rotation). New poloidal field B_{p} can be generated from the toroidal field B_ϕ only by the α effect. New toroidal field $\mathbf{B}_\phi$, however, can be generated from poloidal field B_{p} in two ways: by differential rotation (Ω) and by the α effect (α). The different effects can have very different relative magnitudes, which leads to qualitatively different dynamos. The following jargon has developed.

1. $\alpha\Omega$ *dynamo*: in the generation of the toroidal field, the α term is much smaller than the Ω term. These dynamos tend to have a periodic behavior; in many studies, the solar dynamo is believed to be of the $\alpha\Omega$ type. The $\alpha\Omega$ dynamo is characterized by the dimensionless *dynamo number*,

$$N_{\mathrm{D}} \equiv N_\alpha\,N_\Omega \equiv \frac{\alpha L}{\beta}\,\frac{\Omega_{\mathrm{d}}L^2}{\beta}, \qquad (7.10)$$

where Ω_{d} is the typical difference in the angular rotation rate across the convection zone and L is the typical length scale characterizing the dynamo. The number N_{D} must exceed a critical value for the model dynamo-to operate. The factor N_α measures the ratio between the strengths of the α effect and the diffusion, and N_Ω the ratio between the effect of the differential rotation and that of the diffusion.

2. α^2 *dynamo*: the Ω term is much smaller than the α term, so that in the generation of both the B_ϕ and B_{p} fields, only the α terms remain. These dynamos often have stationary solutions; the dynamo of the Earth could be an α^2 dynamo.

3. $\alpha^2\Omega$ *dynamo*: in the generation of the toroidal field, the α term and the Ω term are of comparable magnitude.

In classical mean-field models, solutions mimicking properties of the 22-yr solar activity cycle have been derived from Eq. (7.8), provided that the parameters α and β are conveniently adjusted; see Fig. 7.1. The toroidal component is much stronger than the poloidal component. The relation with observed features of the solar cycle is established by assuming that toroidal loops emerge because of their buoyancy as soon as the field strength exceeds some ad hoc critical value. The migrating pattern in the toroidal field in Fig. 7.1 is interpreted as the cause behind the butterfly diagram (Fig. 6.3). For an explanation of this dynamo wave, see Yoshimura (1975), Stix (1976), or Hoyng (1992).

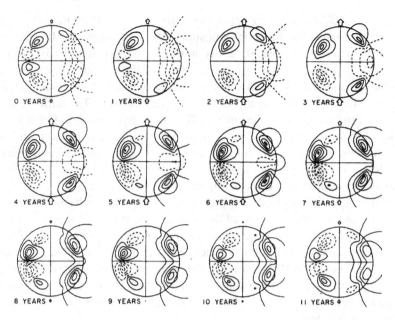

Fig. 7.1. Numerical simulation of a solar $\alpha\Omega$ dynamo (Stix, 1976). Each frame is a meridional cross section; to the left are contours of constant toroidal mean field and to the right are the field lines of the poloidal mean field. Solid curves indicate the toroidal field pointing out of the figure and clockwise poloidal field lines. Arrows indicate the strength and direction of the polar field.

In the kinematic approach, with the local magnetic field passively following the plasma flow, the dynamo equation is linear in the mean magnetic field $\mathbf{B}_0$. Consequently, $\mathbf{B}_0$ is not constrained by Eq. (7.8). In the real dynamo, the effect of the Lorentz force acting back on the velocity field $\mathbf{v}$ is essential in constraining $\mathbf{B}$ and thus $\mathbf{B}_0$; hence the force-balance equation, Eq. (4.1), should be incorporated in the model. The resulting nonlinear dynamo problem has not yet been treated in a consistent manner; instead, several authors have attempted to parameterize some of the nonlinear effects by invoking simplifications. For example, the effect of $\mathbf{B}$ reducing the helicity in $\mathbf{v}_1$ has been mimicked by adopting some ad hoc function for the parameter α that decreases with increasing $\mathbf{B}_0$ (α quenching). Also the Ω effect has been allowed to weaken with increasing $\mathbf{B}_0$. For references to studies dealing with nonlinear effects in the dynamo action, see Schmitt (1993) and Hoyng (1994). Moreover, in some mean-field models, flux losses from the toroidal field caused by buoyancy have been included. Charbonneau and MacGregor (1996) indicated the problem that the α effect is strongly impeded before the mean field has reached energy equipartition with the driving fluid motions. These authors demonstrated, however, that a boundary-layer dynamo (discussed below) can produce equipartition-strength fields, even in the presence of strong α quenching.

Some nonlinear characteristics of dynamos are brought out in recent numerical simulations. For instance, Nordlund *et al.* (1992) have investigated spontaneous dynamo action in turbulent, compressible, hydromagnetic convection above a stable overshoot layer. The model considers a Cartesian box; effects of rotation are included by a Coriolis force corresponding to a rotation of the local coordinate system. If the magnetic Prandtl number $\mathcal{P}_{r,m} \equiv \nu/\eta$ (kinetic viscosity over magnetic diffusivity) exceeds unity, the magnetic energy in the originally weak seed field is found to increase by several orders of magnitude. Then their dynamo saturates as the result of the Lorentz force acting on the fluid flow.

Although the solar differential rotation is most readily observed as a dependence of the angular rotation rate on latitude, initial applications of the mean-field dynamo theory to the convection zone proper (excluding the overshoot layer) usually considered only a radial shear in the angular rotation Ω_r. In order to let the dynamo wave proceed toward the equator, it is assumed that the rate Ω_r increases with depth. From helioseismology it is now known, however, that the rotation rate is virtually constant with depth over the bulk of the convection zone (Section 3.3); hence the Ω effect cannot operate in that way in the convection zone itself.

Whereas in the early mean-field models the dynamo action was supposed to operate on magnetic field throughout the solar convection zone, there is a consensus now that the strong fields responsible for the solar-cycle phenomena are stored below the convection zone. As an example of a drastic modification of the early mean-field dynamo, we consider a dynamo proposed by Parker (1993), which operates at the boundary between two regions (see Charbonneau and MacGregor, 1997, for a model based on that idea). Region A is the convection zone where (radial) velocity shear is zero, but where the α effect operates. Region B is the thin overshoot layer just below the convection zone, where a strong, depth- and latitude-independent velocity shear $\partial\Omega/\partial r$ generates a strong toroidal magnetic field. That strong field suppresses the turbulence, so that $\alpha_B \equiv 0$ in region B, but the turbulent diffusivity is supposed to be much smaller than in region A: $0 < \beta_B \ll \beta_A$. A finite β_B is needed to enable some leakage of the strong, toroidal field from region B into region A, and some downward penetration of the poloidal field from A into B. Parker demonstrated that in a Cartesian geometry the dynamo takes on the character of a surface wave tied to the boundary layer between the regions A and B. He attributed the active regions in the photosphere to flux loops picked up from the strong-field toroidal flux bundles in the overshoot layer.

Much effort is being invested to improve or modify the mean-field dynamo theory, as is indicated by recent review papers; see Hoyng (1992, 1994), the series of papers in the "Solar Dynamo" section of Schüssler and Schmidt (1994), and proceedings of meetings largely or entirely dedicated to the dynamo problem (Tuominen *et al.*, 1991; Krause *et al.*, 1993; Proctor and Gilbert, 1994). One example is the study of symmetry breaking in stellar dynamos by Jennings and Weiss (1991) on the basis of a simple model for a nonlinear $\alpha\Omega$ dynamo (we return to this study in Chapter 15). Ossendrijver and Hoyng (1996) considered the influence of rapid stochastic variations of the dynamo parameters, in an attempt to explain that the sunspot-cycle duration is positively correlated with cycle amplitude.

7.2 Conceptual models of the solar cycle

Whereas the mean-field dynamo models start from assumptions about the magnetic field and the flows in the solar interior, the conceptual model proposed by Babcock (1961) was inspired by the observed time-dependent geometry of the photospheric magnetic field. Babcock's model describes the 22-yr cycle in five stages. During the first stage, approximately three years before the onset of the new sunspot cycle, the global magnetic field to be involved in the new cycle is approximated by a dipole field symmetric about the rotation axis. That poloidal field is identified with the observed field in the polar caps. Low-latitude active regions during that phase are regarded as residuals of the preceding cycle. All the internal poloidal field lines were arbitrarily taken to lie in a relatively thin sheet at a depth of the order of 0.1 $R_\odot$ that extends in latitude from $-55°$ to $+55°$.

During stage 2, the originally poloidal field is pulled into a helical spiral (Fig. 6.10a) in the activity belts, with the result that the field becomes nearly toroidal (Babcock attributes this idea to Cowling, 1953).

During stage 3, bipolar active regions emerge from the nearly toroidal flux ropes. Babcock assumed that the emergence processes start once the magnetic field has reached some specific critical value, mainly because of buoyancy. Because of the $\sin^2 \theta$ term in the Sun's differential rotation, the field intensification proceeds most rapidly around latitudes $|\theta| = 30°$, so the first active regions of the cycle are expected to emerge there. Then the active latitudes gradually shift toward the equator, reaching $|\theta| = 8°$ approximately 8 yr later. In other words, Babcock made his model compatible with Spörer's rule and the Hale–Nicholson polarity rule (cf. Table 6.1).

Stage 4 describes the neutralization and subsequent reversal of the general poloidal field as a result of the systematic inclination of active regions: the following polarity tends to lie at a higher latitude than the leading part (Section 4.5). During the decay of active regions, the following polarity has a slightly larger chance to migrate toward the nearest pole than the leading polarity (Section 6.3.2). In all Babcock-type models it is assumed that wherever magnetic patches of opposite polarities are brought together by flows, equal amounts of opposite flux cancel. More than 99% of the flux emerged in active regions cancels against opposite polarity in adjacent (remnants of) active regions. Less than 1% of the following polarity makes it to the nearest solar pole, first neutralizing the existing polar fields and then replacing them by flux of opposite polarity. The same fraction of leading polarities of the two hemispheres cancels in the equatorial strip.

In stage 5, Babcock assumed that ∼11 yr after the beginning of stage 1 the polar fields correspond to a purely poloidal field opposite to that during phase 1. From here on, the dynamo process has been suggested to continue for the second half of the 22-yr activity cycle, now with all polarities reversed with respect to the previous half of the activity cycle.

In Babcock's model, the statistical effect of the poleward migration of the following polarity toward the nearest solar pole was included simply as an observed fact. Leighton (1969) interpreted the mean flux transport as the combined effect of the dispersal of magnetic elements by a random walk and (like Babcock) the asymmetry in the flux emergence because of the small but systematic tilt of the bipolar regions with respect to the direction of rotation. In an earlier paper, Leighton (1964) had attributed this random walk to erratic motions of the magnetic elements in the magnetic network that are caused by the evolution of the supergranulation, and he modeled the dispersal of the magnetic elements as a diffusion process (see Section 6.3.2). Leighton (1969) included this flux transport in a quantitative, closed kinematic model for the solar cycle, dealing with zonal and radial averages of the magnetic field, which are assumed to vary only with latitude and time.

Leighton's semiempirical model is based on several scaling assumptions; along with parameters deduced from observational data, it contains nine adjustable parameters. For a variety of combinations of parameters, oscillatory modes for the subsurface field are found that reproduce the butterfly diagram (Fig. 6.3) in sunspot emergences.

Recently it became clear that the toroidal flux system cannot reside in the convection zone proper (Sections 4.5 and 5.1.1.2). It is now commonly accepted that a strong toroidal flux is stored in the thin boundary layer below the convection zone. In a series of papers, Durney (1995, 1996a, 1996b, 1997) investigated quantitative models of the Babcock–Leighton type with a realistically thin (thickness $\approx$10,000 km) boundary layer below the convection zone for the generation of the toroidal field. This generation is realized by the prescription of the dependence of differential rotation on depth and on latitude. As in all models of the Babcock–Leighton type, the poloidal field is generated in the top layer of the Sun by the systematic inclination γ (Section 4.5) of the bipole axes of the active regions. A meridional circulation is invoked in an attempt to ensure a proper distribution of magnetic flux across the solar surface, and as the means to transport the poloidal field down to the boundary layer at the bottom of the convection zone. In Durney's studies, observational properties are predicted as depending on various assumptions on input quantities, in particular those concerning the rotational and meridional flows. Although Durney states that in his two-dimensional model there is no mean-field equation relating the poloidal and the toroidal field, he does average over longitude, so that in the generation of the poloidal field, emerging bipolar active regions are replaced by a magnetic ring doublet encircling the Sun. Although necessary to make the dynamo problem tractable, this azimuthal averaging is a serious limitation to Durney's models. The models do not reproduce the butterfly diagram (Fig. 6.3): flux emergences are restricted to latitudes larger than 27°. This defect is caused by the meridional circulation, which is invoked to bring the new poloidal field created in the top of the convection zone down to the bottom of the convection zone. This circulation would dump the poloidal field at high latitudes in the boundary layer.

We conclude that there is not yet a quantitative, closed dynamo model describing the cycle-dependent magnetic topology in the Sun that incorporates all relevant understanding of magnetic structure in the solar interior and of the empirically determined properties of flux transport. A serious limitation of the models by Babcock, Leighton, and Durney is that these do not reproduce a realistic butterfly diagram (Fig. 6.3) that includes the overlaps during which the first active regions of the new sunspot cycle already appear at high latitudes while the last active regions of the old cycle are still emerging near the equator. In an attempt to link the known features of the solar dynamo, Zwaan (1996) composed the scenario summarized in the remainder of this section.

In this scenario, the nearly toroidal flux system (Fig. 6.10a) is also assumed to be generated by differential rotation and stored in the overshoot layer at the bottom of the convection zone. From this toroidal flux system, Ω-shaped loops rise, either to emerge in the atmosphere as (large) bipolar active regions (Figs. 4.8 and 4.10) or to form subsurface arches (Fig. 6.11) whose tops serve as sites from where (also smaller) bipolar regions can emerge.

The erosion of the deep toroidal flux system starts with the succession of emergences of Ω-shaped loops, and it is believed to be completed by the slow emergence of U-shaped loops (Fig. 6.17), as argued in Section 6.4.2. The lateral influx of radiation into slightly cooler flux tubes may also contribute to the weakening of the strong-field system on time scales of many months (Section 4.1.2). Although the initially stiff magnetic root structure is dissolved in the lower part of the convection zone, the tops of flux tubes supporting the patches in the magnetic network remain strong for a longer time for reasons given in Section 5.2. These stiff flux-tube tops are carried about by the meridional circulation and the random walk, as described in Section 6.3.2.

Zwaan (1996) argues that the flux cancellations occurring in the magnetic flux transport across the photosphere are to be interpreted as *field reconnections*, followed by *retraction* of the flux back into the convection zone, as discussed in Section 6.4.1. It is inferred that flux retraction is the common process that returns flux into the Sun, rather than a passive transport caused by the meridional flow, as in Durney's model. In Section 6.4.1 we argue that (some fraction of) the reconnected field tends to move all the way down to the interface below the convection zone, where it arrives being cooler than the ambient plasma.

Closest to the pole, the following polarity of remnants of active regions, which emerged early during the cycle, reconnects with the field in the polar caps, as long as any opposite polarity is left there. Active regions tend to emerge progressively closer to the equator, so that their leading polarity remnants reconnect preferentially with the following polarity of (remnants of) active regions that emerged closer to the equator. Eventually the leading polarities of the northern and southern hemisphere reconnect; this happens close to the equator, if the flux emergences are equally distributed over both hemispheres. The majority of the reconnecting flux loops (see Fig. 6.10b) have the poloidal component of the proper direction for the *next* sunspot cycle. The reconnected flux loops are strong in the photosphere and the top of the convection zone, and they contain relatively cool plasma. Even though the magnetic structure directly underneath is already eroded and rather complex, these sinking poloidal stitches are still linked by tenuous connections deeper in the convection zone. On their way down, the flux loops may meet field lines that are not aligned in the same way, which would lead to field reconnection. Eventually, however, the newly created poloidal component should prevail statistically because of its systematic nature. Upon arrival in the boundary layer, the rotational shear orders and further intensifies the magnetic field into the new, nearly toroidal, helical system.

As in the Babcock, Leighton, and Durney models, the α effect that generates the poloidal field for the next sunspot cycle is introduced in this scenario by the systematic inclination γ of the bipolar active regions with respect to the direction of rotation. Two effects contribute to the systematic inclination γ: (a) the helicity in the toroidal flux system, and (b) the Coriolis effect on emerging flux loops.

Consistent with the butterfly diagram in the flux emergences, the scenario includes a dynamo wave. While the ongoing activity is shifting toward lower latitudes, the new poloidal field component is already generated in its wake. Consequently, the first active regions of the new sunspot cycle can already appear at the high latitudes in the activity belt while the last active regions of the old cycle are still emerging near the equator.

In this scenario, the layers below the bottom and above the top of the convection zone are particularly important in the dynamo action: the overshoot layer harbors the Ω effect and a part of the α effect, whereas the atmosphere plays an additional role in the α effect as the scene for the reconnections and the cooling of the plasma that is retracted with the reconnected field.

7.3 Small-scale magnetic fields

Most dynamo theories deal with the magnetic field on relatively large scales that define the global (outer-)atmospheric properties of the Sun and other cool stars. The convective motions in the envelopes of these stars are likely to generate small-scale fields as well, however. This has been demonstrated by semianalytic studies and by purely numerical experiments. Durney *et al.* (1993), for example, analyze a set of dynamo equations and find that the dynamo process efficiently generates small-scale magnetic fields even in the absence of rotation; this has been confirmed by purely numerical simulations (see Nordlund *et al.*, 1994, and Cattaneo, 1999). This process is frequently referred to as "the turbulent dynamo."

Durney *et al.* (1993) find that the spectrum of the magnetic energy density peaks at a wave number k of about unity, where $k = 1$ is defined to correspond to the eddy dimensions. They argue that the total energy density in the small-scale magnetic field is comparable to the total kinetic energy density in the convective motions on correspondingly small scales. The growth rates for the larger scales increase as helicity increases with faster rotation, but those for the small scales above $k \sim 3$ change little in their simple model.

Most of the magnetic energy is likely to be generated deep in the convective envelope where the kinetic energy density in the convective motions is largest (see Fig. 2.1). The energy in the small-scale field disappears from the small-scale end of the spectrum either through a cascade to larger scales, because of dissipation, or by buoyancy. These three processes limit the increase of the energy density at small scales. The model discussed by Durney *et al.* (1993) does not include the escape of flux from a layer within the envelope as a result of buoyancy. They point out, however, that the balance between this flux leakage and the helicity-dependent growth rate of the large-scale field introduces the possibility of a (weak) dependence of the energy density for the small-scale field on rotation rate.

The convective scales (that are assumed to scale with the pressure scale height) are largest deep in the convective envelope (compare Fig. 2.1). Consequently, what is a small-scale field there would correspond to relatively large scales and to substantial field strengths in the atmosphere, if that field would manage to erupt. The manifestation of this field would depend, however, on its ability to survive the rise through the overlying layers of the convective envelope. Because of pressure equilibrium, rising magnetic field is expected to weaken appreciably (Section 6.4.2). Much of the field probably travels no more than a few pressure scale heights before it is disrupted by diverging flows and eventually dissipated as randomly oriented fields are brought together. The small-scale field in the photosphere is therefore likely to be dominated by fields locally generated at the scale of granular convection, with a rapidly decreasing amount of energy in field on larger scales. In other words, the turbulent dynamo probably is strongly involved in the generation of the internetwork field (Section 4.6), even though there are alternative processes producing weak field, such as emergences of ∪-loops (Section 6.4.2) and the fragmentation and blowup of strong flux tubes (Sections 5.3 and 6.4.2). The various processes that generate internetwork field cannot be discriminated in their results. In fact, the turbulent dynamo is expected to operate immediately on any weak field that becomes available, hence we use the the term turbulent dynamo also to refer to the ensemble of processes leading to the internetwork field.

Magnetic phenomena in the solar atmosphere reveal the presence of two different dynamo aspects. On one hand, the bipolar active regions, which exhibit the characteristic patterns discussed in Chapter 6, find their origin in the solar-cycle dynamo in which the differential rotation in the boundary layer below the convection zone plays an important part. These active regions occupy preferred latitude zones, they exhibit preferred orientations for their dipole axes, and they are part of a distribution of sizes with a smooth size spectrum, ranging from the largest active regions down to and including the ephemeral regions. On the other hand, the turbulent dynamo is probably largely responsible for the internetwork field.

The dichotomy of strong field versus weak field in the photosphere is explained in Section 4.6 as the consequence of the convective collapse during flux emergence that occurs only for concentrations that were already sufficiently strong upon emergence. Because of the convective collapse, strong fields are observed in all active regions, including the ephemeral regions. The tops of Ω loops rising from the toroidal flux system generated by the solar-cycle dynamo emerge as strong fields in the atmosphere. The bulk of the photospheric flux that originates from the turbulent processes associated with the dynamo action on small scales is expected to appear as relatively weak field, although locally some relatively strong bipoles may emerge. In Section 4.6 we point out that a (partial) convective collapse may lead to isolated specks of (moderately) strong field in the internetwork field. Emerging small flux loops that extend over several pressure scale heights may lead to small ephemeral regions. Hence we consider the possibility that in the domain of ephemeral regions there is an overlap of bipoles produced by the solar-cycle dynamo and the turbulent dynamo.

A contribution of the aperiodic turbulent dynamo to the ensemble of ephemeral regions can explain several properties of ephemeral regions (see Section 6.2.2), specifically the low cycle amplitude, the weak E–W preference in the bipole orientations, and the broad distribution in heliographic latitudes.

The suggestion that a part of the small ephemeral regions is produced by the aperiodic turbulent dynamo and a (partial) convective collapse in emergent flux loops is supported by the independence of the minimum strength of the equatorial network of the phase of the solar cycle: the least active areas on scales of a few square degrees show a constant level of activity both in the chromospheric Ca II K intensity (White and Livingston, 1981b, who achieve a measurement accuracy of 0.6% in the 1-Å index) and in moderate-resolution Mt. Wilson magnetograms (LaBonte and Howard, 1980). This network is maintained by flux injection through ephemeral regions that replenish flux that cancels in collisions between concentrations of opposite polarity. Because the replacement time scale is only a few days (Section 5.3), any significant fluctuation in the generation rate of ephemeral regions on time scales of a week or more would result in a significant, persistent change in network strength. Hence we conclude that at least part of the solar surface reflects a constant injection rate of flux in ephemeral regions, which may be attributed to the aperiodic turbulent dynamo.

Note that the chromospheric emission from the quietest equatorial network on the Sun lies significantly above that of the centers of the supergranular cells, and recall that the stellar basal chromospheric emission equals the emission from those cell centers (Section 2.7). Consequently, the energy density in the small-scale magnetic field from the internetwork field up to the ephemeral regions lies above that of a possible turbulent field in extremely slowly rotating stars that might be (in part) responsible for the basal flux. This enhancement of the action of the turbulent dynamo in the Sun over that in the least-active stars could be a consequence of the enhanced helicity in the rotating Sun, which results in a stronger magnetic field even for the turbulent dynamo; we maintain that acoustic processes (Section 2.7) are the dominant cause of the basal emission.

The scales involved in the dynamo action clearly range from the scale size of convective eddies up to the size of the stellar convective envelope. It is often helpful to consider what happens on these scales as separate processes, and in fact to refer to the dominant processes on these vastly different scales as a "turbulent dynamo" and a "cycle dynamo." For the conceptual visualization of the solar-cycle dynamo, such as discussed in Sections 7.1 and 7.2, only the large scales were considered. In a nonrotating star, the turbulent dynamo acting on small scales dominates. This does not mean, of course, that these two "dynamos" operate independently of each other: the small-scale dynamo process probably generates some yet to be determined fraction of the field for the cycle dynamo, while the largest-scale field of the cycle dynamo feeds the turbulent scales through the emergence of ∪ loops, through incomplete cancellation of opposite-polarity fields near the surface, and possibly through subsurface reconnections, to name but several of the processes involved.

7.4   Dynamos in deep convective envelopes

In Chapters 11 and 12 we discuss dynamo action in stars with a large range of fundamental properties. On the coolest side of the Hertzsprung–Russell diagram, stars have very deep convective envelopes, and some are even completely convective. In the latter stars, a boundary-layer dynamo clearly cannot operate. Before we leave the topic of the dynamo, we consider some of the implications of this fundamental change in the internal structure and the change in the dynamo that is likely associated with it.

The action of a dynamo process on small scales – where it is probably only weakly dependent on rotation (Durney *et al.*, 1993) – as well as on large scales – where it depends strongly on (differential) rotation – appears consistent with the decay of activity in evolved stars: as these stars slow down because their moment of inertia increases and the magnetic brake (discussed in Chapter 13) extracts angular momentum, the large-scale dynamo becomes inactive, leaving only a relatively weak turbulent dynamo that itself is no longer strengthened by rotation, but that may play a role in the massive winds of these stars (Section 12.5).

Whereas the effects of decreasing rotation appear to outweigh effects of the deepening envelope in very cool evolved stars, this is not the case in the fully convective main-sequence stars. The disappearance of the boundary layer below the convective envelope while magnetic activity persists led Durney *et al.* (1993) to attribute the magnetic activity in fully convective stars entirely to the turbulent dynamo. It seems doubtful, however, that a turbulent dynamo, with its small-scale magnetic fields, can explain the high level of activity and large-scale atmospheric structure that is observed in the coolest M-type main-sequence stars.

In view of this problem, we speculate that another dynamo mode dominates in rapidly rotating, (nearly) fully convective stars. We propose that magnetic field in the innermost domains of very deep convection zones leaks away (by buoyancy) so slowly that differential rotation can act long enough to order and subsequently strengthen the field appreciably. Such an increase in the time that magnetic field lingers in deep layers is a likely consequence of the high plasma density there: the rate of rise $v_\uparrow$, as estimated from Eq. (4.21), is of the order of the Alfvén velocity v_A [Eq. (4.22)], which becomes very low indeed in a very-high-density environment, as can be estimated for fields of equipartition strength by combining Eqs. (4.11) and (2.29). Given this increased dwelling time, the consequent wrapping up of field could then cause a dynamo to function similarly to the solar boundary-layer dynamo, but within the deep envelope itself. This dynamo mode possibly results in a weaker tendency for the systematic East-West orientation than the solar-type boundary-layer dynamo. This deep-envelope dynamo mode, in all likelihood, becomes gradually more important than the boundary-layer dynamo as a convective envelope deepens.

Before we get too far ahead of ourselves, we now turn to the discussion of magnetic activity in stars other than the Sun. We return to the dynamo problem in Chapter 15, after the discussion of the solar outer atmosphere and of magnetic activity in stars other than the Sun in the next chapters.

8

The solar outer atmosphere

8.1 Topology of the solar outer atmosphere

If there were no magnetic field, the solar outer atmosphere would be featureless except for a stratification in height and time-dependent patterns caused by overshooting convection and gasdynamical waves (Chapter 2). In this chapter we discuss the complex and time-dependent structure of the outer atmosphere as it is controlled by the magnetic field. This first section describes the overall topology in its dependence on its photospheric roots in concentrations and configurations of the strong magnetic field. It provides a general introduction upon which we expand in the following sections.

First we consider the chromospheric and coronal structure issuing from a hypothetical monopolar magnetic network. Recall from Sections 4.2 and 4.3 that the vertical flux tubes constituting the magnetic network occupy only a small fraction of the deep photosphere (small filling factor, f_{ph}) and that these tubes fan outward with height. Hence the filling factor increases with increasing height, until the individual flux tubes merge. The strong flux tubes in a unipolar network thus sustain a cusped vault that overlies the weak-field domain in the photosphere and chromosphere (Figs. 8.1 and 8.2). Spruit (1983) estimates the *merging height* h_m above the photosphere by

$$h_m \simeq -2H_p \ln(f_{ph}), \tag{8.1}$$

where H_p is the pressure scale height and f_{ph} is the photospheric filling factor, $f_{ph} = \langle|\varphi_{ph}|\rangle/|B_{ph}|$, in which $\langle|\varphi_{ph}|\rangle$ represents the mean magnetic flux density in the photosphere and $|B_{ph}|$ is the typical magnetic flux density in a flux tube. Equation (8.1) is based on the monopole approximation of the flux-tube field above the photosphere at a sufficient distance from the tube's axis and on a uniform distribution of the tubes across the photosphere (Spruit, 1981c). This expression shows that the merging height h_m depends only logarithmically on the filling factor: in the quiet solar atmosphere, $h_m \simeq 1,600$ km (i.e., well within the chromosphere), while in active regions, $h_m \simeq 750$ km.

Magnetograms obtained in lines originating sufficiently high above the photosphere show that, away from the center of the disk, plages and network patches exhibit diffuse fringes of reverse polarity on the limbward side and diffuse centerward extensions of the proper polarity. Giovanelli (1980) interpreted these phenomena as being caused by the fanning out of the field with height. The result is the formation of the *magnetic canopy*, which arches over the interiors of the supergranular cells. From an analysis of magnetograms in the Mg I b$_2$ line, Giovanelli concluded that in active regions the bottom of the magnetic canopy is less than 500–600 km above the level where the continuum

Fig. 8.1. The magnetic field issuing from a monopolar area of the photosphere covered with flux tubes of intrinsically strong field, forming a magnetic canopy at chromospheric heights.

optical depth is unity. Although this figure is somewhat lower than the value predicted by Eq. (8.1), both results are in reasonable agreement because both estimates depend on crude model assumptions.

Where the magnetic polarities in the photosphere are mixed, neighboring poles of opposite polarity are connected by field lines. Whatever the precise distribution of the polarities, the photosphere and low chromosphere are completely covered by a magnetic vault consisting of round and cusped arches, supported by flux tubes of strong field rooted in the convection zone (Fig. 8.2). The vault is located within a few thousand kilometers above the photosphere.

The merging level separates the *flux-tube domain* of discrete (bundles of) strong-field flux tubes from the *solar magnetosphere,* where the plasma is everywhere magnetized by the strong fields rooted in the photosphere. The flux-tube domain extends from the

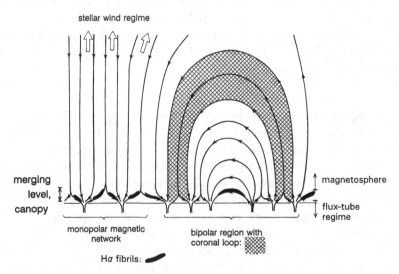

Fig. 8.2. Illustration of concepts of the magnetic structure in stellar atmospheres, showing the field rooted in strong flux bundles in the convective zone. The flux-tube domain of discrete flux bundles, separated by nearly field-free plasma, extends upward through the photosphere and into the low chromosphere. In the upper chromosphere the flux tubes merge, creating the magnetic canopy. The magnetosphere, defined as the domain above the merging level, is pervaded by magnetic field. The stellar wind flows along the magnetic field wherever it is (temporarily) open to interstellar space (figure from Zwaan and Cram, 1989).

photosphere merely to the low chromosphere. The domain of the weak internetwork fields (Section 4.6) is restricted to the space between the flux tubes and beneath the canopy. Hence the magnetosphere comprises the upper chromosphere and all of the corona, and the transition region between chromosphere and corona.

In the pressure balance in the corona, the height-dependence of gravity must be taken into account. In the case of a static equilibrium, the balance equation, Eq. (4.12), with the explicit dependence on gravity, $\mathbf{g}(r)$, reads

$$-\nabla p + \rho \frac{GM_\odot}{r^3}\mathbf{r} + \frac{1}{4\pi}(\nabla \times \mathbf{B}) \times \mathbf{B} = 0. \tag{8.2}$$

Recall that the gas-pressure distribution along a field line obeys simple hydrostatic equilibrium [Eq. (4.15)] because the Lorentz force has no component along the field line. In the photosphere and the chromosphere, where the gravity is almost constant, the gas pressure drops off nearly exponentially with height, with a scale height of only 150 km, which is much faster than the decline with height of the magnetic pressure $B^2/(8\pi)$ within a flux tube. Hence within the tube the plasma β – the ratio of the gas pressure to the magnetic pressure [Eq. (4.13)] – decreases rapidly from approximately unity in the deep photosphere to a very small value in the upper chromosphere and low corona.

Because in most of the chromosphere and in the low corona β is much smaller than unity, the gas pressure plays a minor role in controlling the magnetic structure there. In the nearly horizontal field of the magnetic canopy, relatively large variations in gas pressure and density can therefore be accommodated and supported against gravity by slight adjustments in the (nearly, but not completely, force-free) magnetic field. This property may explain the detailed structure standing out in filtergrams obtained in the H α line core (see Fig. 1.3). It is believed that the dark fibrils visible in H α, which festoon the canopy, are aligned along the magnetic field (Foukal, 1971).

At some distance into the corona, at $r \gtrsim 1.5R_\odot$, the plasma β increases again with height, because from there on the gas pressure drops more slowly with r than the magnetic pressure (the pressure scale height $H_p \approx 400$ Mm for $T = 2 \times 10^6$ K at $r = 2R_\odot$). Thus the plasma β increases again with increasing heliocentric distance, and all terms in Eq. (8.2) are relevant in the description of quasi-static parts of the corona.

Over active regions, large parts of the opposite polarities are connected by closed magnetic loops, forming the coronal condensation (Figs. 1.1*d* and 8.3), well known already from white-light eclipse pictures. There, the gravity and Lorentz force appear to be strong enough to constrain the gas within the loops. Wherever the coronal field is somewhat weaker, as over an aged active region or during some rearrangement of nearby areas of opposite polarities, the lowest coronal loops may still be closed. The higher loops, however, can no longer confine the gas; they burst open and the solar wind (Section 8.9) drags the magnetic field into the interplanetary space.

The part of the solar magnetosphere that is open to interstellar space is called the *wind domain* (Fig. 8.2) – there the field lines are rooted in the photosphere on one side only. In fact, most field lines reaching up to $\sim$2.5 solar radii are open. Locally the wind domain reaches down to the canopy; relatively large corresponding coronal regions are called *coronal holes*. Coronal holes are visible in soft X-rays as relatively dark voids, if viewed in the proper perspective. Large coronal holes are found over the polar caps (as

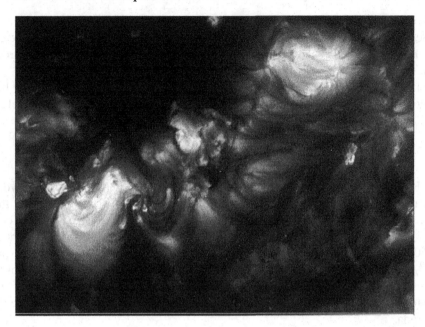

Fig. 8.3. Image taken by the Normal Incidence X-ray Telescope (*NIXT*; see Golub and Herant, 1989), taken on 11 July 1991. The 1.4-Å passband is centered on 63.5 Å, where the main spectral lines are from Fe XVI and Mg X. The area on the diagonal between the active regions NOAO 6,718 (lower left) and NOAO 6,716 (upper right) contains filaments (as seen in H α), plus a newly emerged small active region, NOAO 6,725.

over the solar South Pole in Fig. 1.1*d*) during most of the activity cycle, while the cap is occupied by a unipolar field (Section 6.1.2). Around maximum activity, coronal holes may be found in the activity belts as well. Coronal holes are discussed in Section 8.7.

Observations by satelliteborne telescopes reveal a bewildering complexity of structure, consisting of loops, tubes, and more complicated structures (Figs. 1.1*d*, 1.2*d*, 8.3, and 8.5). The lengths of the loops vary from the angular resolution of ∼1 arc-second of the best telescopes existing to date, up to several hundred thousand kilometers. The width of many of the loops that stand out as relatively isolated structures in active regions do not exceed the instrumental angular resolution by much, so that the resolution required to resolve these structures is not yet reached.

The intricate coronal structure results in part from the localized nonradiative heating, the strongly anisotropic heat transport in the dilute, magnetized plasma in the corona, and from the nearly hydrostatic equilibrium along the field lines (Section 8.6). The heat transport is anisotropic because the heat conduction across the field lines is strongly reduced in comparison with the conduction along the field. The Lorentz force makes charged particles spiral about the field lines with the Larmor radius

$$r_{\rm L} = \frac{m v_\perp c}{|q| B} \equiv \alpha \frac{T_6^{1/2}}{B}, \tag{8.3}$$

where m and q are the mass and the charge of the particle, and $v_\perp$ is its velocity perpendicular to the field of strength B (in gauss). If the perpendicular velocity equals the thermal velocity at temperature T_6 (in megakelvin), the right-hand expression applies with $\alpha = 22$ for electrons, and $\alpha = 950$ for protons. In coronal conditions, the Larmor radii are of the order of centimeters for electrons, and of meters for protons. These radii are many orders of magnitude smaller that the mean-free paths of the particles along the magnetic field, because in the coronal plasma the collision frequencies are much lower than the corresponding gyrofrequencies. The thermal conduction perpendicular to the field is therefore negligibly small. Since the corona is optically thin for almost all radiation, there is virtually no radiative heat exchange between different parts of the corona either. Hence magnetic flux loops are thermally insulated from each other – even neighboring flux loops may differ considerably in temperature.

The fine structure of the corona indicates that the momentary rate of nonthermal heating varies markedly from one magnetic loop to another. Whenever a part of a magnetic loop is heated, thermal conduction by the mobile electrons rapidly smoothes the temperature distribution along the loop. As a consequence, heat flows to the footpoints of the loop, where the temperature rises. Since the pressure and density scale heights increase with increasing temperature, matter from the footpoints ascends. Hence the plasma density in the loop increases, and so does the heat loss from the loop by radiation. If the heating rate is steady, eventually a balance between heating and cooling rates is reached. The heat losses from closed magnetic loops are by radiation and by conduction to the feet, where the denser plasma radiates the energy flow returned by conduction. Solar wind flow provides an additional energy loss from coronal holes. Quantitative models for coronal loops and holes are discussed in Sections 8.6 and 8.7.

It is remarkable that no bright coronal loops end in the dark cores of spot umbrae (Sheeley *et al.*, 1975); some bright loops appear to end in umbral light bridges (Brekke, 1997; Schrijver *et al.*, 1999a). Penumbral fields, in contrast, lie at the base of bright loops.

The high degree of structure in the corona persists a long way out from the photospheric surface (see Fig. 8.11 below): at an enhanced contrast, thin, linear structures extend up to at least a few solar radii above the limb. There are more indications that a similarly small structuring exists even higher: observations of radio Doppler scintillation suggest filamentary structures as small as 5 km to 20 km in the solar wind at 9 solar radii above the surface (Woo and Habbal, 1997).

The coronal magnetic field keeps adjusting to the ever-changing photospheric field with the Alfvén speed v_A. In coronal conditions, the typical Alfvén speed [Eq. (4.22)] ranges from $\sim 1,000$ km/s to 10,000 km/s, hence the coronal magnetic field responds to changes in the photospheric magnetic field within minutes. Movies in soft X-rays obtained with the *SXT* on the *YOHKOH* satellite (Ogawara *et al.*, 1992), with *EIT* (Delaboudiniere *et al.*, 1995) on *SOHO*, and with *TRACE* (Handy *et al.*, 1999) show the frantic activity going on at a great variety of scales in length and time.

During its readjustments, the coronal field tends to keep its large-scale topological structure, that is, the field-line connections within the original bipoles tend to survive. Locally the magnetic configuration is stretched, sheared, and twisted by relative displacements of the poles in the photosphere, and emerging bipolar active regions insert new coronal condensations with magnetic loops initially closed within the region itself. The topology may change when a magnetic loop meets some other loop of a different orientation. At the interface, a current is formed and there reconnection occurs, which links poles of originally separate bipoles. On one hand, most of these reconnections result in little more than altered chromospheric fine structure, minor brightenings, or perhaps small flares. On the other hand, when during the evolution of an activity complex the field is subjected to twisting or shearing, large flares may result (Section 8.3.3).

There is variability in the brightness and shape of features on the time scales of seconds to minutes, particularly in the shortest loops. Long loops over active regions appear to survive for hours up to perhaps days. Many loops brighten and fade gradually. Frequently, shapes of loops change, while the large-scale field slowly evolves. These changes are often so rapid that they suggest substantial changes in the current systems rather than the relatively slow changes in the configuration resulting from large-scale rearrangements of photospheric fields caused by photospheric flows. Entire loop systems are seen to distort, sometimes in association with obvious flares (before as well as after), at other times with much slower intensity changes.

During its adjustments, the coronal magnetic field on observable scales is never far from a force-free configuration [Eq. (4.14)], with

$$(4\pi/c)\,\mathbf{J} = \nabla \times \mathbf{B} = \alpha(\mathbf{r})\,\mathbf{B} ; \qquad (8.4)$$

the parameter $\alpha(\mathbf{r})$ is constant along each field line, which follows from the divergence of Eq. (8.4) leading to $\mathbf{B} \cdot \nabla\alpha = 0$. Different field lines within one coronal configuration may differ in α. The evolution of the coronal field presents a hard problem, even if that field were left alone, without changes being impressed on it from the photosphere underneath. The current picture of coronal field evolution is inspired by theory, with laboratory plasmas in mind. Woltjer (1958) demonstrated that the minimum-energy field within a closed volume for a given helicity is a linear force-free field, that is, a force-free field with a constant α throughout the volume. The question is whether a given system will relax to a minimum energy state. In the case of a small but nonzero resistivity, the Taylor theorem says that, while helicity is approximately conserved, relaxation proceeds toward the minimum energy field, with constant α. The Taylor theorem is confirmed in laboratory plasmas. It cannot be carried over to the entire corona, however, because that is not a closed system. Nevertheless, the Taylor relaxation has been generalized to localized coronal configurations, such as coronal loops. Although the overall helicity is quite nearly maintained, field reconnections on small scales at current–sheet interfaces within the configuration are assumed to redistribute helicity such that a uniform α is approached. It has even been attempted to generalize the Taylor theorem to entire coronal configurations in active regions and activity complexes. A safer approximation is that the coronal field has a piecewise constant-α function; see Berger (1991) for a review on the theory of the coronal magnetic field.

The entire coronal magnetic field never reaches the theoretically lowest possible energy state, that is, with $\alpha = 0$, a current-free (potential) field. The relaxation to a potential field would require a period many orders larger than the sunspot-cycle period, with constant photospheric poles. Instead, the coronal field keeps developing in response to the photospheric magnetic field, with relatively long-lived local structures that appear to be in a quasi-stable state. Within individual configurations, helicity is approximately conserved, with a nearly constant α. Note that locally the outer-atmospheric field departs from a force-free field. For example, in filament-prominence configurations the field supports matter against gravity (Section 8.2). We refer to Low (1996) for a review of the coronal structure and a critical discussion of the role of magnetic helicity therein.

A low-α force-free field resembles a potential field. The first attempt to model the coronal magnetic structure by computing a current-free field fitting the magnetic poles in photospheric longitudinal magnetograms was designed by Schmidt (1964). Altschuler and Newkirk (1969) developed a code for modeling the coronal field above the entire photospheric sphere, in which the effect of the solar wind domain is simulated by forcing the field to become radial at distances larger than $2.5 R_\odot$. The resulting "hairy ball" configurations have been superimposed on eclipse photographs in several cases, and some agreement in the large-scale structure has been found (e.g., Fig. 3.7 in Priest, 1982). Locally, however, there are marked discrepancies between the observed and the computed structures. Good agreement for large-scale active-region fields can be obtained by a force-free magnetic field, provided that a proper α is adjusted. Force-free field extrapolations from photospheric or chromospheric (vector) magnetograms present an ill-posed problem, however (Low and Lou, 1990). For recent three-dimensional computations based on more general MHD assumptions, see Linker *et al.* (1996).

We conclude that the detailed magnetic structure of the corona is not fixed by the momentary photospheric field alone. The preceding evolution of the photospheric field also enters because flux displacement, emergence, and cancellation induce electric currents in the corona, some of which decay very slowly.

Some large-scale coronal configurations that depart strongly from the potential field live for days to weeks with little change. Such configurations are, for example, found in and around prominences (Section 8.2). Another conspicuous feature is the *helmet streamer*, named after its appearance at the solar limb in white light during a solar eclipse (Fig. 8.4). Its magnetic field is rooted in two, usually elongated, photospheric strips of opposite (dominant) polarity. The lower part consists of closed loops forming the *coronal arcade* running over and along the polarity-inversion zone in the photosphere. The top

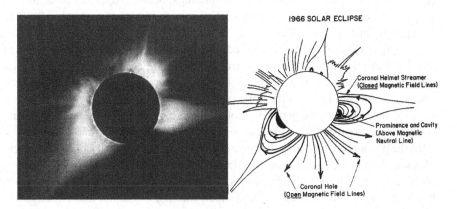

Fig. 8.4. White-light corona during a solar eclipse, showing helmet streamers, with prominences and cavities underneath. The sketch (Hundhausen, 1994) shows some of the inferred magnetic field lines in streamers and coronal holes.

of the helmet tapers into a thin, very elongated streamer. At the base, below the coronal arcade, usually there is low-density *cavity*, in which in most cases a quiescent prominence is found. Despite its magnetic complexity, the configuration of prominence/filament, cavity, coronal arcade, and streamer is a common feature in the solar outer atmosphere. It may last longer than a solar rotation period; it may get (partly) disrupted (see Section 8.3), but often it returns in nearly the same configuration. For a review on coronal structure, including helmet streamers and prominences, we refer to Low (1996). Note that in a seemingly steady outer-atmospheric structure the magnetic footing may change: see the discussions of filaments (Section 8.2) and of coronal holes (Section 8.7).

Soft X-ray images of the Sun show many small emission features that have been called X-ray bright points (Golub *et al.*, 1974; Golub, 1980), but since these features are also visible in extreme-ultraviolet emission lines (Habbal and Withbroe, 1981) we refer to them as *coronal bright points* (the term bright points dates back to the *Skylab* era, when they appeared at the limit of the instrumental resolution, rather than as extended emission patches with looplike substructure on top of small bipoles, as we see them now). The size of the emission areas is less than ~20,000 km; their lifetimes are typically some hours, but some live longer, up to ~48 h. They are widely distributed in latitude, but with a preference for the activity belts. A fraction of the coronal bright points overlay ephemeral active regions, but most of them are found over encounters of magnetic grains of opposite polarity, engaged in flux cancellation (Sections 5.3 and 6.4). The brightness of these points has been interpreted as a result of flux reconnection in the corona preceding flux cancellation in the photosphere (Webb *et al.*, 1993; Harvey *et al.*, 1994). This idea is incorporated in a three-dimensional modeling of coronal bright points by Parnell *et al.* (1994). The comparability of emerging and canceling features, however, suggests that much of the heating is caused by processes other than reconnection of the structure's large-scale field.

In the corona over the quiet photosphere, many changes are induced by the emergence of ephemeral active regions and flux cancellation events (Section 5.3), which cause continual reconnection in the coronal field by adding and removing footpoints. For the rest, the coronal structure in the quiet Sun comprises a multitude of faint connections (either dark or bright relative to their surroundings) between opposite polarities that form, evolve, and disappear as flux concentrations move about, embedded in an unresolved background haze of emission.

The large-scale corona varies markedly during the sunspot cycle because of its response to the photospheric magnetic field. Around maximum activity, it is extremely complex, with many condensations and streamers. Around minimum activity, it is less bright and it seems much simpler: over the polar caps the thin polar plumes are seen, whose patterns inspired the concept of a general dipole field in the Sun. At lower latitudes, the corona shows a bulge consisting of closed loops and an occasional helmet streamer.

In the physics of the chromospheric structure, the high opacity in strong spectral lines plays an important role; hence the simplifying coronal approximation (Section 2.3.1) in the computation of radiative losses is not applicable. Moreover, radiative exchange between neighboring flux tubes makes their thermal insulation less effective than is the case between their counterparts in the corona. Yet the observed intricate chromospheric structure suggests a very uneven heating. Spruit (1981c) pointed out that even a small temperature variation low in the atmosphere, if it extends over several pressure scale heights, can lead to large pressure variations in the chromosphere. Consider two chromospheric columns, labeled 1 and 2, which differ in temperature by a height-independent factor $1 + \epsilon$, with $0 < \epsilon \ll 1$:

$$T_2(h) = (1 + \epsilon) \, T_1(h). \tag{8.5}$$

If hydrostatic equilibrium holds, the gas pressures in the two columns are related by

$$\frac{p_2(h)}{p_1(h)} = e^\epsilon \, N(h), \tag{8.6}$$

where $N(h)$ is the total number of scale heights for column 1 above $h = 0$. Even when ϵ is quite small, the pressure ratio becomes large because the chromosphere spans ~ 10 pressure scale heights. So, although at the photospheric level the temperature, gas pressure and density within a flux tube are lower than they are outside, at several scale heights up into the chromosphere, where the temperature is higher in the tubes than outside, the internal pressure and density may be substantially *higher* than in the external plasma.

Within the patches of the magnetic network or in plages, the fanning tubes merge at a low level in the chromosphere, as discussed in the beginning of this section. The fanning shape of the resulting flux-tube bundles is reflected in the emission features visible in strong resonance lines such as the Ca II H and K lines: whereas in the far line wings the photospheric magnetic patches correspond to aggregates of extremely fine bright points, the emission dots become progressively coarser toward the line cores that are formed in the chromosphere.

Recently, some progress has been made in modeling the photospheric and low-chromospheric magnetic and thermodynamic structure in the flux-tube domain in the solar atmosphere; see Section 8.5. The structure in the upper chromosphere, that is, above the merging level, has not yet been modeled quantitatively; the main observational features and a qualitative interpretation in terms of magnetic topology are summarized in the paragraphs that follow.

Magnetic patches in the network and in plages are seen as clusters of bright "fine mottles" in H α (Fig. 1.3a), both in the very line core and in the far wings (at $|\Delta\lambda| \simeq 2\,\text{Å}$). The bright H α mottles correspond to the bright mottles seen in the Ca II H and K line cores. Most of the dark mottles, seen in the H α line core and inner wings, are elongated; they tend to be directed radially away from (bright) magnetic structure nearby (Fig. 1.3). Near the center of the disk, around an isolated network patch, the dark mottles form a rosette centered on the patch of bright fine mottles, like needles on pin cushions. Along chains of bright mottles in the network, dark mottles are arranged in a double row on both sides (see Fig. 8.2 for a schematic representation). Near the limb, the dark mottles are seen in bushes (or, rather, hedges) along the network. Where the dark mottles appear very elongated, as for instance along the edges of a plage, they are called fibrils. Their typical width is between approximately 1 Mm and 2 Mm, and their typical length is between 10 Mm and 15 Mm; they exhibit internal velocities up to ± 10 km/s. Their individual lifetimes are between 10 and 20 min, but their patterns show very little change over periods of hours. For descriptions, references, and figures, see Gaizauskas (1994), Section 9.1 in Foukal (1990), and Chapters 4 and 7 in Bruzek and Durrant (1977).

Foukal (1971) makes a distinction between fibrils, which are seen as rooted at one end only, and threads, which are long loops, apparently rooted at both ends in opposite polarities. From a study of fibrils across the disk, Foukal determined the shape of a typical fibril: its root is nearly vertical, whereas at a height of a few thousand kilometers it bends over into a nearly horizontal section, which is up to ~15 Mm long.

Particularly long fibrils are found arranged radially in the *superpenumbra* around a mature or decaying sunspot (Fig. 1.3). The individual fibrils are seen from within the outer boundary of the white-light penumbra, but the pattern extends far beyond the white-light penumbra, by about the spot radius. Apparently the superpenumbra corresponds to a part of the fanning magnetic field above the photosphere. In the superpenumbral fibrils, plasma is flowing inward and downward into the spot (inverse Evershed effect; see Section 4.2.1), with velocities of some 20 km/s in H α.

Spicules are rapidly changing, spikelike structures in the chromosphere that are observed as emission features outside the solar limb in chromospheric lines and in absorption in the extreme ultraviolet (Fig. 8.5). There is a range of inclinations up to, say, 40° about the vertical. They appear to shoot up from the low chromosphere, reaching speeds between ~20 and 30 km/s, and a height of about 9,000 km, and then they fall back and/or fade. The total lifetime is between 5 and 10 min. They exhibit substantial rotational velocities about their axes (Beckers, 1977).

The correspondence between spicules and features seen on the disk is a matter of debate; see Bray and Loughhead (1974) and Athay (1976) for reviews. Beckers (1977) and Foukal (1990) associate spicules with dark mottles that happen to be nearly vertical. Gaizauskas (1994) questions this hypothesis on the grounds that this identification cannot be quantitatively demonstrated, and that there is a discrepancy between the dynamical properties of spicules and those of dark mottles. Nearly all workers in the field agree, however, that all the elongated features – spicules, dark mottles, fibrils, and threads – are aligned along the magnetic field.

In deducing information about the magnetic structure from H α features, one should bear in mind that H α, as any spectral line, can show up only where the plasma densities and excitation conditions are appropriate. The observational data suggest that the spicules and all the dark mottles are in the upper chromosphere, in the magnetosphere above the merging level (Fig. 8.2). The good visibility of the low-lying fibrils may be explained by the limited vertical extent of these nearly horizontal features: in the low-β conditions in the chromosphere, relatively large mass densities can be supported against gravity, so that the structures have a substantial optical thickness. The observed flows and limited lifetimes indicate a mildly dynamical MHD force balance in the support of fibrils. Well-developed fibrils are found in active regions and enhanced, unipolar network, where the canopy is a cusped vault.

In static or mildly dynamic cases, the plasma settles to hydrostatic equilibrium; hence nearly vertical flux tubes contain little mass. Viewed near the center of the disk, there is little plasma above the cores of network clusters and of plages, so the observer can look down to the bright magnetic structure in the low chromosphere, without dark mottles obstructing the view. This explains why plages and plagettes are seen as bright in the H α line core; see Fig. 1.3a.

Spicules extend over many scale heights above the low chromosphere. In order to make such flux tubes visible in H α, sufficient matter must be pushed up against gravity.

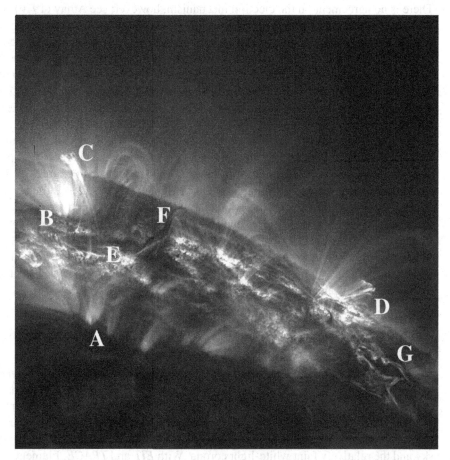

Fig. 8.5. Image taken by the Transition Region and Coronal Explorer, *TRACE*, in the 171-Å passband centered on Fe IX and Fe X lines emitted by plasma at temperatures around 1 MK. The field (400 arcsec to a side) shows two active regions at the northwestern limb of the Sun on 19 August 1998 at 06:04 UT; North is to the right and West to the top. Most of the visible loops are substantially higher than the associated pressure scale height, and as a result the emission is concentrated in the lower segments of the loops (as near A and B), but the fainter top parts of the loops can still be seen above the limb. Fans of relatively cool loops are often seen emanating from the umbral–penumbral interface (B) or umbral light bridges. Some material at 1 MK can be seen high in the corona in rapidly cooling loops (C), particularly in postflare loops (D). The bulk of the coronal plasma over the magnetic plages is at temperatures well above 1 MK, and therefore not visible at 171 Å, but the conduction of thermal energy downward results in a low-lying, dynamic pattern called moss (covering much of the magnetic plages, particularly clearly near E), in which short-lived ejections of material at chromospheric temperatures absorb part of the extreme-ultraviolet radiation. At the limb, spicules are seen in absorption. The cooling material in the tops of the loops near C also absorbs extreme-ultraviolet radiation. Cool material is being thrust up into a long-lived spray (near F) and spills over the loop's top. Hot and cool material move along the spray following different, intertwined paths. In the lower-right corner (G) part of a filament can be seen, temporarily wrapped in a rapidly evolving, fleeting pattern of brightenings.

It is not surprising, then, that spicules appear as ejections from the lower chromosphere. There is no agreement on the ejection mechanism, however; see Athay (1976).

According to Zirin (1974), spicules occur exclusively over the network; he claims that they are absent over plages. Note, however, that Zirin's statement is based on the appearance of dark H α fibrils on the disk, which he identified with bright spicules, but this identification has been questioned. At any rate, the presence of a fine structure containing neutral H and He, and He$^+$ in the upper chromosphere is indicated by multitudes of dynamical dark features seen against coronal emission observed with *TRACE* (see Fig. 8.5 and Section 8.6).

As a conclusion to this section on outer-atmospheric topology, we recommend the reader to scrutinize Fig. 8.5 and its caption. Since such an image in the 171 Å passband does not show the bulk of the coronal plasma (which is at temperatures well above 1 MK), the sparser structure emitting at about 1 MK stands out in all its intricacy. Another advantage is that in the extreme ultraviolet the extinction in the bound-free continua from the ground states of H, He, and He$^+$ makes cool chromospheric structure (at $T \lesssim 10^4$ K) stand out as dark against the 1 MK emission. As a result, Fig. 8.5 brings out that plasmas of widely different temperatures and densities are found closely next to each other at the same altitude in the solar atmosphere. Time series of such images show transient features in the outer atmosphere.

8.2   The filament-prominence configuration

A *prominence* is a bright feature above the solar limb when observed in strong chromospheric spectral lines. A *filament* is a similar feature projected as a dark, usually elongated structure on the disk (see Figs. 1.1*c* and 1.3). The *filament-prominence configuration* contains relatively dense and cool (5,000 $\lesssim T \lesssim$ 8,000 K) plasma, which is partly visible in the light of H α. The correspondence of filaments to prominences appears when the configuration is observed while being carried across the solar limb by the solar rotation: what appears as dark against the disk appears as bright against the sky and the relatively faint white-light corona. With *EIT* and *TRACE*, filaments near the solar limb are seen in the extreme ultraviolet as dark features against the bright coronal background (see the frontispiece and Fig. 8.5).

For a classical observational study we refer to d'Azambuja and d'Azambuja (1948) and to Kiepenheuer's (1953) review, which contains a synopsis of the classical data. Since then there has been a flood of both observational and theoretical papers; we refer to the monograph by Tandberg-Hansen (1995) and to proceedings edited by Priest (1989) and by Webb *et al.* (1998). Chapter 10 in Bruzek and Durrant (1977) provides illustrations and brief descriptions of the phenomena.

There is no consensus on a theoretical model for the filament-prominence configuration. Here we summarize what appear to be their main observational features, and we try to sketch their place in the geometry and the evolution of the solar magnetic structure.

All filaments reside over zones of polarity inversion, between two strips of (dominant) opposite polarity. There is a great variety in sizes and shapes, however, because the appearance of the filament depends on the magnetic environment. Filaments within activity complexes, over the polarity inversions between dense plages of opposite polarities, are very thin, with virtually no transverse structure. The corresponding prominences are low, with heights of less than 10 Mm. The other extreme is formed by the filament-prominence configurations found over polarity inversion zones between weak magnetic network: these are higher ($\approx$50 Mm), with an intricate, dynamic fine structure. Despite the variety in shapes and sizes, it is believed that all these configurations share basic patterns in their magnetic structure and form a continuity in the appearances, which suggests common processes of formation and support. In the following paragraphs, we describe a so-called quiescent filament-prominence configuration, as found in a moderately dense magnetic environment away from active regions – such a system shows all the characteristics best.

The overall shape of a quiescent prominence appears as a bridge, supported by one or more arches. The main body of the configuration is a nearly vertical blade, typically 40 Mm high but no more than 6 Mm thick. Usually the filament axis, i.e., the centerline of the filament projected onto the disk, shows at least some curvature; a typical length is 200 Mm, but it may be smaller or larger by a factor of 4. The pillars of the bridge structure are broader than its body. Whereas its overall shape may vary little over weeks, a prominence shows a dynamic fine structure down to the observational resolution. Beneath and close to its filament, the Hα fine structure exhibits a striation parallel to the polarity inversion line; this phenomenon has been called a *filament channel* (see Fig. 1.3*b*).

On each hemisphere, there are two latitude belts where filaments are most frequently found. One is called the polar crown, which is the polarity-inversion strip between the polar cap and the following polarity in remnants of active regions; this belt shifts poleward during the sunspot cycle. The other preferred belt is $\sim$10° poleward of the sunspot belt at that time; during the sunspot cycle, this filament belt shifts toward the equator, together with the sunspots. Once formed, however, individual filaments show a mean, systematic *poleward* drift that is compatible with the meridional surface flow discussed in Sections 3.2 and 6.3.2 (see Fig. 34 in d'Azambuja and d'Azambuja, 1948).

Filaments exhibit a well-defined differential rotation: Eq. (3.1) describes the data very well in the latitude range up to 68°. Figure 3.1 demonstrates that the magnetic pattern underlying a filament must move with respect to the photospheric magnetic feet, because at higher latitudes the latitude-dependent rotation rate for filaments is found to differ markedly from the rotation rate for small magnetic features. This indicates that filaments are dynamic structures, with their magnetic structures changing such that they move ahead with respect to the underlying photospheric field configuration.

Studies of the Hα structure in and around filaments and measurements of the magnetic field in prominences indicate that in filament channels and in the filaments/prominences proper the strongest field component is horizontal and nearly parallel to the polarity-inversion line and to the filament axis (see Bommier *et al.*, 1994). Note that this direction is nearly orthogonal to that expected for a potential field. In the majority of the cases, even the small field component perpendicular to the filament axis is found to be directed against the direction expected in a potential field.

In many, if not all cases, a filament prominence is overarched by a *coronal arcade*. Such an arcade is rooted in the approximately vertical fields in the opposite-polarity strips on either side of the filament channel. Together with the helmet streamer (Section 8.1) topping the structure, the coronal arcade and the underlying filament-prominence system form one magnetic configuration.

The dense and cool prominence matter, at elevations many times higher than the scale height of a few hundred kilometers corresponding to its temperature, must be supported against gravity and be insulated from the hot surrounding corona by the magnetic field. The dense plasma must hang or, rather, move in magnetic slings; wherever it is visible in $H\alpha$, the plasma must be able to maintain its temperature at $T \approx 5 \times 10^3$ K. Note that only a part of the complex magnetic-field system is visible in $H\alpha$.

Filaments may suddenly disappear, but generally their filament channels remain visible, at least for a while. In about two-thirds of the sudden disappearances (also known as "disparitions brusques"), the filament is seen to reappear within a period ranging from less than an hour to a few days, in many cases in a similar shape. Such a reappearance indicates that the eventual change in the field configuration has been rather slight. In Section 8.3 we return to the phenomena of filament disappearances.

The component of the magnetic field along the filament axis can have two orientations with respect to the dominant polarities along the polarity inversion zone. If the observer, viewing along the filament in the direction of the field, finds the strip of positive polarity on the right-hand side, then the chirality (handedness) of the axial component is called *dextral*. In the opposite case, the chirality is called *sinistral*. The chirality of filaments exhibits hemispheric and latitudinal patterns (see Martin *et al.*, 1994). For a discussion of observational aspects of chirality in the solar atmosphere and a tentative interpretation, see Zirker *et al.* (1997).

Filaments indicate polarity-inversion zones, but some of the long polarity-inversion zones are never occupied by filaments and filament channels. Apparently flux cancellations across the polarity inversion zone are a prerequisite for filament formation (Martin, 1990a), so surface flows converging on polarity-inversion zones favor the formation and presumably also maintenance of filaments and filament channels (see Gaizauskas *et al.*, 1997). In other words, polarity-inversion zones across which appreciable amounts of flux are canceled are likely to be (partly, and part of the time) occupied by filaments, and the other way around. Hence filaments probably indicate zones that are important in the flux removal from the solar atmosphere (Section 6.4.1) and in processes in the solar dynamo mechanism (Section 7.2).

8.3   Transients

In response to the changing photospheric magnetic field, the coronal field generally evolves gradually through a sequence of quasi-stable configurations (Section 8.1). During this process, local instabilities develop, which lead to a great variety of transient phenomena in the solar atmosphere. These transients show large ranges in complexity, in dynamical effects, and in energy release.

8.3.1   Ejecta

Ejecta are transient phenomena that appear to be dominated by ordered, large-scale bulk motion of plasma. For brief descriptions and illustrations of ascending prominences, surges, and sprays, we refer to Chapter 10 in Bruzek and Durrant (1977). More comprehensive discussions and references are found in Section 3.6.2.3 of Tandberg-Hansen (1995).

Section 8.2 alludes to the *sudden disappearance* of filaments. First the filament becomes activated; in other words, the internal motions increase. As a filament it becomes darker and as a prominence it becomes brighter.* Thereupon the filament/prominence may lose part or all of its matter through downflows with speeds of ≈ 100 km/s. Another possibility is that (part of) the structure ascends with increasing speed, up to more than 100 km/s, sometimes more than the escape velocity of 600 km/s. A fraction of the matter in such an ascending (erupting) filament (prominence) disappears high in the corona, while the rest falls back onto the chromosphere. In approximately two-thirds of the cases, filaments reappear in roughly the same shape, usually within an hour to a few days. These features suggest a destabilization of the magnetic structure, possibly by some change in the photospheric magnetic field.

A *surge* is a straight or slightly curved spike that is shot out of a small luminous mound in the chromosphere at velocities of 100–200 km/s. Most of the observations are in Hα, but extreme-ultraviolet observations also show them well. On the solar disk, they appear usually in absorption but, in their initial phase, sometimes in emission. Surges reach heights up to 200 Mm and typically last 10–20 min. The surge matter either fades or returns to the chromosphere, usually along the trajectory of ascent, but sometimes the matter loops over and descends elsewhere. Surges originate in small brightenings, which are Ellerman bombs (Section 8.3.2) or tiny flares, close to spots and pores. Many (small) surges start near penumbral borders and are directed radially away from the spot. From a study of surges in emerging flux regions, Kurokawa and Kawai (1993) find that surges occur from the very first stage of flux emergence. Surges are seen at sites where emerging field collides with preexisting field of opposite polarity, which the authors take as an indication that magnetic reconnection is involved. Small surges tend to recur at a rate of ≈ 1/h. The trajectories of the moving plasma and their collimation indicate that surges are confined to magnetic loops.

Sprays are vigorous ejections of plasma that frequently appear with a substructure of bright clumps. After an initial acceleration of a few kilometers per square second, the material decelerates slightly while following trajectories spreading over a large volume; the maximum velocities, generally $\gtrsim 400$ km/s, often exceed the velocity of escape. A fraction of the material is seen to fall back into the chromosphere; the rest fades. Sprays are associated with flares in their neighborhood; see Tandberg-Hansen (1995) for references. For an example of a very long-lived spray at the interface of two active regions observed in the extreme ultraviolet, see Fig. 8.5.

YOHKOH/SXT has revealed that frequently small-scale ejecta can be observed in soft X-rays; see Shibata (1996) for a review. The *X-ray jets* (coronal jets may be a better term) are bright ejections, most of which are associated with microflares (i.e., transient coronal-loop brightenings, which are discussed in Section 8.3.2) and subflares at their footpoints. The lengths range from 10^4 to 4×10^5 km, and their apparent velocities from 10 to 10^3 km/s; the smallest are the most frequent. The lifetimes range from a few minutes to 10 h; see Shimojo *et al.* (1996) for a summary of the properties of coronal jets.

Many jets are ejected from sites where emerging flux regions interact with preexisting field, or from coronal bright points. From a comparison with NSO/Kitt Peak magnetograms, it appears that more than 70% of the coronal jets are ejected from locations where the polarities are seen to be mixed (Shimojo *et al.*, 1996), which is compatible with the suggestion that coronal jets are related to flux cancellation.

---

* In the EUV, rapidly evolving brightenings envelop the dark core.

The coronal jets and the Hα surges show similarities in shape and in dynamics. Although many Hα surges are not associated with coronal jets, there are cases in which Hα surges are adjacent to coronal jets, with ejections in the same direction. Canfield *et al.* (1996) found indications that reconnection is involved in cases of associated chromospheric surges and coronal jets.

Kundu *et al.* (1995) found a Type III radio burst (Section 8.3.4) occurring simultaneously and cospatially with a coronal jet. This implies the presence of relativistic electrons, which suggests a generation by magnetic reconnection, such as occurs in major flares.

X-ray plasma ejections or plasmoid ejections occur well above the soft X-ray loops in some impulsive flares observed close to the solar limb; see Shibata (1996). These ejections are looplike, bloblike, or jetlike; the velocities range from 50 km/s to 400 km/s, and typical sizes are in the range $(4-10) \times 10^4$ km. The strong acceleration of the ejections occurs nearly simultaneously with the hard X-ray impulsive peaks.

The Hα surges, coronal jets, and plasma ejections are interpreted as caused by magnetic reconnection; see Shibata (1996) for references. From high-resolution observations of surges, Gaizauskas (1996) concludes that the dynamical phenomena are consistent with reconnection theory; recurrent surges appear to be promoted by the cancellation of flux rather than by flux emergence.

With *TRACE*, two types of fairly common ejections have been found: (a) ejections of relatively cool matter obscuring the bright coronal background, which are thrown up as streams of some length and subsequently fall down again, either to the same point, to an opposite end of a loop, or to both ends if the deceleration caused by gravity matches the height of the loop; and (b) plasmoids, small bright blobs forming streaks on images as they move during the exposure, which travel along the coronal field with speeds of several hundred kilometers per second. The first type probably corresponds to surges, and the second one is reminiscent of sprays, except that here matter appears hot; the intricate coexistence of hot and cool material in many dynamic processes allows such ejections to have a wavelength-dependent appearance.

The importance of *coronal mass ejections* (CME), once called coronal transients, was established by means of coronagraphs carried by *Skylab* (Tousey, 1973; MacQueen *et al.*, 1974), but earlier dramatic transient events had been observed in the 5,303-Å coronal line with a Sacramento Peak Observatory coronagraph (DeMastus *et al.*, 1973). Since early 1996 the *SOHO/LASCO* coronagraphs have been providing unprecedented views of the outer corona and CMEs up to 30 solar radii. For reviews and illustrations of CMEs we refer to Wagner (1984), Hundhausen (1993a), and Webb (1998).

A CME appears in white light as an expulsion of mass from the outer atmosphere; most of its mass leaves the Sun. Many CMEs suggest planar loops but they may as well be ellipsoidal bubbles. Others are rather amorphous clouds or radial tongues. The most common type of CME displays a bright loop or bubble, expanding with a speed of several hundred kilometers per second, which is followed by a relatively dark cavity; in many but not all cases, this cavity leads an erupting prominence. Since a CME moves with a supersonic but sub-Alfvénic speed, it is inferred that a slow MHD shock is formed that moves ahead of the bright structure.

Most CMEs are large in extent, much larger than one active region, often involving more than one polarity inversion zone with the overlaying coronal arcade. They arise in closed magnetic structures, most of which are seen as preexisting coronal streamers

(Hundhausen, 1993a). Many CMEs appear as a "blowout" of a streamer, which increased in brightness and in size for days before the eruption. The streamer usually reforms after one day or so.

CMEs can be followed in the heliosphere by remote sensing and by satellites in situ. The transient but strong interplanetary phenomena include shocks, density and temperature changes, and looplike magnetic topologies. CMEs reaching the Earth after some 4 days can cause major geomagnetic storms.

A CME carries a total mass in the range between $\sim 10^{15}$ and 10^{16} g, which is of the order of the estimated mass in a helmet streamer, or roughly 10% of the entire mass of the corona (see Section 10.3 in Golub and Pasachoff, 1997). It involves an energy release of $\sim 10^{31}$–10^{32} erg in lifting the mass against gravity and in producing the kinetic energy of the expelled mass. This energy is comparable to that of the largest flares (Section 8.3.3).

The latitude distribution of the CMEs corresponds to that of coronal streamers and of filaments – which includes equatorial belts broader than the sunspot belts plus the domains of the polar-crown filaments. The rate of CMEs varies in phase with the solar cycle. During low activity, the rate is approximately one CME per day; this rate increases abruptly shortly after sunspot minimum once on each solar hemisphere two dominant zones of filaments are formed (see Webb, 1998).

A fraction of one-third to one-half of the CMEs is not associated with transient phenomena in the lower solar atmosphere. Roughly half of the CMEs are coupled with erupting filaments. Only a small fraction of CMEs is associated with optical flares. Moreover, a study of CMEs associated with flares showed that such CMEs typically begin 20 min *before* the flare. In other words: flares are neither sufficient nor necessary to produce CMEs.

Low (1984) suggests that a CME results from the instability in the coronal magnetic system: once magnetic tension and gravity can no longer contain the magnetized plasma, an MHD outflow develops. Low's solution for the nonlinear, timedependent MHD equations of such an outflow shows features resembling observed CME phenomena.

Since plasma clouds expelled in CMEs and erupting filaments carry tangled magnetic fields, the Sun may shed a large amount of its magnetic helicity through these ejections.

8.3.2 *Small-scale, transient brightenings*

The solar atmosphere shows various small-scale, transient brightenings that hold clues to small-scale MHD processes. In the low atmosphere, the Ellerman bombs are visible in the wings of strong spectral lines. A profusion of small-scale coronal brightenings of short duration became apparent with the advance in spatial and temporal resolution offered by the *YOHKOH/SXT* instrument.

Ellerman bombs appear as bright points, with diameters from $2''$ (1,500 km) down to the resolution limit, on filtergrams taken in the wings of the Hα and Ca II K lines (for illustrations and references, see Section 7.14 in Bruzek and Durrant, 1977, and Section 5.4.2 in Bray and Loughhead, 1974). We encourage the reading of the delightful, brief discovery paper by Ellerman (1917). The Ellerman bombs have been called *moustaches* by Russian observers because of their characteristic appearance in strong spectral lines: emission in the extended wings, out to 5 or 10 Å, and absorption in the line center. Ellerman bombs are visible in the Balmer lines up to H_{10}, and in a number of moderately-strong and strong metal lines. In Hα and Hβ, linear polarization has been measured by

Firstova (1986). The typical time scale of a bomb is 20 min; some may last for a few hours. Usually they brighten within 2 to 3 min, keep their brightness (maybe with fluctuations) during most of their lifetime, and then disappear suddenly within 2 to 3 min.

During part of their lifetime, some bombs are topped by a surge (Section 8.3.1), and conversely, many surges have an Ellerman bomb at their roots.

When the seeing is excellent, numerous bombs are seen in active regions that are still growing or that have just reached maximum development. The pattern of bombs is particularly dense at the edges of young sunspots and pores. The density of bombs decreases rapidly during the decay of active regions – no bombs are found in remnants of decayed regions.

Bombs do not obviously correspond to features in medium-resolution Kitt Peak magnetograms; their occurrence in young active regions suggests that they are related to the complexity of the magnetic field. They are probably sitting on tight, small bipoles; there are indications that bombs coincide with small, isolated magnetic peaks of the polarity that is opposite to that of the surrounding field (Rust, 1968; Roy, 1973). This suggestion agrees with the indication that surges mark sites of flux cancellation (Section 8.3.1).

From the linear polarization of the Hα and Hβ emission, Firstova (1986) inferred that the emission is produced by beams of energetic particles. Ding *et al.* (1998) explored this possibility by computing Hα line profiles resulting from energetic particles bombarding the atmosphere. They find that the characteristic line profiles in Ellerman bombs can be reproduced in two extreme cases: either by injection from the corona of high-energy particles ($\gtrsim$60 keV electrons or $\gtrsim$3 MeV protons), or by injection of less energetic particles ($\sim$20 keV electrons or $\sim$400 keV protons) in a low-lying site located in the middle chromosphere or deeper. The intricate magnetic structure in growing active regions suggests that the particle acceleration is caused by reconnection of magnetic field quite low in the atmosphere. This impression is supported by the frequent occurrence of surges over Ellerman bombs.

Ellerman bombs pose a practical problem in the calibration of Ca II K filtergrams as magnetograms (see Section 8.4): with a $\sim$1-Å passband one cannot tell Ellerman bombs apart from faculae. Consequently, in scatter plots of Ca II K line-core intensity against magnetic flux density (Fig. 8.8 below) the bombs are seen at high Ca II K intensity, but low magnetic flux density.

The solar corona is the scene of numerous *transient loop brightenings*, as observed because of the spatial and temporal resolution of the *YOHKOH/SXT* (Shimizu *et al.*, 1992), *SOHO/EIT*, and *TRACE*. Small loops within some active regions are seen to brighten and fade within a few tens of minutes; because of their peak temperatures ($T \approx 10^{6.7-6.9}$ K) and other properties they are called *microflares*. Their typical energy output, $\sim$10^{27} erg, places them at the low end of the energy spectrum of solar flares (Section 8.3.3), and well above the hypothetical nanoflares that are invoked to explain the coronal heating (Chapter 10).

Microflares are often associated with X-ray jets (Section 8.3.1); Shibata (1996) estimates the fraction to be larger than 60%. The location of microflares suggests that they are caused by magnetic reconnection, as are the X-ray jets. In fact, Shibata (1996) reports numerical simulations indicating that a magnetic reconnection may cause a cool surge, a coronal jet, or a microflare; the product depends on the initial field configuration.

The coronal bright points, which we discussed already in Section 8.1 because they catch the eye in any high-resolution coronal image, are also features to be classified as small-scale transient brightenings. During their lifetimes, these bright features fluctuate in brightness, and some even flare.

8.3.3   Flares

Flares come in a bewildering variety of spatial extent, temporal behavior, and spectral shape. An essential aspect of any flare is that much of its energy output is in electromagnetic radiation; this characteristic sets it apart from phenomena that we called ejecta (Section 8.3.1). A complication is that in many cases flares and ejecta overlap in space and in time.

For a brief summary of features, terms, and illustrations, see Chapter 9 in Bruzek and Durrant (1977). More detailed discussions are found in Zirin (1988) and in Golub and Pasachoff (1997). Švestka (1976) gives a classical monograph on solar flares; both solar and stellar flares are discussed in Haisch *et al.* (1991b) and in proceedings published by Haisch and Rodonò (1989). More recent developments are given in proceeding published by Uchida *et al.* (1996), and they are condensed in Brown *et al.* (1994).

Flares were first observed in 1860 in white light – we now know that only the most intense flares produce observable brightenings in the visible continuum. Solar white-light flares offer a comparison for stellar flares observed with broadband photometry (Section 12.4). For a review of observational data and an interpretation of white-light flares, see Neidig (1989).

The development of the spectroheliograph, and particularly the application of mono-chromatic birefringent filters in movie cameras, boosted observational flare research, mainly in the light of the $H\alpha$ line. Observations in the radio spectrum reveal both thermal and nonthermal components; nonthermal phenomena stand out at meter and decameter wavelengths (Section 8.3.4). Extreme-ultraviolet and X-ray observations indicate coronal temperatures in excess of 10^7 K – these emissions are attributed to bremsstrahlung produced by beams of energetic electrons. Many flares emit gamma rays (see Section 11.6 in Zirin, 1988; Section 9.3.3 in Golub and Pasachoff, 1997). Spectral lines in the gamma-ray spectrum indicate nuclear reactions, which require the acceleration of protons and heavier ions in the flare. A small number of flares, particularly those associated with coronal mass ejections, emit high-energy electrons and nuclei that reach the Earth; see Section 11.5 in Zirin (1988) and Hudson and Ryan (1995) for reviews.

The large range in flare sizes and intensities makes the $H\alpha$ flare-importance classification cover more than an order of magnitude in surface area, and the X-ray classification more than 3 orders of magnitude in X-ray intensity (see Section 9.1.1 in Golub and Pasachoff, 1997). Small flares release energies of $\lesssim 10^{30}$ erg in $\lesssim 10^3$ s; a very energetic solar flare emits $\approx 10^{32}$ erg in $\approx 10^4$ s. Note that the energy carried by a CME is often larger than the energy liberated in any accompanying flare. There is no lower limit to flare energies: the term flarelike brightening is used for transient and very local brightenings in $H\alpha$ plages. In the corona, there are numerous transient loop brightenings (Section 8.3.2). Microflares emitting $\sim 10^{27}$ erg have been detected in hard X-rays and in the extreme ultraviolet; nanoflares liberating $\sim 10^{24}$ erg have been proposed as a substantial source of coronal heating (Chapter 10).

Observed in Hα, most moderately-large and large flares develop as a pair of two bright strands on either side of the main polarity-inversion line of the magnetic field in the active region; these are called *two-ribbon flares*. Often the polarity-inversion line is overlain by a filament before the flare onset; such a filament may disappear early during the flare event. During the flare, the ribbons expand and move apart. The coronal emissions originate somewhere above the polarity inversion line.

Figure 8.6 sketches the development of a large flare in several spectral passbands. The following phases can be distinguished:

The *precursor* covers phenomena that precede the main flare event by some tens of minutes (see Gaizauskas, 1989). There is a variety of such phenomena, such as filament

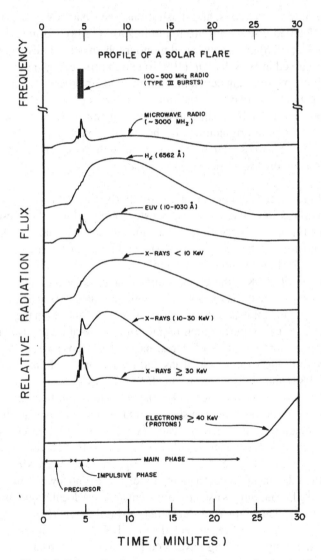

Fig. 8.6. Schematic picture of the development of a large solar flare in different spectral windows (from Golub and Pasachoff, 1997).

activation, often followed by filament eruption, surges, and in some cases the onset of a coronal mass ejection (Section 8.3.1). There are cases of gradual preflare heating, which are visible in microwaves, ultraviolet, and X-rays, as sketched in Fig. 8.6.

During the *impulsive* or *flash phase*, energy is suddenly released. During this phase, a collection of emission spikes is observed in the radio, microwave, extreme-ultraviolet, X-ray, and gamma-ray spectrum. Much of this radiation is nonthermal, which is attributed to particle acceleration in small sites of magnetic field reconnection. If energetic particles produced during the impulsive phase can reach the Earth, they arrive after approximately 20 min, traveling at almost half the speed of light.

A discussion of models for the emissions during the impulsive phase is outside the scope of this book. For detailed descriptions of how and where electrons are accelerated, and for the interpretation of the succession of spikes in various impulsive emissions in terms of times of flight of accelerated electrons we refer to Benz *et al.* (1994) and Aschwanden and Benz (1997); these are but two out of a range of outstanding papers in this field.

The bulk of the flare emission is radiated during the *main phase*; this is largely thermal radiation from plasma heated in the flare, with coronal temperatures of $\gtrsim 10^7$ K. Mass falling back from flare-associated ejecta adds to the chromospheric flare emission.

During the main phase, rising loops become visible in Hα. Initially their tops are very bright in coronal emission lines. After major flares, such loops remain visible for hours as *postflare loops*. Postflare loops are also observed in the extreme ultraviolet, see Fig. 8.5, and in X-rays.

In addition to two-ribbon flares, there are *compact* or single-loop *flares* (Pallavicini *et al.*, 1977). Compact flares have a small coronal volume (10^{26}–10^{27} cm^3; compared to more than 10^{28} cm^3 for two-ribbon flares) and a short duration (tens of minutes). In soft X-rays, compact flares are seen as an intense brightening of one or a few small coronal loops. X-ray ejections are seen from well above the tops of compact flares. The distinction between compact flares and two-ribbon flares is an early attempt to categorize the diversity of flares on the basis of simple shapes. Recognizing that merging and reconnection of field lines hold the key to flare mechanisms, recent schemes try to focus on differences in the physical processes. In an *eruptive* flare, overlying field is first blown apart and then reconnected. Thereby particles are accelerated and mass is ejected along open (or at least very long) field lines. The process then settles down to dissipate the remaining excess energy by gradual thermal processes on newly connected field loops. In a *confined* flare, field lines merge without blowing open the overlying field; the rich menu of dynamic phenomena displayed by eruptive flares is absent or severely restricted.

The large energy output of major flares indicates that the released energy had been stored in the stressed coronal magnetic field, to be suddenly released. The stored energy may have been there straight from the emergence of constituents to the active complex, or be built up by shearing motions of magnetic feet in the photosphere, or both. The instability may be triggered when the evolution causes the field parameters to exceed a threshold; the process might involve canceling flux, or shearing motions. Alternately, a trigger may be provided by a wave issuing from a nearby flare. Such waves have been invoked as a means to explain sympathetic flares, occurring nearly simultaneously in different, sometimes remote, active regions. Indeed, occasionally large-scale fronts are seen to move through the corona like shock waves: Thompson *et al.* (1997) found such a Ramsey–Moreton wave emanating from a flaring active region, traveling at a speed of approximately 300 km/s, spreading across the solar disk radially from the flare site. Many more have been seen since then.

Although the coronal magnetic field must change during a flare, in many cases the overall structure remains nearly the same, including the involved disappearing and reappearing filament. The preservation of the basic magnetic structure explains the occurrence of homologous flares, that is, successive Hα flares at the same position, that show a remarkably similar structure and development; such a succession may last for days. The rapid repetition of homologous flares, sometimes with increases in flare strength, suggests that the magnetic configuration passes through a series of quasi-stable states.

The complexity and strength of the photospheric magnetic field in active regions favors the occurrence of flares. Major flares tend to occur in large magnetic complexes (Section 6.2.1). The most violent events have been observed over magnetic complexes containing so-called δ spots, that is, spots having umbrae of opposite polarity enclosed by one penumbra, which is the most extreme in magnetic complexity; see Section 6.2.1.

8.3.4   Radio bursts

Transient processes in the corona lead to a variety of enhancements and bursts of radio emission that provide basic information on the physical processes involved. We limit our sketch to a few salient features. For more comprehensive introductions we refer to Dulk (1985), to Sections 8.5 and 11.7 in the monograph on solar physics by Zirin (1988), and to Bastian *et al.* (1998).

The various enhancements in coronal radio emission, which occur in young active regions or are associated with ejecta and flares, are pictured in dynamic spectra, as shown in Fig. 8.7. This diagram shows the radio emision enhancements that may be produced by large flares, but in many flares, some types of radio emission may be weak or altogether missing.

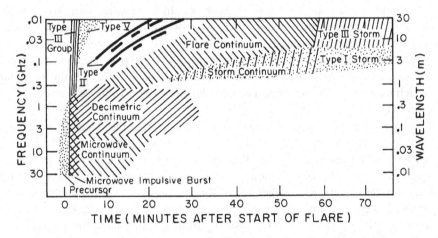

Fig. 8.7. Schematic dynamic spectrum of radio bursts such as those that might be produced by a large flare (from Dulk, 1985). For comparison, the plasma frequency (Eq. 2.61) equals $\nu_p \approx \sqrt{n_{e,10}}$ GHz.

Most of the terminology used to describe radio bursts dates from the early years of radio astronomy (Wild and McCready, 1950):

1. *Type I bursts* are narrow-band (3–5 MHz) bursts that last less than $\sim$1 s; usually they are strongly circularly polarized. Generally they appear in great numbers, superimposed on a variable continuous background. They make up the *Type I noise storms* that last from hours to days. The brightness temperature of a burst may reach 10^{10} K, which indicates that the origin is not thermal.

2. *Type II bursts*, or slow-drift bursts, are observed at $\lambda \gtrsim 1$ m; they drift to lower frequencies with $\approx$1 MHz/s. At a given frequency, a burst lasts for approximately 10 min. Usually Type II bursts are not polarized. Harmonics at twice the frequency are often observed. Brightness temperatures of these intense bursts may reach 10^{13} K, indicating a nonthermal origin.

3. *Type III bursts*, or fast-drift bursts, also observed at $\lambda \gtrsim 1$ m, are short-lived (1–2 s) bursts characterized by a rapid frequency drift of $\sim$20 MHz/s toward lower frequencies.

4. *Type IV bursts* consist of broadband emission, covering the radio spectrum from centimeter to decameter wavelengths. There are large-scale and fine-scale structures in Type IV emission; hence Type IV emission has been split into flare continuum, storm continuum, decimetric continuum, and microwave continuum. At meter wavelengths, there are Type IV bursts with moving sources (see Fig. 11.12 in Fokker, 1977) that last for approximately half an hour; these are partly circularly polarized. Type IV bursts with stationary sources may last for hours. *Microwave impulsive bursts*, at $\lambda \lesssim 10$ cm, are associated with the impulsive phase of flares.

5. A *Type V burst* is a broadband continuum emission at wavelengths ranging from several meters into the decameter domain that follow a Type III bursts for $\sim$1 min.

Type I bursts and noise storms are observed in young active regions, particularly over and around their sunspot groups. It is suggested that these bursts are associated with the restructuring of the coronal magnetic field during and shortly after flux emergence. The mechanism is not yet clear, however. Type I emission following a flare is called a Type I storm continuum; such a continuum may last a few hours.

The drifts toward lower frequencies of Type II and Type III bursts are caused by an outward movement of the source, exciting emission at the local plasma frequency; interferometric imaging confirms these outward motions. For Type II bursts, the outward velocities amount to $\sim$1,000–1,500 km/s; for Type II bursts, velocities of $\sim 1 \times 10^5$ km/s are found.

Type II bursts are associated with coronal mass ejections and possibly with other ejecta as well; the source of such a burst is attributed to the MHD shock preceding the CME.

Type III bursts may occur during the impulsive phase of flares; frequently they come in groups. The excitation of these bursts is attributed to streams of fast electrons created in reconnection sites. Type III bursts occur almost exclusively in active regions, but for many of them no (sub-)flares have been reported as counterparts. Conversely, many X-ray flares occur without Type III bursts.

Type IV emission is associated with flares; for the various components, different emission mechanisms are invoked.

The time profiles of impulsive microwave bursts are remarkably similar to those of impulsive hard X-ray bursts; both emissions are attributed to beams of electrons accelerated in the flare (see Bastian *et al.*, 1998, for a comprehensive review).

8.3.5   The role of outer-atmospheric transients in the development of magnetic activity

We conclude that individual outer-atmospheric transients as such affect the large-scale photospheric magnetic structure only slightly – in many cases, even the resulting changes in the outer-atmospheric structure from before to after the transient seem rather small. Yet transients are important events in the gradual evolution of the magnetic structure during the activity cycle, because they indicate topological changes through reconnections in the magnetic field. Reconnection is a prerequisite for the flux retraction (Section 6.4) that is needed to remove magnetic flux from the atmosphere during the course of the magnetic activity cycle.

According to the site of the inferred magnetic reconnection, the transients indicating reconnection can be tentatively listed as follows:

Ellerman bombs, with their associated surges, may indicate magnetic reconnections deep in the atmosphere, between pairs of opposite polarity at a very small scale (too small to be resolved in present standard magnetograms). We suggest that these reconnections are followed by flux retractions that simplify the small-scale complexity in young active regions. Also Type I radio bursts may be associated with the restructuring of a recently emerged magnetic field.

The majority of *coronal bright points* mark the sites in the low corona where poles of opposite polarity reconnect, before they cancel and the recently formed loops are retracted. In this case, the separation between the poles is large enough to be visible in high-resolution observations; hence the height of the reconnection sites is also larger than in the case of Ellerman bombs.

Flares indicate magnetic reconnections at sites situated still higher in the corona. Major flares point to the reconnection of field lines between different bipoles constituting an activity complex.

It has been suggested that the Sun sheds much of its magnetic helicity through coronal mass ejections. If a disrupted coronal streamer reforms, its reconnected magnetic field is assumed to be of a simpler geometry, with less helicity. In this way, CMEs are suggested to lead to a more simple topology of the magnetic field in the outer atmosphere.

8.4   Radiative and magnetic flux densities

The multitude of associated events at different temperatures in the solar outer atmosphere reveal the coupling of the different domains by the magnetic field. That same link exists in (apparently) quiescent conditions. In this section we discuss the relationships between the emissions from the various domains in the solar outer atmosphere, and that between radiative and magnetic flux densities.

8.4.1   Around the temperature minimum

For the low outer atmosphere, just above the photosphere, projection effects between the location of the photospheric magnetic field and the location where the corresponding outer-atmospheric emission originates are negligible. This allows point-by-point correlative studies, even at a fairly high angular resolution.

Cook and Ewing (1990) studied the relationship between the magnetic flux density $|\varphi|$ and the brightness temperature T_b of solar fine-structure elements observed at 1,600 Å within a 20-Å FWHM bandpass. At those wavelengths the classical Vernazza-Avrett-Loeser models (Section 8.5) place the height of formation of the ultraviolet continuum very close to the temperature minimum region (which, at least in the classical static models, coincides with the formation height of the Ca II K_1 minima). The formation of the continuum at 1,600 Å is near local thermodynamic equilibrium. The 1,600-Å continuum shows a quiet photosphere with a network pattern of supergranular cells with bright patches of emission corresponding to the magnetic elements in the network. Within the cell interiors, there are transient brightenings that are related to acoustic phenomena, discussed in Section 2.8.

Cook and Ewing (1990) analyzed data obtained during a rocket flight of the High-Resolution Telescope and Spectrograph (*HRTS*), which has a resolution of 1 arcsec. They calibrate the observed brightness to an effective brightness temperature T_b, assuming blackbody radiation. The comparison with magnetic flux densities $|\varphi|$ measured simultaneously at Kitt Peak yields the relationship

$$T_b = 6|\varphi| + 4,380 \, (\text{K}), \tag{8.7}$$

which has been derived to be statistically valid for $4,400 \lesssim T_b \lesssim 5,700 \, \text{K}$, or for magnetic flux densities up to $\sim 220 \, \text{G}$.

Foing and Bonnet (1984) showed that network elements stand out as bright specks in the ultraviolet continuum near 1,600 Å. They attribute this relative brightness in part to the Wilson depression (Section 4.3). Flux-tube brightening near the temperature minimum can be related either to unknown local heating or to effects associated with the poorly measurable Wilson depression in a complicated magnetic environment. This ambiguity makes Eq. (8.7) hard to interpret in terms of magnetic activity. This is one of the reasons why the behavior of the continuum emission associated with magnetic activity at the temperature minimum has received relatively little attention.

8.4.2   The Ca II resonance lines as a magnetometer

The qualitative correlation between persistently enhanced chromospheric emission and the presence of photospheric magnetic flux had been known for several decades (e.g., Leighton, 1959; Howard, 1959) when the relationship was first explored quantitatively by Skumanich *et al.* (1975) for a quiet solar region with mean flux densities between 25 and 120 G as observed with a 2.4-arcsec aperture. Schrijver *et al.* (1989a) extended this relationship to higher magnetic flux densities in a study of an active region and its surrounding network (confirmed for another data set by Schrijver *et al.*, 1996a, and for a large data set by Harvey and White, 1999). They excluded spots from their description. Figure 8.8 relates the observed line-core intensity and the magnetic flux density. The scatter in this diagram is surprisingly large, which is in part associated with the method of observation, because of (a) small-scale polarity mixing which leads to flux cancellation within resolution elements, and (b) the short integration time per pixel within a dynamic environment in which waves modulate the intensity. Moreover, the coarse spatial resolution and the single snapshot ignore the intrinsic dependence

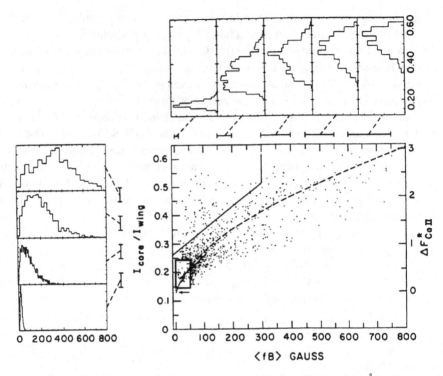

Fig. 8.8. The Ca II K-intensity ratio I_{core}/I_{wing} (I_{core} is the intensity in a 1.04-Å band centered on the K line, and I_{wing} is the intensity in a 2.08-Å band 7.39 Å to the red) plotted against the absolute value of the magnetic flux density $|\varphi| = \langle fB \rangle$ measured with a 2.4-arcsec resolution (figure from Schrijver *et al.*, 1989a). Pores and spots are excluded. Only 10% of the data used to derive Eq. (8.8) is plotted, and an even smaller number in the box in the lower left corner to avoid crowding. Solid line segments outline an area of peculiar points, mostly Ellerman bombs (Section 8.3.2). A short line segment in the lower-left domain of the data shows the range studied by Skumanich *et al.* (1975). The dashed curve represents the power-law fit of Eq. (8.8). Four histograms on the left show the distribution of magnetic flux densities for pixels with a comparable value of I_{core}/I_{wing}. Five histograms on the top show the distribution of I_{core}/I_{wing} within intervals of $|\varphi|$.

of the emission on the detailed magnetic topology (discussed in Section 9.5.1) and the role of the detailed history of the field dynamics that is expected in the case of intermittently occuring reconnection. The most strongly deviating pixels are found in the upper-left corner of Fig. 8.8. Some of the corresponding pixels are located near line-of-sight polarity inversions, where small-scale reconnection, a large inclination of the magnetic flux, or subaperture mixing of opposite polarities may be responsible for the peculiar behavior. Many of the peculiar pixels are likely associated with Ellerman bombs (Section 8.3.2). All of the strongly deviating pixels are located within the magnetic plage.

The intensity ratio I_{core}/I_{wing} shows an empirical minimum value that significantly exceeds zero. This minimum value is reached only over pixels where the magnetic flux density $|\varphi|$ does not differ significantly from zero. This minimum is in part associated with

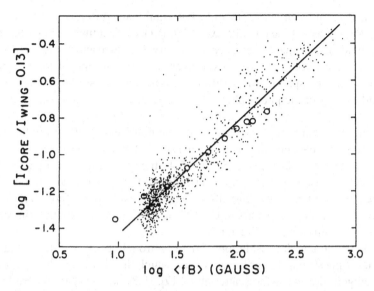

Fig. 8.9. The Ca II K line-core intensity ratio $I_{\text{core}}/I_{\text{wing}}$ (see Fig. 8.8), after subtraction of the minimal line-core intensity ratio, as a function of the magnetic flux density $|\varphi| = \langle fB \rangle$ for an angular resolution of 14.4 arcsec (figure from Schrijver *et al.*, 1989a). The power-law fit of Eq. (8.8) has an exponent of 0.6. The circles show the interval-averaged values from Skumanich *et al.* (1975).

line-wing emission that is transmitted in the bandpass, and in part with the *basal emission* that is probably acoustic in origin (Section 2.7). If these nonmagnetic components are subtracted from the measured radiative intensities, and after the intensity is transformed to a flux density, the relationship between the remaining *excess flux density* $\Delta F_{\text{Ca II}}$ in the Ca II HK lines and the magnetic flux density $|\varphi|$ is a power law:

$$\Delta F_{\text{Ca II}} = 6.4 \times 10^4 \, |\varphi|^{0.6 \pm 0.1} \equiv \gamma'(f \, B_0)^\beta \tag{8.8}$$

(see Fig. 8.9 for the effect of the subtraction of the minimal signal). To obtain the right-hand expression, we have assumed a characteristic intrinsic magnetic field strength B_0 (Section 4.3) for flux tubes that cover a fraction f of the photospheric surface area. Cast into an equivalent form for flux tubes with a characteristic circular cross section πr_0^2 in the photosphere, it reads

$$\Delta F_{\text{Ca II}} = \gamma'(\pi r_0^2 B_0)^\beta n^\beta \equiv \gamma n^\beta, \tag{8.9}$$

where n is the spatial number density of flux tubes. Apparently, the chromospheric emission does not increase in proportion to the number density of flux tubes threading the chromosphere. The weaker-than-linear increase with the magnetic flux density appears to be an effect of the field geometry and of radiative transfer, as discussed in Sections 8.5 and 9.5.1.

Harvey and White (1999) investigate the relationship between Ca II K brightness and magnetic flux density based on a large sample of full-disk observations. They find power-law relationships for flux densities between ~30 G and ~500 G that agree with

Eq. (8.8), but they draw attention to mild but significant differences between mean relationships for subsets of the data that were defined to reflect the local environment of the magnetic field. The primary difference is one between plage and network, which they attribute to the occurrence of emission over polarity inversions, and different heights of formation of the chromospheric line in different environments associated with the divergence of field lines over the photosphere. These differences are yet to be interpreted in detail.

We point out one peculiarity associated with the nonlinearity of the relationship between Ca II H+K and magnetic flux densities in Eq. (8.8): when a plage decays following the phase of a coherent, "adult" plage (Section 5.4), the spreading of the flux can in principle lead to an increase in the emission, depending on the environment into which the flux disperses and in which it eventually cancels against opposite-polarity flux. This would make the plage signal persist in Ca II H+K more prominently than in magnetic flux (or in coronal emission; see Section 8.4.5).

A reduction of the angular resolution of the spectroheliogram and magnetogram by binning adjacent pixels into larger boxes, within which the absolute values of flux densities are averaged, decreases the scatter about Eq. (8.8) markedly (compare Figs. 8.8 and 8.9). The correlation coefficient reaches its highest value when boxes of ≈ 15 arcsec are used. Surprisingly, this reduction of the angular resolution leaves the power-law index unchanged. Although this insensitivity to spatial binning suggests the existence of a characteristic length scale at or above ≈ 15 arcsec, much of the explanation of this scale invariance lies in the random positioning of flux tubes on small scales, discussed in Section 5.3, which makes the mean relationship between radiative and magnetic flux densities independent of angular resolution on scales that do not exceed $\sim 10,000$ km.

For the Sun as a whole, Oranje (1983b) analyzed the emission profile in Ca II K as a function of time through a 3-yr period (1979–1982). He finds that the Ca II K profile can be modeled by a time-invariant baseline profile $C(\lambda)$ on top of which there is an emission profile $A(\lambda)$ that varies only in strength but little, if at all, in shape:

$$I(\lambda, t) = C(\lambda) + a(t) A(\lambda). \tag{8.10}$$

This *plage emission profile*, $A(\lambda)$, which measures the blend of network and plage emissions, can be considered as the mean profile characteristic for all elements emitting chromospheric radiation. It has only a very shallow K_3 depression between the K_2 peaks. The red peak is slightly stronger than the blue peak, contrary to that in regions on the Sun of very low magnetic activity. With this emission profile it would be possible to estimate surface filling factors for activity on Sunlike stars, but this method has never been applied, either to stars very much like the Sun in spectral type, or indeed for others by establishing the emission profiles for these stars as they vary through a cycle.

8.4.3 The C IV doublet

One of the most prominent spectral diagnostics for the transition-region domain (between several tens of thousands of Kelvin up to just below a million Kelvin) is the C IV resonance doublet at 1,548.2 Å and 1,550.8 Å. The formation temperature of the C IV lines is approximately 100,000 K. This line pair and the C II line at 1,335 Å are of particular importance to the study of solar and stellar activity, because both have been observed extensively for both Sun and stars.

The height and spatial extent of the C IV emitting regions (Section 8.8) leads to substantial projection offsets between the photosphere and transition region, so that a point-by-point correlation is not meaningful. Instead of such a comparison, Schrijver (1990) compared distribution functions of brightness to derive a relationship. His method does not prescribe a specific functional dependence and assumes only that there is a statistically valid, monotonic function

$$F_{\rm CIV} \equiv f(|\varphi|) \tag{8.11}$$

between the radiative and magnetic flux densities when atmospheric regions are mapped back onto the photosphere along field lines. Under that assumption, the normalized distribution functions $\Psi(|\varphi|)$ and $\Xi(F_{\rm CIV})$ can be used to derive the relationship: since at a C IV flux density $F_{\rm CIV} = f(|\varphi|)$ the same fraction of points is brighter in C IV than stronger in $|\varphi|$ the following relationship holds:

$$\int_{|\varphi|_1}^{|\varphi|_2} \Psi(x)\,dx = \int_{f(|\varphi|_1)}^{f(|\varphi|_2)} \Xi(x)\,dx. \tag{8.12}$$

Schrijver (1990) applied this method to C IV spectroheliograms observed by the Ultra-Violet Spectro-Polarimeter onboard the Solar Maximum Mission (*SMM*) to establish the relationship between the C IV radiative flux density and the corresponding photospheric magnetic flux density $|\varphi|$. Figure 8.10 shows this relationship as derived from distribution functions for quiet-Sun and for typical active regions:

$$F_{\rm CIV} = 1.6 \times 10^3\,|\varphi|^{0.70} \tag{8.13}$$

(the constant of proportionality has been raised by a factor of 1.6 relative to that given by Schrijver, 1990; see Schrijver, 1995). The presence of a basal flux density of the order of 5×10^3 erg cm^{-2} s^{-1}, as expected from the stellar data listed in Table 2.7, on this relationship would be significant. Schrijver (1990) pointed out that there may be a substantial calibration offset between the C IV fluxes predicted by Eq. (8.13) for the Sun and the C IV fluxes for stars, however. This uncertainty hampers a direct comparison of the solar and stellar data for this line. In view of this and other problems, the solar data presented in Fig. 8.10 should not be used to constrain the stellar C IV basal flux density. The power-law character of the relationship over more than 3 orders of magnitude in average magnetic flux densities may be taken to imply, however, that the C IV basal flux is negligible compared to magnetically related radiative losses, at least in the solar outer atmosphere.

Relationship (8.13) also applies to radiative and magnetic flux densities averaged over active regions as a whole, as can be seen from the data plotted in Fig. 8.10. More surprisingly, it even holds for solar and stellar disk-average data. This peculiar transformation property of disk-averaging appears to hold in general; see Section 9.5.2.

The nonlinear dependence of the transition-region C IV emission on the local magnetic flux density may be intrinsic, or it may be the result of a mixed origin: C IV emission may reflect both local heating and heat conduction down from the corona. This is discussed in more detail in Section 9.5.1.

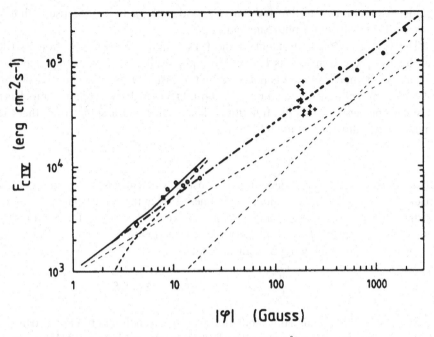

Fig. 8.10. Flux densities F_{CIV} in the C IV λ 1,548+1,551 Å doublet plotted against magnetic flux densities $|\varphi|$. The full-drawn line is the relationship for solar quiet regions, using Eq. (8.13). The thick-dashed curve results after correction for a possible basal flux of 10^3 erg cm^{-2} s^{-1}. The power-law fit to active-region data is shown by a thick, short-dashed line segment between 50 G and 300 G; its dashed-dotted extension above 300 G matches stellar data. Symbols: $\diamond$, mean for solar quiet regions; $\circ$, the Sun as a star (as observed by the Solar Mesosphere Observer and by Kitt Peak); +, average values for entire active regions; •, stellar values. The two thin-dashed lines represent the two terms in Eq. (9.11). Figure from Schrijver (1990).

8.4.4   *Other chromospheric and transition-region activity measures*

The study of magnetic activity in the solar atmosphere is not limited to comparisons of outer atmospheric radiative losses to magnetic flux densities: intercomparisons of radiative losses among themselves also provide useful results. Schrijver (1992), for example, studied spectroheliograms of quiet and active solar regions, observed in spectral lines originating in the upper chromosphere and transition region (using data obtained with the spectroheliometer developed by the Harvard College Observatory for *Skylab*; Reeves *et al.*, 1972). A set of six simultaneously observed wavelengths covers the temperature range from the middle chromosphere (10^4 K) up to the corona (2×10^6 K). Schrijver argues that the slight curvature that is observed in scatter diagrams formed by a pixel-to-pixel comparison is consistent with a model comprising two components: (a) a basal emission, $I_{b,i}$, and (b) a component ΔI_i, referred to as the excess intensity, that depends on the mean flux density of the intrinsically strong magnetic fields in the photosphere. The excess intensity is assumed to be related to other excess intensities

Table 8.1. *Average power-law exponents* b_{ij} *of best fits*
$\Delta I_j = a_{ij} \cdot (\Delta I_i)^{b_{ij}}$ *as in Eq. (8.14)*[a]

Line	C II	C III	O IV	O VI	Mg X
Ly α	1.19 ± 0.04	1.23 ± 0.16	1.26 ± 0.18	1.15 ± 0.11	(1.09 ± 0.07)
C II	1	1.10 ± 0.06	1.07 ± 0.05	1.01 ± 0.05	(0.96 ± 0.11)
C III		1	1.01 ± 0.01	0.95 ± 0.05	(0.84 ± 0.11)
O IV			1	0.95 ± 0.04	(0.80 ± 0.14)
O VI				1	(0.89 ± 0.08)

[a] Data from Schrijver (1992). For temperatures of formation, see Table 2.2.

through power laws:

$$\Delta I_j = a_{ij} (\Delta I_i)^{b_{ij}}. \tag{8.14}$$

The basal fluxes are then estimated as average results from the power-law fits for all pairs of intensities. The subtraction of a fixed basal intensity is, of course, a simplification, because the basal atmosphere shows a range of intensities as waves travel through the atmosphere (see Section 2.8), but as long as the observed intensities are sufficiently larger than the basal flux, subtraction of a single value is statistically allowed, provided that low-lying points are excluded from the analysis.

The power-law indices of the various relationships that came out of this analysis are listed in Table 8.1. The slopes differ only slightly from linearity, but these small differences are significant. The deviation from linearity increases with an increasing difference in the formation temperature of the lines (see Fig. 9.6). The nearly linear character of these relationships reflects a stunning property of the solar atmosphere: at a spatial resolution of several arcseconds, the run of average density as a function of temperature above $\sim 50,000$ K appears to be of the same shape in a variety of network and plage features, differing only by a multiplicative factor. This similarity does not extend to the low chromosphere or hot corona.

8.4.5 Coronal radiative losses

Establishing a quantitative relationship between coronal soft X-ray and magnetic flux densities is greatly hampered by the intrinsically nonlocal response of the coronal brightness to the photospheric boundary conditions. This is particularly true where the magnetic topology is complex or dynamic, such as in quiet Sun, but will be difficult even in relatively simple environments, such as coherent, relatively large bipolar active regions, because of the interaction between neighboring fields and the evolution of internal current systems. In view of these problems, we should look for ensemble-averaged, nonlocal relationships in which we average over entire active regions and substantial areas of the quiet Sun. This procedure is aided by the fairly well-defined perimeter of the active region associated with the plage state (Section 5.4). The coronal volume seen in soft X-rays, traditionally referred to as the coronal condensation, can also be outlined in projection on the disk with fair objectivity (compare Fig. 1.1*d*), although the fainter outlying loops complicate this.

Fisher *et al.* (1998) – confirming an earlier result by Schrijver *et al.* (1985), based on a much smaller sample – report a nearly linear dependence between the integrated X-ray brightness L_X as measured by the *YOHKOH/SXT* and the total absolute magnetic flux Φ, using a sample of 333 active regions of various sizes and in different evolutionary phases; there is no other residual dependence on other parameters derived from the vector magnetograms that can improve the fit that covers 87% of the variance in the data:

$$L_X \approx 10^{26} \left(\frac{\Phi}{10^{22}} \right)^{1.2}. \tag{8.15}$$

This dependence is marginally steeper than the index 1.1 found by Schrijver *et al.* (1985), which probably reflects the fact that the *YOHKOH/SXT* instrument responds to hotter plasma than the Mg X line used in the earlier study.

Deriving a relationship for the quiet Sun is difficult, because coronal structures, or loops, respond to the emergence, movement, and cancellation of photospheric magnetic flux over distances from a few arcseconds to more than several supergranules (e.g., Schrijver *et al.*, 1997b). Loops longer than that are also expected to respond, but these are generally very faint. An early constraint on such a relationship between integrated properties comes from Schrijver *et al.* (1985): the contrast in the soft X-ray flux density between a typical magnetic plage and the typical quiet Sun is a factor of approximately 30, whereas the magnetic flux densities show a contrast of a factor of ≈ 25. This suggests that the large-scale relationship between coronal radiative losses in the range of 1 MK up to ~ 4 MK and the average photospheric magnetic flux densities is roughly linear. This near-linearity is supported by stellar observations, discussed in Section 9.4.

8.5   Chromospheric modeling

Our understanding of the complex and dynamic chromosphere is severely limited by our lack of insight into its heating mechanism and by the complexity of the three-dimensional, non-LTE radiative transfer in the chromospheric spectral lines. The first target for modeling is the simplest lower part, in the flux-tube domain where the structure may be approximated by vertical flux tubes, separated by virtually field-free plasma. Although the individual magnetic flux tubes move about in response to convective flows (Section 5.3), the plasma flows within and around the flux tubes appear to be subsonic, so magnetohydrostatic equilibrium is a plausible approximation.

The first models for the solar chromosphere are embodied in the one-dimensional, static, semiempirical models that list the stratifications of temperature and gas pressure for the photosphere and chromosphere. In order to accommodate various chromospheric conditions, sets of models have been constructed with different chromospheric extensions, each one adapted to special conditions, such as a plage, the average quiet Sun, and so on. The VAL models (Vernazza *et al.*, 1976, 1981) and their revisions are still frequently used.

Solanki *et al.* (1991) constructed models that comprise two components: the magnetic component is an axially symmetric, vertical flux tube, and the space around the flux tube is filled by the nonmagnetic component. The authors present a wine-glass model for a single axially symmetric flux tube with its canopy bounded by a vertical circular cylinder.

Their approximation that the space in and above the canopy would be completely filled with identical, axially symmetric flux tubes is not consistent with the observations, of course, but is used as a first approximation for the typical magnetic configuration in the photosphere and chromosphere. This simple model is adjusted such that it reproduces the profiles observed in the Ca II K line at the center of the solar disk, that is, viewed parallel to the axis of the tube. The main controlling parameters are the magnetic field strength at the axis of symmetry $B(z = 0)$, and the filling factor $f(z = 0)$, both at the height $z = 0$ of unit optical depth in the nonmagnetic photosphere.

Each two-component model is specified by the selection of one (semiempirical VAL) model for the thermal stratification in the flux tube, and one for that in the nonmagnetic atmosphere. Gas-pressure distributions and the geometric height scales follow from the assumption of hydrostatic equilibrium; the Wilson depression, that is, the shift between the height scales, is determined such that the tube is in magnetohydrostatic equilibrium with the surrounding deep photosphere (see Section 4.3). Static equilibrium (except for a possible allowance for turbulent pressure) applies within each component; the thermodynamic quantities are supposed not to vary horizontally in each of the domains. All tubes have the same diameter of 200 km at $z = 0$.

Solanki *et al.* (1991) adopted a fairly sophisticated treatment for the non-LTE formation of the Ca II K line, but in the calculation of the source function for the Ca II K line they used a horizontally infinite one-dimensional atmosphere in which radiative exchange between the two model components was ignored. For particular choices of empirical models and magnetic parameters the computed profiles reproduce the observed profiles fairly well. For one thing, the observed curved relationship between the Ca II K line-core intensity and the magnetic flux density (see Section 8.4) can be qualitatively reproduced.

As admitted by Solanki *et al.* (1991), the assumption of constant temperatures across horizontal areas within flux tubes is not compatible with the observations that the network patches remain well visible in ultraviolet spectral lines formed in the middle and upper chromosphere, up to the transition zone, with a limited increase in the horizontal size with height (Reeves, 1976). Apparently the warmest and most dense parts of the chromosphere are concentrated near the axes of the flux-tube bundles threading network patches. Hence chromospheric heating must be intense near the flux tubes' axes, and reduced over the canopy.

The picture of a chromosphere consisting of a large volume of relatively cool gas and a small fraction of hot plasma is supported by observations in the range of submillimeter, millimeter, and centimeter wavelengths. This spectral region presents interesting diagnostics for chromospheric temperatures because that radiation is free-free emission, which reflects the local electron temperature without non-LTE effects intervening. High-resolution maps of the chromosphere obtained at $\lambda = 3.6$ cm at the Very Large Array (*VLA*) show that the chromospheric network stands out with a temperature of $\sim 16,000$ K, whereas the rest of the chromosphere is barely detectable, at a temperature of $\sim 7,000$ K (Gary *et al.*, 1990).

During the 1991 solar eclipse, Ewell *et al.* (1993) determined that at $\lambda = 870\,\mu$m the solar limb is situated at $3,380 \pm 140$ km above the visible limb; the central brightness temperature is $6,400 \pm 700$ K. This low mean temperature confirms that the substantial volume of the nonmagnetic component and much of the volume immediately above the canopy is cool. In submillimeter waves, the relatively small volume of the heated parts of the flux tubes contributes little to the emission of the mean chromosphere, because the source function, when approximated by the Rayleigh–Jeans law [Eq. (2.39)], depends linearly on temperature.

The extent of several thousand kilometers for the "mean" chromosphere above the limb, which is confirmed in all truly chromospheric diagnostics, cannot be explained by static models, because this extent exceeds the pressure scale height substantially. In fact, as summarized in Section 8.1, the observed structure in the upper chromosphere is transient: mildly dynamic in the nearly horizontal $H\alpha$ fibrils and highly dynamic in the nearly vertical spicules. There are no quantitative models yet for spicules because our understanding of their physics is too incomplete. It has been suggested that spicules are the dynamical response to overheating of the chromosphere by the conductive heat flux from the corona, which is expected to occur in the magnetic network concentrations where much of the coronal volume is magnetically rooted (Kuperus and Athay, 1967; Kopp and Kuperus, 1968).

For the low atmosphere below the canopy, the following picture emerges. The flux tubes define the catacomblike geometry of the canopy. Between the columns of the intrinsically strong field, the field is relatively weak (Fig. 8.1), and acoustic processes dominate. Well below the canonical canopy height, much of the area is covered by cool CO-containing material, exhibiting predominantly 5-min or 3-min oscillations, depending on height. The propagating waves turn into shock waves below halfway between the photosphere and the canopy, starting to dissipate low in the outer atmosphere. Nonradiative energy is being dumped into the essentially nonmagnetic atmosphere in a quasi-continuous manner: although clearly fluctuating, the basal emission never drops (close) to zero at a resolution of a few arcseconds. This may be caused by more than one shock dissipating its energy within the formation height of the line (see Figs. 9–12 in Carlsson and Stein, 1994), while alternatively or perhaps additionally several small-scale wave fronts may be dissipated within resolution elements.

Whenever the photospheric piston produces exactly the right boundary conditions – where three-dimensional interference is probably important – some particularly strong brightenings may occur, leading to $Ca\,II\,K_{2V}$ bright grains (Section 2.7). The simultaneous existence of material emitting in CO lines and a continually emitting basal chromosphere has not been conclusively resolved, but this may point to length scales of only a few arcseconds or less, or – which appears more likely in view of the numerical simulations discussed in Section 2.7 – to relatively brief heating intervals that are not long enough to dissociate (all of) the carbon monoxide. The time scales for the formation and dissociation of CO appear to be of the same order of magnitude as the intervals between successive strong shocks, so that the chemical balance is continuously shifting (Ayres, 1991a). The coherence of the waves decreases with increasing height, being destroyed by partial reflection and scattering by inhomogeneities and by interaction of wave fronts originating from different locations.

The chromospheric topology may explain the extremely low field strength found by Landi Degl'Innocenti (1998) in the Na I D lines (see Section 4.6 for the methods used). The cores of these lines are formed in the low chromosphere. In the warm plasma in the flux tubes, Na is virtually completely ionized, so the Na I D lines are nearly exclusively formed in the cool part of the chromosphere. Presumably the formation region is just below the canopy, yet well above the weak and turbulent photospheric internetwork field of some 10 G, whose typical strength is expected to drop off rapidly with height.

8.6   Solar coronal structure

Characteristic coronal radiative flux densities in the soft X-ray band (typically 6–60 Å, or 0.2–2 keV) in nonflaring, or "quiescent," conditions range from 3×10^3 erg cm^{-2} s^{-1} for coronal holes (Schrijver *et al.*, 1985), through 6×10^4 erg cm^{-2} s^{-1} during cycle minimum and 1.6×10^5 erg cm^{-2} s^{-1} during cycle maximum (Vaiana *et al.*, 1976) averaged over the solar disk, up to 6×10^5 erg cm^{-2} s^{-1} averaged over active regions (Schrijver *et al.*, 1985), thus covering a range of a factor of 200 even at such a low angular resolution. In contrast to this large range in radiative flux densities, the quiescent solar corona shows a relatively narrow range of temperatures, ranging from $\sim$1 MK to 5 MK (see, for example, Rosner *et al.*, 1978; Schrijver *et al.*, 1999a) covering most, if not all, of that range within both quiet and active regions. Observations with *YOHKOH*'s Soft X-ray Telescope and Bent Crystal Spectrometer have shown that relatively rapidly evolving structures associated perhaps with microflares, but not with obvious flares, have plasma up to $\sim$7 MK (Yoshida and Tsuneta, 1996; Sterling *et al.*, 1997). Measuring the temperature of plasma in the open-field environment of coronal holes is difficult, because there is always some mixed polarity within the photospheric boundaries of these regions, which leads to a contamination of the open-field emission by emission from compact loops from deeper down. Near the limb (even at the poles) there is contamination by emission from the brighter surrounding regions in the line of sight. Estimates based on Thompson scattering of white light suggest an average density stratification low in coronal holes compatible with a temperature of $\sim$1.2 MK (Esser and Habbal, 1997).

The radiative loss curve $\mathcal{P}(T)$ for an isothermal plasma at a given density varies by only approximately a factor of 3 over the range from 1 MK to 7 MK (Fig. 2.4). Hence the large contrast in soft X-ray flux densities predominantly reflects differences in plasma density; see Eq. (2.59). The function $\mathcal{P}(T)$ is sensitive to chemical abundances, the ionization balance, and to atomic oscillator strengths (which can differ substantially between codes; compare spectral codes developed by Kato, 1976; Mewe *et al.*, 1985; Brickhouse *et al.*, 1995; and Landini and Monsignori–Fossi, 1991, and references therein). It is often assumed to depend only on temperature, so that $E_{\text{rad}} \propto n_{\text{e}}(T)n_{\text{H}}(T)\mathcal{P}(T)$; compare Eqs. (2.58) and (2.59). The characteristic coronal electron density $\hat{n}_{\text{e}}$ over quiet regions is of the order of 10^8 cm^{-3} near $T \approx 1$ MK (e.g., Laming *et al.*, 1997, measuring off limb). From the characteristic flux densities given above, we conclude that $n_{\text{e}} \sim 2 \times 10^7$ cm^{-3} for coronal holes (see also Koutchmy, 1977), provided that we assume that the same fraction of the volume is emitting as in the quiet corona. Bright loops in active regions have measured electron densities near temperatures of 2–4 MK of $\hat{n}_{\text{e}} \sim 2 \times 10^8$–$4 \times 10^9$ cm^{-3} (e.g., Bray *et al.*, 1991) for high, faint loops to relatively short, bright loops, respectively. The average density within the coronal condensation is substantially lower than that because of the partial volume filling, keeping the overall emissivity of the coronal condensations down.

The general assumption is that stellar coronae are so tenuous that extreme-ultraviolet and soft X-ray photons escape without further interaction with plasma in the stellar outer atmosphere. This assumption is too simple for the low corona (in which spicules and other ejecta as well as dense, low-lying loops at chromospheric temperatures absorb soft X-ray and extreme-ultraviolet emission), and also for regions containing filaments. But the assumption is undoubtedly correct in the high corona for the entire continuum and for the bulk of coronal spectral lines. It has, however, been questioned for a small subset of the strongest lines. Acton (1978) already investigated the problem of resonance scattering of X-ray emission lines in the solar corona. More recently, others have addressed this issue, reaching the conclusion that some of the strongest lines (such as the 15.01-Å Fe XVII line, O VIII at 18.97 Å, and Ne IX at 13.44 Å) have an optical thickness reaching up to 3 in active regions. There is still controversy about the details of these effects, and even on whether the maximum optical thickness is reached for regions viewed from above at disk center or edge-on on the limb (compare Rugge and McKenzie, 1985; Schmelz *et al.*, 1992; Waljeski *et al.*, 1994; Phillips *et al.*, 1996; and Saba *et al.*, 1999). These differences may well depend on the structure of the region. Resonant scattering is also important in the quiet corona for a few strong spectral lines of Fe IX, X, and XII just below 200 Å (Schrijver and McMullen, 1999).

A crude estimate of the optical thickness of the truly coronal plasma can be made as follows. In the case of thermal Doppler broadening, which probably dominates in coronae (unless turbulent line broadening is important, such as may be the case for an ensemble of microflaring loops), the optical thickness at line center is given by (e.g., Mariska, 1992):

$$\tau_0 = 1.2 \times 10^{-17} \left(\frac{n_i}{n_{el}} \right) A_Z \left(\frac{n_H}{n_e} \right) \lambda f \sqrt{\frac{M}{T}} n_e \ell \equiv C_d \left(\frac{A_Z}{A_{Z,\odot}} \right) \left(\frac{n_{e,10}\ell_9}{\sqrt{T_6}} \right), \qquad (8.16)$$

where λ is the wavelength in ångstroms, f is the oscillator strength, M is the atomic weight, T is the temperature, ℓ is a characteristic dimension, $A_Z = n_{el}/n_H$ is the abundance of element "el," $n_H/n_e \approx 0.85$ is the ratio of hydrogen to electron number densities, and (n_i/n_{el}) is the ion fraction [the final expression is given for easy combination with the scaling law in Eq. (8.22) below]. The constant C_d is of the order of unity for the strongest spectral lines. The product $n_{e,10}\ell_9$ (with $n_{e,10}$ in units of 10^{10} cm^{-3}, and ℓ_9 in units of 10^9 cm), the column density, depends on the coronal geometry. Mewe *et al.* (1995) and Schrijver *et al.* (1995) argue that the line-center optical thickness can be of the order of unity for a small set of the strongest coronal lines at X-ray and extreme-ultraviolet wavelengths.

The nonnegligible optical thickness in a few spectral lines by itself hardly affects the total radiative losses from the solar corona (or stellar coronae, for that matter; see Section 9.6), because τ_0 is not large enough to result in a measurable photon destruction within the corona itself by collisional deexcitation. Deviations in line-flux ratios for strong resonance lines introduced by the coronal geometry provide a diagnostic to estimate coronal electron densities and the indirect measurement of properties of the high corona, and perhaps the detection of a Sun-like stellar wind (Section 9.6). Whether resonant scattering affects level populations to a significant degree (e.g., Bhatia and Kastner, 1999) remains to be determined.

TRACE and *SOHO/EIT* and *CDS* observations frequently show dark matter in filaments, low-lying loops, and other dynamic, high-chromospheric structures (see the frontispiece). The extinction in the forest of spicules, fibrils, and so on, is so strong that the limb seen in the extreme ultraviolet is offset by 4,000 km to 9,000 km relative to the white-light limb (Section 8.8) which is attributable to the Lyman continuum of H, and the bound-free continua of He I and He II. Kucera *et al.* (1999) demonstrate that this same continuum causes prominences to show up as dark features in lines at several hundred ångstrom in the *SOHO/CDS* data. It is likely that absorption in the neutral hydrogen and helium continua causes the multitude of dynamic, dark features well above the spicule-like phenomena that are seen in *TRACE* observations at 171 Å and 195 Å, in addition to the filament/prominence configurations.

Despite the dynamic nature of the solar corona, the relationship between length, pressure, and characteristic coronal temperature of isolated coronal loops has been approximated by a quasi static model (starting with models developed in parallel by Craig *et al.*, 1978; Rosner *et al.*, 1978; and Vesecky *et al.*, 1979; see also a monograph on solar coronal loops by Bray *et al.*, 1991, and a discussion on the sensitivity of loop appearance to the details of the physics by Van den Oord and Zuccarello, 1996). These models assume that changes occur on time scales longer than the dynamic, radiative, and conductive time scales. These time scales are estimated on the basis of scaling arguments. Deviations from pressure equilibrium are restored on the *dynamical time scale*,

$$\hat{t}_{L,d} = \frac{L}{c_s} = 1.1 \frac{L_9}{\sqrt{T_6}} \text{ (min)}, \tag{8.17}$$

for loop half-length L, sound speed c_s, and temperature T, all in cgs units with L_9, for instance, in units of 10^9 cm.

Thermal perturbations are restored either by radiation or by conduction. The *conductive time scale* is given by the ratio of the thermal energy to the rate at which conduction across a temperature gradient transports energy:

$$\hat{t}_{L,c} = \frac{3n_e kT \ell^2}{\kappa_c T^{7/2}} = 86 \frac{n_{10} \ell_9^2}{T_6^{5/2}} \approx 10 \left(\frac{\ell_9}{L_9} \right)^2 \hat{t}_{L,d} \text{ (min)}, \tag{8.18}$$

where ℓ is the length scale of the disturbance and κ_c is the coefficient of the classical heat conductivity (see below). The final expression is obtained by inserting the loop scaling law Eq. (8.22). The ratio ℓ_9/L_9 shows that small fluctuations are erased very efficiently, leaving those spanning the entire loop for last. Equation (8.18) should be applied to the loop as a whole with some care, because the temperature gradient in a loop is strongest low in the legs, whereas the top part of the arch has a much weaker gradient. Consequently, the time scale for energy to be conducted out of the top part of a quasi-static loop upon the cessation of nonthermal heating, for example, is substantially larger than the value estimated from Eq. (8.18) by setting $\ell \sim L_9$; $\hat{t}_{L,c}$ should not be interpreted as the time scale to evolve from one state to another.

The *radiative time scale* is given by the ratio of the internal energy of the plasma (here taken to be fully ionized hydrogen) to the rate of radiative cooling, that is,

$$\hat{t}_{L,rad} = \frac{3n_e kT}{n_e n_H \mathcal{P}(T)} \approx 5 \frac{T_6^{5/3}}{n_{10}} \approx 35 \, T_6^{1/6} \hat{t}_{L,d} \text{ (min)}, \tag{8.19}$$

for electron density n_e. The radiative loss curve $\mathcal{P}(T)$ has been approximated by $1.5 \times 10^{-18} T^{-2/3}$ erg cm^3 s^{-1} (valid roughly between 0.4 MK and 30 MK; compare Fig. 2.4).

The dynamical time scale and conductive time scale for gradients on the largest scales are of the order of tens of minutes in typical solar active-region loops of some 10^5 km in length. The radiative loss time scale is substantially longer, reaching about half an hour to an hour. Many bright structures on length scales of a few thousand kilometers or more survive for at least several tens of minutes, so that the quasi-static approximation seems applicable. Even loops of the order of 1,000 km, such as in very compact coronal bright points, may not deviate from the quasi-static conditions much of the time, because for these loops the dynamical and conductive time scales are some 10 s or so. But the final verdict on this is not yet in: the radiative time scale is so much longer than the other time scales, that the radiative inertia of the loops can mask strong fluctuations in the heating rates inside the observed structures. Numerical simulations discussed in Chapter 10 reveal that such intermittent heating is to be expected, but also that it may not lead to strong differences between the properties of static loops and of dynamic loops.

In quasi-static models, only three terms are involved in the energy balance at any point at a distance s above the base, measured along the loop: the local volume energy deposition, ϵ_{heat}, the conductive flux density, F_c, along the loop, and the radiative losses per unit volume $\epsilon_{\text{rad}} = n_e n_H \mathcal{P}(T, \ldots)$ for radiative losses from an optically thin plasma [Eq. (2.59)]. Then:

$$\epsilon_{\text{heat}}(T, n_e, s, \ldots) = \frac{1}{A(s)} \frac{d}{ds} \left[A(s) \kappa_c \frac{dT(s)}{ds} \right] + n_e n_H \mathcal{P}(T, n_e, A_i, \ldots). \qquad (8.20)$$

In general, the loop cross section $A(s)$ is expected to change along the loop. The classical heat conductivity κ_c equals 8×10^{-7} erg cm^{-1} s^{-1} K$^{-7/2}$ (Spitzer, 1962).

Equation (8.20) explains why coronal loops are hot: wherever nonthermal energy is deposited high in the outer atmosphere, this energy can only be lost again through conduction or radiation. Conduction is only efficient if the plasma is hot or if the temperature gradient is large, which indirectly requires the corona to be hot. Radiation is only efficient if the plasma density is high enough or the collisional excitation rate high enough, which also requires the corona to be hot (see Section 2.3.1). In hydrostatic equilibrium, sufficient matter to radiate efficiently can only fill coronal loops if the pressure scale height [Eq. (2.9)] is large. In the photosphere the pressure scale height H_p is only ~ 150 km. At 10^5 K it is 5,000 km, but at 10^6 K, $H_p \approx 50,000$ km, which is large enough to fill coronal loops to radiate sufficiently. Some of the longest loops at temperatures close to 1 MK show the settling of material near their footpoints, whereas hotter ones are filled completely (see the frontispiece and Fig. 8.5).

Both the heating and the radiative loss functions may vary along a loop. The efficient thermal conduction reduces the sensitivity of the structure to the local value $\epsilon_{\text{heat}}(s)$, so that relatively little can be deduced about this function from observations alone. The radiative loss curve $\mathcal{P}$ is generally assumed to be a function of temperature only (thus implicitly assuming ionization balance, Maxwellian velocity distributions, and so on), but in recent years the chemical homogeneity of the solar corona has been questioned (see the review by Haisch et al., 1996). Systematic differences of approximately half an order of magnitude have been reported between elemental abundances in the photosphere and in the corona (and solar wind) related to the first ionization potential (FIP) of the elements involved [possibly related to Lyman α (10.2 eV) photoionization and the consequent selective coupling of ions to the magnetic field, e.g., Von Steiger and Geiss, 1989; also see Section 9.6 for FIP effects in stellar coronae]. Moreover, abundances also have been found to be different for different environments. For instance, the FIP bias appears to be weak in the fast solar wind emanating from high solar latitudes and coronal holes (Von Steiger *et al.*, 1997; and Feldman *et al.*, 1998). In closed-field regions, the chemical composition appears to depend on the history of the magnetic field (see the review by Haisch *et al.*, 1996); for example, the deviation from photospheric abundances has been reported to be stronger in "old" loops than in loops newly emerged into the outer atmosphere. The enhancement of the elements with a FIP below 10 eV (e.g., Fe, Si, Mg) over others (e.g., C, N, O), excluding hydrogen, is well established (see the review by Von Steiger *et al.*, 1997). Where hydrogen – itself a high-FIP element – fits in this picture is less clear, because it has no coronal spectral lines. Indirect arguments suggest, however, that the abundances of low-FIP elements are actually enhanced relative to hydrogen in the solar corona. Here we use the approximation that coronae are chemically homogeneous, although not necessarily of the same composition as the photosphere. We note that Cook *et al.* (1989) explored the sensitivity of the radiative loss curve to the chemical composition: (high-FIP) iron dominates $\mathcal{P}$ above 5×10^5 K, whereas (low-FIP) oxygen and carbon are the main contributors in the transition region near 10^5 K (Fig. 2.4).

The condition for hydrostatic equilibrium [Eq. (4.15)] completes the pair of equations describing quasi-static loops. The assumption that hydrogen is the dominant electron donor, and that all hydrogen is ionized in the corona, leads to the approximation

$$\frac{d}{ds}nT = \frac{m_H}{2k}n(s)g(s). \qquad (8.21)$$

A set of boundary conditions is required for unique solutions to the pair of Eqs. (8.20) and (8.21). Different sets can be used, made up from combinations of the loop apex temperature, $T_a(s = L)$, a zero conductive flux $F_c(s = L)$ at the loop top, and a negligible conductive flux at the base, the base temperature $T(s = 0)$, or the base pressure $p_b(s = 0)$. The detailed structure of the loop is not very sensitive to the boundary conditions at its base. The stability of the model loops, however, appears to be quite sensitive to the detailed specification of the boundary conditions; the difficulty of treating this boundary condition rigorously has left loop-stability criteria controversial.

Rosner *et al.* (1978) derived scaling laws for quasi-static coronal loops, subject to a set of assumptions. First, they assume that the vertical extent of the loop is small compared to the coronal pressure scale height $H_{p,9} \approx 5T_6(g_\odot/g)$ [Eq. (2.9)], where g is the gravitational acceleration and T_6 is the temperature in megakelvin. They further assume that loops have a constant cross section A, a uniform volume heating ϵ_{heat}, and solar photospheric abundances, that $T(s)$ is a monotonically increasing function of s only, that the maximum temperature T_a is at the apex, and that the conductive flux at the base is negligible. The scaling law relating pressure p, loop half-length L, and apex temperature T_a reads

$$T_a = \frac{1400}{\Gamma^{0.1}}(pL)^{1/3}\mathcal{H}(L) \Leftrightarrow T_{a,6} = \frac{2.8}{\Gamma^{0.15}}(n_{a,10}L_9)^{1/2}\left[\mathcal{H}(L)\right]^{3/2}, \qquad (8.22)$$

where $n_{a,10}$ is the density at the loop apex in units of 10^{10} cm^{-3}. The relationship on the left is generally referred to as the Rosner, Tucker, and Vaiana, or RTV, scaling law, for $\Gamma \equiv 1$ and $\mathcal{H}(L) \equiv 1$. The scaling with the expansion parameter Γ was derived by Schrijver *et al.* (1989c), using the code developed by Vesecky *et al.* (1979), but for uniform heating and neglecting pressure stratification. The factor Γ is the ratio of the cross sections at the loop top to that at the footpoint near the photosphere, assuming that the loop geometry matches that of a line dipole, that is, described by two circles of slightly different radii, both lying in the same plane perpendicular to the surface, and touching at some depth below the photosphere, where their separation is taken to equal the loop diameter. The function $\mathcal{H}(L)$ describes the dependence of the scaling law on the pressure scale height H_p and the scale height H_h of the volume heating function; Serio *et al.* (1981) find $\mathcal{H}(L) = \exp[-L(2/H_h + 1/H_p)/25]$ for $\Gamma \equiv 1$. Note that we have condensed various studies into Eq. (8.22); the case for which $\Gamma \neq 1$ as well as $\mathcal{H} \neq 1$ has not yet been studied, so that that combination should not be made when applying Eq. (8.22).

A second scaling law, also from Rosner *et al.* (1978), is

$$T_a = 60(\epsilon_{\text{heat}} L^2)^{2/7}. \tag{8.23}$$

By studying a collection of loops observed with the *Skylab* S-054 Soft X-ray Telescope, Rosner *et al.* (1978) argued that Eq. (8.22) is likely to be successful for a wide variety of solar coronal loops with lengths $2L$ from $10,000\,\text{km}$ up to 30 times as much (i.e., with heights up to the pressure scale height), and pressures from $0.1\,\text{dyn/cm}^2$ up to 2,000 times as much, although their verification diagram showed a very substantial scatter, presumably largely because of the poor angular resolution of the telescope, and the associated difficulties in estimating electron densities. Note that with the scaling law in Eq. (8.22), the time scales given by Eqs. (8.17), (8.19), and (8.18) imply that conduction and pressure adjustments occur far more rapidly than those associated with radiative cooling, as required by the quasi-static assumption: $\hat{t}_{\text{L,d}} \lesssim \hat{t}_{\text{L,c}} \ll \hat{t}_{\text{L,rad}}$.

Not all bright structures are compact compared to the pressure scale height. The Fe IX/X and Fe XII passbands of *SOHO/EIT* and *TRACE* show that the outer loops of active regions are frequently relatively cool, with a temperature of 1–1.5 MK. For such loops, bright emission is only seen in the lower section of the loop, in keeping with the pressure scale height. Over the solar limb, the line-of-sight integration and the relatively dark background reveal the faintly emitting top segments of these loops.

Even though the loops in the solar corona have been observed with angular resolutions down to an arcsecond, the dependence of the loop cross section on height remains unclear. The observation by Schrijver *et al.* (1985) that the projected area of the coronal condensation over an active region of average size exceeds the underlying plage area by approximately a factor of 5 (see Section 5.4, Table 5.2, line 5), suggests that at least the average field strength decreases with height by that same factor. Imaging observations show that loops maintain a fairly constant cross section along most of their length, but these results are obtained for structures near the angular resolution of the telescopes (Golub *et al.*, 1990, Klimchuk *et al.*, 1992), even with the substantial improvement of the resolution of *TRACE* just below an arcsecond. Using a different approach, Kano and Tsuneta (1996) model the brightness profile along loops observed with the *YOHKOH/SXT*. They can model the profile for their sample of 16 loops if they assume that each loop is composed of filamentary strands with varying cross section with height. Their model requires the strands to be approximately six times larger at the loop top than low in the loop. This value is quite close to the typical expansion suggested by the data of Schrijver *et al.* (1985).

A proper interpretation of loop expansion factors requires determination of the coronal field geometry precisely where the emission occurs. Since the coronal field cannot readily be measured locally, it is to be determined by an extrapolation based on the observable photospheric or lower-chromospheric magnetic field. The general assumption that the magnetic field is force free [Eq. (4.14)] is invalid in the photosphere whenever field and flows interact, but a good approximation in the chromosphere above $z \approx 400$ km (Metcalf *et al.*, 1995) and in the inner corona (except very locally in the thin current systems where reconnection occurs between strongly sheared fields). Yet even if high-resolution chromospheric (vector-)magnetograms would be available, force-free extrapolations would prove difficult because of the mathematically ill-posed nature of the problem (Section 8.1). The intrinsic substructure of the field below currently available angular resolutions associated with small-scale currents introduces further uncertainties. Nevertheless, force-free extrapolations have been quite successful for the large-scale active-region fields, with appropriate adjustments of α [Eq. (4.14)] as a function of position within the photospheric boundary value. These extrapolations often result in loop cross sections that vary with height substantially less than would be the case in potential fields (e.g., McClymont and Mikic, 1994; Wang and Sakurai, 1998). We return to the problem of loop cross sections when we discuss constraints on loop geometries based on the spectroscopy of stellar coronal emission in Section 9.6.

The RTV scaling law leads to a somewhat counterintuitive scaling of loop brightness and length: for loops shorter than the pressure scale height, and that have the same relative variation of cross section with height, the total X-ray power P_X emitted by a loop scales like

$$P_X \propto n_e^2 \mathcal{P}(T) d^2 L \sim \frac{d^2}{L} T^{7/2}, \tag{8.24}$$

for a loop (top) diameter d. If loops of a given geometry were simply subjected to a volume scaling factor of $d^2 L$, shorter loops would be dimmer than their larger counterparts, as might be expected from simple volume arguments. But for ensembles of loops with the same surface filling factor of their footpoints (or for loops with unit cross section), the contribution to the total coronal radiative loss by compact loops exceeds that for longer loops.

The applicability of the RTV scaling law of Eq. (8.22) to solar loops suggests it is also a useful tool to interpret stellar coronal spectra (see Section 9.6). Unfortunately, this general applicability results from its insensitivity to many of the details that determine the loop's structure and the observable radiative losses. The insensitivity to loop geometry, as measured by the ratio Γ of cross sections at top and footpoint, is already shown in Eq. (8.22). Chiuderi *et al.* (1981) discuss the effect of a dependence of a hypothetical heating function on temperature, $\epsilon_{\text{heat}} \propto a T^\gamma$, on the scaling law, and they argue that measurement errors preclude an experimental determination of γ to a useful degree of accuracy without detailed observations of loops on time and length scales well below those of their entire volume. It appears impossible to constrain the heating function beyond the requirement that $\epsilon_{\text{heat}}(T)$ cannot intersect the radiative loss curve $n_e n_H \mathcal{P}(T)$ more than once (Van den Oord and Zuccarello, 1996).

On the basis of earlier studies, Antiochos and Noci (1986) point out that, in addition to the hot coronal loops, with $T > 10^5$ K, under certain conditions cool stable loops, with $T < 10^5$ K, are possible. The loop stability depends on the ratio of the pressure scale height $H_{p,5}$ at $T = 10^5$ K to the loop height L. If $H_{p,5}/L \ll 1$, only hot loops are possible, but if $H_{p,5}/L \lesssim 1$ then both hot and cool loops would be possible. From an involved stability analysis by Antiochos *et al.* (1985), Antiochos *et al.* (1986) deduced that for loops with heights close to $H_{p,5}$ hot loops are not stable; hence short loops prefer a cool state.

8.7   Coronal holes

The corona over large regions of unipolar regions of quiet Sun sometimes shows a remarkably low emission; such regions are referred to as coronal holes. They exist over the polar regions during most of the solar cycle, disappearing only during a cycle maximum. Low-latitude coronal holes are seen less frequently. Coronal holes can be as wide as 300,000 km, or as narrow as 40,000 km (then sometimes referred to as coronal dark channels). Low-latitude coronal holes can be either isolated, entirely surrounded by a relatively bright quiet corona, or they can be connected to the polar coronal holes, occasionally extending across the equator.

Coronal holes are found where the field in relatively large areas over the quiet Sun opens up into interplanetary space. Their location can often be predicted from a potential-field extrapolation of the photospheric magnetic field (starting with Levine *et al.*, 1977; see also, for instance, Hoeksema, 1984, and Wang and Sheeley, 1993); see Section 8.1. If the model-field lines that extend out to beyond $2.5 R_\odot$ from the Sun's center are traced back to the solar surface, then large coherent areas of such field lines outline the base of coronal holes. The location of a coronal hole tends to coincide with the local maximum in the absolute flux density in very low-resolution magnetograms. However, locating such an extremum is insufficient to predict the occurrence of coronal holes, because they develop well before a coronal hole materializes and persist some time after it disappears (Hoeksema, 1984).

Nonradiative heating in coronal holes results in a temperature that does not seem to be very much – although significantly – lower than in the closed, bright loops, but the electron density is much lower. Remarkably, Withbroe (1981) concludes that the total energy flux density into the quiet corona appears to be a factor of almost 3 smaller than that into a coronal hole: whereas the losses through radiative emission and downward thermal conduction are substantially smaller for a coronal hole, the energy dissipation in open-field regions is apparently used primarily to drive the outflow of the solar wind (Section 8.9). Withbroe finds that the energy flux density in the solar wind exceeds the radiative plus conductive losses in a coronal hole by an order of magnitude, being 8×10^5 and 7×10^4 erg cm^{-2} s^{-1}, respectively.

Whereas the large-scale geometry of the magnetic field within coronal holes is that of field lines that have opened up into the solar wind, closed loops do occur low down. Occasionally a small active region emerges into a coronal hole, or drifts into the hole if the two rotate at different velocities (see later in this section); if the region is not too large, this intrusion does not much affect the hole surrounding it. The ubiquitous ephemeral regions also inject flux into coronal holes. Coronal bright points that are associated with the emergence of ephemeral regions or coronal bright points that develop because of chance encounters of magnetic concentrations, appear to exhibit the same morphology and variability as in the predominantly closed-field environment of the quiet corona

(Habbal *et al.*, 1990). The appearance of coronal bright points does not depend on whether they lie below closed- or open-field line regions. This indicates that their heating is mainly determined by their internal magnetic field.

Although the coronal holes stand out very clearly in soft X-rays, the contrast decreases rapidly toward lower temperatures: in H α and Ca II K, for instance, the quiet-Sun unipolar network does not differ significantly from the base area of coronal holes (Vaiana *et al.*, 1973). Apparently, the chromospheric heating and structuring is largely unaffected by the coronal field geometry. The He I λ 10,830 Å line is an exception to this: its formation is affected by either irradiation from the low corona or, more probably, by downward-conducted thermal energy. The result is that observers can trace the outlines of coronal holes as the perimeters of relatively bright regions in He I λ 10,830 spectroheliograms.

Polar coronal holes disappear only during a cycle maximum when the predominant polarity within the polar caps is being reversed (Section 6.2). Large low-latitude and midlatitude coronal holes have an average lifetime of five to seven rotations, whereas some live as long as nine rotations (Hoeksema, 1984). The average lifetime of smaller holes is much shorter.

At low latitudes, coronal holes are rare. The large unipolar regions that are a prerequisite for the formation of low-latitude coronal holes require a fortuitous positioning of active regions, both present and past. The formation of a coronal hole can be associated with the emergence of an active region: if the regions are in a favorable space–time configuration, a coronal hole may appear simultaneously with the emergence. Large coronal holes are rare near the time of a solar minimum, because there are fewer active regions in that phase.

Low-latitude coronal holes often show rotation profiles that differ significantly from that of the photosphere. Some holes rotate almost completely rigidly whereas others show a considerable, yet modified differential rotation. The rigid rotation is, at first glance, similar to the peculiarly rigid rotation of the poleward "streaks" of flux escaping from decaying large active regions that was discussed in Section 6.3.2.2. The cause of the rigid rotation of some coronal holes, however, requires more than that. Wang and Sheeley (1993) demonstrated that coronal holes are associated predominantly with distortions of the large-scale bipolar and quadrupolar components of the global solar field by the occurrence of a sizable active region. If that region lies at a different latitude than the bulk of the coronal hole, the hole moves at a rate that differs substantially from that of the underlying photospheric plasma and the field embedded within it. As a result, field concentrations drift through the hole, both in longitude because of this difference in local rotation rate, and in latitude because of the meridional flow. As a concentration moves into the hole, its field frequently, but not necessarily, opens up, and as it leaves a hole on the other side, any open field reconnects to form closed loops. This reconnection forces other field connections apart, and their reconnection initiates a progression of field-line rearrangements over large distances. Wang and Sheeley (1993) discuss this process, and they point out that this continuous reconnection in the corona, merely the result of differential rotation (and meridional flow), must also be occurring if there are no coronal holes at all. Schrijver *et al.* (1998) point out that the frequent emergence and cancellation of the photospheric field associated with the evolution of ephemeral regions may be an important mechanism enabling these frequent reconnections anywhere in the corona.

Given the mechanism of the formation of coronal holes, their rotational characteristics are expected to depend on details of the surrounding field and, as a result, a trend depending on the phase of the cycle is expected. Indeed, in the declining phase of the cycle, polar-hole extensions tend to rotate rigidly when active regions emerge near the equator and the polar fields are strong, whereas they are strongly sheared during the rising phase of the cycle when active regions erupt at higher latitudes. Isolated coronal holes over remnants of decayed active regions near the cycle's maximum, when the polar coronal holes are absent, rotate differentially at the local rotation rate (Wang and Sheeley, 1993).

8.8   The chromosphere–corona transition region

The temperature domain between that of the chromosphere and the corona, that is, between $\sim$50,000 K and $\sim$0.5 MK, is referred to as the *transition region*. This definition is often (implicitly) narrowed further to a definition based on a physical model, in which the transition region is defined as the region with a steep, conductively determined temperature gradient between chromosphere and corona. Mariska (1992) wrote a monograph dealing exclusively with the transition region. Here we discuss only a few of its properties regarding its geometry and dynamic character.

Both the classical, spherically symmetric models of the solar outer atmosphere and models for static outer-atmospheric loops include a geometrically thin transition region which was inferred to be heated primarily by thermal conduction from the corona above it, and cooled through fairly efficient radiation. The transition region is geometrically thin because of the rapidly decreasing conductivity with temperature, strengthened by the efficiency of radiative cooling that increases with decreasing temperature down to $\sim$10^5 K for optically thin plasma (Fig. 2.4; see, for instance, Bray *et al.*, 1991, for the argument). As observations in these few diagnostics have become more detailed, however, the temperature domain of the transition region presents and increasingly complicated picture. *TRACE* observations, for example, confirm that the low-lying, conduction-dominated transition region actually extends to temperatures in excess of 1 MK underneath the 3–4 MK quiescent corona over active regions (Berger *et al.*, 1999). Hence, it includes temperatures traditionally referred to as truly coronal. Perhaps we should redefine the transition region to be a coronal domain (in terms of its generally negligible optical thickness) in which there is a steep temperature gradient along the magnetic field.

The magnetic canopy above which the entire outer atmosphere is filled with a magnetic field is located in the high chromosphere (Section 8.5). The pattern of the magnetic network is seen in the transition region up to a temperature of at least 0.3 MK (O VI 1,032 Å); this requires that much of the transition-region emission originates within the regions of relatively strong field on or in the vicinity of the axes of the magnetic concentrations, that is, somewhere over the pillars of the canopy (Fig. 8.1). At temperatures of $\gtrsim$0.5 MK, the network structure is no longer visible but is replaced by a hazy background containing looplike structures.

The general correspondence between bright areas in coronal and transition-region images (see Mariska, 1992) results in a good correlation of patterns (and in the relationships between intensities discussed in earlier sections) at a low angular resolution and in a well-defined mean trend between the transition-region brightness and the underlying magnetic flux density (Section 8.4.3) that allows matching of the corresponding patterns.

The detailed correlation between the brightness of the corona and that of the transition region, however, appears to be poor at high spatial resolution, just like that between the chromosphere and transition region (Reeves *et al.*, 1976; Berger *et al.*, 1999). Feldman and Laming (1994), for example, searched *Skylab* NRL S082a spectroheliograph observations for the transition-region footpoints below bright coronal loops. They concluded that these footpoints are less bright than they were expected to be from loop-model computations. Conversely, a bright region at transition-region temperatures does not always correspond to a coronal brightening. These clashes with the predictions from quasi-static loop models (Section 8.6) point to a much more dynamic picture than envisioned in these models.

Within the complex and dynamic geometry of the solar outer atmosphere, transition-region emission originates from a range of heights, overlapping with both the cooler and hotter domains in the solar atmosphere (see Dowdy *et al.*, 1986, on the quiet-Sun transition region). Loops at transition-region temperatures occur up to heights well over the associated pressure scale height, up to heights where they are surrounded by coronal structures. This applies to postflare loops, but also to other loops not obviously associated with large flares. The raggedness of the transition region emphasizes the inadequacy of the simple layered models because the corona and chromosphere are interspersed over a height interval of thousands of kilometers. Daw *et al.* (1995) have used high-resolution *NIXT* data to show that cool spicular material extends up to heights of 4,000 km up to 9,000 km above the photosphere that absorbs the coronal emission from surrounding hot loops. *TRACE* images show cool material being thrown up and falling back down in spiculelike structures that show up as rapidly evolving, absorbing features. The statistical effect of this ensemble of dynamic phenomena is that limb-brightening curves suggest an overall mean stratification with a substantial range in height for the transition region (Mariska *et al.*, 1978): the peak within the broad brightness distributions shifts from ∼2 arcsec above the limb for Si III (1,892 Å; 80,000 K) up to 4 arcsec for O V (1,218 Å; 0.3 MK).

At an angular resolution of an arcsecond or worse, the profiles of the emission lines from the transition region often show multiple components, substantial net velocities, and turbulent broadening. Multiple components (e.g., Kjeldseth-Moe *et al.*, 1993) suggest velocities from subsonic up to the supersonic value of the order of 100 km/s, both up and down. The mean linewidth for C IV for all features suggests a nonthermal velocity of some 20 km/s (Dere *et al.*, 1987; Cheng *et al.*, 1996). The complex line profiles require a dynamic structure below the angular resolution.

Frequently, the transition-region lines with formation temperatures between ∼20,000 K and at least up to 0.2 MK show a general broadening during short periods of time. These events are referred to as transition-region explosive events (see the review by Dere, 1994). They are ∼2 arcsec in size, with broadenings up to ≈200 km/s, lasting for 20 s up to 200 s, and with the maximum line-of-sight velocities independent of the viewing angle (Dere *et al.*, 1989). These are likely the product of magnetic reconnection in the mixed-polarity network. Some 600 form every second on a completely quiet Sun.

Another common phenomenon is the *jet*: brightenings in spectral lines with a blue-shifted component as far out as 400 km/s. They are most easily seen in C IV, but they are also seen in other lines formed from some 20,000 K up to several hundred thousand degrees.

The volume filling factor at transition-region temperatures can be estimated by combining measurements of the electron density, the height extent of the line-forming layer as observed at the limb, and high-resolution observations at the center of the disk. With the use of different methods, filling factors have been reported of 0.2 (see Mariska, 1992) down to between 10^{-5} and 10^{-2} (Dere *et al.*, 1987); the latter range suggests a substructure down to scales as small as 3–30 km, compatible with similar filling factors found for the corona (e.g., Martens *et al.*, 1985). Wikstøl *et al.* (1998) point out that these estimates are sensitive to the assumption that the transition region is static, and that far larger filling factors are found based on dynamic models, in which impulsive heating and propagating pressure fluctuations lead to an equally acceptable paradigm.

8.9   The solar wind and the magnetic brake

8.9.1   *Solar wind and mass loss*

The Sun continually loses mass through the solar wind, despite the fact that the coronal plasma is gravitationally bound to the Sun, with an average energy per proton that is only approximately a tenth of the gravitational binding energy near the solar surface. But because of the efficient thermal conductivity, the temperature stays high out to large distances, and near $10R_\odot$ the thermal energy exceeds the binding energy. The outer corona therefore evaporates, and matter from below streams up to replace the lost mass (Parker, 1997). Although the pressure scale height [Eq. (2.9)] in the low corona is only some 50,000 km to 100,000 km, it increases rapidly with height (scaling close to $r^{12/7}$, which would be expected if only thermal conduction were important in a spherical geometry) as $g(r)$ decreases rapidly, whereas $T(r)$ decreases close to $r^{-2/7}$.

Parker (1958) was the first to propose a model for a thermally driven outflow from the solar corona into interstellar space that accommodates the transition from a subsonic flow near the surface to a supersonic flow further out. He argued that the gravitational field of the Sun, aided by the geometry of the outflows, forces a solution in which the solar wind acts as if it were flowing through a Laval nozzle. The thermally driven wind as proposed by Parker necessarily becomes supersonic at some distance from the Sun. This is readily seen for a steady, isothermal, radially expanding wind by combining the equation of continuity, $\rho v r^2 = \text{const}$, and the momentum balance:

$$\rho v \frac{dv}{dr} = -\frac{dp}{dr} - \rho \frac{GM_\odot}{r^2}. \tag{8.25}$$

The assumption that the wind is isothermal is a rather restrictive simplification, but the highly efficient thermal conduction in the solar corona makes this assumption not fundamentally inappropriate; in fact, similar solutions are found for other models, such as a polytrope with $T \propto \rho^\alpha$ for $\alpha < 1/2$ (Parker, 1963b).

Equation (8.25) formally has a hydrostatic solution in which $v(r) \equiv 0$, and a solution that corresponds to an outflow that is described by

$$\left(v - \frac{c_s^2}{\gamma v} \right) \frac{dv}{dr} = \frac{2c_s^2}{\gamma r} - \frac{GM_\odot}{r^2}, \tag{8.26}$$

where c_s is the sound speed given by $\sqrt{\gamma p / \rho}$, where $\gamma = c_p/c_v$, the ratio of specific heats. Because the acceleration dv/dr should be finite everywhere, this expression defines a critical point at $r = \gamma GM_\odot/2c_s^2$ where the flow velocity equals the isothermal sound speed, $v = c_s/\sqrt{\gamma}$, unless dv/dr were to vanish there. In his pioneering work, Parker argued that the boundary conditions allowed neither a hydrostatic nor the solution with $dv/dr = 0$ at the critical point, and that the solar wind should consequently be slow in the low corona and supersonic far enough away from the Sun.

This model was sufficiently compatible with what was then known about the corona, the causes of aurorae, and the properties of cometary tails (see Parker, 1997). More recent *in situ* observations from spacecraft again proved Parker right, at least to a first approximation: the wind at the Earth's orbit is indeed supersonic. As it turns out, the simple model places the critical point at $\sim 7R_\odot/T(\mathrm{MK})$ above the photosphere, that is, at approximately ten solar radii for a plasma that has cooled to somewhat below 1 MK. The sonic point probably lies at different heights above the solar surface in different solar wind environments, such as above streamers and above the polar caps.

To date, all *in situ* probes of the solar wind have been limited to the domain well above the critical point. An indirect way of tracking the subsonic solar wind near the Sun has been used by Sheeley *et al.* (1997): they track intensity features moving away from the Sun (as obverved with the *SOHO/LASCO* instrument; see Fig. 8.11 for an example), presumed to be passively blown away in the solar wind, on *SOHO* coronagraphic image sequences. They find that the wind maintains a nearly constant acceleration of ~ 4 m/s^2 between $2\,R_\odot$ and $30\,R_\odot$.

It is primarily the very high thermal conductivity that is responsible for the solar wind in the model by Parker. This allows the transport of thermal energy across large distances, thus maintaining a relatively high pressure that can sustain the wind. In the actual solar wind, substantial heating and acceleration are required in addition to what can be provided in the model for the thermally driven wind (e.g., Axford and McKenzie, 1991). The details of these processes are not yet fully understood, but it appears likely that dissipation of Alfvénic fluctuations associated with a cascade of turbulent energy is involved (e.g., the review in Goldstein *et al.*, 1995, and references therein). For the Sun, but also for cool stars in general, the radiation pressure is in general negligible as a driver of the wind.

In reality, the solar wind is neither steady, isothermal, nor precisely radially expanding. It is affected dramatically by the magnetic field, as discussed to some degree in the model described in Section 8.9.2 below. The wind exhibits a quiescent outflow that is frequently interrupted by transient disturbances. The quiescent solar wind in the ecliptic is composed of intermittent high-speed and low-speed streams (see, e.g., the reviews by Gosling, 1996, and Axford and McKenzie, 1997). The high-speed flows, which originate in coronal holes, typically have velocities of 700–900 km/s, a particle density of ~ 3 cm^{-3}, and a mean proton temperature of 230,000 K at the Earth's orbit. It is this high-speed wind that is to first order described by the Parker model, and its modifications that include heating and momentum deposition. The low-speed flows originate wherever the field lines are largely closed low in the corona. These flows have a velocity of typically somewhat over 300 km/s, a particle density that is approximately three times higher than in the fast streams, and a mean proton temperature of only $\sim 40,000$ K and an electron temperature of 130,000 K at 1 AU. This part of the solar wind poses a challenge to modelers, as it does not correspond to an equilibrium configuration along field lines, but instead originates from coronal domains where the connectivity appears to evolve on a time scale of only a few hours.

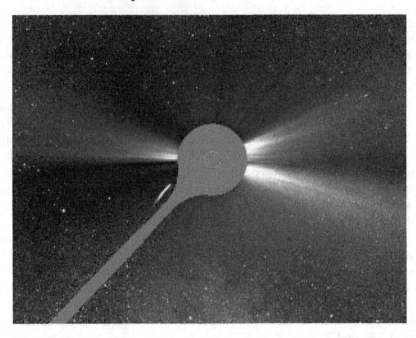

Fig. 8.11. The white-light outer corona as seen on this cutout by the *LASCO*-C3 coronagraph on *SOHO* on 23 December 1996, at 01:07 UT. The circle in the occulting disk shows the size of the photospheric disk. The image shows low-latitude streamers, as well as structure in the polar field (some of which can be traced back to polar plumes). Note the background stars and the Sun-grazing comet SOHO-6. The *SOHO/LASCO* data used here are produced by a consortium of the Naval Research Laboratory (USA), Max-Planck-Institut für Aeronomie (Germany), Laboratoire d'Astronomie (France), and the University of Birmingham (UK). *SOHO* is a project of international cooperation between ESA and NASA.

The matter in these slow and fast streams moves almost strictly radially (compare Fig. 8.11). The deviation from a radial outflow, caused by the interaction with the magnetic field, is only ∼8 km/s at the Earth's orbit. The result of this pattern of the flow is that the fast and slow streams necessarily interact: as the Sun rotates, a source region for high-speed wind can move to a location where it lies beneath a slow-speed stream that set out earlier, enabling the eventual interaction of plasma in these two types of wind. This interaction produces large-scale compressive structures in the solar wind. The low-speed solar wind is highly filamentary, and neighboring filaments may originate in distant regions on the Sun. These origins likely evolve rapidly from closed to open and conversely, so that one cannot refer to these origins as "roots."

These quiescent streams are frequently disrupted by transient events, such as the dramatic coronal mass ejections (see Section 8.3.1), that occur in a broad latitude belt, with a preference for relatively low latitudes. The range of latitudes within which mass ejections occur appears to increase with the solar activity level (Hundhausen, 1993a). During these ejections some 10^{15} to 10^{16} g of solar matter is propelled into the wind. After an initial acceleration phase, coronal mass ejections eventually reach about the average solar wind speed near the Earth, at 470 km/s. The frequency changes from once every 5 days around a cycle minimum up to 3.5 per day at a cycle maximum.

The essentially radial outflow from a rotating Sun leads to the characteristic Archimedean spiral pattern of magnetic field lines [use the same ducks trained for the explanation of the rigid rotation of poleward streaks between Eqs. (6.8) and (6.9), but now released into interplanetary space]. As a result, the field at the Earth's orbit is deflected from the radial direction by on average 45°. The average value of this angle changes with the solar cycle, in response to the mean solar wind speed, by approximately 10° (see, e.g., Smith and Bieber, 1991).

At a given large height above the solar surface, the time-averaged magnetic field in the wind reflects the polarity patterns of the photospheric magnetic field on the largest scale. In the ecliptic plane, the field direction exhibits a sector structure, with alternating polarities. These sectors (often four, sometimes two circling the Sun) are separated by a warped current sheet that cuts through the ecliptic at the sector boundaries (Wilcox, 1968; Schatten *et al.*, 1969; Schulz, 1973). This current sheet is the extension of the large-scale polarity-inversion lines in the photosphere, defining the polarity-inversion surface in the solar wind. The sector structure disappears at sufficiently high latitudes, where only a high-speed wind is observed (Marsden and Smith, 1996).

The *Ulysses* spacecraft explored the solar wind out of the plane of the ecliptic. After it moved south of a latitude of 36° S it found only a relatively uniform high-speed wind throughout the high latitudes. This wind originated in the large coronal hole over the southern pole. At midlatitudes, the interleaving of high- and low-speed streams was found to be much simpler than near the equator, consistent with a tilt in the global dipole moment of the Sun in this phase near the minimum of the sunspot cycle.

The interaction between streams at different velocities and transient streams that are associated with coronal mass ejections results in a multitude of shocks, both traveling away from the Sun and toward it. In addition, waves and turbulence in the solar wind may be generated by the dynamic character of the wind base, including filaments, plumes, and coronal bright points. The solar wind is consequently rather gusty (see reviews by Goldstein *et al.*, 1995, and Gosling, 1996).

The net mass loss through the solar wind is $\approx 2 \times 10^{-14}\, M_\odot/\mathrm{yr}$ (an Earth mass every 150 million years, or $10^{-4} M_\odot$ in the time the Sun has spent on the main sequence). In this total budget, coronal mass ejections contribute a nonnegligible fraction; an early estimate put it at 3% at cycle minimum up to 10% at maximum (Howard *et al.*, 1985), but more recent observations from *SOHO* suggest that this fraction is likely to be larger. The evolution of the Sun is not affected by this outflow, but its rotational history is, as illustrated in the next section.

8.9.2 *Angular-momentum loss by the solar wind*

Parker's original work on the solar wind did not include a magnetic field. The magnetic field introduces the possibility of carrying away more angular momentum than associated with the outflowing matter itself, resulting in magnetic braking of the Sun. We follow the model of Weber and Davis (1967), as adapted by Belcher and MacGregor (1976), which can be used to approximate the rates of mass loss and angular-momentum loss associated with a rotating, magnetized wind. Instead of a single critical point, as in Parker's model, a magnetized wind shows two more, namely where the radial velocity equals the phase speeds of the fast magnetosonic and Alfvén wave modes (with the slow magnetosonic mode replacing pure sound waves).

The Weber–Davis model assumes a fully ionized hydrogen corona extending from some radius r_0 somewhat larger than the stellar radius. It also assumes that the magnetic field at the base at r_0 is radial. Only steady-state, axisymmetric solutions are allowed for the equatorial plane, assuming that neither the flow nor the magnetic field have components out of that plane, that is, $\mathbf{v} = [v_r(r), 0, v_\phi(r)]$ and $\mathbf{B} = [B_r(r), 0, B_\phi(r)]$. The energy equation is replaced by a polytropic relationship of the form $p/p_0 = (\rho/\rho_0)^{\alpha'}$, with $1 < \alpha' \leq 3/2$. The base density ρ_0, temperature T_0, and field strength B_{r_0} are prescribed as boundary conditions.

From these definitions it follows that at the Alfvénic point, where $v_r(r = r_A) \equiv B_r/(4\pi\rho)^{1/2}$, the specific angular momentum $\mathcal{L}$ transported by the wind is given by the ϕ component of the induction equation,

$$\mathcal{L} = rv_\phi - \frac{rB_\phi B_r}{4\pi\rho v_r} = \Omega r_A^2. \tag{8.27}$$

Hence, interestingly, the value of $\mathcal{L}$ transported by the magnetized wind equals the specific angular momentum estimated using the stellar angular velocity and the Alfvén radius as the equivalent arm over which the torque is applied. This provides a simple scaling for stellar studies for an otherwise complicated problem; see Section 13.3. (Note, by the way, that the angular momenta that are being carried away by the slow and fast streams in the solar wind are comparable, because the velocity contrast is largely offset by the density contrast.)

Figure 8.12 depicts two solutions characteristic of the present-day Sun and of a young Sun, as it may have appeared at an age of some 50 Myr, respectively. The Alfvén point at r_A (solid dot on these figures) moves inward as the rotation rate and surface field strength decrease, resulting in a decreased rate of angular-momentum loss. Note that although the rate of angular-momentum loss in these two cases is very different, the mass loss rate differs only by a factor of $\sim$3. This is because momentum deposition by magnetic and centrifugal forces occurs mostly beyond the sonic point. Hence, the effect on the flow is primarily an increase in the asymptotic flow speed rather than in the mass flux. Note also that these two solutions differ markedly in their dynamics: the wind in the current Sun is mostly driven by thermal energy redistribution in the highly conducting wind, whereas in the rapidly rotating young Sun the magnetocentrifugal driving by the stirring by the magnetic field dominates at large distances.

The assumption of a monopolar magnetic field is a serious shortcoming of the Weber–Davis (WD) model. And even with that assumption, the model cannot be extended out of the equatorial plane consistently. A more realistic field geometry, however, can only be investigated numerically. The main difficulty in complex field geometries is of a topological nature: the flow speed for a thermally driven wind is expected to increase monotonically with distance, whereas the field strength of a multipole field decreases rapidly with distance. Close to the star, the flow is effectively channeled by the magnetic field, whereas beyond some critical distance the energy density of the flow becomes greater than the magnetic energy density, and the magnetic field lines are forced open, allowing the outflow of a wind. Such a configuration has a magnetospheric dead zone, in which magnetic field lines are closed onto the surface, and a wind zone, in which field lines are open to interstellar space. The existence of the dead zone reduces the

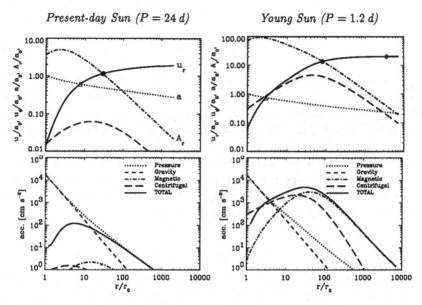

Fig. 8.12. Weber–Davis models for a rotating, magnetized wind emanating from a G2 V star. The left-hand panels are for the following boundary conditions: a radial field strength at the wind base at $r_0 = 1.25 \, R_\odot$ of $B_{r_0} = 2 \, \mathrm{G}$, a rotation period of $P_{\mathrm{rot}} = 2\pi/\Omega = 24.3$ days, a polytropic index of $\alpha = 1.13$, a base temperature of $T_0 = 1.5 \times 10^6$ K, and base density of $\rho_0 = 5 \times 10^7$ protons/cm^3. The right-hand panels are for a surface field strength 20 times larger, that is, of $B_{r_0} = 40 \, \mathrm{G}$, and a rotation period of $P_{\mathrm{rot}} = 1.21$ days. The top panels show the variations with distance of the radial velocity u_r (solid curve), the tangential velocity u_ϕ (dashed curve), the sound speed c_s (for which the symbol a is used in the figure), and the radial Alfvén speed $A_r \equiv B_r/(4\pi\rho)^{1/2}$ (dashed-dotted curve). The triangle, solid dot, and diamond indicate the location of the slow, Alfvén, and fast points, respectively (the Alfvén and fast point nearly coincide for the slow rotator). The lower panels show the force balance in the wind (figure from Charbonneau *et al.*, 1997).

total mass flow and the net angular momentum carried by it. Approximate wind solutions constructed within this framework (Mestel, 1968; Mestel and Spruit, 1987; Collier Cameron *et al.*, 1991) suggest that in the limit of high rotation rate and surface field strength, the effective angular momentum loss may be substantially less than that predicted by WD-type solutions: the WD solution probably provides an *upper limit* to the true angular-momentum loss rate, for a given coronal temperature, gravity, rotation rate and surface field strength. Because a comprehensive set of numerical solutions for the axisymmetric MHD wind problem is lacking, only crude approximations have been used in stellar models (Section 13.3).

The model for the magnetized wind of the present Sun is characterized by mass- and angular-momentum loss rates $\mathrm{d}M/\mathrm{d}t = 2.9 \times 10^{-14} \, M_\odot \, \mathrm{yr}^{-1}$ and $\mathrm{d}\mathcal{L}/\mathrm{d}t \simeq 2 \times 10^{31}$ dyn cm. Both compare well to *in situ* observations at the Earth's orbit (Pizzo *et al.*, 1983). The time scales for mass- and angular-momentum loss for the present Sun are then

$$\hat{t}_{\mathrm{M}} = M_\odot \left(\frac{\mathrm{d}M}{\mathrm{d}t} \right)^{-1} \sim 10^{13} \, \mathrm{yr}, \qquad (8.28)$$

and

$$\hat{t}_\mathcal{L} = \mathcal{L}_\odot \left(\frac{d\mathcal{L}}{dt} \right)^{-1} \sim 10^9 \, \text{yr}, \tag{8.29}$$

where we use the approximation that the present-day Sun rotates rigidly (Section 3.3) so that $\mathcal{L}_\odot = \Omega I_\odot$, for a moment of inertia $I_\odot \simeq 7 \times 10^{53} \, \text{g cm}^2$ (Claret and Giménez, 1989). The time that the Sun spends on the main sequence is $\hat{t}_\odot \simeq 10^{10} \, \text{yr}$; compare Eq. (2.5). The present time scale $\hat{t}_M$ for solar mass loss is far too large for the Sun's structure to be affected by mass loss over the remainder of the Sun's lifetime (and in fact it has been ever since the Sun reached the main sequence). The braking time scale $\hat{t}_\mathcal{L}$ for the present Sun, however, is less than the time the Sun will remain near the main sequence, so that even now substantial angular-momentum loss occurs. In the past, as we discuss in Chapter 13, $\hat{t}_\mathcal{L}$ was substantially shorter.

9

Stellar outer atmospheres

9.1 Historical sketch of the study of stellar activity

It has long been known that emission components are present in the cores of the Ca II H and K resonance lines in the spectra of many stars of spectral type G and later. The discovery paper by Schwarzschild and Eberhard (1913) was followed by a stream of usually brief papers from which lists of stars with references were collected by Joy and Wilson (1949) and Bidelman (1954). In the early 1950s, a vigorous study of the Ca II H and K resonance lines in stellar spectra was started by Olin C. Wilson at the Mt. Wilson Observatory; see Section 9.3. This research was boosted once more when in 1977 the efficient Ca II HK photometer (Vaughan *et al.*, 1978b) was installed at the Mt. Wilson 60-in. telescope. The productivity of the Ca II HK photometer was enhanced by letting researchers (including Ph.D. students) from other institutes use the instrument. The Catania Observatory has a long tradition of pursuing solar and stellar observations in parallel, also aimed at a better understanding of magnetic activity (Godoli, 1967).

From the early investigations onward, the researchers were aware that the mechanisms causing the stellar Ca II H and K emission are probably similar to those of the solar chromosphere. In addition, it was realized that the large range in Ca II H+K emission flux densities from stars of identical effective temperature $T_{\rm eff}$, surface gravity g, and metallicity indicates that the standard stellar classification does not fully describe stellar atmospheres.

In this and the following chapters we indicate that fundamental insights into stellar magnetic activity had been obtained from studies of stellar Ca II H and K emission before the first observations from space became available. Moreover, in addition to data from space, the Mt. Wilson Ca II HK photometer keeps providing fundamental data that are hard to obtain in some other way. Yet, the study of the magnetic activity of stars gained momentum when in 1978 two extremely successful satellite observatories were launched: the International Ultraviolet Explorer (*IUE*) enabling the observation of chromospheres and transition regions, and HEAO-2 *EINSTEIN* for coronal soft X-ray emission.

9.2 Stellar magnetic fields

9.2.1 Measurements of stellar magnetic fields

Attempts to measure magnetic fields of active cool stars by means of the polarization of Zeeman components in spectral lines have failed. The explanation is suggested by the example provided by the Sun: apparently the polarities are strongly mixed, so the polarization signals cancel. Strong fields with sufficiently large filling factors are detectable, however, by an additional broadening of Zeeman-sensitive spectral lines in the unpolarized spectrum; see Section 2.3.2. This fact is used in the Unno (1956) theory, which was developed by Robinson (1980) and applied for the first time by Robinson *et al.* (1980) in an attempt to measure fields in active cool stars.

Usually the analysis assumes a two-component model: magnetic patches of a well-defined mean field strength B embedded with a filling factor f in the quiet stellar atmosphere. The radiative flux spectrum is

$$F = f F_{\rm m}(B) + (1 - f) F_{\rm q}(B \equiv 0), \tag{9.1}$$

where $F_{\rm m}$ is the flux spectrum in the magnetic regions and $F_{\rm q}$ is that of the quiet (or, more precisely, the nonmagnetic) atmosphere. It is assumed that the fields are radial, and that the patches are nearly uniformly distributed across the stellar disk. In many applications, B is assumed to be constant with height, and the atmospheres in magnetic and quiet patches are supposed to be identical. Lines of different Zeeman sensitivity are incorporated in the analysis in order to separate magnetic from nonmagnetic line broadening.

Among the various investigations, there is appreciable diversity in the choice of the set of spectral lines, the treatment of radiative transfer in the line formation, the choice of the flux-tube and atmospheric models, and the method to compare the observed line profiles with the theoretical ones; for detailed descriptions and discussions, see Saar (1988) and Basri *et al.* (1990); see also Gray (1984). From tests and a comparison of different schemes applied to the same stars it follows that the mean flux density $\langle |\varphi| \rangle = f B$ is better determined than f and B individually, because errors in f and B partly cancel. The filling factor f is particularly sensitive to the treatment of the line formation and to the flux-tube and atmospheric models. Measurements in infrared lines (Valenti *et al.*, 1995) improve the accuracy of the determination of f and B because the magnetic sensitivity increases linearly with wavelength, and the emergent radiative intensities depend less sensitively on the atmospheric models. Note that measurements in the visible spectrum severely underestimate the contributions of starspots to the total magnetic flux.

Saar (1996) compiled a critically selected sample of magnetic measurements published up to that date, which contains 17 G-, K- and M-type main-sequence stars and one K2 subgiant component in an active spectroscopic binary. The moderately active G- and early K-type stars have an intrinsic field strength B ranging from 1.0 to 1.9 kG, and filling factors f ranging from $\sim$1.5%–35%. The very active Ke and Me stars (with chromospheric emission lines and starspots) exhibit B up to 4.2 kG and f up to 70%. We discuss these results below.

Although the number of field measurements is small, the results are fundamental in the interpretation of outer-atmospheric phenomena. The results confirm that the stellar magnetic structure fits into the framework of the solar magnetic structure discussed in the preceding chapters. Moreover, the magnetic flux densities correlate well with the outer-atmospheric radiative flux densities (Section 8.4, particularly Fig. 8.10; Section 11.3), which provides calibrations of radiative flux densities as activity measures.

9.2.2   *Scaled flux-tube models and structural hierarchy of stellar fields*

From the solar analog, we infer that the photospheres of stars with convective envelopes comprise two components (see Section 4.2): (1) a strong field in dark spots and bright faculae, consisting of flux tubes (Section 4.3), separated by (2) a photosphere with an intrinsically weak field (Section 4.6) that does not noticeably contribute to the Zeeman line broadening. Following Zwaan and Cram (1989), we attempt to estimate properties of the magnetic concentrations in cool stars by scaling the concentrations observed in the Sun by rules inferred from flux-tube models (Section 4.3) and observed solar features.

The geometrical and thermodynamic structure of flux tubes and flux-tube bundles depends on three length scales: (a) the flux-tube radius R, (b) the Wilson depression Δz (Fig. 4.5), and (c) the mean-free path of photons ℓ_R [Eq. (2.55)]. These scales depend strongly on the height z in the atmosphere; hence for representative use we take the values at the geometrical height z_0 corresponding to the monochromatic optical depth $\tau_5 = 1$ at 5,000 Å, which is deep in the photosphere, directly above the convection zone. The tube radius R determines the amount of magnetic flux Φ contained in the tube (bundle), and its appearance as either a bright grain, a dark pore, or a complete spot. The Wilson depression Δz is connected with the field strength in the tube; it depends on the thermodynamic stratification within the tube, which is set by the energy transport in the tube. As long as this energy transport is not understood, the Wilson depression Δz must be treated as an adjustable parameter. The photon mean-free path ℓ_R is involved in the radiative exchange between the flux tube and its surroundings, which, for one thing, determines whether a flux tube is observed as bright or dark. These three structure parameters are not independent.

Zwaan and Cram (1989) suggested that the structural parameters $\ell_R(\tau_5 = 1)$ and $\Delta z(\tau_5 = 1)$, if expressed in units of the pressure scale height $H_p(\tau_5 = 1)$, depend only slightly on T_{eff} and gravity g. For the mean-free path $\ell_R(\tau_5 = 1)$ this is readily verified: from the Carbon–Gingerich (1969) models and Rosseland opacity tables published by Kurucz (1979) it follows that for main-sequence stars ranging from F5 to M0 and giants from G0 III to K5 III the ratio $\ell_R(\tau_5 = 1)/H_p(\tau_5 = 1)$ differs by less than a factor of 2 from the solar value of 0.22. The dependence on metal abundance is also weak: for a G-type dwarf with $T_{eff} = 5,500$ K, ℓ_R drops from 117 km (solar abundances) to 53 km if the metal abundance is reduced by a factor of 100.

Spruit (1976) deduced that the relative Wilson depressions $\Delta z/H_p$ range from 0.6 to 1.3 for tiny facular or network grains up to small pores in the Sun. The Wilson depression measured in sunspots indicates that in umbrae $\Delta z/H_p$ equals 3 to 4. Zwaan and Cram (1989) assumed that these results depend only slightly on T_{eff} and gravity g. The increase of $\Delta z/H_p$ with increasing flux-tube radius R is expected to be associated with an increase of the field strength of the tube, as is the case in the Sun (Sections 4.2 and 4.3).

The fanning out of the flux tubes or flux-tube bundles in the photosphere depends on the ratio of the photospheric tube radius R to the pressure scale height H_p. For the Sun, the division between pores without penumbrae and complete spots with umbrae and penumbrae occurs around $R(\tau_5 = 1)/H_p(\tau_5 = 1) = 13$, and this value may apply to a pore–spot transition in all stars with convective envelopes, because the effect is a direct consequence of magnetostatic equilibrium.

With Zwaan and Cram (1989) we assume that the small-scale magnetic structure can be described by the slender-flux-tube approximation. A characteristic field strength B^* is estimated by

$$B^* \equiv [8\pi p(z_0)]^{1/2}, \tag{9.2}$$

where z_0 is the geometrical height where the monochromatic optical depth τ_5 at 5,000 Å in the quiet photosphere equals unity, and p is the photospheric gas pressure. At a given geometrical height, the optical depth inside a solar flux tube is smaller than in the photosphere outside by a factor somewhat larger than 10 because of the Wilson depression Δz (Fig. 4.5), which for facular and network grains is approximately equal to the pressure scale height $H_p \simeq 150$ km. The geometrical level z_0 is crudely estimated to correspond to the level within the tube where the magnetic signal for standard Zeeman measurements originates. In other words, we estimate that the characteristic level of line formation is situated higher than the level of continuum formation by an amount that happens to correspond to the Wilson depression Δz. In any case, for the Sun Eq. (9.2) yields $B^* = 1{,}800$ G, which corresponds to the upper limit of intrinsic field strengths measured in solar plages and network, and somewhat below the field strengths measured in pores (Section 4.2). We assume that B^* yields a similarly typical measure for bright magnetic structures in cool stars.

From the grid of model atmospheres published by Carbon and Gingerich (1969), Zwaan and Cram (1989) computed B^* for a range of effective temperatures T_{eff} and surface gravities g, and by interpolation and extrapolation, for main-sequence stars and for giants; see Fig. 9.1. Along the main sequence, B^* ranges from 1.4 kG in the mid-F-type stars to 3.2 kG in the late-K-type dwarfs. For giants, B^* decreases from $B^* \simeq 0.65$ kG for early G-type giants to $\simeq 0.4$ kG at K5 III.

Presumably the relative Wilson depression $\Delta z / H_p$ depends on R / H_p because the normalization of the structural parameters by H_p largely eliminates the dependence on the photon mean-free path ℓ_R. We conjecture that, by analogy with the Sun, the strong magnetic-field patches in any cool star form a hierarchy consisting of four classes (see Table 4.1). The radii R are scaled from the solar values by the pressure scale height H_p, and the ranges of field strengths B are taken from the solar values and scaled with

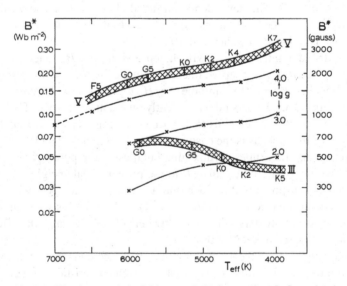

Fig. 9.1. The characteristic field strength B^* for small-scale flux tubes in stellar photospheres as a function of effective temperature T_{eff} and gravity g. The cross-hatched bands indicate the estimates for main-sequence stars (LC V) and giants (LC III) (from Zwaan and Cram, 1989).

the characteristic field strength B^* according Eq. (9.2). These four classes are described below.

1. *Network patches and faculae* consist of bright grains with radii $R \lesssim 1.5H_p$, with field strengths B in the range $0.6B^* \lesssim B \lesssim B^*$.
2. *Micropores or magnetic knots*, which are flux tubes with $R \approx 2H_p$ and field strength $B \approx B^*$; these knots are expected to appear as bright Ca II H and K faculae.
3. *Dark pores*, which are small starspots without penumbrae, with $3.5H_p \lesssim R \lesssim 13H_p$ and $1.1B^* \lesssim B \lesssim 1.3B^*$.
4. *Starspots*, which presumably consist of umbrae surrounded by penumbrae, with total radii $R \gtrsim 13H_p$ and umbral field strengths B_u in the range $1.4B^* \lesssim B_u \lesssim 1.7B^*$.

The properties of the inferred stellar magnetic configurations depend on the classical stellar parameters effective temperature T_{eff} and gravity g through $p(z_0)$ in Eq. (9.2) and the scaling of the structure parameters with the pressure scale height H_p. The filling factor of the magnetic structure depends on an additional "activity parameter" which describes the dynamo strength, as discussed in Section 11.3.

Presumably starspots contribute little to the flux spectrum in the visible; hence it is expected that the magnetic signature of the visible spectrum is largely determined by the network patches, the faculae, and the magnetic knots. Thus we expect that the measured field strengths B turn out to be nearly equal to the characteristic strength B^*. The measurements compiled by Saar (1996) for the G- and early K-type dwarfs confirm this expectation, including the expected trend of the intrinsic field strength to increase with decreasing effective temperature. For the very active, "spotted," Ke- and Me-type dwarfs, the measured value of B turns out to be somewhat higher than B^*; Saar (1996) explained that the majority of these observations were obtained in the infrared where the contribution of the sunspots to the flux spectrum probably is significant and hence is expected to lead to a higher mean B.

We conclude that the speculative extrapolation from solar applications of flux-tube modeling to that of stars finds support in observational data. It should be pointed out, however, that the simple scaling of solar structure to stars, as proposed above, probably fails in cases with large magnetic filling factors; for the Ke- and Me-type stars, Saar (1996) lists filling factors between 50% and 70%. For instance, in closely packed magnetic structures, starspots would have no room to unfurl their penumbrae. More fundamentally, the side-by-side coexistence of convection and strong magnetic structure (Section 4.1.2) may change in character when the convective envelope is largely filled by a strong field.

9.3 The Mt. Wilson Ca II HK project

The great importance of the Mt. Wilson Ca II HK program for the study of magnetic activity lies in the large and homogeneous data sets that have been and are being collected. During the first program (starting in 1938), 10-Å/mm photographic spectrograms of G-, K- and M-type stars were collected at Mt. Wilson and at Mt. Palomar; this data set has been used in the determination of the Wilson–Bappu effect (discussed in the paragraphs that follow), and in the first systematic studies of the magnetic activity of evolved stars (Section 11.1).

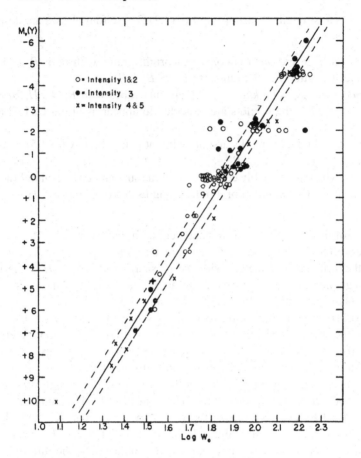

Fig. 9.2. Absolute visual magnitude M_v plotted against the logarithm of the Wilson–Bappu width $W_\circ = W_{WB}$, expressed in kilometers per second and corrected for instrumental broadening. The stars are divided into three ranges of their activity level (from Wilson and Bappu, 1957).

9.3.1   The Wilson–Bappu effect

The width W_{WB} of the emission feature in the core of each of the Ca II H and K resonance lines (see Fig. 2.6) correlates remarkably closely with the absolute visual magnitude M_v of the star. This effect is called the *Wilson–Bappu effect* after the first detailed investigation by Wilson and Bappu (1957). The Wilson–Bappu width W_{WB} is defined by the outer edges of the emission feature as seen on a suitably exposed, photographic spectrogram. Wilson and Bappu found a linear relation between the absolute magnitude M_v and log W_{WB}, covering the entire range of luminosity classes, from main-sequence stars to supergiants (Fig. 9.2). According to Wilson and Bappu, the shape of this relation does not depend on the stellar spectral type and not even on the activity level, as indicated by the strength of the Ca II emission. In their investigation of the dependence of the Wilson–Bappu effect on stellar atmospheric parameters, Lutz and Pagel (1982) found

$$W_{WB} \sim T_{\text{eff}}^{1.65} \, g^{-0.22} \, [\text{Fe/H}]^{0.10}. \tag{9.3}$$

Considering the small range of effective temperatures in the cool-star domain, the Wilson–Bappu width W_{WB} is found to depend primarily on the surface gravity g, somewhat on the effective temperature T_{eff}, and only very slightly on the metal abundance [Fe/H].

Solar data indicate a small but significant increase of the width of the Ca II emission cores with increasing activity (White and Livingston, 1981b). In his analysis of Ca II K profiles measured for the Sun as a star, Oranje (1983a) used the wavelength separation W_0 between the half-intensity points between the K2 peak and the K1 minimum on both sides as a proxy for the Wilson–Bappu width W_{WB}. One of his findings is that $W_0 \simeq W_{WB}$ increases only slightly with an increasing activity level. The reason turned out to be that the variable Ca II K line profile $I(\lambda, t)$ is readily described by the sum of a line profile for minimal activity, $C(\lambda)$, plus the product of an activity factor $a(t)$ and a virtually constant "plage emission profile" $A(\lambda)$ [Oranje, 1983b; see Eq. (8.10) in Section 8.4.2]. Apparently this standard profile $A(\lambda)$ represents the mean emission profile for all elements emitting chromospheric radiation. It is a surprising finding that this profile remains virtually constant during the activity cycle, from the minimum of the cycle when the emission comes exclusively from the quiet magnetic network, to the maximum of the cycle when there is an appreciable contribution from plages and enhanced network in active regions.

The Wilson–Bappu effect is also found in the widths of ultraviolet emission lines, including the Mg II resonance lines and other saturated chromospheric and transition-region emission lines (see McClintock *et al.*, 1975, and Engvold and Elgarøy, 1987). The Mg II h and k lines display a well-defined Wilson–Bappu relation; in addition, the Mg II line widths are seen to increase somewhat with increasing activity (Elgarøy *et al.*, 1997).

The solar line profile for minimal activity, $C(\lambda)$ in Eq. (8.10), does not show a significant emission core, which could suggest that the Wilson–Bappu effect is a property of the chromospheric structure pervaded by a strong magnetic field only. In the spectra of virtually all K- and M-type giants and supergiants, however, conspicuous emission cores are present in the Ca II H and K line cores, with well-defined widths satisfying the Wilson–Bappu relation. Since for the majority of these evolved stars the chromospheric emission consists nearly completely of basal emission (Section 11.1), the Wilson–Bappu effect appears to apply to the basal emission as well. Hence the Wilson–Bappu effect is a *general* chromospheric property of all cool stars.

There is not yet a convincing quantitative explanation of the Wilson–Bappu effect. It is argued that the main effect, namely the broadening of the emission line cores with decreasing surface gravity, results from the increase of the column mass density in the chromosphere (see Ayres, 1979). Numerical simulations of a restricted problem, based on some assumed turbulent velocity for the nonthermal line broadening, support this notion (Cheng *et al.*, 1997).

9.3.2 *Spectrophotometry: Ca II H and K variability*

Does the magnetic (chromospheric) activity of main-sequence stars vary with time, and if so, how? This question drove Olin C. Wilson during most of his career. Although the Ca II H and K line-core emissions are the most sensitive activity criterion in the visible spectrum, Wilson became convinced that photometric accuracies of $\sim 1\%$ are required to pinpoint time-dependent effects. Thus Wilson (1968) engaged in a program using the Coudé scanner of the 100-in. telescope at the Mt. Wilson Observatory as a two-channel photometer, centering with one channel on the Ca II H or K line with a 1-Å passband, and with the other channel counting in two 25-Å (R and V) windows, separated by 250 Å, on either side of the Ca II H and K region. The Ca II H+K index is defined by the ratio S of the counts in the H and K windows over those in the R and V channels.

This sensitive and stable photometer proved the variability of the Ca II H+K signal from cool stars with an emission level above the minimum level, on time scales ranging from a fraction of a day to many years. In 1966, Wilson started his search for variations in main-sequence stars with periods over several years, with stellar analogs of the solar activity cycle in mind; we discuss results in Section 11.7. Wilson (1968) published some first results, among them that for main-sequence stars of some specific Strömgren color $b - y > 0.290$ (spectral type later than F5), there is a range of Ca II H+K emission indices. In other words, there is an *activity parameter*, supplementary to the standard parameters effective temperature T_{eff} and surface gravity g, that determines the structure of atmospheres of cool stars. Another striking feature is the color-dependent lower boundary to the Ca II H+K emission index.

Virtually all stars in the sample, including the 18 "standard stars" with the smallest Ca II H+K flux, show apparently intrinsic, short-term variability during each of their observing seasons – even the feeblest chromospheres of main-sequence stars appear to be variable on time scales of less than a year. The amplitude of the short-term fluctuations increases (with little scatter) with the mean Ca II H and K flux of the star as averaged across the entire observing period.

In 1977, the survey program was continued using the Mt. Wilson 60-in. telescope with the Ca II HK photometer. Since then, this instrument has been used nearly every night in various activity programs, such as the continuation of Wilson's "cycle program." The measurements are presented in terms of the Ca II H+K emission index:

$$S = \alpha \, \frac{H + K}{V + R},\tag{9.4}$$

where $H + K$ are the counts in the Ca II line cores, and where V and R are the simultaneous counts in the violet and red "continuum bands" on both sides of the Ca II doublet, and α is the nightly calibration factor that is determined by means of standard sources. Note that the normalization by the flux in the "continuum bands" makes the emission index S blow up with decreasing effective temperature. For the conversion of the index S into the line-core flux density F, see Rutten (1984a).

An important application of the Ca II HK photometer is the determination of stellar rotation periods from a modulation of the Ca II H+K signal, which is an improvement over the spectroscopic determination of the projected rotational velocity $v \sin i$, with the unknown projection factor $\sin i$. Rotation periods have been determined ranging from 2 days up to $\sim$50 days (Vaughan et al., 1981; Baliunas et al., 1983). For main-sequence stars with a moderate to large emission index $\langle S \rangle$, rotation modulation is nearly always observed. Rotation periods longer than $\sim$1 month are hard to detect, because the activity level $\langle S \rangle$ tends to decrease with increasing rotation periods (Section 11.3), and because evolutionary changes in activity complexes decrease the coherence of the signal (Section 6.2). Of the subgiants and giants in the Baliunas *et al.* (1983) sample, only two rapidly rotating stars show detectable rotation modulation on the time base of ∼100 days. From 8 yr of monitoring of 12 late G- and early K-type giants, Choi *et al.* (1995) derived rotation periods between 112 and 183 days for eight stars and two more with 68 and 59 days; in two cases, no modulation was found. In 10 out of 12 cases, the modulation is clearly visible, which indicates that in active giants the global distribution of the magnetic structure varies little over time spans of 100–200 days.

9.3.3   The solar-neighborhood program

Vaughan and Preston (1980) presented results of their solar-neighborhood survey with the Ca II HK photometer for more than 400 stars within 25 parsec from the Sun. Figure 9.3 shows the results for the stars with known $B - V$ colors. This sample confirms the large variation in the activity level of otherwise identical (main-sequence) stars. Vaughan and Preston (1980) called attention to the two fairly distinct strips of increased

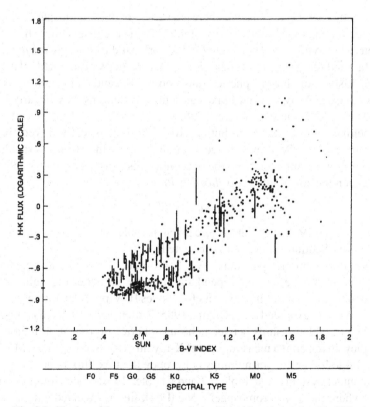

Fig. 9.3. The logarithm of the Ca II emission index S against color $B - V$ for (main-sequence) stars from the solar-neighborhood survey (Vaughan and Preston, 1980). The bars indicate the excursions of the index S for stars monitored by Wilson (1978) for long-term activity variations. The Vaughan–Preston gap is visible for $B - V < 1.1$ (figure from Hartmann and Noyes, 1987).

density of data points in the range $0.4 < B - V < 1.0$, separated by a strip of low density, which is often called the *Vaughan–Preston gap*. Note that there is no trace of such a gap for $B - V > 1.0$.

9.4   Relationships between stellar activity diagnostics

The clear correlation between the strength of the atmospheric magnetic field and the outer-atmospheric emission in the solar atmosphere formed the rationale behind the search for similar trends in stars other than the Sun. The explorations of stellar chromospheric activity by O. C. Wilson using Ca II H+K measurements revealed a substantial range of activity in stars, even at a given surface gravity and effective temperature: the observed range in emission strengths (see Fig. 2.17) by far exceeds the short-term fluctuations that would be expected as the result of rotational modulation and active-region emergence and decay in a star like the Sun. Wilson (1968) already pointed out, for instance, that there was a strong increase in the range of Ca II H+K fluxes in the F-star domain, an increase which he termed the "chromospheric bulge." Later, it turned out that the primary factor that controls the activity level in stars was the rotation rate (Section 11.3).

In addition to the Ca II H+K emission, there is a variety of outer-atmospheric emissions that serve as measures of magnetic activity (Table 2.2). These diagnostics show similar, and interrelated, behavior, as demonstrated in the early studies by Ayres *et al.* (1981b), Zwaan (1981a), and Oranje *et al.* (1982a). Since then, rather well-defined relationships have been established for chromospheric, transition-region, and coronal emissions from early-F- through early-M-type stars. Fluxes at radio wavelengths vary linearly with the soft X-ray flux (see the review by White, 1996).

If the sometimes substantial contributions from the (acoustically driven) basal atmosphere (introduced in Section 2.7; see Section 11.4 on the strong basal fluxes in early-F-type stars) are subtracted from the observed emissions, power-law relationships are found between the so-called *excess flux densities*:

$$\Delta F_{\mathrm{j}} = a_{\mathrm{ij}} \cdot \Delta Fi^{b_{\mathrm{ij}}} \tag{9.5}$$

(see the example in Fig. 9.4; exponents b_{ij} are given in Table 9.1). The operator Δ implies that any basal flux (Section 2.7) is to be subtracted.

These power-law relationships extend over 2–4 orders of magnitude, depending on the diagnostic used, and yet do not depend significantly on either the stellar effective temperature or on the stellar luminosity class for classes II–V, provided the very cool stars beyond mid-M type are excluded (e.g., Oranje, 1985; Rutten *et al.*, 1989). The supergiants of LC I do obey the relationships between chromospheric and transition-region fluxes, even though they do not emit a measurable soft X-ray flux (see Section 11.1). Most active binaries (see Table 9.2 on the terminology) also obey the flux–flux relationships. Only the W UMa contact systems (see Table 9.2 and Chapter 14) deviate from the flux–flux relationships where the low-chromospheric Mg II h+k flux is involved, but not in other diagnostics. The very rapidly rotating, apparently single FK Com stars appear to deviate somewhat from the main flux–flux relationships (Bopp and Stencel, 1981; Zwaan, 1983), as do the very young T Tauri stars with accretion disks (Section 11.6). But with those few exceptions, the flux–flux relationships hold for all cool stars.

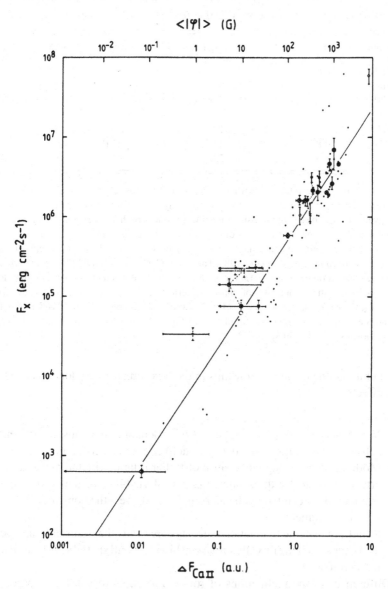

Fig. 9.4. Relationship between the soft X-ray flux density, F_X (0.05–2 keV; observed with *EXOSAT*'s *CMA*/L1 assembly), and the Ca II H+K excess flux density $\Delta F_{\text{Ca II}}$ (one unit equals 1.3×10^6 erg cm^{-2} s^{-1}; see Section 2.3.4). The line is a power law with slope 1.5. *EXOSAT* and Mt. Wilson Ca II H+K measurements were obtained within 3 days for 13 of the stars, and within 3 months for the others. Small dots are used for the comparison of *EINSTEIN*/*IPC* and Mt. Wilson Ca II H+K flux densities that were measured up to several years apart. The scale for the mean absolute magnetic flux densities at the top of the diagram is determined with Eq. (8.8). Typical solar soft X-ray fluxes (Section 8.6): 3×10^3 erg cm^{-2} s^{-1} for coronal holes, 6×10^4 erg cm^{-2} s^{-1} during cycle minimum (for the quiet network), and 6×10^5 erg cm^{-2} s^{-1} as a typical average for nonflaring active regions (figure from Schrijver *et al.*, 1992a).

Table 9.1. *Exponents b_{ij} of power-law fits to stellar magnetic and radiative flux densities, $\langle|\varphi|\rangle$ and $(\Delta)F_i$, respectively[a]*

Diagn.	$\langle|\varphi|\rangle$	Ca II	Mg II	Si II	C II	Si IV	C IV
Ca II	0.6						
Mg II		1.1	–				
Si II		1.4	1.2	–			
C II		1.8	1.5	1.3	–		
Si IV		2.2	1.6	1.3	0.9	–	
C IV	0.8	1.9	1.6	1.4	1.0	1.1	–
X-ray	1.0	2.2	2.0	1.8	1.3	1.1	1.3

[a] The exponents are taken from Rutten *et al.* (1991) for relationships between radiative losses. These values were derived by optimizing a set of relationships using rather conservatively low basal flux densities leading to numbers that differ noticeably from other published values based on studies of pairs of radiative diagnostics. Exponents for relationships between stellar average magnetic flux densities and Ca II H+K or C IV and soft X-rays were taken from Schrijver *et al.* (1989a) and Saar (1996), respectively. The uncertainties in the exponents range from ∼10% for the relationships between chromospheric diagnostics up to 30% in the comparison of chromospheric to coronal diagnostics (compare, for example, the exponent for the relationship between soft X-rays and Ca II H+K in this table, and the value for the relationship plotted in Fig. 9.4). The formation temperatures of the diagnostics are given in Table 2.2.

The scatter about the mean relationships in the flux–flux scatter plots is caused by the following effects:

1. Variability on different time scales:
   - Variations caused by (a) cyclic or (quasi-)chaotic dynamos and (b) the birth, subsequent evolution, and decay of active regions (see Section 9.5.2).
   - Modulation of the signal because of stellar rotation (including center-to-limb effects in optically thick lines, the partial visibility of the coronal emission from regions occulted by the disk, and – as above – the nonsimultaneity of the flux measurements).
   - Rapid, small-scale fluctuations such as flares and other transients, and possibly deviations from the flux–flux relationships on small spatial scales (as discussed in Section 8.4).
2. Differences in the inclinations of rotation axes, so that different parts of any activity belts are seen, weighted with center-to-limb effects.
3. Binarity of the source, with both stars active, or both stars contributing to a reference bandpass.
4. Interstellar extinction, and possible circumstellar emission.
5. Calibration uncertainties related to the details of the source spectrum (see Section 2.3.4), and other instrumental uncertainties.

The best-determined relationships largely avoid the scatter related to item 1 by comparing nearly simultaneously observed fluxes. Schrijver *et al.* (1992a), for instance, study the relationships between Ca II H+K and Mg II h+k flux densities that were obtained within

Table 9.2. *Definitions of some common classes of stars and binary systems in the cool star domain[a]*

Name	Definition
Algol	Semi-detached binary system with a cool subgiant and an early-type companion, with mass transfer from the Roche-lobe filling secondary to the primary star, which has resulted in the more evolved star being the least massive.
ζ Aur	Eclipsing binary consisting of a bright K (super) giant and a hot B star.
BY Dra	The most active main-sequence emission-line stars of type dKe or dMe; many are binary systems.
FK Com	Very active, rapidly rotating, late-type single giants, possibly coalesced binaries.
RS CVn	In the strict definition, a close binary with stars of nearly equal masses, with the hotter component of spectral type F, G, or K, and luminosity class V or IV (but with the subgiant not filling its Roche lobe), with orbital periods between 1 day and 2 weeks. In the expanded definition, any tidally interacting binary system with at least a single cool star. RS CVn stars have negligible mass exchange.
UV Ceti	Red-dwarf flare stars, variable, either single or members of binaries.
T Tau	Irregularly variable pre-main-sequence stars.
W UMa	In the strict definition: eclipsing contact binaries of main-sequence stars. In the relaxed definition, semidetached systems are sometimes included.

[a] Compare Fig. 14.1 for the definition of (semi) detached and contact systems.

36 h of each other, and between Ca II H+K and soft X-ray flux densities obtained within a few days up to a few months of each other (see Fig. 9.4). For the dwarf and giant stars in their sample, ranging from F5 to K3, the scatter about the average relationship is entirely compatible with calibration and instrumental uncertainties.

In spectroscopic binaries (or binaries too narrow for X-ray telescopes to resolve) the presence of a companion may introduce offsets from the mean relationships too: the addition of the fluxes of the companions offsets fluxes along a linear relationship in a flux–flux relationship, and thus off the main relationship if that itself is nonlinear. Additional offsets are introduced if the basal fluxes differ for the two components, and if the companion contributes to a reference channel, such as would be the case in Mt. Wilson Ca II H+K measurements (Section 2.3.4).

The studies quoted above suggest that the scatter about the mean flux–flux relationships can be understood without leaving room for a dependence on fundamental stellar parameters. Hence we conclude for the early-F through early-M type (bright) giant, subgiant, and main-sequence stars, that whenever a magnetic field manages to deposit nonradiative energy in the outer atmosphere, the radiative losses from different parts of the atmosphere adjust to reach a well-defined distribution. This is apparently the case both statistically speaking on relatively small length scales (see Section 8.4), as well as on the scale of the star as a whole, including (tidally interacting) detached binaries (Chapter 14). Some of these relationships, such as that shown in Fig. 9.4 for soft X-rays versus Ca II H+K, hold for over a factor of 10,000 in soft X-ray flux densities! This is a factor of $\sim$50 beyond the contrast between a solar coronal hole and the average for a nonflaring active region (Section 8.6).

This astounding adherence to power-law relationships regardless of stellar age, size, or luminosity suggests that one single *activity parameter* determines the structure of the outer atmospheres (Zwaan, 1983). This parameter Υ varies along the flux–flux relationships, itself being determined primarily by the stellar rotation rate; see Section 11.3.

Most of the power-law flux–flux relationships deviate significantly from linearity, which means that the changing conditions do not merely reflect changing numbers of some (ensemble of) active features, but that a fundamental change in the response of the outer atmosphere to the mean magnetic flux density is required (this is further discussed in Section 9.5.2). The power-law index differs increasingly from unity as the difference in formation temperatures of the diagnostics used increases; see Table 9.1. The power-law indices listed in that table are derived by iteratively determining a best color-dependent basal flux density that is to be subtracted from the net observed flux densities. Because of the rather crude parameterization of the color dependence of this basal flux that was used by Rutten *et al.* (1991), the values of the exponents are approximations only. For instance, Schrijver *et al.* (1992a) find that the relationship between Ca II H+K and Mg II h+k excess flux densities for dwarf stars with $B - V$ exceeding $\sim$0.45 is perfectly linear:

$$\Delta F_{\mathrm{MgII}} = (0.78 \pm 0.05)\Delta F_{\mathrm{CaII}}. \tag{9.6}$$

For stars hotter than $B - V \approx 0.45$, the relationship changes (Oranje and Zwaan, 1985), probably because Ca$^+$ begins to ionize to Ca^{2+} as the photospheric temperature increases. The relationships involving soft X-ray fluxes depend on the passband and thus on the instrument: instruments sensitive to hotter plasmas yield a higher exponent.

Despite the substantial difference in formation temperature (Table 2.2), the emission in the C II and C IV lines is strictly proportional. The C II and C IV lines are particularly useful for studies of magnetic activity of relatively warm "cool stars," such as the early-F-type main-sequence stars, because these lines lie in a part of the ultraviolet wavelength range where the stellar continuum is very low for stars cooler than approximately mid-F. The Si II and Si IV lines are very similar to the C II and C IV lines in their response to atmospheric activity.

The relationships between flux densities discussed thus far only compare radiative losses from the different outer atmospheric domains. The direct comparison between radiative and magnetic flux densities for stars is far more difficult than for the Sun, because magnetic fields on stars are not easy to measure (Section 9.2). Schrijver *et al.* (1989a) and Saar (1996), for example, have made comparisons of $\langle |\varphi| \rangle$ and radiative losses, such as the Ca II H+K, C IV, and coronal soft X-ray flux densities. These values are also shown in Table 9.1. Saar (1996) lists a power-law index of 0.5 for the relationship between the Ca II H+K excess flux density and the magnetic flux density, close to the value of 0.6 found by Schrijver *et al.* (1989a), but notes that there seems to be a saturation for $\langle |\varphi| \rangle \gtrsim 400$ G. This observation is based on a sample of only 11 stars, however, and Saar warns that it should be viewed with caution.

The issue of an apparent saturation in stellar activity for the fastest rotators has been raised in numerous studies. It appears that the power-law relationship as shown in Fig. 9.4 does not hold for the most active stars with $F_X \gtrsim 10^7$ erg cm^{-2} s^{-1}. In that figure, the most active stars lie well above the mean power-law relationship, indicating a saturation in the chromospheric emission. Other studies now suggest that both diagnostics saturate, but that chromospheric diagnostics reach saturation before the coronal ones do, causing an upturn in the flux–flux relationship. In Ca II H+K, the emission at saturation coincides roughly with the brightest nonflaring regions in solar plages, where a similar saturation is observed (Schrijver *et al.*, 1989a). We return to the issue of dynamo saturation in the discussion of the rotation–activity relationship (Section 11.3) and of stellar angular-momentum loss (Chapter 13).

9.5 The power-law nature of stellar flux–flux relationships

9.5.1 Solar data with angular resolution

The origin of the nonlinearity of relationships between radiative losses from the outer atmosphere and photospheric magnetic flux densities, such as the example for Ca II K emission and $|\varphi|$ in Eq. (8.8), lies ultimately in the dependence of the internal atmospheric structure on the (dynamic) magnetic geometry. The crude model developed by Solanki *et al.* (1991), mentioned in Section 8.5, involves the computation of the field geometry given the photospheric values of the intrinsic field strength B_0 and the flux-tube radius r_0, and the radius out to which the field can expand before becoming vertical into the high atmosphere. They find a nonlinear dependence of the tube-integrated Ca II K flux on the flux-tube density, primarily because of the dependence of the merging height of the magnetic field (Section 8.1) on the filling factor. For small filling factors the merging of the fields of neighboring tubes occurs so high that the surface-averaged emission in the core of the K line comprises a significant contribution from the cool material underneath the canopy, resulting in a low core intensity. For progressively larger filling factors the merging height decreases, which results in a weaker than linear increase in the relative contribution from the flux-tube atmosphere. Once the merging height has dropped below the formation height of the line core, the emission saturates and becomes independent of the filling factor.

Unfortunately, empirical relationships such as that between Ca II K emission and $|\varphi|$ cannot be compared directly with these model results for individual flux tubes, because observations at moderate angular resolution average over a number of flux concentrations of which the brightness depends nonlinearly on the local flux density. To bridge the gap between models for individual tubes and observations, Schrijver (1993b) started from the approximation that the total chromospheric emission observed from an area (say, a resolution element of a few arcseconds) can be seen as the sum of the emissions from all of the tubes contained within that pixel in which neighboring tubes merely determine the area into which flux expands, disregarding effects on the internal stratification. He then uses the small-scale positioning of these footpoints, as discussed in Section 5.3, to demonstrate that spatial smoothing over small length scales hardly affects the mean relationship between radiative and magnetic flux densities, by the following argument.

Let $N(a, n)$ da be a histogram of the number of identical flux tubes of which the field expands to an area between a and $a + $da at the merging height given an average flux-tube density n. This function $N(a, n)$ serves as the parent distribution function from which areas are drawn for any realization of flux-tube positions within some observed area. Assume that the integrated chromospheric flux L_i (e.g., in Ca II K) of an individual flux tube in excess of any nonmagnetic emission is a function of the expansion area a, which is determined by the distance to a few nearest neighbors, with the set of expansion areas covering the plane in a surface-filling tesselation (in a crude approximation in which only the forces exerted by the nearest neighbors are considered, this would be described by the Voronoi tesselation discussed in Section 2.5.1). With these assumptions, the expectation value of the radiative flux density ΔF from a unit area with average flux-tube density n is

$$\Delta F = \int_0^A N(a, n) L_i(a) \, da. \tag{9.7}$$

The upper limit A is to be sufficiently large so that ΔF_i becomes insensitive to its precise value, because of the rapid decrease of the function $N(a, n)$ with a (see below).

Surprisingly, it is not necessary to know $N(a, n)$ in detail to understand the transformation properties of flux–flux relationships. As discussed in Section 5.3, magnetic flux tubes, well after emergence, appear to be distributed randomly in the solar photosphere on scales up to at least 3 arcsec. This implies that $N(a, n)$ is statistically homogeneous, that is, invariant under rotations and translations, and with simple scaling properties. Schrijver (1993b) argues that therefore

$$N(a, n) \, da = \alpha n^2 h(a \cdot n) \, da, \tag{9.8}$$

where $h(a \cdot n)$ is a function of the product of a and n only, and α is a normalization constant. The quadratic dependence on n follows from the combination of the scaling of the total number of expansion areas with n, and the transformation properties of distribution functions. Note that this same scaling behavior holds in the case in which tubes have a range of strengths that is independent of n (which is roughly analogous to the multiplicatively weighted Voronoi tesselation).

A combination of Eqs. (9.7) and (9.8) and the power-law relationship between a radiative flux density and the magnetic flux density [such as Eq. (8.9)] with power-law index β_i yields

$$\int_0^\infty n^{1-\beta_i} L_i(a) h(a \cdot n) d(a \cdot n) = \frac{\gamma}{\alpha}. \tag{9.9}$$

Because this should hold for all number densities n, it follows that to a good approximation the product $L_i(a) \cdot n^{1-\beta_i}$ is a function of $a \cdot n$, or $L_i(a) \propto a^{1-\beta_i}$, so that for a single flux tube

$$\Delta F_i^* \equiv \frac{L_i(a)}{a} = ka^{-\beta_i} = kn_i^\beta, \tag{9.10}$$

with k as a constant of proportionality. This relationship shows the same exponent as the relationship observed with moderate spatial resolution, such as Eq. (8.9), which was derived from solar observations with an angular resolution of 2.4 up to 15 arcsec.

Now that we have demonstrated the (approximate) equivalence of the relationships for individual flux tubes and for averages at an intermediate resolution, we make a direct comparison of observed relationships with numerical models. The qualitative agreement of the models by Solanki *et al.* (1991) and the empirical relationship in Eq. (8.8) is encouraging. It suggests that a combination of traditional one-dimensional static models is compatible with the observations, but a fully self-consistent model including multidimensional radiative transport, waves, and magnetic geometries is yet to be developed. Note, moreover, that this discussion is limited to moderate to high filling factors of magnetic flux. In areas of the Sun where there is very little flux, the expansion of the field is so strong that the approximation that the internal structure is a simple stratified chromosphere is untenable: the observations in Ca II H or K show that the emission is strongly concentrated toward the cores of flux bundles.

Whereas radiative transfer effects appear to cause the nonlinearity of flux–flux relationships for the chromosphere, this cannot be the case for the transition region where radiative diagnostics such as C II and C IV lines are (at least effectively) optically thin anywhere on the disk except perhaps very near to the limb. In Section 8.4.3 we showed that the amount of energy radiated in the C IV lines varies weaker than linearly with the magnetic flux density (and the same is, by inference, true for, e.g., the C II lines because the C II and C IV fluxes are strictly proportional; see Section 9.4). Schrijver (1990) discusses this nonlinearity. He argues that it might reflect that plasma at transition-region temperatures is heated by two distinct processes: heat conduction from the coronal part of high loops, and some local heating that occurs either in that same region or in lower-lying loops that do not have truly coronal segments. The radiative losses in C IV can, for instance, be predicted quite accurately from the following relationship reflecting coronal and local heating:

$$F_{\mathrm{C\,IV}} = 80|\varphi| + 900|\varphi|^{0.6} = \frac{F_{\mathrm{X}}}{40} + \frac{\Delta F_{\mathrm{Ca\,II}}}{70} \quad (\mathrm{erg\,cm}^{-2}\,\mathrm{s}^{-1}). \qquad (9.11)$$

This possible separation is certainly not a unique solution.

9.5.2 Hemisphere-averaged stellar data

The next step in understanding the relationships between disk-averaged stellar fluxes is to relate the relationships for solar data with intermediate angular resolution to those observed for the Sun as a star in comparison to other cool stars. To do this, it is necessary to integrate the intensities $I_i(|\varphi|)$, as expected from relatively small areas (see Section 9.5.1), over appropriate histograms $h_t(|\varphi|)$ of flux densities occurring on the visible hemisphere of the stellar surface (described for the Sun in Section 6.3.1.2). The angular resolution to be used for the determination of the histograms $h_t(|\varphi|)$ need not be better than a few arcseconds, according to the discussion in Section 9.5.1. A lower angular resolution may suffice, however, and may even be desirable. As discussed in Section 8.4.2, the relationship between the Ca II K line-core intensity and the magnetic flux density does not change when the resolution is changed from $\sim 2.5''$ up to $\sim 15''$. Using the lower resolution of some $15''$ to determine $h_t(|\varphi|)$ reduces the scatter about the relationship between radiative and magnetic flux densities. Formally, the resolution at which the relationship between radiative and magnetic flux densities is determined should be the same. This led Schrijver and Harvey (1989) to use the synoptic magnetograms discussed in Section 6.3.1: their resolution of one heliocentric degree matches the resolution of $15''$ to within 12% at disk center.

For the transition-region and coronal emissions an additional argument lends support for the use of data at even poorer angular resolution: the large emission scale height and the curvature of magnetic field lines results in substantial projection offsets (Section 8.4) and ambiguities on what features are being compared to each other in the magnetogram and spectroheliogram. For coronal losses the angular resolution should be coarse enough to encompass entire active regions or large patches of quiet Sun (Section 8.4.5).

In accordance with the solar results discussed in Section 8.4.2, Schrijver and Harvey (1989) assumed that there is a monotonic correspondence between the magnetic flux density $|\varphi|$ in the photosphere and the emission from the upper atmosphere in any given diagnostic at the resolution of the synoptic magnetograms extending up to sunspots, which are excluded from this discussion. They parameterize these as power laws between the mean magnetic flux density, $|\varphi|$, and some radiative diagnostic of activity expressed as a flux density F_i:

$$F_i = a_i |\varphi|^{b_i}. \tag{9.12}$$

The magnetic flux density, averaged over the stellar surface at some time t, is

$$\langle |\varphi| \rangle_t = \int_0^{|\varphi|_{\max}} |\varphi| h_t(|\varphi|) d|\varphi|, \tag{9.13}$$

where $h_t(|\varphi|)$ is the histogram of magnetic flux densities on the stellar disk at that time (compare Fig. 6.13 for the Sun). The integration up to $|\varphi|_{\max}$ excludes sunspot data, both for observational and theoretical reasons. Observationally, on one hand, the highest values of the flux density should be excluded because the synoptic maps do not properly accommodate the flux in sunspots, nor are these properly measured in the original high-resolution magnetograms. On the other hand, atmospheric emissions are in general reduced over sunspots, so that their exclusion from the integration in Eq. (9.13) is warranted. Schrijver and Harvey (1989) use a value of $|\varphi|_{\max} = 512\,\mathrm{G}$.

If center-to-limb effects are ignored, the disk-averaged stellar radiative flux density, $\langle F_i \rangle$, equals

$$\langle F_i \rangle_t = \int_0^{|\varphi|_{\max}} a_i |\varphi|^{b_i} h_t(|\varphi|) d|\varphi|. \tag{9.14}$$

Center-to-limb effects introduce an offset factor of the computed surface-averaged properties. The precise value of this offset depends, among other things, on the details of the geometry of the activity belts, but these effects are of limited magnitude compared to the scatter about the flux–flux relationships. Other time-dependent and line-of-sight effects listed in Section 9.4 also result in a limited scatter about the average value.

The relationship between the $\langle F_i \rangle_t$ and $\langle |\varphi| \rangle_t$ for the Sun throughout the solar cycle is obtained by performing the integrations in Eqs. (9.13) and (9.14) at different times in the solar cycle. Figure 9.5 shows such relationships as a function of the assumed value of the power-law index b_i. Schrijver and Harvey (1989) approximate these relationships by

$$\langle F_i \rangle_t = 2.5^{b_i(b_i-1)} a_i \langle |\varphi| \rangle_t^{b_i}. \tag{9.15}$$

If, for example, $b_i = 0.6$, as is the case for the Ca II K line-core emission [see Eq. (8.8)], then $2.5^{b_i(b_i-1)} = 0.80$. This explains the seemingly surprising result that the relation for mean stellar fluxes and the relation for solar fluxes with a moderate angular resolution are nearly the same (see Fig. 8.10; also, for example, Schrijver *et al.*, 1989a).

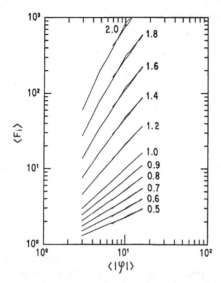

Fig. 9.5. Simulated relationships between surface-averaged flux densities, determined by convolving power-law relationships between fluxes determined from data with a moderate angular resolution with the flux-density distribution $h_t(|\varphi|)$ for the stellar disk [as in Eqs. (9.13) and (9.14)]. The values of b_i from the input relationships Eq. (9.12) are shown as labels to the curves (figure from Schrijver and Harvey, 1989).

We point out that there is an inconsistency in this argument for coronal radiative losses: these do not merely reflect the local photospheric flux density, as discussed in Section 8.4.5, but they also respond to fields in the surrounding area. The resolution of approximately 12,000 km at which $h_t(|\varphi|)$ was determined is too small for such an averaging at coronal heights, although it is appropriate for the chromosphere and most of the transition region. However, the nearly linear response of the soft X-ray flux to the large-scale photospheric field leaves the above conclusion valid even for coronal emissions.

With the above results, the transformation between any pair of radiative diagnostics with power-law indices b_i and b_j in the relationship with magnetic flux densities can be written as

$$F_i = A_{ij} F_j^{B_{ij}} \leftrightarrow \langle F_i \rangle = 2.5^{b_i b_j (B_{ij}-1)} A_{ij} \langle F_j \rangle^{B_{ij}}, \qquad (9.16)$$

where $B_{ij} = b_i/b_j$ and $A_{ij} = a_i/a_j^{B_{ij}}$. Apparently, the properties of $h_t(|\varphi|)$ preserve the nonlinearity of relationships between radiative flux densities derived from solar observations by using some intermediate angular resolution. One rather convenient consequence of this is that we can now combine the solar and stellar relationships discussed in Sections 8.4 and 9.4 (see also Cappelli *et al.*, 1989, and Cerruti–Sola *et al.*, 1992). Figure 9.6 demonstrates the increase in the steepness of the relationship between radiative and magnetic flux densities with an increasing temperature of formation: at chromospheric temperatures the relationship is nearly a square root, while the coronal emission is nearly proportional to the total amount of flux in the photosphere.

The transformation in Eq. (9.16) not only allows us to combine solar and stellar data in the study of stellar atmospheric structure, but it also validates to a limited extent the

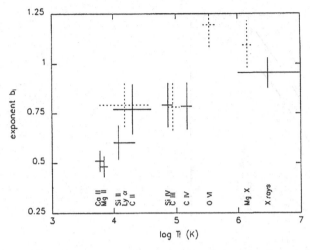

Fig. 9.6. Exponents b_i in the power-law relationships $F_i \propto |\varphi|^{b_i}$ between solar and stellar radiative flux densities and the photospheric magnetic flux density, as a function of the temperature of formation T_f of the radiative diagnostic. The exponents were derived from data either observed with some angular resolution or disk-averaged for the Sun as a star. Dotted lines show values derived from data averaged over entire active regions (figure from Schrijver, 1991).

modeling of stellar atmospheres by a single-component atmosphere. The problem with this kind of modeling has its roots in the way nonlinearities and filling factors combine. Figure 9.7 shows the simulated contributions of areas of different flux densities to the total emission from the Sun as a star. Near the maximum of the sunspot cycle only ~40% of the total coronal emission comes from the quiet network with flux densities below 50 G (cf. Section 5.4), whereas 75% of the total chromospheric emission comes from

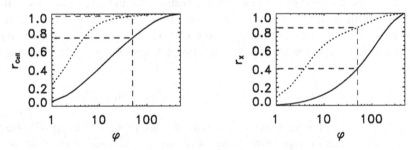

Fig. 9.7. Simulated cumulative contributions in the chromospheric Ca II H+K (left) and coronal soft X-ray (right) emissions. The curves show the fractional contribution to the total emission from areas with absolute magnetic flux densities φ (at an angular resolution of the order of a heliocentric degree) or less. The emission strength was estimated by using the distribution functions $h_t(|\varphi|)$ for the magnetic flux (shown in Fig. 6.13) for the minimum of the sunspot cycle (dotted curves) and for the maximum (solid curves), and multiplying these with the expected emission (i.e., with $\varphi^{0.6}$ for the chromospheric emission and $\varphi^{1.2}$ for the coronal emission; see Fig. 9.6). Pixels with flux densities above 500 G, generally in sunspots, were ignored. The dashed lines identify the levels corresponding to a threshold of 50 G, which is a suitable contour level for plage regions (Section 5.4), although it also includes some active network.

these regions (and even more if we would include the basal component). Hence if one takes the emission from the Sun as a star and models it by a simple spherically symmetric atmosphere, one tends to line up chromospheric emission that is weighted preferentially toward the network with coronal emission that is weighted preferentially toward plages. Despite this, the transformation from spatially resolved to disk-averaged radiative losses is apparently such that disk-averaged emissions do correspond to some "mean" magnetic feature somewhere on the disk, so that a single-component model does model an existing feature, provided the quasi-static approximation is valid (which is not always the case, as has been demonstrated for the basal atmosphere; see Section 2.7). This rough matching does not make up for inherently three-dimensional effects that photons are subjected to as they leave and enter distinct environments on their way into interstellar space, so three-dimensional (and probably dynamic) radiative transfer models, of course, are still required.

9.6 Stellar coronal structure

9.6.1 Coronal spectroscopy

The structure of stellar coronae can only be unraveled indirectly through the use of spectroscopic tools. Our understanding of stellar coronae is helped by the negligible optical depth of stellar coronae (except maybe in a few strong spectral lines; see below and Section 8.6, and ignoring blockage by cool matter such as that in filaments), so all emissions simply contribute directly to the total observed spectrum. Despite that, it is not easy to recover the coronal temperature structure from such a disk-integrated spectrum. A key complication is the intrinsic mathematical instability of the kind of inversion involved.

Let the plasma in a stellar corona with some given temperature T, unit electron and proton densities, and a given set of abundances emit a spectrum $\mathbf{f}'(\lambda, T)\Delta\lambda$ (which can be represented as a high-dimensional vector for a discrete and finite $\Delta\lambda$). This spectrum is the sum over all densities involved, but we ignore the limited set of density-sensitive lines in this discussion. Prior to being subjected to a detailed analysis, this spectrum is affected by interstellar absorption, weighted by the wavelength-dependent effective area of the detecting instrument, and convolved with the instrumental dispersion profile and the subsequent wavelength smoothing associated with the finite spatial resolution of the detector, and so on. The observed isothermal spectrum is represented by $\mathbf{f}(\lambda, T)\Delta\lambda$. For a composite plasma with temperatures ranging from $T_{\min}$ up to $T_{\max}$, the net expected spectrum $\mathbf{g}(\lambda)\Delta\lambda$ is given by the total contribution of all temperatures, weighted by their emission through $n_e n_H$ [Eq. (2.59)]:

$$\mathbf{g}(\lambda)\Delta\lambda = \int_{T_{\min}}^{T_{\max}} \mathbf{f}(\lambda, T) n_e n_H(T) dV(T) \, \Delta\lambda \tag{9.17}$$

$$= \int_{T_{\min}}^{T_{\max}} \mathbf{f}(\lambda, T) n_e n_H(T) \frac{dV(T)}{d\log(T)} d\log(T) \, \Delta\lambda.$$

When this is represented numerically for wavelength intervals $(\lambda_i, \lambda_i + \Delta\lambda]$, with $i =$

$1, \ldots, N$, and temperature intervals $(T_j, T_j + \Delta T]$, for $j = 1, \ldots, M$, we find

$$\mathbf{g}_i(\lambda_i) \approx \sum_{j=0}^{M} \mathbf{f}(\lambda_i, T_j) \mathbf{D}(T_j) \Delta \log(T). \tag{9.18}$$

This equation introduces the *differential emission-measure distribution* $\mathbf{D}(T) \equiv n_e n_H$ $dV/d \log T$. This is the weighting function that measures the relative contribution to the total spectrum by plasma of a given temperature. This definition can be combined with the model equation for quasi-static loops, Eq. (8.20), to yield a relationship between $\mathbf{D}(T)$ and the conductive flux density F_c along a loop with constant pressure p that passes through the loop's cross section A (Van den Oord and Zuccarello, 1996):

$$\mathbf{D}(T) \propto \frac{A(T) p^2 T^{3/2}}{|F_c(T)|}. \tag{9.19}$$

Within an isobaric loop (i.e., with a length well below the pressure scale height), therefore, $\mathbf{D}(T)$ is large wherever T is high and $|F_c|$ is low, that is, particularly near the top of the loop. Alternatively, in any part of the loop where the conductive flux is independent of temperature, $\mathbf{D}(T) \propto A T^{3/2}$ (e.g., Jordan, 1991).

Derivation of $\mathbf{D}(T)$ from an observed spectrum involves an inversion of Eq. (9.18). The inversion of such a Fredholm equation of the first kind is an ill-posed problem and solutions tend to show large, unphysical oscillations (e.g., Craig and Brown, 1986; Press *et al.*, 1992; and Harrison and Thompson, 1992). A number of distinct methods have been developed to regularize this inversion, that is, to dampen these oscillations. Each of these methods smoothes the recovered $\mathbf{D}(T)$ to a point where further smoothing would reduce the quality of the fit of the observed to the theoretical spectrum. The algorithm can be formulated to allow only positive definite solutions, but although it is a requirement for any physically valid solution, statistically speaking this is not mandatory; inversion results should merely not have statistically significant negative excursions. Some inversion algorithms are guaranteed to find the best solution, such as those based on matrix inversions in the least-squares problem. Others converge to a solution by finding the extremum in some measure of quality. In principle, all methods should converge to the same solution within the uncertainties, and in fact they appear to do just that, as shown in a comparative study edited by Harrison and Thompson (1992). Unfortunately, a consistent treatment of precisely what these uncertainties are is lacking at present. Such a treatment is difficult because the uncertainties are often dependent on uncertainties at other temperatures.

9.6.2   *Observed temperature structure*

The highest spectral resolutions prior to the *AXAF* and *XMM* missions for stellar soft X-rays have been achieved with the grating instruments on board *EXOSAT* (e.g., Lemen *et al.*, 1989) and the Extreme-UltraViolet Explorer, *EUVE* (Bowyer and Malina, 1991). In Fig. 9.8 we have collected some of the published $\mathbf{D}(T)$ curves for a sample of cool stars as observed with *EUVE*. Its wavelength range makes the *EUVE* spectrometer particularly sensitive to plasma with temperatures between a few times 10^5 K to $\sim 10^7$ K. Within that range, the (effectively) single main-sequence stars of moderate activity show a dominant emission component at ~ 2–4 MK. This component is generally assumed to

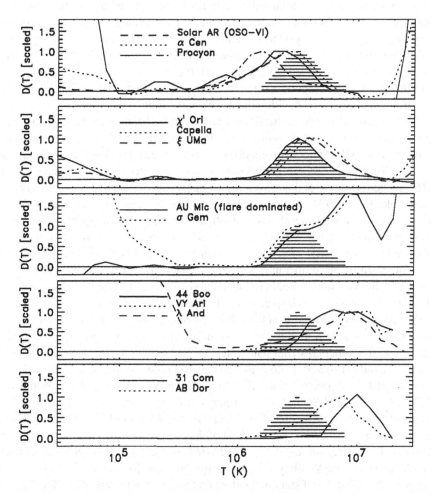

Fig. 9.8. Comparison of normalized emission-measure distributions $\mathbf{D}(T)$ [Eq. (9.18)]. The curves are normalized to unity at a temperature near 3 MK for the stars in the top three panels, and at a temperature near 10 MK for the stars in the lower two panels; they are smoothed over $\Delta \log(T) = 0.2$. The dashed region repeated in all panels for the purpose of cross reference is $\mathbf{D}(T)$ for χ^1 Ori between 1.5 MK and 8 MK. Top three panels: stellar data from Mewe *et al.* (1995), Schrijver *et al.* (1995), and (dashed-dotted) from Schmitt *et al.* (1996a); the solar curve is for active regions as observed with *OSO-VI* (Dere, 1982). The difference in temperature scales between the two curves for Procyon demonstrates the sensitivity of inversion results on both the details of the spectral codes and the inversion algorithm, both of which differ between these two inversions of the same *EUVE* data set. Bottom two panels: data from Dupree *et al.* (1996). Figure modified from Schrijver *et al.* (1996b).

α Cen	G2 V + K2 V	Procyon	F5 IV–V	χ^1 Ori	G0 V
Capella	G5 III+G0 III	ξ Uma B	G0 V	AU Mic	M0 Ve
σ Gem	K1 III	44 Boo	G0 V	VY Ari	K3–4 V–IV
λ And	G8 IV–III	31 Com	G0 III	AB Dor	K0 V*

* Dupree *et al.* (1996) list this star as K1 IIIp.

be directly comparable to the most common component in the quiescent solar corona, namely a combination of relatively stable active-region and quiet-photospheric loops. The function $\mathbf{D}(T)$ for this component peaks at somewhat different values for different stars, but a significant dependence on any stellar parameter has yet to be formulated.

A hotter component exists for the more active stars shown in Fig. 9.8: for more active dwarf stars and for evolved stars in general, the emission measure rises with T above $\sim$5 MK. This hotter component is not well constrained by the *EUVE* data, because only a few strong lines are formed within the wavelength range to which it is sensitive, while the peak in the continuum has shifted shortward of the *EUVE* range at these temperatures, so that even the continuum has no diagnostic value to constrain the temperature distribution above some 10 MK. Other instruments, such as on board *ASCA* or the combination of the *EXOSAT* Transmission Grating Spectrometer (*TGS*) with its Medium Energy Experiment (*ME*), are better suited for this wavelength range. From that work (e.g., Lemen *et al.*, 1989, and Tanaka *et al.*, 1994) it appears that a separate, hot component exists in the coronae of these stars, even in the absence of obvious flaring. The temperature of this hotter component lies between $\sim$7 MK and $\sim$15 MK. Dupree (1996) points out a particular enhancement near 6 MK in active stars with $P_{\text{rot}} \lesssim 13$ days as part of the general trend of an increasing high-temperature emission measure at higher activity. The origin of this spike, which is not so prominent in inversions of some of the same stars by, for example, Mewe *et al.* (1995) and Schrijver *et al.* (1995), remains unclear.

The number of stars for which the hot coronal component has been accurately measured is limited to less than a dozen, so little can be said in a quantitative sense about trends with fundamental stellar parameters. If we assume that this two-component structure is generally sufficient to describe the entire cool-star domain, the spectroscopic data obtained at lower spectral resolution by *EINSTEIN*'s Imaging Proportional Counter, *IPC* ($\Delta E/E \approx 1$), *EXOSAT*'s Low-Energy (*LE*) instruments, and *ROSAT*'s Position-Sensitive Proportional Counter, *PSPC* ($\Delta E/E \approx 0.45$ at 1 keV) allow the study of trends based on a much larger data set than that of the high-resolution data alone. Schmitt *et al.* (1990), for instance, survey all *EINSTEIN/IPC* sources in the Bright Star Catalog (and Supplement) and the Woolley Catalog with more than 200 detected photons. From this it appears that cool dwarf stars in the domain from mid-F to early-K all have a dominant 3-MK component. The observed M-type dwarf stars (generally rather active) display a component at 10 to 20 MK with an emission measure that exceeds that of the 3-MK component. The detected giant stars all show a dominant component within the range of 10 to 20 MK. Schrijver *et al.* (1984) found the same trends based on a substantially smaller sample of stars. They study the change in a mean "characteristic coronal color temperature" T_X in more detail, however, and show that T_X increases with activity for the mid-F to early-K stars. It does not deviate strongly from 3 MK, however, for main-sequence stars with rotation periods above some 10 days. Wood *et al.* (1995) analyze *EINSTEIN/IPC*, *EXOSAT/LE*, and *ROSAT/PSPC* data for 18 main-sequence stars ranging in spectral type from F5 to M4.5. They confirm this dichotomy: the temperatures allowed by their count-rate ratios lie in the ranges from 2.5 to 6 MK and above 25 MK. They caution that these very-low-resolution data are to be interpreted with great care, and that test input data allow a substantial family of one- and two-temperature fits that are all statistically acceptable.

Coronae of giant stars differ markedly from those of main-sequence stars: all observed spectra of giant stars show the signature of a hot component. Even giants that have an activity level comparable to that of the Sun (in terms of the coronal flux density in soft X-rays) have this hot component as the dominant contribution to the coronal color temperature. Note, however, that the composition of the dwarf and giant samples differ: giant stars near the red-giant branch that are as active as the Sun must be members of relatively close binary systems, because otherwise stellar evolution and magnetic braking would have slowed down the rotation of the stars and thus have quenched their activity (see Section 13.3).

The rather poor spectrum obtained with the *EXOSAT/TGS* of Procyon suggested a third coronal component with a temperature of about half a million kelvin (Lemen *et al.*, 1989), but the higher-quality *EUVE* spectrum of Procyon, covering the more appropriate longer-wavelength regime, contradicts this: the $\mathbf{D}(T)$ shown in Fig. 9.8 shows a gradual decrease for decreasing temperature, much like the Sun and α Cen. All other stars in that figure have emission-measure distributions that decrease much more rapidly below approximately 1–2 MK, with only very weak low-temperature tails.

9.6.3 Loop geometry

Spectroscopy can constrain loop geometry in stellar coronae: a changing cross section along the loop directly impacts the emission-measure distribution $\mathbf{D}(T)$ [Eq. (9.19)], resulting in a change in the emission measure at that temperature. Because we cannot confidently extrapolate the coronal magnetic field even from the solar photospheric boundary conditions, there is no a priori model for loop cross sections in stars, nor can all of them be expected to be the same. Schrijver *et al.* (1989c) have shown that in spite of this some constraints can be put on loop cross sections: the two good coronal spectra of cool stars that were obtained with the *EXOSAT/TGS* cannot be reconciled with coronae comprising only loops of constant cross section, because the differential emission measure (as derived by Rosner *et al.*, 1978) is not steep enough. Schrijver *et al.* (1989c) compute models for loop families with changing loop cross sections (similar to field lines in line dipoles) using the model by Vesecky *et al.* (1979). From this, they conclude that the spectra of σ^2 CrB (F8 V+G1 V) and Capella (G5 III+G0 III) are compatible with coronae consisting of two distinct ensembles of quasi-static loops. The relatively cool loops with temperatures near 5 MK appear to expand markedly with height: the best fits are obtained for ratios Γ in cross section at the loop top and base of 30 and 50 for Capella and σ^2 CrB, respectively, being larger than 5 and 2 at the 99% confidence level. If the relatively hot component is also modeled by loops, then their expansion with height appears to be much less, with best-fit values of Γ of 2.5 and 4, respectively. The spectra allow that the expansion factors for cool and hot loop families could be $\Gamma \sim 10$ for both families at the 90% confidence level. Schrijver *et al.* (1989c) argue that these expansion factors are underestimated if there is a spread in loop-top temperatures, which is likely to be the case.

These results on loop expansion factors find support in more recent high-resolution *EUVE* spectra of some cool stars. Van den Oord *et al.* (1997) discuss the analysis of *EUVE* spectra of four cool stars: α Cen (G2 V), Capella (G5 III+G0 III), χ^1 Ori (G0 V), and ξ UMa B (G0 V+···). They find that the minimally required expansion factors lie between 2 and 5. These more recent limits on these expansion factors are compatible with the factor of ~5 inferred in some studies of the solar corona (Section 5.4). The older values based on the much lower-resolution stellar *EXOSAT* spectra and those for individual solar loops, however, do not agree; perhaps this is due to the lower spectral resolution, perhaps due to the different spectral window.

9.6.4   Loop lengths and area coverage

The next step in the analysis of stellar coronae is the determination of the fractional area filled by the loop footpoints (the surface filling factor) and the characteristic loop lengths. Some constraints can be extracted by combining the observed (characteristic) temperatures and radiative losses with the quasi-static loop models discussed in Section 8.6. Unfortunately, unless the electron density is known by some independent means, the RTV scaling laws, Eqs. (8.22) and (8.23), only lead to estimates that combine loop length and surface filling factor. This allows some estimates of loop lengths in the following way: following Schrijver *et al.* (1984), let us define the specific emission measure ζ as the total emission measure per unit area:

$$\zeta \equiv \frac{1}{4\pi R_*^2} \int n_e n_H dV \equiv \frac{1}{4\pi R_*^2} \int D(T) d\log T. \tag{9.20}$$

For a corona consisting of N loops of constant electron density n_e, proton density n_H and temperature T, and fixed loop half-length L and cross section A,

$$\zeta = G f \left(\frac{p}{2kT} \right)^2 \ell, \tag{9.21}$$

where G is a geometry factor ($1 - G$ is the fraction of the corona occulted by the stellar disk; below we take $G = 0.7$), and f is the surface filling factor $2NA/4\pi R_*^2$. The emission scale length ℓ is limited either by the loop half-length L if $L < H_p/2$, or by $\ell \approx H_p/2$ for longer loops. The combination of Eq. (9.21) with the RTV scaling law in Eq. (8.23) yields

$$\zeta_{27} \approx 0.12 \frac{f}{L_{10}} T_6^4 \qquad (L < H_p/2) \tag{9.22}$$

$$\approx 0.03 \frac{f}{L_{10}^2} \frac{g_\odot}{g} T_6^5 \quad (L > H_p/2), \tag{9.23}$$

where, for example, ζ_{27} is in units of 10^{27} cm^{-5}. For the scaling law in Eq. (8.22), involving the volume heating rate ϵ_{heat}, we find:

$$\zeta_{27} \approx 39 f \, \epsilon_{heat}^{1/2} T_6^{9/4} \qquad (L < H_p/2) \tag{9.24}$$

$$\approx 3,200 f \, \epsilon_{heat} \frac{g_\odot}{g} T_6^{3/2} \quad (L > H_p/2). \tag{9.25}$$

Schrijver *et al.* (1984) have applied the above relationships to a sample of low-resolution *EINSTEIN/IPC* spectra for which they derived characteristic coronal temperatures T_X and estimates of the specific emission measure ζ. They show that the data for the average Sun and for α Cen A and α Cen B are compatible with stars that are covered mostly with long loops with $L \approx H_p/2 \approx 50,000$ km. Vaiana *et al.* (1976)

already argued that long loops indeed emit the dominant contribution to the quiescent solar coronal emission. For moderately active dwarfs, with $10^{28} < \zeta < 10^{30}$ cm^{-5}, $L_{10}/f \approx 0.1$ and $L \ll H_\mathrm{p}$. For these stars, apparently most loops are compact, probably more compact than what are referred to as compact active-region loops seen on the Sun. For the most active stars, both giants and dwarfs with $\zeta > 10^{30}$ cm^{-5}, $L_{10}/f \approx 10$–100, so that $L \approx H_\mathrm{p}/2 \approx R_*$ if $f \approx 1$, or much more compact if $f \ll 1$.

9.6.5 Coronal densities

A much better constraint on loop lengths can be derived if the characteristic mean coronal electron density within the loops contributing most to the observed spectrum is known. Some *EUVE* spectra contain sets of spectral lines that are strong enough to allow the comparison of the permitted (radiatively deexcited) and forbidden (mostly collisionally deexcited) lines. Table 9.3 summarizes the results of such analyses. The derived electron densities are quite uncertain, not only because of measurement uncertainties but also largely because of uncertainties in the spectroscopic models and the expected inhomogeneity of coronae. First, substantial differences in density occur along coronal loops because of the temperature gradient from loop top to footpoint. Second, solar data show that substantial density contrasts also occur across field lines, because otherwise no loops could stand out at all.

The characteristic coronal electron densities derived from diagnostics formed near 3 MK are typically $\sim$3 orders of magnitude lower than those found for the hotter

Table 9.3. *Electron densities, n_e, for quiescent stellar coronae, derived from spectra obtained with the Extreme Ultraviolet Explorer, EUVE*[a]

Source	n_e(cm^{-3})	Fe Lines	Ref.[b]
Procyon (F5 IV–V)	$(1$–$10) \times 10^9$	X–XIV	1
	$(1$–$30) \times 10^9$	X–XIV	2
α Cen (<u>G2 V</u>+<u>K2 V</u>)	$(2$–$20) \times 10^8$	X–XIV	3
Capella (G5 III+<u>G0 III</u>)	10^9	XII–XIV	4
(P $=$ 104d, d $=$ 167$R_\odot$)	$(1$–$20) \times 10^{11}$	XXI–XXII	4
	$(1$–$20) \times 10^{12}$	XIX–XXII	1
σ Gem (<u>K1 III</u>+...)	10^{12}	XXI–XXII	1
(P $=$ 19.6d, d $=$ 59$R_\odot$)	$< 3 \, 10^{12}$	XXI–XXII	5
ξ UMa B (<u>G0 V</u>+...)	5×10^{12}	XXI–XXII	1
(P $=$ 3.98d, d $=$ 12$R_\odot$)			

[a] For binary stars, the components expected to dominate the observed spectrum have been underlined. For close binaries, the orbital period P and mean separation d are also listed. For comparison, the electron density in bright loops in solar active regions is $\sim$3 $\times 10^8$ cm^{-3} (Section 8.6).
[b] References: 1, Schrijver *et al.* (1995); 2, Schmitt *et al.* (1996a); 3, Mewe *et al.* (1995); 4, Brickhouse (1996); 5, Monsignori-Fossi and Landini (1994).

components with temperatures of 10 to 15 MK (Table 9.3). For Capella this contrast is found even within the same corona. The associated high volume emissivity (proportional to n_e^2) of the hot component implies that the volume filling factors of the hot components are much smaller than those associated with the 3-MK component. If the quasi-static loop scaling law in Eq. (8.22) is applied to these loops, then they are some 2 orders of magnitude shorter than a typical solar coronal loop, namely of the order of 100 km to 1,500 km for, for example, Capella's hot component. This is surprisingly and uncomfortably compact and may imply that these structures are not in equilibrium; perhaps we are looking at the signature of a multitude of small, compact, short-lived brightenings occurring over a substantial part of the stellar surface but with a small filling factor (consistent with the paucity of radio detections of nearby, active M dwarfs; see White, 1996). Dupree (1996) points out that this high-intensity, dense coronal component coincides with the domain in activity where polar spots are found (Section 12.2), but whether these phenomena are directly related remains to be seen.

9.6.6   *Abundances and radiative transport*

As the quality of spectroscopic observations of stellar coronae has improved over the years, unexpected deviations from the simple models have come to light. Two problems are particularly puzzling: the chemical composition and the radiative transport. The problem of the chemical composition of the solar corona was briefly discussed in Section 8.6. The awareness of this problem in the field of solar coronal spectroscopy has triggered a search for similar effects in stellar coronae, with some surprising results. Laming *et al.* (1995) have simulated the coronal spectrum of the Sun as a star, and they conclude that the solar first-ionization-potential (FIP) effect is limited to material hotter than ~ 1 MK. Drake *et al.* (1995) do not find any indication of a significant FIP effect in the *EUVE* spectrum of Procyon (F5 IV–V, with a cool white-dwarf companion). In a review of other results, Drake (1996) concludes that coronae of the most truly active stars are metal deficient according to the FIP division of elements. Although this is in general referred to as the FIP effect, it is important to note that for these stars – most of which are much more active than the Sun – the fractionation tends to be opposite to that found in the solar corona: metal deficiency in stellar coronae in contrast to metal richness in the solar corona. For ξ Boo A, a star of intermediate activity, Laming and Drake (1999) find a sunlike coronal composition. No clear trend has yet emerged describing the composition of stellar coronae relative to their photospheres.

Schrijver *et al.* (1994; see also Mewe *et al.*, 1995; and Schrijver *et al.*, 1995) have raised another problem in stellar coronal spectroscopy: they argue that radiative transfer effects caused by the finite optical thickness may not be negligible in some of the stronger coronal lines in stellar environments, as was recognized earlier for some solar coronal lines (Section 8.6). They argue that if a somewhat more tenuous medium of a comparable temperature overlies the brightly emitting coronal loops, then scattering of the photons in the strongest spectral lines can affect the observed line strengths noticeably even in disk-integrated spectra, because in such a geometry more line photons will be scattered down (to be destroyed on impact in the lower atmosphere) than back up; this inelastic scattering is effectively absorbing line photons. The resulting decreased line-to-continuum ratio is indeed compatible with what appears as a continuum between ~ 70 Å and 140 Å in the *EUVE* spectra of most notably α Cen and Procyon. The nature of this apparent continuum is still being debated, however: Schmitt *et al.* (1996b), for instance, argue that possibly a large number of spectral lines not included in the spectral codes conspire to result in a thermal continuum. It remains unclear whether enough lines are missing, and whether together they can indeed mimic an apparent continuum that can in other stars, including the Sun, be absent. Another alternative is that cool matter embedded high within the corona affects the spectrum through the strong wavelength dependence of the optical depths in the bound-free continua of hydrogen and helium, which scatter photons not only in different spatial directions (which would not affect the disk-integrated spectra very much) but also in wavelength throughout the continuum (thus removing them from the coronal lines and hiding them in the weak coronal continuum).

On one hand, resonance scattering in the highly structured stellar coronae complicates the interpretation of spectra tremendously. On the other hand, such scattering opens up new means to detect a dimly emitting domain surrounding bright loops, or perhaps even a hot stellar wind: Schrijver *et al.* (1995), for instance, propose that in the case of Procyon, a coronal 1- to 3-MK envelope with a density scale height some five times the pressure scale height and a base density of approximately 3×10^8 cm^{-3}, which is only twice that thought characteristic of the base density of the solar wind, can model Procyon's *EUVE* spectrum successfully, including the destruction of backscattered photons in some strong spectral lines. Thus coronal resonant scattering in only a few spectral lines could open up a window to stellar winds in a region of the H–R diagram where they are notoriously hard to detect by other means (compare Section 11.4).

Stellar and solar coronal spectroscopy are providing us with evidence that coronae differ markedly from star to star, even in apparently quiescent situations. Nevertheless, all stars obey the relationships between radiative flux densities from very different temperature regimes in their outer atmospheres, as was discussed in Section 9.4. The implications for coronal heating mechanisms are discussed in Chapter 10.

10

Mechanisms of atmospheric heating

After our discussion of the observed radiative emissions from solar and stellar outer atmospheres, we now turn to the mechanisms heating the outer-atmospheric domains to a temperature that even in apparently quiescent conditions is up to 3 orders of magnitude higher than that of the photosphere. Over the years, a multitude of mechanisms has been proposed to transport energy from the stellar interior into the outer atmosphere and to dissipate it there (see the compilations by Narain and Ulmschneider, 1990, 1996, and the summary in Table 10.1). It has become clear that the real question is no longer *how* the atmosphere can be heated, but *which* mechanisms dominate under specific conditions. The problem of outer-atmospheric heating can be separated into two parts: (a) what is the source of the energy, and (b) how is it transported and dissipated? Although at first sight the second part seems to be separable into the problems of transport and dissipation, this is not always the case, because some mechanisms intricately link these two processes through cascades or critical self-regulation, as we mention below.

Our discussion concentrates on coronal heating, with a substantial bias to current-based mechanisms. Space limitations do not allow an in-depth discussion of the many distinct wave-heating processes. Chromospheric heating follows similar principles as coronal heating, but the added complexity of radiative transfer and nonforce-free fields complicate studies in this area. In contrast, the larger viscosity and resistivity of the chromosphere make it easier to dissipate both currents and waves. Large flares fall outside our purview.

The first question is how long the dissipation of the magnetic field itself could sustain the radiative losses of the corona (which in all but the most active stars is smaller than the chromospheric radiative losses). If we ignore heat conduction, the field-dissipation time scale that would meet the heating rate ϵ_{heat} of a quiescent coronal loop of unit cross section can be estimated by using the RTV scaling law [Eq. (8.23)]:

$$\hat{t}_{\text{B, diss}} = \frac{E_{\text{mfe}}}{\epsilon_{\text{heat}}} \approx 7.7 \frac{B_2^2 L_9^2}{T_6^{7/2}} \text{ (days)}, \tag{10.1}$$

where E_{mfe} is the magnetic free-energy density, that is, the energy in the field in excess of the potential-field energy. This energy is associated with currents and waves, and is only a fraction of the total energy density $B^2/8\pi$. If we use $B^2/8\pi$ for an order-of-magnitude estimate of the upper bound of the available energy, then for active regions, with $B_2 \approx 1$, $L_9 \approx 5$ and $T_6 \sim 3$, this time scale is only a few days, which is much less than the lifetime

Table 10.1. *Sources and mechanisms of nonradiative heating for outer atmospheres of cool stars*[a]

Heating	Source	Dissipation
Hydrodynamic:	acoustic waves	shock dissipation
	pulsational waves	shock dissipation
AC/wave:	slow-mode MHD waves, longitudinal tube waves	shock dissipation
	fast-mode MHD waves	Landau damping
	Alfvén waves (torsional, transverse)	resonance heating, viscous heating, Landau damping
	magnetoacoustic surface waves	mode coupling, phase mixing, resonant absorption
DC/current:	electric currents	reconnection (turbulent heating, wave heating) associated with resistive or viscous dissipation

[a]Table after Ulmschneider (1996).

of coronal condensations over active regions. For the high corona over quiet Sun, with $B_2 \approx 0.02$, $L_9 \approx 0.2$, and $T_6 \approx 1.5$, the time scale is only a fraction of a second. Brightly radiating loops fill only a fraction of the total volume, however, so that only that fraction of the field has to be dissipated at any one time. Correcting for this effect increases the dissipation time scale for the total coronal volume. According to a study by Fisher *et al.* (1998), the dissipation time scale for active regions ranges from a few days to months, with a average value of ~ 1 month. For the quiet Sun the dissipation time scale may be of the order of a few hours, which is far too short given the flux-replacement time scale there of some 40 h in the quiet Sun (Section 5.3). A similar argument has been made for the quiet-photospheric field below the coronal holes: to power the high-speed solar wind, the dissipation time scale of the field is only approximately half an hour (Axford and McKenzie, 1997). Consequently, after the magnetic field reaches up into the corona, a substantial amount of nonradiative energy is to be supplied subsequently to compensate for the radiative losses. The energy requirements for the chromosphere are much larger, so this conclusion holds there a fortiori.

The stellar luminosity dominates the radiative emission from the outer atmosphere by at least some 2 orders of magnitude (Section 11.3), so in principle there is sufficient energy to heat that atmosphere, which doesn't come as much of a surprise, of course.

Within the convective envelope, the large-scale kinetic energy density associated with convection represents a small but nonnegligible fraction of the thermal energy density (see Section 2.2 and Fig. 2.1): the kinetic energy density $\rho \hat{v}_{ML}^2 / 2$, as computed from a mixing-length model, reaches a few percent of the stellar luminosity at approximately one pressure scale height below the stellar surface (Vilhu, 1987). Transport mechanisms that can deposit some of this energy into the outer atmosphere involve the magnetic field and acoustic waves, where fields, flows, and waves may interact to heat the stellar outer atmosphere.

Theoretical studies show that acoustic waves can certainly reach chromospheric heights (Section 2.7). The enormous density contrast between the photosphere and corona, however, makes it very unlikely that a sufficient acoustic flux survives shock damping to reach coronal heights (Hammer and Ulmschneider, 1991). Moreover, acoustic waves by themselves cannot explain the correlations between fields and radiative emission discussed in Section 8.4. Nor do they contain sufficient power to heat the outer atmospheres of the most active stars (Vilhu, 1987). The conclusion by elimination is that nonradiative heating is driven primarily by convection that couples to the magnetic field.

Traditionally, the mechanisms of energy transport and dissipation that involve magnetic fields are grouped into two classes. If the field perturbations occur fast compared to the Alfvén crossing time along the outer-atmospheric part of a field line, the mechanism is referred to as an alternating-current, or AC, mechanism. If it is slow, it is a direct-current, or DC, mechanism. In AC mechanisms, wave energy is dissipated. The waves are either picked up from the convective motions directly or from the associated sound waves in the convective envelope, or they may be generated within the outer atmosphere itself as the result of rapid field reorganizations that occur at sites of fast reconnection. In DC mechanisms, slower processes lead to more persistent currents. These currents are induced by moving the footpoints of coronal field lines in an interaction with the photospheric flows, as well as by the injection of new flux.

Before addressing details of the dissipation processes, we make a rough estimate of the electromagnetic energy flux into the outer atmosphere, using the Poynting flux density **S**, associated with both flows and waves, through some area of the solar photosphere, represented by a vector **ds** normal to the surface (e.g., Jackson, 1975):

$$\mathbf{S} \cdot \mathbf{ds} = \frac{c}{4\pi}(\mathbf{E} \times \mathbf{B}) \cdot \mathbf{ds} = -\frac{c}{4\pi}[(\mathbf{v} \times \mathbf{B}) \times \mathbf{B}] \cdot \mathbf{ds}. \tag{10.2}$$

This flux density contributes to the total outer-atmospheric internal energy in three different ways, as demonstrated by Poynting's theorem (e.g., Priest, 1993):

$$\oint \mathbf{S} \cdot \mathbf{ds} = \frac{\partial}{\partial t}\underbrace{\int \frac{B^2}{8\pi}\,dV}_{\text{internal energy}} + \frac{1}{c}\underbrace{\int \mathbf{v} \cdot (\mathbf{j} \times \mathbf{B})\,dV}_{\text{work}} + \underbrace{\int \frac{j^2}{\sigma}\,dV}_{\text{ohmic dissipation}}. \tag{10.3}$$

The three terms on the right-hand side of the equation reflect the change in the internal magnetic energy of the outer atmosphere, the work done by the field through the Lorentz force, and the ohmic heating through current dissipation. The internal energy of the field represents stored energy: the complexity of the field allows energy to be stored or released, depending on the boundary motions. The tension and pressure associated with the field opposes the movement of the photospheric footpoints of field lines. If forced to move against these forces, work is being done on the outer-atmospheric field. Currents are unavoidably generated, because the outer atmosphere tries to keep the magnetic fluxes through closed paths constant by inducing counteracting currents (e.g., Longcope, 1996).

Poynting's theorem can be applied to the entire outer atmosphere, with the photosphere as the boundary for the surface integral of the left-hand side of Eq. (10.3). We disregard energy losses through the solar wind. When averaged over the entire photospheric surface, the internal energy of the outer-atmospheric field does not show systematic changes on a time scale of up to months, but merely statistical fluctuations around some mean. These fluctuations may be small when associated with footpoint motions or new flux injection in ephemeral regions, or large when new active regions emerge. Such fluctuations are relatively small, however, if averaged over a sufficiently large area and over a sufficiently long time. Temporal changes in the total internal magnetic energy as measured by the first term on the right-hand side of Eq. (10.3) may then be neglected. In a (very nearly) force-free field in which $\mathbf{j} \parallel \mathbf{B}$, such as in a corona in which large-amplitude waves would be absent, the next term, measuring the work, is often neglected (but important because this work is done in jets associated with current-sheet reconnection that "messes up the surrounding field, and thus contributes to creating a hierarchy of scales;" Nordlund and Galsgaard, 1997). To sustain the outer-atmospheric radiative losses, the term measuring ohmic dissipation should therefore equal the integrated radiative emission (plus, if we restrict ourselves to the corona, the total conductive losses to the lower-temperature regions of the atmosphere).

The dissipation term in Eq. (10.3) is hard to determine quantitatively, however, because very small-scale currents develop embedded within the large-scale field. Recently, numerical experiments have reached a sufficient level of sophistication to be of help. These experiments are difficult, however, so that in many discussions that focus on order-of-magnitude estimates, the dissipation term is often approximated by an estimate based on the dimensions of the quantities in the Poynting flux density in Eq. (10.2):

$$\langle |\mathbf{S}| \, ds \rangle \sim \frac{\hat{B}_{\text{cor}}^2}{4\pi} Q(\hat{\ell}, \hat{t}, \hat{L}, \eta, \hat{f}, \ldots) \, ds, \tag{10.4}$$

where $\hat{B}_{\text{cor}}$ is a suitably chosen value characterizing the coronal field strength. The efficiency function Q depends on characteristic values of the length scale $\hat{\ell}$ and the time scale $\hat{t}$ of the photospheric driver, the length scale $\hat{L}$ of the coronal field lines, the plasma conductivity η, the filling factor $\hat{f}$ for the photospheric field, and possibly other parameters; Q has the dimension of a velocity.

The efficiency function Q is difficult to estimate. We illustrate this by discussing three different estimates for DC processes. Parker (1972) proposed that continual braiding of the footpoints of the outer-atmospheric field leads to the formation of strong currents. This braiding results from the wrapping of field lines around each other by the turbulent motions in the photosphere. In simplified models, braiding is often approximated by twisting two field lines around one another. It is important to distinguish between twisting and braiding, however. Simple twists of pairs of field lines are readily reversed, because twists can be undone as easily as they are created in a convective medium that has no clearly preferred direction of helicity. Braiding, which involves three or more field lines, in contrast, is irreversible in a statistical sense: to undo a braid requires something akin to time reversal. It has been shown that when many field lines are moved around each other randomly, the complexity of the field increases steadily, while the currents increase exponentially (e.g., Berger, 1994; Antiochos and Dahlburg, 1997).

Parker argued that the field develops discontinuities in the form of current sheets (see references in Parker, 1994), and ohmic dissipation then leads to outer atmospheric heating. If the corona is indeed heated in this way, then coronal loops may preferentially form at interfaces of the field from neighboring flux tubes or perhaps simply randomly throughout the volume. As a result of this the coronal footpoints may not correspond closely to the position of the photospheric flux tubes when observing at very high angular resolution; this is compatible with high-resolution observations by Berger *et al.* (1999).

In Parker's model, the efficiency function Q is approximately

$$Q \sim \hat{v}^2 \hat{t} / \hat{L} \tag{10.5}$$

for field lines of length $\hat{L}$, and for photospheric velocities $\hat{v}$ that act for a time $\hat{t}$ before dissipation occurs. Parker estimated $\hat{t}$ by matching the Poynting flux density to the observed characteristic radiative losses, which corresponds to a characteristic pitch angle over which the field lines are to be wrapped around each other of $\sim 20°$.

Van Ballegooijen (1985, 1986) argues that the convective braiding of the footpoints does not lead to discontinuities, but instead results in a cascade in which energy at large scales is transformed into currents on very small scales. He estimates

$$Q \sim \left(\frac{q}{3\pi} \log \mathcal{R}_m \right) D / \hat{L}, \tag{10.6}$$

where q is a correction factor close to unity. Just as Parker's result, this expression involves the diffusion coefficient $D \propto \hat{v}^2 \hat{t}$ associated with the convective flows in the photosphere (compare Section 6.3.2, and Table 6.2). The magnetic Reynolds number $\mathcal{R}_m$ (see Table 2.5 for the definition) enters into this expression because of the assumption that magnetic diffusion and hence reconnection become important when the resistive time scale equals the cascade time scale.

More recently, numerical simulations with three-dimensional MHD computer codes that include resistivity with moderately high magnetic Reynolds numbers have become possible, although the Reynolds numbers for the smallest simulated scales are still many orders of magnitude lower than coronal values inferred for the expected dissipation scales. Galsgaard and Nordlund (1996) and Nordlund and Galsgaard (1997) find that current sheets form throughout the simulated volume. These currents form quickly, typically within only a few correlation time scales of the boundary flows. These current sheets form a hierarchical pattern in which successively smaller scales are associated with larger ones. The system quickly reaches a state in which currents dissipate energy without a further increase in the internal magnetic energy (thus coupling the dissipation and the transport mechanisms). Nordlund and Galsgaard find that this state of statistical equilibrium is largely independent of the resistivity of the system, and that the final state is determined by a limitation on the winding number, that is, the number of times neighboring field lines are wrapped around each other. Their numerical experiments suggest that reconnection limits the (signed) winding number to zero on average, as expected when stellar rotation is neglected so that there is no preferred sense of twisting, with a dispersion of approximately unity, independent of the resistivity.

These numerical simulations suggest that rapidly evolving, highly intermittent currents exist throughout stellar outer atmospheres. As a result, the heating is expected to reach very different magnitudes at different locations within the coronal volume and probably even within individual loops as observed at present-day spatial resolution. This rapid variability reflects a process that Parker (1988) refers to as nanoflares. Although this may appear to be at odds with the concept of the quasi-static loop as discussed in Section 8.6, this is not necessarily the case: if the heat input occurs in a multitude of small bursts (relative to the thermal energy content of the loop) then the loop will settle into a quasi-steady state (e.g., Peres, 1997), in which the fluctuations in the heating input are smoothed out by rapid adjustments by conduction and pressure adjustments; see Eqs. (8.17) and (8.18). The relatively long radiative time scale $\hat{t}_{L,rad}$ [Eq. (8.19)] helps the loop to continue radiating on its thermal energy content without substantial changes on very short time scales.

The numerical experiments by Nordlund and Galsgaard suggest that

$$Q \sim \tan(\varphi_w) \sim \frac{\hat{v}\hat{\ell}}{\hat{L}}. \tag{10.7}$$

Here the winding angle φ_w is similar to the pitch angle that Parker estimated from the coronal energy requirements. If these numerical experiments are sufficiently precise, then we know not only the magnitude of the characteristic pitch angle (close to the value originally estimated by Parker) but also how the field manages to limit this pitch angle efficiently, settling at a maximum corresponding to a remarkably low winding number.

The small-scale reconnection events that lead to nanoflaring can be observed only to a limited extent, because most of these events occur at energies below the detection thresholds. Only the largest of these events can be observed. Krucker and Benz (1998) find an energy distribution $f(E) = f_0 E^{-\delta}$ for δ in the range of 2.3 to 2.6 for flare energies E from 8×10^{24} to 160×10^{24} erg cm^{-2} s^{-1}. If $F(E)$ extends down to about 0.3×10^{24} erg cm^{-2} s^{-1} in that same way (requiring some 28,000 nanoflares to go off per second over the entire Sun), then such flaring could be responsible for all coronal heating in the quiet corona they observe. That the exponent δ does not hold over the entire energy range, or may differ between active and quiet regions, is exemplified by Shimizu (1995), who found δ to lie between 1.5 and 1.6 for flare energies exceeding 10^{27} erg cm^{-2} s^{-1} for flares observed by the *YOHKOH/SXT* in active regions. These larger flares, by the way, amount to no more than 20% of the total X-ray brightness of active regions.

Strong shear in the field, and thus strong electric currents, can also be formed by the injection of new flux into an existing configuration by flux emergence, which can lead to the short-term heating of loops (including flares). Note that flux injection in the network in the form of ephemeral regions not only introduces shear on short time and length scales, but also contributes to the braiding of long loops because of the footpoint displacements associated with the processes of emergence and subsequent cancellation (see, Schrijver *et al.*, 1998). The sustained brightness of many coronal loops indicates a (possibly intermittently) sustained Poynting flux through waves or displacements rather than flux injection as the dominant mechanism providing energy to the outer atmosphere.

Rapid convective motions and acoustic waves drive slow-mode and fast-mode MHD waves, longitudinal tube waves, transverse and torsional Alfvén waves, and magnetoacoustic surface waves, all of which can result in AC heating (see the compilations by Narain and Ulmschneider, 1990, 1996). Coupling between the various wave modes can occur particularly where there are strong gradients in density, temperature, or magnetic field. Waves dissipate efficiently only if large gradients are formed, which can be the result of, for example, phase mixing or resonances (see Table 10.1).

The dissipation of wave power through shock formation relies on significant density gradients, so this mechanism does not seem to be very promising for the corona, but it may well play an important role in chromospheric heating. Resonance heating in an isolated coronal structure depends on the formation of large amplitudes through the constructive interference of waves bouncing back and forth in coronal resonance cavities formed by loops that match specific frequencies, as well as on the accumulated weak damping during repeated passes of the wave through the same volume. Global resonances, in which a substantial coronal volume acts as an electromagnetic resonator, may also play an important role, but these are not yet understood well enough. In other heating models, phase mixing is used. Such models invoke small differences in propagation speeds of nearby field lines to allow large gradients to build up as waves propagate. In a very tangled field, waves starting in phase at some location do not have enough time to build up strong phase differences before the wave paths diverge, but in such a complicated field geometry this appears, in fact, to lead to enhanced wave absorption and therefore heating (see, for example, Similon and Sudan, 1989). The fanning out of magnetic field with height in the chromosphere directly affects the wave amplitudes because of energy conservation (see, for example, Cuntz *et al.*, 1998), and therefore wave heating of the chromosphere is sensitive to the details of how flux is bunched at the photospheric level.

AC and DC mechanisms cannot be as easily separated as the definition suggests, because waves can affect reconnection rates, whereas rapid reconnection generates waves. Such AC/DC interactions make the problem of coronal heating difficult to treat in a unified way. It appears that many of the proposed AC, DC, and composite mechanisms hold sufficient promise to act at significant levels in stellar outer atmospheres. Hence, it seems likely that several mechanisms function side by side, possibly depending in relative magnitudes on the local environment.

Each of the three expressions in Eqs. (10.5), (10.6), and (10.7) for the heating function Q above, based on the field-line braiding models by Parker, Van Ballegooijen, and Nordlund and Galsgaard, involves an inverse proportionality to loop length L. The RTV scaling laws, Eqs. (8.22) and (8.23), indicate that the total heating rate required for a loop with unit cross section, $\epsilon_{heat} L$, also is roughly inversely proportional to L (e.g., Porter and Klimchuk, 1995), because otherwise the apex temperature would show a substantial range in values. Apparently, the energy requirements of coronal loops are such that the entire spectrum of photospheric convective flows is important for coronal heating: whereas granular braiding is efficient for relatively short loops, long loops are wound up efficiently by large-scale flows.

Note that the fine-scale mixing of polarities in the quiet solar photosphere results in a range of loop lengths, existing side by side (Fig. 10.1). The efficient heating of loops of a broad range of lengths is compatible with the corona over these regions, in which a haze of X-ray emission contains a few compact loops that stand out brightly, but contribute only a fraction of the total coronal emission (one estimate by Habbal and Grace, 1991, puts the contribution of coronal bright points to the total quiet-Sun coronal emission at a mere 10%).

The corona must have a temperature of the order of ~ 1 MK or more in order for the pressure scale height H_p [Eq. (2.9)] to be comparable to the loop length: a virtually empty loop could not radiate energy deposited by nonradiative heating, nor could a relatively cool loop lose that energy through conduction (see Section 8.6). Hence, the

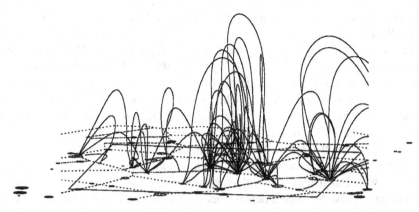

Fig. 10.1. Potential-field model for a mixed-polarity area. Flux concentrations are positioned on a Voronoi tesselation that mimics the supergranular network (Section 2.5). The flux distribution extends over an area nine times larger than shown in the plot to reduce edge effects. The polarities (open and filled circles) are assigned randomly, with the fluxes drawn from an exponential distribution (as is observed for the quiet solar magnetic network; Section 5.3). Only field lines starting over flux concentrations in a strip around the center of the area are shown (figure from Schrijver *et al.*, 1997b).

coronal temperature must be high enough to fill at least the lower parts of the loops with matter. The impulsive nature of the heating may be instrumental in pushing the loop temperature to a value higher than needed to fill it with matter (following an argument made by R. Rosner, private communication). Let us assume an initial heating to occur at a rate of E_{init} for the entire loop. This rate can be computed by dividing the thermal energy content of the loop by the onset time scale $\hat{t}_{\epsilon,\text{onset}}$. With Eq. (8.19), we find

$$\frac{E_{\text{init}}}{E_{\text{rad}}} > \frac{3n_e kT \, L/\hat{t}_{\epsilon,\text{onset}}}{n_e^2 \mathcal{P}(T) \, L} = \frac{\hat{t}_{\text{L,rad}}}{\hat{t}_{\epsilon,\text{onset}}} \approx \left(\frac{80}{\hat{t}_{\epsilon,\text{onset}}} \right) \frac{L_9}{\sqrt{T_6}}, \qquad (10.8)$$

for $\hat{t}_{\epsilon,\text{onset}}$ in minutes, and where E_{rad} is the radiative cooling rate. Loop onset times are observed to be as short as a minute or even shorter (Golub *et al.*, 1990), comparable to the sound transit time scale $\hat{t}_{\text{L,d}}$ [Eq. (8.17)]. Hence the initial heating rate apparently exceeds the steady-state heating rate by 1–2 orders of magnitude, depending on loop length and temperature, which pushes the loop to relatively high coronal temperatures. Thus the temporal and spatial intermittency of the heating that is expected from the theoretical and numerical models may well be the reason for the corona to be typically at several million kelvin.

For some coronal loops in active regions to stand out clearly requires spatial intermittency: apparently only a subset of the field lines is heated to coronal temperatures or at least filled to substantial densities at any one time. Recent *TRACE* observations show that heating can start or terminate within minutes over coronal volumes encompassing loop bundles with cross sections up to a few arcseconds (Schrijver *et al.*, 1999a). Higher-resolution observations are required before empirical statements on the intermittency of coronal heating below ∼1 arcsec can be made. The spatial intermittency of the outer-atmospheric heating is directly supported by transition-region observations. These indicate very small spatial filling factors, well below the angular resolution of the instruments (Section 8.8).

The estimate for the Poynting flux density in Eq. (10.4) for DC heating predicts a proportionality to $\hat{B}^2_{\rm cor}$ that is at first sight not affected by the estimates of Q in the models discussed above. Observations, in contrast, have shown that coronal radiative losses are very nearly proportional to the mean photospheric flux density φ, as measured by the characteristic value $\hat{B}$ (for solar data, see Section 8.4; and for stellar data, see Sections 9.4 and 11.3). It thus appears that the product $B^2 Q$ varies linearly with φ. One may speculate that this reflects a shift in the predominant heating mechanism from the quiet photosphere to active regions such that an apparent relationship appears that has $Q \propto \hat{B}^{-1}$. But it may also reflect the dependence of Q on the mean flux density. One potential cause for such a dependence is that the abnormal granulation in magnetic plages and patches of strong network (Section 5.4) is likely to reduce the value of Q as $\hat{B}$ increases. This reduction in mobility is strongest in spot umbrae, and in fact loops emanating from umbrae are always dark in the corona, despite the strong field strengths. Why the movements at the other end of those loops do not (in part) compensate for this is not known, but may for instance be a consequence of a limited scale height for coronal heating. Nor is it clear why coronal loops in the umbra–penumbra interface appear to be particularly bright for plasma at 1–2 MK.

Another reason why Q may depend on $\hat{B}$ is that the power supplied to the outer atmosphere is likely to depend on the number density of the photospheric footpoints of the outer-atmospheric magnetic field. In Section 5.3 we discussed a model that explains why the number density of photospheric concentrations of flux is not simply proportional to the flux density: the balance of collisions and fragmentations causes the average size of the concentrations to increase through interactions with neighboring concentrations when the average magnetic flux density in a region increases.

Another dependence that plays into the apparently linear response of the Poynting flux density $B^2 Q$ to the magnetic flux density φ is the difference in the characteristic loop length between active and quiet regions.

Note that the inverse scaling of supplied energy flux density with the loop length is incompatible with the linear scaling of coronal emission with total absolute flux in active regions (Section 5.4): despite the appeal of Eq. (10.4), much work still needs doing to prove its validity and to clarify how it might be applied in general.

Once the energy is brought into the corona, it has to be dissipated through resistivity or viscosity. Ulmschneider (1996) quantifies the need for short length scales in these processes by comparing rough estimates of the viscous and Joule heating rates:

$$\epsilon_{\rm visc} = \eta_{\rm v} \left(\frac{{\rm d}v}{{\rm d}z}\right)^2 \sim \eta_{\rm v} \frac{\delta \hat{v}^2}{\hat{\ell}^2}, \tag{10.9}$$

$$\epsilon_{\rm Joule} = \frac{J^2}{\sigma} \sim \frac{c^2}{16\pi^2\sigma}|\nabla \times \mathbf{B}|^2 \approx \frac{c^2\delta\hat{B}^2}{16\pi^2\sigma\hat{\ell}^2}, \tag{10.10}$$

where $\eta_{\rm v}$ is the viscosity and σ is the conductivity, and δv and δB are measures for the small-scale changes in the velocity and field strength, respectively. These heating rates approach observed values only if the scales $\hat{\ell}$ are less than some kilometers for the chromosphere and much less than that for the corona.

In the corona, the energy dissipation need not occur in proportion to the local radiative losses, because the thermal conduction there is very efficient. This efficient conduction makes it hard to put observational constraints on the location of coronal heating along loops.

The transition region between the chromosphere and corona derives much of its heat from the efficient conduction of heat from the corona. Nevertheless, local nonthermal energy deposition is also likely to occur (see Mariska, 1992, for an extensive discussion focused on the transition-region properties). Local heating in the transition region may well be associated with transition-region explosive events (Dere, 1994) that are a likely consequence of small-scale flux emergence. In more active stars, one expects a higher injection rate. It may well be that increased rates of transition-region explosive events are the mechanism behind the broad wings seen in stellar C IV line profiles of active stars (Wood *et al.*, 1997); the fraction of the emission in the broadened component of the lines increases with an increasing level of activity (Linsky and Wood, 1996).

Chromospheric heating rests on the same physical principles as coronal heating, although different mechanisms may dominate in these different atmospheric environments: AC mechanisms may be more efficient in the chromosphere with its strong density gradients, whereas DC mechanisms may dominate in the more tangled field of the tenuous corona. The much higher density of the chromosphere allows much lower temperatures, so that temperature difference may but need not imply differences in heating mechanisms. Heating depends on density and temperature explicitly through resistivity and viscosity, but this has little direct consequence for the discussion of the mechanisms of chromospheric versus coronal heating as long as the scales involved are not known.

At the chromospheric level, there seems to be at most a negligible difference in radiative losses when network underneath coronal holes is compared to a quiet network. Apparently the energy deposition in the chromosphere does not depend on whether the field is closed or not on length scales larger than some tens of thousands of kilometers. Note that the marked difference in the brightness of a coronal hole and of the quiet-Sun corona does not reflect a similar difference in the energy deposition rates: these regions have similar coronal energy requirements (Section 8.7), but in coronal holes much of the energy goes into the large-scale kinetic energy of the outflow rather than into thermal energy that is then lost radiatively and conductively (see Table 2.1). One characteristic difference between low-coronal and chromospheric emission patterns emerging from high-resolution observations, is that whereas chromospheric emission primarily comes from over the core of magnetic flux concentrations (Sections 8.1 and 8.5), the emission from the high-temperature transition region ($\sim$1 MK) appears to be largely uncorrelated in position with the magnetic footpoints at scales near 1 arcsec (Berger *et al.*, 1999). What clues this holds remains to be explored in detail, but some ideas are emerging; see, for example, the paragraph preceding Eq. (10.5).

Stellar observations suggest two domains of preferred coronal temperatures (Section 9.6): one lies near 3–5 MK, and the other near 15–20 MK. The cause of these preferences remains unknown. A connection of loop stability with the shape of the radiative loss curve has been proposed, but numerical studies to date do not support this unambiguously, whereas effects of possible deviations from solar abundances through the radiative loss curve $\mathcal{P}(T)$ (Sections 2.3.1 and 8.6) have not been included in these studies. Moreover, *TRACE* observations show long-lived, stable loops with a temperature of only 1–2 MK.

An alternative explanation for the existence of dominant temperature intervals is that the relatively hot component, which weakens with decreasing activity and probably also with increasing surface gravity (Section 9.6), reflects an increased frequency of (relatively small) flares or, more precisely, intermittent phases of impulsive heating. The multitude of weak flares and the cooling phases of strong flares could then result in the 3 to 5 MK component, whereas a much smaller number of relatively large flares could result in a substantial hot component (Guedel, 1997).

Finally, let us turn to global constraints on outer-atmospheric heating requirements. In Chapter 11 a reasonably well-defined rotation–activity relationship is discussed, which includes the substantial spread about any of the mean relationships that is caused by long-term activity variations (Section 11.3). This implies that nearly all the parameters controlling both the dynamo and the energy dissipation mechanisms have little impact on the relative distribution of energy over the outer-atmospheric domains, despite the large range of the radiative flux densities covered in the stellar sample. The potentially strong dependence of the net heating rates on these processes is apparently, at most, weak. There is only one significant exception: the stellar rotation rate. To quote Rosner (1991): "We shall have to understand the physics of . . . heating at a sufficiently deep level to comprehend why the details which matter to our present understanding of these theories ultimately do not matter." The same has been said of the acoustic energy-deposition mechanism discussed in Section 2.7, where self-regulating mechanisms have been shown to operate. The numerical experiments by Nordlund and Galsgaard point to a remarkable independence of heating rates on coronal conditions, also suggesting such a self-regulation, but the physical principles behind it are not yet entirely clear.

11

Activity and stellar properties

11.1 Activity throughout the H–R diagram

This section provides an overview of the magnetic activity of single stars across the H–R diagram; see Fig. 11.1. Activity in binary stars is discussed in Chapter 14. The observational data are summarized and qualitatively interpreted in terms of the empirical activity–rotation relation and the evolution of the rotation rate. The quantitative analysis follows in Section 11.3 and in Chapter 13.

Plots of Ca II H+K flux density against color (see Figs. 2.17 and 9.3) display a broad distribution of emission fluxes. Young stars from open clusters, such as the Pleiades and Hyades, occupy the upper strip in such plots, that is, above the Vaughan–Preston gap (Section 9.3.3). Older main-sequence stars, such as the Sun, and many subgiants and giants are found in the lower part of such plots.

Dedicated studies proved that a convective envelope is a necessary but not a sufficient condition for significant magnetic activity. A convective envelope turned out to be a necessary condition, because indications for magnetic activity are exclusively found in stars with convective envelopes (Fig. 11.1), which corresponds to the domain of the H–R diagram to the right of the granulation boundary (see Fig. 2.10), including pre-main-sequence stars, such as the T Tauri stars (these objects are discussed in Section 11.6).

A convective envelope is not a sufficient condition for significant magnetic activity, however: magnetic activity is found to depend on the stellar rotation rate (the activity–rotation relation is discussed in quantitative detail in Section 11.3). Among the main-sequence stars, the most active, single stars are the young and relatively rapidly rotating stars that have recently arrived on the zero-age main sequence (Section 2.4). Thereupon the stars spin down because of magnetic braking (Section 13.3), and consequently the activity level declines. The Sun is among the aged, slowly rotating stars with a low level of activity.

Along the main sequence, magnetic activity is found all the way from the F-type stars down to the coolest M-type stars. The onset of magnetic activity in the range of the F-type stars, with their shallow convective envelopes, is discussed in Section 11.4. The coolest M-type stars are fully convective; Section 11.5 scrutinizes activity signatures of stars with very deep convection zones and of fully convective stars.

In the domain of stars that have evolved from the main sequence, the patterns in the outer-atmospheric signatures appear to be complicated at first sight. The study of such stars started with Ca II H and K data, but this field of research boomed after the

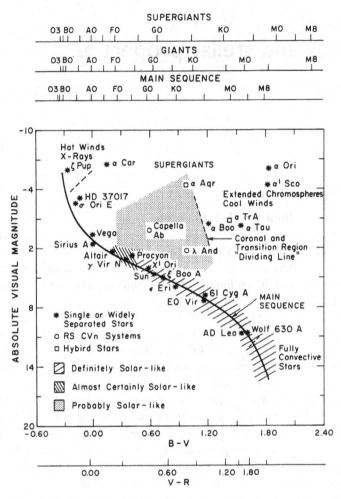

Fig. 11.1. An H–R diagram showing stars with magnetic activity, which in the original paper (Linsky, 1986) are distinguished in groups of solar likeness. Also indicated is the region where massive winds occur and hot coronal plasma is apparently absent. Some frequently studied stars (both magnetically active and nonactive) are identified by name.

space observatories *IUE* and *EINSTEIN* became available. Wilson (1976) published a study of the chromospheric activity of ~500 evolved stars, based on eye-estimated classes of the Ca II H and K line-core emission strengths. These data were reanalyzed by Middelkoop and Zwaan (1981), and Middelkoop (1982a) complemented these studies with his quantitative measurements of 335 evolved stars with the Mt. Wilson Ca II HK photometer (Section 9.3).

The Ca II H+K emission from evolved single stars with masses of $M_* \lesssim 1.2 M_\odot$, that is, subgiants and light giants, is weak. In fact, in diagrams of chromospheric flux densities against effective temperature or color, the data points of these evolved low-mass stars are located very close to the lower limit defining the level of the basal emission (Section 2.7, Fig. 2.17).

Giants with masses of $2M_\odot \lesssim M_* \lesssim 3M_\odot$ show appreciable chromospheric activity during their F- and early G-type phase, with $0.35 < B-V < 0.85$. The chromospheric emissions from single K-type giants ($B-V > 1.0$) are so small that the data points border the lower-limit flux level in flux–color diagrams. A fraction of the more massive bright giants, with $3M_\odot < M_* < 5M_\odot$, still show appreciable chromospheric emission as early K-type giants, with $1.00 < B-V < 1.30$ (Middelkoop, 1982a).

The supergiants (LC I), with $M_* > 5M_\odot$, cover a broad strip in diagrams of chromospheric flux density against color $B-V$, both in Ca II H+K and in Mg II h+k (Oranje and Zwaan, 1985). Pasquini *et al.* (1990) found that supergiants show appreciable Ca II H+K emission throughout spectral range G up to K3, that is, for $B-V \lesssim 1.4$. However, some apparently inactive G-type bright giants and supergiants, with $B-V \leq 0.9$, show chromospheric flux densities far below the basal emission level as indicated by giants and subgiants. Two possible explanations for such extremely dark line cores come to mind: a strong departure from LTE in the rarefied outer atmospheres, or additional extinction by circumstellar matter.

Giants and bright giants showing a chromospheric emission in excess of the basal emission satisfy the flux–flux relations as defined by main-sequence stars – weakly active, single giants define those relationships down to the smallest measured flux densities (Fig. 9.4). Note that the mean surface brightness in soft X-rays of such weakly active giants is less than that of a solar coronal hole!

The drop in chromospheric emission from giants with decreasing effective temperature over the late G-type domain corresponds to an even sharper drop in emissions from the transition region and the corona, as we already noted in Section 9.4. This has lead to the discovery of so-called *dividing lines* in the H–R diagram (see Fig 11.1). One such line was proposed by Linsky and Haisch (1979) from a study of *IUE* ultraviolet spectra: to the higher effective temperature side of this dividing line, G-type giants are found that show both transition-region (mainly C IV) and chromospheric emissions, whereas toward the lower temperatures the K-type giants display no outer-atmospheric matter hotter than a chromosphere. From soft X-ray measurements with the *EINSTEIN* Observatory, Ayres *et al.* (1981a) found a coronal dividing line, which for giants and bright giants virtually coincides with the C IV dividing line. From the *ROSAT* All-Sky Survey, the coronal dividing line for subgiants (LC IV), giants (LC III) and bright giants (LC II) is found to be well defined at spectral type K3 (Haisch *et al.*, 1991a). In contrast with the C IV dividing line, the coronal dividing line does not penetrate in the domain of the supergiants (LC I), because none of the cool supergiants emits detectable X-ray emission.

Earlier, a dividing line was discovered by Reimers (1977) in his study of the circumstellar Ca II H and K absorption lines in spectra of late-type giants and supergiants. For many late-type, evolved stars, the Ca II H and K line-core profiles exhibit an extra absorption component that is shifted to a shorter wavelength. In many cases, this component is quite sharp; it is described as circumstellar, and its shift is interpreted as being caused by the outflow of a massive, cold stellar wind. The outflow velocities range from 10 to 100 km/s. Reimers found a dividing line in the H–R diagram, with stars with strong, cold stellar winds located to the cooler effective-temperature side, and stars without such a strong and cool wind to the warmer side. Particularly near the boundary line, the circumstellar absorption component is variable in strength and somewhat in wavelength over time intervals of approximately 1 to several months.

The overall shape of the Ca II K line-core emission profile (see Fig. 2.6 for the nomen-clature) changes across the H–R diagram, with a dividing line that virtually coincides with the mass-loss dividing line (Stencel, 1978). To its cool side, the K2V emission peaks in most of the Ca II K line cores are much lower than the K2R peaks, while in the domain on the warm side the peaks are approximately equal or the K2V peaks are slightly higher. The asymmetry with the K2V peak being the lowest has been interpreted as a relative violet shift of the K3 component caused by an outflow in which the speed increases with height (see Stencel, 1978); this interpretation is compatible with the conclusions from the circumstellar Ca II lines. The mass loss from cool stars is further discussed in Section 12.5.

Scrutiny of the observational data bears out the idea that the dividing lines are narrow strips rather than sharp lines. In the domain of the giants (LC III) and bright giants (LC II), the dividing strips found from the chromospheric emission, from transition-region (C IV) emission, from soft X-ray emission, and from cool winds practically overlap. In other words, a (bright) giant either exhibits all the outer atmospheric emissions but no massive cool wind, or it displays a strong cool wind but no observable transition-region and coronal emission. Within and near these overlapping dividing strips, the emissions and winds are quite variable. There we also find the so-called *hybrid* bright (LC II) giants that sport both a cool, massive wind and transition-region emissions; see Fig. 11.1 (Hartmann *et al.*, 1981b).

All G-, K- and M-type supergiants (LC I) exhibit massive cool winds, and none show soft X-ray emission. Whereas most of the G- and early K-type supergiants, with $B-V$ up to 1.0, display conspicuous chromospheric and transition-region emissions, the late K- and M-type supergiants emit small amounts of chromospheric radiation only.

Note that the dividing lines refer to (effectively) single stars only. Many components in close binaries show an activity level that is much higher than that of an otherwise identical single star. For instance, a K-type giant in a binary with an orbital period less that 100 days emits copious amounts of chromospheric, transition-region and coronal radiation (see Zwaan, 1981a).

The variation of the levels of outer-atmospheric emissions across the H–R diagram may be interpreted as a decrease in magnetic activity with decreasing stellar rotation rate (see Section 11.3) as stars evolve away from the main sequence (Middelkoop and Zwaan, 1981; Zwaan, 1991). During that evolution, the stellar rotation rate decreases because of the increasing moment of inertia (Section 13.1) and because of "magnetic braking" caused by the streaming out of a stellar wind along the coronal and circumstellar magnetic field (Section 13.3).

Because stars with a mass of $M_* \lesssim 1.2 M_\odot$ have had convective envelopes ever since their arrival at the zero-age main sequence, they have been losing angular momentum by their magnetized stellar winds during their entire, relatively long existence as hydrogen-burning stars (see Table 2.3). Consequently, they leave the main sequence as very slowly rotating stars, and they develop into subgiants and giants with an extremely low activity level, hardly above the level of the basal emission.

Stars with a mass of $M_* \gtrsim 2 M_\odot$ do not have convective envelopes during the main-sequence phase, hence these stars leave the main sequence rotating rapidly. While moving to the right in the H–R diagram, such a star develops a convective envelope (see the gran-ulation boundary in Fig. 2.10), which rapidly deepens while the star develops into a truly giant star. Some time after the onset of convection in the envelope, the star is still rapidly rotating, so the dynamo action is strong, and the star shows up as an active G-type giant once the convective envelope is deep enough. Thereupon the stellar rotation rate drops rapidly, largely because of the rapidly increasing moment of inertia (Section 13.2). Hence, while the star passes rapidly across the dividing line, the outer-atmospheric emissions drop to the low basal level, leaving detectable emissions only from the chromosphere.

Whereas the evolution of the stellar rotation rate as sketched above (and discussed in detail in Chapter 13), together with empirical activity–rotation and flux–flux relations, provides a promising scenario for the distribution of the activity level for single stars across the H–R diagram, some problems remain. How do giant stars without measurable X-ray emission or transition-region emission, or both, produce such conspicuous cool winds? Why do G-type supergiants ($B-V < 1.0$) show transition-region emissions but no soft X-ray emission? What is the nature of the hybrid stars? In short, how does an outer atmosphere react to the photospheric magnetic field forced upon it? We explore these questions in following chapters, notably in Section 12.5.

11.2   Measures of atmospheric activity

Throughout this book, we express stellar atmospheric activity in terms of a radiative surface flux density F_i in some outer-atmospheric emission. Here we discuss its merits relative to other measures that are frequently used in the literature, specifically the surface-integrated value of F_i, that is, the luminosity L_i, and the luminosity relative to the bolometric luminosity $R_i \equiv L_i/(4\pi R_*^2 \sigma T_{\text{eff}}^4) \equiv F_i/\sigma T_{\text{eff}}^4$. Each of these has its own intuitive foundation: R measures the fraction of the total energy generated by a star that is converted into nonradiative heating, F is a direct measure for the energy budget per unit area of the star, and L measures the global stellar budget. Given the lack of a solid theoretical understanding of the series of processes involved from the generation of the magnetic field to the dissipation of nonradiative energy into heat in the outer atmosphere, there is no argument that favors any particular activity measure over another *a priori*. The question to ask is the following: Which unit yields the simplest relationships with the least scatter about them for the largest sample of cool stars?

In the preceding chapters we may have given the impression that the uncovering of the remarkably tight relationships between flux densities from different outer-atmospheric temperature domains was rather straightforward following the subtraction of the basal flux densities. Historically, the idea of the existence of a basal atmosphere was developed in an interplay with the definition of the flux–flux relationships themselves. Because the choice of the measure of activity itself rests in part on that development, we dwell on some of the historical details for a moment.

Most of the flux–flux relationships discussed in Section 9.4 are nonlinear. This is fundamental to the identification of the appropriate unit for the activity measure, because transforming units from, say, the surface flux density F to L, or R, introduces functions on both sides of such a relationship that do not cancel against each other. Specifically, a relationship between diagnostics A_i and A_j that is approximated by

$$A_j = c_{ij}(A_i)^{e_{ij}} \tag{11.1}$$

is sensitive to the units in which it is expressed if $e_{ij} \neq 1$, because if the units are changed into those of A'_j by a color- or radius-dependent scaling,

$$A'_j f(T_{\text{eff}}, R, \ldots) = c_{ij}[A'_i f(T_{\text{eff}}, R, \ldots)]^{e_{ij}}$$

$$\Updownarrow \qquad\qquad (11.2)$$

$$A'_j = c_{ij}(A'_i)^{e_{ij}} f(T_{\text{eff}}, R, \ldots)^{(e_{ij}-1)}.$$

The resulting transformed relationship shows a residual dependence on basic stellar parameters.

Schrijver (1983) subjected a sample of dwarf and giant stars to a multidimensional principal component analysis. This statistical method can be used to find the lowest dimensional power-law relationship between any set of parameters that incorporates all of the variance in the data except for that associated with uncertainties (here including the effects of stellar variability caused by nonsimultaneous measurements). Schrijver finds that the simplest relationship between measures for coronal and chromospheric emission, independent of stellar parameters, is found when relating flux densities. He finds, for example, a relationship between the soft X-ray flux density F_X measured by the *EINSTEIN/IPC* and the Mt. Wilson Ca II H+K flux density $\Delta F_{\text{Ca II}}$ in excess of the minimal level (compare Section 2.7), which is given by

$$F_X \propto \Delta F_{\text{Ca II}}^{1.67}. \qquad\qquad (11.3)$$

This relationship holds over four decades in F_X (compare Fig. 9.4, which shows a power-law index of 1.5; this difference is probably caused both by the differences in the stellar sample and by the differences in the soft X-ray passbands). Rewriting Eq. (11.3) to a relationship between luminosities results in a clearly noticeable radius dependence (e.g., Rutten and Schrijver, 1987a), which excludes L as a proper measure of activity. The relationship between normalized fluxes R is almost as tight as that between flux densities F, however, and also virtually independent of color and luminosity. A transformation yields

$$F_X \propto \Delta F_{\text{Ca II}}^{1.67} \Leftrightarrow R_X \propto (\Delta R_{\text{Ca II}})^{1.67} T_{\text{eff}}^{2.67}. \qquad\qquad (11.4)$$

Some 80% of the stars in his sample are contained in a range in $\log T_{\text{eff}}$ of merely 0.09, so that the differentiation by stellar color is hidden in the scatter associated with stellar variability associated with cyclic and rotational modulation, and measurement and calibration uncertainties (e.g., Schrijver *et al.*, 1992a). In a comparison of near-simultaneous measurements of *ROSAT* soft X-ray and Mt. Wilson Ca II H and K fluxes of a sample of effectively single stars, Piters (1995) slightly favors R over F. Basri (1987) compared correlations in scatter diagrams relating L, R, and F for various activity diagnostics for a sample containing single stars as well as tidally interacting binaries. He finds the best correlation for the surface flux density F. Given this evidence, we do not feel we can unambiguously distinguish between F and R once and for all as the most suitable measure for flux–flux studies in single stars, but the above studies favor F if both single stars and close binaries are involved.

Let us now extend this discussion beyond the atmospheric properties and include the driving force behind it, namely the dynamo process itself. Although the stellar rotation rate is the primary parameter determining the dynamo activity, other parameters may also play a role. The surprising tightness of the stellar flux–flux relationships, and the large range over which they are valid, led to the introduction of an *activity parameter* in Section 9.4: it appears likely that there is a single parameter Υ that specifies the distribution of radiative losses over the temperature domains in stellar outer atmospheres. In order to be consistent with the local and global power-law relationships between radiative and magnetic flux densities discussed in Section 9.5, we expect that to a good approximation the magnetically driven (excess) flux density depends on Υ through a power-law relationship,

$$\Delta F_i = a_i \Upsilon^{b_i(T_f)}, \tag{11.5}$$

where $b_i(T_f)$ is a function of only the formation temperature T_f of the diagnostic i under consideration. The activity parameter Υ is itself a function of at least stellar rotation rate and stellar structure, but also of binarity (Section 14.2). This expression shows that the radiative losses from any domain in the outer atmosphere generally are not simply proportional to the activity level of the star, however that is expressed. This conclusion is a restatement of our earlier finding that the disk-averaged radiative flux density depends on the details of the distribution of magnetic flux over the surface (Section 9.5.2).

11.3   Dynamo, rotation rate, and stellar parameters

The mean level of magnetic activity of a cool star is a function of its rotation rate. This connection has been explored ever since the pioneering work by Wilson (1966) and the study by Kraft (1967) that he inspired (see Noyes, 1996, for a historical review). At first, the rotation–activity relationship was considered to be an indirect one through stellar age: a (primordial) magnetic field was thought to be slowly decaying because of the finite resistivity and the large scales involved [cf. Eq. (4.10)], while rotation was slowing down in response to angular-momentum loss through the magnetized stellar wind. Skumanich (1972) suggested that the activity–rotation relationship was a causal consequence of dynamo action.

When comparing any radiative measure of stellar magnetic activity with stellar angular velocity for stars of comparable effective temperature and gravity, one finds that activity increases monotonically with increasing angular velocity for stars with rotation periods exceeding 1–2 days. Before we discuss this relationship in more detail, let us address the problem of how to determine which parameters are involved.

Dynamo theory aims to explain the properties of stellar magnetic fields. It focuses in particular on the dependence of the mean activity level on stellar parameters, but also on the spatial and temporal patterns in stellar magnetic activity. At present, there is no consistent dynamo model (see Chapters 7 and 15), so that empirical studies are without clear guidance when looking for trends. Some qualitative trends have been extracted from crude theoretical models, however. For a kinematic dynamo, in which flows are not affected by the magnetic field, the dynamo problem is purely linear, so that there is no constraint on the strength of the magnetic field contained in the dynamo equation. The growth rate for the field is, however, specified by the model. If we let $\langle \mathbf{B}_0 \rangle \propto \exp{(\Lambda t)}$, the temporal and spatial parts of the dynamo equation, Eq. (7.8), can be separated, resulting in an eigenvalue problem:

$$\Lambda \langle \mathbf{B}_0 \rangle = \nabla \times (\mathbf{u}_0 \times \langle \mathbf{B}_0 \rangle + \alpha \langle \mathbf{B}_0 \rangle - \beta \nabla \times \langle \mathbf{B}_0 \rangle). \tag{11.6}$$

This equation can be made dimensionless by multiplying it with the time scale for turbulent diffusion

$$\tau_\beta \approx L^2/\beta_0, \tag{11.7}$$

where L is the scale of the region within which the dynamo operates and the index 0 indicates that some typical value should be chosen. This eigenvalue problem suggests two dimensionless *dynamo numbers*, $N_\alpha \equiv \alpha_0 L/\beta_0$ and $N_\Omega \equiv \Omega_d L^2/\beta_0$ (where L is the scale of the region within which the dynamo operates, and Ω_d is a characteristic value of the differential rotation). These two are often combined into a single number as in Eq. (7.10). Durney and Latour (1978) propose that both N_α and N_Ω are comparable to $\ell_{\rm ML}\Omega_d/v_{\rm ML}$, for a mixing-length velocity $v_{\rm ML}$. The resulting joint dynamo number is

$$N_{\rm D} \equiv N_\alpha N_\Omega = \left(\frac{\ell_{\rm ML}\Omega_d}{v_{\rm ML}}\right)^2 \propto \left(\frac{\hat{t}_c}{P}\right)^2 = \mathcal{R}_o^{-2}, \tag{11.8}$$

where P is the rotation period of the star, $\mathcal{R}_o$ is the Rossby number, and $\hat{t}_c$ is the characteristic turnover time scale of convection. Rosner and Weiss (1992) use slightly different arguments to arrive at

$$N_{\rm D}' = \mathcal{R}_o^{-2}\left(\frac{\tilde{d}_{\rm CE}}{\ell_{\rm ML}}\right)^2, \tag{11.9}$$

which suggests a strong direct dependence of the dynamo number on the size of the convective cells relative to the convecting volume with depth $\tilde{d}_{\rm CE}$ (plotted in Fig. 2.11 for some main-sequence stars).

Equation (11.6) has a discrete spectrum of complex eigenvalues Λ_i. For a certain critical dynamo number $N_{\rm D,cr}$, there is one marginally stable solution of Eq. (11.6), with growth rate $\Re e(\Lambda_0) \equiv 0$, called the fundamental mode. Higher modes are not excited because they have a negative growth rate. If $N_{\rm D} > N_{\rm D,cr}$, then $\Re e(\Lambda_0) > 0$ and at least one mode will grow. If, in contrast, $N_{\rm D} < N_{\rm D,cr}$, then all modes have negative growth rates and $\langle \mathbf{B}_0 \rangle$ eventually decays away. This does not necessarily mean that *all* field decays away, but merely that the mean field vanishes! In principle, this allows a transition from a global to a turbulent dynamo for a decreasing rotation rate (see Chapter 15).

The Rossby number, or rather the dynamo number, does not necessarily inform us about the dynamo strength, because growth rate and amplitude need not be proportional, while in any kinematic model the amplitude is a completely free parameter. Moreover, when stars of different convective turnover times are compared, many other parameters will be different as well, which may or may not affect the dynamo number directly, depending on which approximation is used. Another complicating factor in the empirical search for the relevant parameters in the rotation–activity connection is that the field that is generated in the stellar interior is transported to the stellar photosphere, where it channels the injection of nonradiative energy into the outer atmosphere, which is subsequently radiated away and used to measure dynamo action. Each of these steps introduces its own dependence on stellar parameters. Without unambiguous theoretical guidance about the functional dependence on the parameters involved, the description of what is generally referred to as the *rotation–activity relationship* can only be determined empirically with large samples covering much of the parameter space. Assembling suitably large data bases is still a major problem.

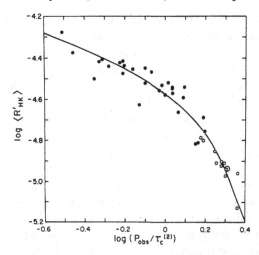

Fig. 11.2. The ratio R'_{HK} of the emission from the chromosphere in the cores of the Ca II H and K lines, after subtraction of a photospheric contribution (*not the basal flux*), to the total bolometric emission of these main-sequences stars vs. the ratio of the rotation period to the color-dependent function $\tau_c^{(2)} \equiv g(B-V)$ from Eq. (11.11). Open and closed circles indicate stars below and above the Vaughan–Preston gap, respectively (Section 9.3; figure from Noyes *et al.*, 1984).

Despite the numerous links in the chain leading from internal dynamo to atmospheric radiative losses, a tight rotation–activity relationship was found for main-sequence stars. Noyes *et al.* (1984) have argued that the activity level of main-sequence stars is primarily determined by a parameter that closely resembles the Rossby number. They gave an approximate parameterization for the ratio R'_{HK} of the emission from the chromosphere in the cores of the Ca II H and K lines, after subtraction of a photospheric contribution (not including the basal signal), to the total bolometric emission of main-sequences stars:

$$R'_{HK} \approx 6 \times 10^{-5} \exp\left[-0.9P/g(B-V)\right]. \tag{11.10}$$

Thus they found that when the Ca II H+K emission relative to the bolometric flux is compared to the rotation period scaled by a color-dependent function $g(B-V)$, the main-sequence stars used in their study collapse onto a single relationship (Fig. 11.2). The best-fit function $g(B-V)$ approximates the convective turnover time of convection just above the bottom of the convective envelope as estimated from mixing-length theory, that is, $g(B-V) \approx \hat{t}_c(B-V)$ for

$$\log g(B-V) = 1.362 - 0.166x^2 - 5.323x^3, \quad x > 0, \tag{11.11}$$
$$= 1.362 - 0.14x, \quad x \le 0,$$

for $x = 1 - (B-V)$ (with $g \approx 20.6$ days for the Sun). That scaling function was later extended from $B-V \approx 0.4$ down to $B-V \approx 0.25$, that is, into the early F-type domain; see Section 11.4. The study by Noyes *et al.* subsequently prompted the search for a similar dependence using other activity diagnostics and expanded stellar samples. Several hundred research papers cite this pioneering empirical study. For example, Stępień (1994), finds essentially the same relationship but presents a slightly different normalization for his empirically determined equivalent of the convective turnover time; Hempelmann *et al.* (1995), derive an exponential relationship for the coronal radiative losses, which also transforms nearly into Eq. (11.10) when using a power-law index between soft X-rays and Ca II H+K of $\sim$1.7; see Eq. (11.3). Figure 11.4 shows the rotation–activity relationship for soft X-rays for cluster and field stars of ages as short as 28 Myr for Rossby numbers of about 2 and below.

The support for the role of $\hat{t}_c$ in the dependence of activity on rotation rests on the invariance of the shape of the rotation–activity relationship as a function of spectral type. Rutten and Schrijver (1987a) point out that the relationship, $A(P_{rot}, T_{eff}, \ldots)$, between rotation period, P_{rot}, and activity, A, is not simply a separable function of rotation rate and effective temperature; in other words, $A(P_{rot}, T_{eff}) \neq A_0(P_{rot}) \times A_1(T_{eff})$. Rutten and Schrijver (1987a) emphasized that the color-dependent scaling of the rotation period may well be limited to the relatively active stars. For rotation periods exceeding $\sim$10 days, the functional dependence of the relationship between the chromospheric Ca II H+K flux density and surface rotation period is found to change significantly with color: the cooler the star, the slower the decrease in activity with increasing period (Fig. 11.3). This implies that *no* color-dependent scaling of activity measure or rotation rate can bring all of the main-sequence stars into line. Although based largely on the same data for the main-sequence stars, the two primary differences with the study by Noyes *et al.* (1984) are that (a) Rutten and Schrijver (1987a) subtract a higher empirical minimum flux

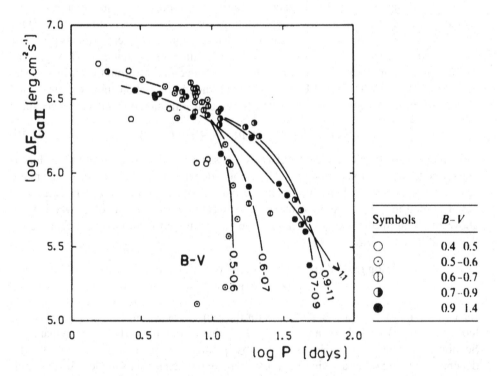

Fig. 11.3. The chromospheric excess flux density ΔF_{HK} versus the stellar rotation period in main-sequence stars. Figure from Rutten and Schrijver (1987a).

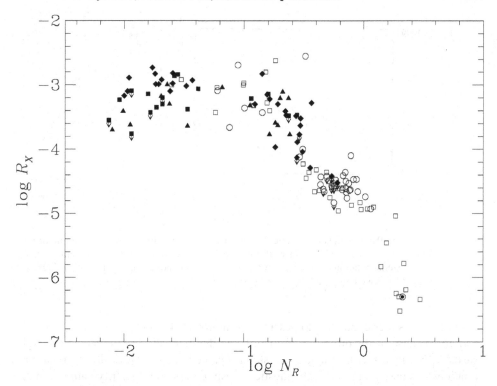

Fig. 11.4. Soft X-ray relative emission $R_X = L_X/L_{bol}$ as a function of Rossby number $\mathcal{R}_o \equiv N_R$ for F5 through M5 main-sequence stars in four clusters and a selection of field stars. Symbols: filled triangles for IC 2391 (28 Myr), filled squares for α Per (51 Myr), filled diamonds for Pleiades (78 Myr), open circles for Hyades (660 Myr), and open squares for field stars. The Sun ($\odot$) is located near the low end of the activity measure. Figure from Patten and Simon (1996).

(incorporating the photospheric line-wing flux as well as the line-core basal flux), which makes the color dependence of the rotation–activity relationship more pronounced; and (b) effectively single giant stars are included in the sample of Rutten and Schrijver. The giant stars particularly contribute to the sample at long periods, where they strengthen the suggestion already present in the main-sequence sample. The color dependence in the rotation–activity relationship is also apparent in coronal soft X-ray data (Hempelmann *et al.*, 1995, their Fig. 3a) for stars with Rossby numbers exceeding unity. Rutten and Schrijver (1986) point out that the onset of the color dependence in terms of activity lies near the Vaughan–Preston gap, a band of intermediate activity in the diagram of Ca II H+K flux versus color containing relatively few stars (Fig. 9.3). The significance of this observation is yet to be determined.

In contrast to the differences in rotation–activity relationships for slowly rotating stars, which do not support a single scaling of activity with some stellar parameter, more rapid rotators do allow such a scaling. It is somewhat disappointing that an ambiguous conclusion is reached for these stars: for main-sequence stars with periods below approximately 10–15 days, it seems that $F - P$ and $R - \mathcal{R}_o$ diagrams do about equally well, with the scaling by effective temperature and convective turnover time leading to comparable results.

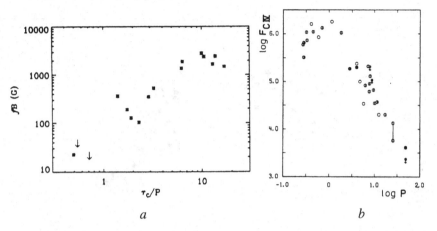

Fig. 11.5. Panel *a*: mean magnetic flux density vs. the Rossby number in a sample of main-sequence stars (figure from Saar, 1996). Panel *b*: the C IV transition-region emission F_{CIV} vs. the stellar rotation period in main-sequence stars, including short-period W UMa binaries (figure from Rutten and Schrijver, 1987a).

Outer-atmospheric radiative losses are commonly used as indirect measures of stellar activity because of the difficulty of quantifying the magnetic flux threading the stellar surface. Yet any direct information on stellar magnetic activity is desirable to calibrate relationships based on indirect diagnostics. Despite the mixed-polarity nature of a stellar disk, which results in a cancellation of any polarization signature, the presence of magnetic fields on stars leads to a broadening of disk-integrated magnetically sensitive lines (Section 9.2). This signature in principle allows a direct measurement of the surface filling factor f and of the intrinsic field strength B. The small magnitude of the Zeeman splitting, however, limits the use of this method to moderately active stars with on one hand substantial filling factors and on the other hand not an excessive rotational Doppler broadening of the line profiles. Access to the infrared has increased the sensitivity of this method (the Zeeman separation relative to the Doppler width scales with λ; see Section 2.3.2). These measurements (summarized by Saar, 1996; see Fig. 11.5a) show a rapid and monotonic increase of the surface-averaged magnetic flux density with rotation period decreasing from $\sim$25 to 4 days. For periods between 1.5 and 4 days, the activity increase appears to level off at a surface-averaged magnetic flux density of $\langle |\varphi| \rangle \approx$ 2–3 kG. That plot alone may not be convincing evidence in favor of dynamo saturation, but there is other evidence for this, as we discuss in the paragraphs that follow. In the most active stars, at the onset of saturation, some 60% of the surface appears to be covered by strong fields, not even including dark starspots and pores (see Section 12.2).

A fit to the data in Fig. 11.5a for $\hat{t}_c/P \lesssim 10$ yields

$$\langle |\varphi| \rangle \propto (\hat{t}_c \, \Omega)^{1.7}. \tag{11.12}$$

This relationship is at odds with another that can be derived by combining other relationships. For the period range from a few days up to 20 days, the relationship between coronal radiative losses and rotation periods, for instance, can be approximated by $\langle F_X \rangle \propto \Omega^{0.8 \pm 0.2}$ (e.g., Schrijver and Zwaan, 1991). Combining this approximation with the relationship $\langle F_X \rangle \propto \langle \Delta F_{\text{Ca\,II}} \rangle^{1.5 \pm 0.2}$, found for stars, and $\Delta F_{\text{Ca\,II}} \propto |\varphi|^{0.6 \pm 0.1}$ found for the Sun observed with a moderate angular resolution, one finds

$$\langle |\varphi| \rangle \propto \Omega^{0.9 \pm 0.3}, \tag{11.13}$$

which is, roughly speaking, a linear dependence. Clearly, this is a very indirect derivation, but although Eq. (11.12) compares magnetic flux densities and rotation rates directly, the derivation of the flux densities depends on several critical model assumptions (Section 9.2). With both relationships suffering from substantial uncertainties, Eq. (11.13) is favored frequently in models of magnetic braking; see Section 13.3.1.

For stars with rotation periods below a few days, depending on spectral type, the outer-atmospheric radiative losses reach a maximum, suggestive of dynamo saturation or quenching when the photospheric magnetic filling factors approach unity. In the shortest-period systems the emission actually *decreases* with a decreasing rotation period, now referred to as "supersaturation." This quenching was first reported by Vilhu and Rucinki (1983) for contact W UMa systems but is now confirmed (Fig. 11.4) for the most rapidly rotating single stars in the young clusters α Per, IC 2391, and IC 2602 for stars with rotational velocities exceeding $\sim$100 km/s (e.g., Stauffer *et al.*, 1997; Jeffries, 1999, and references therein). The saturation seems to occur for chromospheric emission at higher rotation periods than for coronal emission, resulting in an upturn in the flux–flux relationships for the most active stars (which shows up in Fig. 9.4, but for too few data points to be significant). Hall (1991) points out that there is no indication of a saturation in the modulation of the visual magnitude resulting from spot coverage down to the shortest rotation periods in his sample, which reaches down to stars with $\mathcal{R}_o \approx 0.01$. But that sample may not extend to sufficiently rapid rotators: O'Dell *et al.* (1995) point out that optical modulation reaches an observed maximum around a rotation period of approximately half a day, but they caution that this is not unambiguous evidence for saturation. Indirect support for dynamo saturation comes from the study of the evolution of stellar rotational velocities under the influence of magnetic braking; see Sections 13.3.1 and 13.3.4.

The maximum power that is emitted from the nonradiatively heated outer atmosphere from stars is $\sim$$10^{-2}$ of the total luminosity (Vilhu and Walter, 1987), and 10^{-3} for the coronal fraction (Haisch and Schmitt, 1996). The former is so high that we can legitimately ask the question whether more power can be extracted from the convective motions.

We note that the different saturation levels for different diagnostics may be reflected in a property of solar activity: Schrijver *et al.* (1989a) pointed out that the Ca II K emission of solar plages suggested saturation of the emission for a level equivalent to $|\varphi| \approx 600$ G. The radiative flux density at which this occurs is equivalent to the disk-integrated Ca II H+K emission for the most active stars of solar spectral type. Using the solar analogy, we would then expect that once a star is completely covered with plage fields found only in the densest cores of solar plages, the Ca II H+K emission saturates, while more flux could still be added (apparently up to several kilogauss, on average; see Fig. 11.5a), leading to differences in saturation behavior between photosphere, chromosphere, and corona. Although this is appealing, there are stars whose surface-averaged level in the chromospheric Mg II h+k lines exceeds that of plages on the Sun (Section 12.2), which contradicts the above idea. We should allow, of course, for the possibility that the saturation level is different for the Ca II and Mg II resonance doublets, but there is no observational evidence for that at present.

The details of this activity saturation contain an important clue to the activity parameter: Rutten and Schrijver (1987a) pointed out that all main-sequence stars in their sample with periods between half a day up to more than 10 days fell on the same relationship (Fig. 11.5*b*) between surface flux density in transition-region emission and rotation period; plotting activity versus the Rossby number $\mathcal{R}_o \equiv P/\hat{t}_c$ rather than rotation period P would disrupt the color-independence of that relationship.

Another serious problem with the use of the Rossby number as the stellar parameter to determine activity emerges when the stellar sample is expanded to include evolved stars. There is a consensus that the Rossby number does not result in a rotation–activity relationship for the sample of all single, cool stars that is as tight as that for the dwarf stars only. But whereas, for example, Rutten and Schrijver (1987a) and Stępień (1994) conclude that the Rossby number is not the proper parameter, Basri (1987) finds that its use improves the rotation–activity relationship for the entire data set somewhat, although the relationships for each of the subclasses in his sample are somewhat worse when Rossby numbers are used. His sample includes tidally interacting binaries, which are known to deviate from the relationships for single stars (Chapter 14); this may have affected his conclusion.

Since the appropriate parameters to measure stellar activity in a rotation–activity relationship are not yet determined, the analytical shape of these relationships is even less well established. Our problem is that the samples are small and the intrinsic stellar variability substantial. It seems that no single power law or exponential can describe the relationship accurately even for subsets that include only stars in a narrow range of spectral type and of a given luminosity class. For rather crude estimates, however, Eqs. (11.10), (11.12), or (11.13) can be used.

Finally, let us look at the slowest rotators. The crude formulation of the eigenvalue problem for the dynamo in Eq. (11.6) suggests that if the rotation rate is slow enough, no positive eigenvalue and hence no growth in field amplitude should be expected. Consequently, there should be some minimal rate of rotation that is required to excite a global dynamo. Observational studies rule out, however, that this minimal rotational velocity is measurable with present techniques. Schrijver *et al.* (1989b), for instance, argue that for main-sequence stars the rotation rate drops to zero at or very near the empirical lower-limit flux in a flux–color diagram. They also point out that the long-term Ca II H+K variability as well as the coronal soft X-ray emission disappear at or very near this lower-limit flux. Hence they conclude that magnetic activity A_m increases smoothly and monotonically with a decreasing rotation period for the relatively inactive stars.

Verifying whether there is a minimally required nonzero rotational velocity v_c for a global dynamo to operate is more difficult for giants, because of the generally long rotation periods of single giants. Observed $v \sin(i)$ values set an upper limit to v_c of approximately 1–2 km/s (Gray, 1989), which corresponds to rotational periods of a number of months because of the large stellar radii. In general, therefore, all stars with rotation periods below approximately 2–3 months show magnetic activity, that is, all lie above any such minimally required rotation period. This period range includes essentially the entire sample of cool stars, however, so establishing whether there is a period at which the dynamo ceases for even slower rotators will be very difficult.

Flux–color diagrams contain valuable information on the excitation of dynamos in slow rotators: whereas dwarf stars may not spin down sufficiently to reach v_c while on the main sequence, evolving single giant stars continue to slow down as long as their moment of inertia increases (see Section 13.1). If the activity would discontinuously change from $A_m(v_c + \epsilon, T_{eff}, \ldots) > 0$ to $A_m(v_c - \epsilon, T_{eff}, \ldots) = 0$ (for small ϵ) at some critical rotation rate, the evolution of the moment of inertia should yield stars on the cool side of flux–color diagrams at the basal level, separated from the "active" stars by a gap. Such a gap is not observed, so activity apparently decreases smoothly to (near) zero with a decreasing rotation rate. This means that if in fact the critical rotation rate differs from zero, then the activity generated by a dynamo that is just excited, $A_m(v_c + \epsilon, T_{eff}, \ldots)$, is at most small compared to the basal level A_b at which $A_m \equiv 0$. Hence any transition in the dynamo mode, as discussed in Section 12.5, to explain the corona-wind dividing line should not require a measurable discontinuity in activity.

11.4 Activity in stars with shallow convective envelopes

We now take a closer look at a few critical corners of the H–R diagram: F-type stars in this section, then M-type dwarfs and the young T Tauri stars in subsequent sections.

The efficiency of the dynamo appears to increase rapidly yet smoothly with decreasing effective temperature around the temperature at which envelope convection sets in. It is not easy to determine the relation between rotation rate and magnetic activity in F-type stars because of the substantial basal contribution to chromospheric emissions (up to 50% of the C II emission in even the most active F0 V stars, for instance), and the reduced usefulness of particularly Ca II H+K but also Mg II h+k as a result of double ionization in these warm atmospheres (Section 9.4). The high basal fluxes are associated with high convective velocities [see the crude estimate in Eq. (2.75)], which – in a mixing-length approximation – peak at about $T_{eff} = 8,000$ K, or $B–V \approx 0.11$ at spectral type A3 (Renzini *et al.*, 1977) on the main sequence. At that effective temperature the characteristic convective velocities are estimated to be $\approx$60% of the sound speed just below the photosphere. In even warmer stars, convection becomes rapidly less efficient as a means of energy transport. Convective velocities drop until convection disappears altogether at $T_{eff} \approx 9,000$ K (Bohn, 1984) or $B–V \approx 0.06$ at a spectral type of approximately A2. Hence, in order to study magnetic activity, the basal emission of these stars should be studied too, for which an upper chromospheric diagnostic at ultraviolet wavelengths, such as C II (Section 2.3.4), is well suited.

Schrijver (1993a) argued that after the proper basal correction, there is no significant deviation of F-type main-sequence stars from the flux–flux relationships defined by cooler stars (Section 9.4). In the rotation–activity diagrams, however, F-type stars exhibit a severely reduced emission compared to cooler stars at the same rotation period. It turns out that for a stellar sample with rotation periods limited to the range from $\sim$1 up to 10 days, the magnetic activity of main-sequence stars as measured in C II fluxes in excess of the basal level can be approximated by

$$\Delta F_{CII} = 3 \times 10^3 \, \bar{g}(B–V)(v \sin i)^{1.5}, \tag{11.14}$$

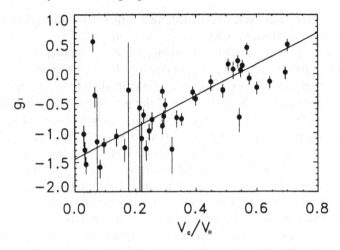

Fig. 11.6. The efficiency function g defined in Eq. (11.14) for magnetic activity in main-sequence stars plotted as a function of the fractional volume of the convective envelope for a mixing length of two pressure scale heights (from Schrijver, 1993a).

where $\bar{g}(B-V)$ is introduced as a color-dependent efficiency function (Fig. 11.6). This efficiency function was determined by iteratively improving the flux–flux and flux–rotation diagrams by optimizing the basal flux. Schrijver (1993a) found that $\bar{g}(B-V) \approx 1$ for $B-V \geq 0.6$. Below $B-V = 0.6$, the value of $\bar{g}(B-V)$ decreases rapidly, and is down to one hundredth at $B-V = 0.22$. Note that a description such as in Eq. (11.14) fails to describe a sample of stars covering a larger range of periods, as discussed in Section 11.3.

Not only C II emission shows a decrease with increasing depth of the convective envelope. The magnitude of variability in broad passbands in the visible is a direct signature of photospheric activity that is due to spots and faculae. Lockwood *et al.* (1997) found that stars with $B-V < 0.4$ showed no significant variability above the measurement uncertainties (at a remarkably low 2–3 millimagnitude rms). The observed amplitude range then increases rapidly with $B-V$. The sample of some 40 cool stars (in a total of 140 program and comparison stars) is not large enough to draw an envelope, but the increase appears to continue at least up to $B-V \approx 0.6$, consistent with the outer-atmospheric variability. That variability was observed to increase with an increasing Ca II H+K emission, consistent with the assumption that photospheric variability is caused by magnetic activity. Hall (1991) shows a compilation of results on stellar variability, compatible with these results. He finds support for a similar trend for subgiants; his sample is too small to study the effects in giants.

The rapid change in dynamo efficiency in the F-star range is also reflected by the flux–color diagram of soft X-ray fluxes, which have the advantage that they are not significantly affected by a basal contribution. A large sample of soft X-ray flux densities compiled by Vilhu (1987; his Fig. 5) shows an increase in the maximum observed flux densities for single stars of ~ 2 orders of magnitude between $B-V = 0.2$ and 0.6, despite the decrease in the maximum rotation rates of single stars in that range. Schrijver (1993a) shows this increase to be compatible with the parameterization in Eq. (11.14). Note that he also demonstrates that the magnetic brake in F-type stars is reduced by a factor that is consistent with the dynamo efficiency, and that a very similar color-dependent factor appears to be required to describe the braking in subgiants with shallow convective envelopes; we return to this in Section 13.3. There is no evidence for dynamo saturation for F-type stars; possibly the lowered dynamo efficiency in these stars keeps even the most active among them from reaching the level required for the activity to saturate.

The steep decrease in the acoustically driven basal component and increase in the dynamo efficiency with decreasing effective temperature for the early F-type main-sequence stars creates a rather sudden transition from stars in which the chromospheric and low-transition-region emissions are acoustically dominated to where they are magnetically dominated. The acoustic dominance in the warmest stars predicts that the short-term variability of chromospheric and transition-region emissions should be low, which is confirmed for the three warmest stars in a study by Ayres (1991b). Such a transition from acoustic to magnetic dominance has also been proposed by Simon and Drake (1989) based on data for giant and subgiant stars; they put this transition at $B-V \approx 0.6$ at G0 IV and $B-V = 0.7$ at G0 III. But although they proposed basically the same change from acoustic to magnetic dominance that was proposed by Schrijver (1993a), their model differs in detail. They suggested the transition to be far more sudden (introducing a dividing line). Moreover, they attributed the deviation of warm stars (defined as late-A and early F-type stars) in diagrams comparing total rather than excess flux densities to a deficient soft X-ray flux that would be associated with a massive coronal wind rather than a magnetically dominated corona. Within the context of the study by Schrijver (1993a) this is not supported, however: the apparent X-ray deficiency of stars with shallow convective envelopes is likely an artifact of the strong basal emission in atmospheric domains with lower temperatures.

The physical foundation behind the color dependence of the dynamo efficiency $\bar{g}(B-V)$ remains veiled, primarily because many physical parameters vary monotonically along the main sequence, so that arguments favoring any functional dependence are necessarily ambiguous. In the literature, however, there is a strong preference for the convective turnover time $\hat{t}_c$, as in the mixing-length concept, at the bottom of the convection zone to be a key parameter (see Section 11.3, and Chapter 15). In main-sequence stars, $\hat{t}_c$ increases sharply along the spectral sequence up to near G0 and then levels off toward cooler stars (Fig. 2.11). Despite the good agreement (reflecting what was already known for a narrower color range from the work by Noyes *et al.*, 1984) this does not uniquely identify $\hat{t}_c$ as the (sole) controlling parameter: the convection zone depth $\tilde{d}_{CE}$, the fractional volume V_{CE}/V_* of the convective envelope, and the mass contained in it – to name but a few examples – also vary in a similar manner as a function of color.

Let us briefly explore possible consequences of a shallow convective envelope on the outer-atmospheric field. Little is known about sizes of stellar active regions (Section 12.2), but one may argue that the maximum size of bipolar regions is limited to (some fraction of) the depth of the convective envelope (Section 4.5). Schrijver and Haisch (1996) point out the consequences for the coronal structure of F-type main-sequence stars. This effect has its origin in the balance of gas and magnetic pressures in the corona: if the magnetic pressure is low enough, coronal field lines can be forced open (Section 8.1). The (mean) size of the bipolar regions is an important factor in this balance: for an isolated region, the height at which the field is expected to open up because of the disrupting effect of the gas pressure depends strongly on the separation of the opposite poles. For a bipole in which the total flux Φ in one polarity varies as $\Phi \propto d_{AR}^2$ (see Section 5.4) – with d_{AR} as the separation between the poles – the on-axis field varies with height h like

$$B(h) \propto \frac{d_{AR}^3}{(h^2 + d_{AR}^2)^{3/2}}. \qquad (11.15)$$

At heights sufficiently exceeding d_{AR}, the plasma β [Eq. (4.13)] is therefore proportional to

$$\beta \propto p_g \left(\frac{h}{d_{AR}} \right)^6, \qquad (11.16)$$

for a gas pressure p_g. Hence, for a given atmospheric stratification the height h_{crit} at which $\beta \approx 1$ is substantially lower for small regions than for large regions. Since the loop length scales with d_{AR}, a combination with the loop scaling law, Eq. (8.22), yields

$$\frac{h_{crit}}{d_{AR}} \propto T^{-1/2} d_{AR}^{1/6}. \qquad (11.17)$$

Hence, when scaled to the size of the region, the height at which the plasma β equals unity moves down slightly faster, so that it can be expected that there is an increasing fraction of the field that opens up to a stellar wind, thereby losing this flux from the X-ray emitting coronal ensemble.

A topological shift in the coronal magnetic field from predominantly open to predominantly closed that is related to a maximum size of bipolar regions imposed by the depth of the convective envelope is compatible with stellar data, but the evidence is inconclusive. Despite this potential shift, F-type stars do adhere to the flux–flux relationships between excess fluxes.

11.5   Activity in very cool main-sequence stars

The color independence of the power-law relationships between radiative losses from the outer atmosphere, discussed in Section 9.4, breaks down in the domain of the M-type main sequence stars. Rutten and Schrijver (1987b) and Rutten *et al.* (1989) showed that dwarf stars with a given rotation period are deficient in their chromospheric emissions beyond $B-V \gtrsim 1.3$. The deficiency increases rapidly with decreasing effective temperature, weakening the Ca II H+K excess flux density for stars beyond $B-V \approx 1.5$ by a factor of 60 relative to that expected for warmer stars of the same rotation period. The deficiency decreases with increasing temperature in the atmosphere: soft X-rays and C IV do not seem to be deficient, Si II, Si IV, and Hα are marginally deficient, but the low-chromospheric emissions are strongly suppressed. There is, moreover, a dependence of the deficiency on metallicity. The departures appear to be independent of the level of magnetic activity of the M dwarf.

Rutten *et al.* (1989) argue that the flux in the Balmer lines does not compensate for the reduced emission in lines such as Ca II H and K and Mg II h and k, and that therefore there is a real net deficiency in the chromospheric emission from M-type dwarf stars. This does not necessarily imply a reduced heating rate: probably the heating shifts to the upper photosphere where the H$^-$ continuum and the myriad of weak lines serve as radiative coolants. This is consistent with the large values of the minimum atmospheric temperature over the photosphere that are required to explain the Ca II K-line profiles of these very cool stars (Giampapa *et al.*, 1982).

The M-type main-sequence stars are expected to be fully convective beyond spectral type M5 and a mass of less than $0.3 M_\odot$. If a stellar dynamo requires the existence of such a boundary layer to operate efficiently, as is argued for the solar dynamo (see Chapter 7), then one might expect a significant change in activity, or even a disappearance of activity around spectral type M5. Direct measurements of magnetic fields in the photospheres of M-type dwarfs through Zeeman effects on line profiles extend down to M5 V (Johns–Krull and Valenti, 1996, and references therein), and thus stop just short of the point on the main sequence where the radiative core is expected to disappear. Stars beyond that spectral type are very faint, which explains at least part of this bias in the data. Using atmospheric diagnostics of activity, Liebert *et al.* (1992), for example, show that all cool stars down to $M_v \approx 19.5$, that is, well beyond M5 V, have H α in emission, with no obvious discontinuity around M5. Basri and Marcy (1995) find that the average integrated emission in H α decreases smoothly toward later spectral types.

The most sensitive measure for stellar magnetic activity is the soft X-ray flux. Barbera *et al.* (1993) report on a drop in mean flux levels near $M_v = 13.4$ or just blueward of M5, based on a limited sample of *EINSTEIN/IPC* observations. More recently, however, Fleming *et al.* (1995) refute these conclusions on the basis of a volume-limited sample of the *ROSAT* All-Sky Survey: not only does the mean coronal activity change smoothly across the boundary in the Hertzsprung–Russell diagram where stars become fully convective, but even the distribution of coronal activity as a function of color shows no discontinuous change. Nor are there significant discontinuous differences in the coronal temperature structure, as far as can be inferred from the low-resolution data that are available (Giampapa *et al.*, 1996).

An explanation for the apparent similarity in outer-atmospheric activity across the transition from stars with bounded convective envelopes to fully convective stars has been proposed, in fact well before the observational data became available. Rosner (1980) and Durney *et al.* (1993), for example, argue that convection always drives a distributive or turbulent dynamo that generates field locally even in nonrotating and fully convective stars. If such a dynamo would dominate in stars later than about spectral type M5, then observations require the transition between the turbulent and boundary-layer dynamo to be smooth. In Section 7.4 we argue that yet another dynamo mode may dominate in these stars, in which the main driver is the differential rotation in the deep interior, rather than only turbulent motions.

The relationship between rotation rate and activity level for M-type dwarf stars remains unknown. These stars do tend to be relatively rapid rotators, however, which suggests that their magnetic brake is less efficient than in warmer stars (even when incorporating the effects of the long-lasting phase during which contraction counteracts spin down; see Section 13.3.4).

At the cool end of the main sequence, we find the brown dwarfs, that is, stars in which there is no stable hydrogen burning. Theory predicts that for masses below $\sim 0.09 M_\odot$ (Burrows *et al.*, 1993) no stable hydrogen burning occurs (and that in stars with masses below $0.065 M_\odot$ no lithium burning occurs, which provides a test for the sub-stellar nature of the least massive stars; see, e.g., Rebolo *et al.*, 1996, on Pleiades brown dwarfs). These light stars emit their thermal energy under the influence of gravitation, with some support of nuclear fusion of deuterium and limited hydrogen burning. For objects with masses below $\sim 0.013 M_\odot$ not even deuterium burning occurs (Hubbard *et al.*, 1994).

The domain of the brown dwarfs cannot be reached by current X-ray detectors: they are simply too small and distant to be detectable even if they were as active as their warmer counterparts, the M-type red dwarfs. Among the coolest stars observed to have activity is VB 10, an M8 Ve star that was observed to have flared in transition-region lines (Linsky *et al.*, 1995), but that star is expected to have nuclear fusion in its core. Basri and Marcy (1995) report on a study of the coolest single star known at the time, BRI 0021, with a spectral type of M9.5+. At $v \sin(i) \approx 40$ km/s, this star has no detectable Hα feature, which they argue to mean that there is no significant opacity in the Balmer transitions, rather than that the line happens to be just filled in. They find a similar star, PC 0025, at M9.5, with an Hα equivalent width of 275 Å. Whether this reflects possible magnetic activity in these very cool objects remains to be seen.

11.6   Magnetic activity in T Tauri objects

The T Tauri stars were discovered as a class of irregularly variable, late-type stars with prominent emission lines resembling those of the solar chromospheric spectrum (Joy, 1942). These stars are pre-main-sequence stars that are associated, both spatially and kinematically, with dark and dusty molecular clouds. They display near-ultraviolet and infrared continuum excesses (see Gahm *et al.*, 1974). In addition to the "classical" T Tauri stars, so-called weak-line or "naked" T Tauri stars are found in star-forming regions; their spectra indicate pre-main-sequence stars that have evolved from the classical T Tauri stage to closer to the zero-age main sequence (see Walter, 1987). The T Tauri stars do not rotate very rapidly: typically with $10 < v \sin i < 25$ km/s (see Hartmann and Noyes, 1987). For reviews of T Tauri stars, we refer to Bertout (1989) and Hartmann (1990). For some weak-line T Tauri stars, magnetic fields of kilogauss strength have been measured (Basri *et al.*, 1992; Guenther and Emerson, 1996). Attempts to measure fields in classical T Tauris have not yet been successful.

The shape of the spectral continuum of the classical T Tauri stars can be explained by a model of a star, with $T_{\text{eff}} \lesssim 6{,}000$ K, typically $T_{\text{eff}} \approx 4{,}000$ K, surrounded by an extended accretion disk (see Bertout *et al.*, 1988). Basri and Bertout (1989) worked out such a model, extending an earlier study by Lynden-Bell and Pringle (1974). In these models, the near-ultraviolet continuum excess is attributed to the warm inner part of the disk, and the infrared excess to the cool outer part. The star outshines the disk in the continuum in the visible. Below we use the term T Tauri object instead of T Tauri star when we refer to the composite consisting of a spherical star, a disk, and possibly other circumstellar structure.

The classical T Tauri objects stand out by their very large emissions in the chromospheric and transition-region lines: the fluxes in the ultraviolet emission lines are larger than in those of the most active main-sequence and evolved stars by factors up to 40 (see Fig. 11.7). Note that also for the T Tauri objects the flux densities F are normalized per unit area of the *stellar* surface. In the relations between the flux densities F in chromospheric and transition-region lines (Section 9.4), the data points of the T Tauri objects extend these relations for main-sequence stars and evolved stars toward higher emission.

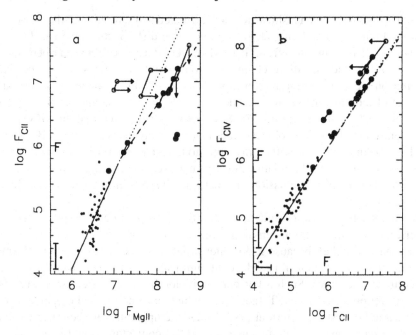

Fig. 11.7. Relationships between flux densities in ultraviolet lines for T Tauri objects (large circles), compared with those for active main-sequence stars and giants (small dots); the solid line segments and their dotted extrapolations are from Rutten *et al.* (1991). The dashed line segment in each panel indicates a linear relationship, starting from the most active main-sequence star. The T Tauri objects are represented by sets of two symbols, each one resulting from a specific correction for interstellar and circumstellar extinction. Filled symbols represent accurate measurements; open symbols with arrows indicate upper or lower limits. Typical C II and C IV flux densities in solar active regions are indicated by bars, and those in solar flares by F (from Lemmens *et al.*, 1992).

Figure 11.7 illustrates that the chromospheric and transition-region fluxes of classical T Tauri objects exceed those from the most active main-sequence stars, solar active regions, and even solar flares by appreciable factors. It is hard to believe that these emissions originate from the stellar disk from which continuum emission in the visible spectrum originates. It seems more plausible that the bulk of the chromospheric and transition-region emission originates from hot circumstellar matter, possibly attached to accretion disks invoked to explain the near-ultraviolet and infrared continuum excesses. Scrutinizing flux–flux relations, such as in Fig. 11.7, Lemmens *et al.* (1992) noticed that the T Tauri extensions do not follow those power-law relations that are nonlinear: there is a kink in the relationship where the T Tauri extension meets the relationship defined by main-sequence and evolved stars.

For all combinations of chromospheric and transition-region lines, within the observational accuracies, the relationships between the fluxes from T Tauri objects are linear. This supports the notion that the bulk of the chromospheric and transition-region emission from T Tauri objects originates from somewhere outside the stellar surface – recall that for older stars the nonlinearity of the relations between transition-region flux densities and chromospheric flux densities has been attributed to a saturation of the chromospheric fluxes at high magnetic filling factors on the stellar disk (Sections 9.5 and 11.3).

The soft X-ray emission from T Tauri objects does not follow the behavior of the chromospheric and transition-region emissions: the X-ray fluxes do not correlate with chromospheric or transition-region fluxes. Some T Tauri coronae are as bright as the coronae of the most active main-sequence and RS CVn systems, but not brighter; see Bouvier (1990), particularly his Fig. 8. It has been suggested that virtually all the soft X-ray emission of a T Tauri object comes from the stellar corona; indeed Bouvier (1990; his Fig. 4) finds that the soft X-ray emission increases with increasing rotation rate in a fashion similar to the relationship for main-sequence and evolved stars. Thus we conclude that the X-ray emission originates exclusively in the corona attached to the stellar photosphere.

Apparently the source outside the stellar disk that provides the bulk of the chromospheric and transition-region emission contains plasma of temperatures up to a few times 10^5 K, but not higher, because it does not produce X-ray emission. This plasma must be optically thin in the far-ultraviolet continuum yet with substantial emission measures in the ultraviolet lines. Such a hot plasma is not predicted by classical models for T Tauri accretion disks around T Tauri stars, which have no internal dynamo. The model constructed by Basri and Bertout (1989) yields temperatures in the inner boundary region not exceeding 10^4 K. Lemmens *et al.* (1992) conjectured that the hot plasma is in a magnetosphere enveloping the inner part of the disk and corotating with it. The large velocity dispersions of $\sim$150 km/s deduced from the large widths of the ultraviolet lines (Brown *et al.*, 1986) cannot be explained by the moderate stellar rotation rates, but this dispersion is compatible with Keplerian velocities at a distance of a few stellar radii. Observations indicate that, in addition to very broad C IV line profiles, occasionally also relatively narrow and mostly redshifted profiles are seen, with shifts of approximately +30 km/s (Calvet *et al.*, 1996). The latter authors suggest that the accretion flow may be channeled to the star along magnetic field lines, so that a localized accretion shock at the stellar surface may serve as an (additional) source of emission at transition-region temperatures.

Although the combination of a central star plus an accretion disk may explain several spectral characteristics, some extreme T Tauri objects present additional phenomena pointing to the presence of an extended halo, winds, and collimated flows. For descriptions and references, we refer to Bertout (1989) and Hartmann (1990). Presumably these additional features are located around the rotation axis of the star–disk system and on both sides of the accretion disk; see Shu *et al.* (1987).

An evolutionary scenario has been sketched as follows (for a discussion of various phases in the formation of pre-main-sequence stars, see Shu *et al.*, 1987). In the cold molecular cloud, cores (concentrations) form. Such a core collapses, and at its center a protostar is formed, which is surrounded by a disk that contains most of the angular momentum in the concentration. At first, this protostellar object is obscured by dark clouds. After the star has accreted sufficient mass, deuterium ignites in the core, which creates a star that is nearly completely convective. It has been proposed that this convection and the resulting differential rotation produce a dynamo, which allows the star to develop a stellar wind. This wind is supposed to be strongest along the rotational poles, which leads to collimated jets and bipolar outflow. These outflows blow off obscuring matter out of increasingly wider cones around the rotation axis, with the result that the star and the accretion disk become visible as a T Tauri object. In the mean time, more rotating matter keeps falling in onto the disk rather than directly onto the star, and eventually through viscous dissipation in the disk onto the star.

An extreme T Tauri object, with a large and dense accretion disk, presumably starts somewhere in the upper-right corner of the flux–flux diagrams (Fig. 11.7). With a gradually decreasing accretion disk, the object approaches the top of the standard flux–flux relation for main-sequence and evolved stars. Its accretion disk decreases further, and then through the stage of a "naked" T Tauri star it eventually enters the standard flux–flux relation as a young main-sequence star whose emission flux densities are completely dictated by the stellar rotation rate. During the evolution from an extreme T Tauri object to a young main-sequence star, the shrinking star spins up because of its decreasing moment of inertia and the angular momentum transferred from the disk to the star (see Chapter 13).

11.7   Long-term variability of stellar activity

O. C. Wilson searched for chromospheric variations on time scales of several years, having stellar analogs of the solar cycle in mind. His program (see Section 9.3) encompassed 91 stars on or near the main sequence from approximately F5 to M2, which were measured roughly once per month during the observing seasons from March 1966 until his retirement in 1977. Wilson (1978) found that within those 11 years approximately a dozen stars had completed a Ca II H+K flux variation remarkably similar to the solar chromospheric flux variation during one sunspot cycle.

In 1977 the survey program was continued using the Mt. Wilson Ca II HK photometer; see Section 9.3. Baliunas *et al.* (1995) published plots of the data, including Wilson's 1966–1977 measurement and the Ca II HK photometer measurements covering 1977–1992. In this survey, Wilson's sample has been enlarged to 111 stars, all on or near the main sequence. Examples of these long-term records are shown in Fig. 11.8.

Baliunas and colleagues grouped stars in four categories depending on their long-term variability: *cyclic*, *flat* (no significant variability on time scales longer than a few years), *long* (long-term trend, variability on time scales longer that 25 yr), and *variable* (significant variability without pronounced periodicity on time scales longer than 1 yr but much shorter than 25 yr). In several cases, two cyclic periods were attributed (see HD 78366 in Fig. 11.8).

According to the authors, roughly 52 stars, including the Sun, show indications of a cyclic variation, 31 have a flat or more or less linear dependence on time, and 29 exhibit variability without a clear periodicity between a few years and 25 yr. The measured cycle periods range from 2.5 yr up to 25 yr (the period covered by observations). If the sample of stars is restricted to the cases of cycle period determinations graded as "good" or "excellent," only 21 stars remain, with a shortest cycle period of 7 yr. Hence only 20% of the program stars exhibit a pronounced cyclic behavior. Baliunas *et al.* (1995) find that the cycle profiles are asymmetric in the sense that the rise is followed by a slower decline, as is also typical for the solar cycle.

In Fig. 11.9, showing the mean Ca II H+K emission index $\langle S \rangle$ [Eq. (9.4)] against color $B-V$, there is a division along a diagonal strip for $B-V < 1.1$. This division corresponds

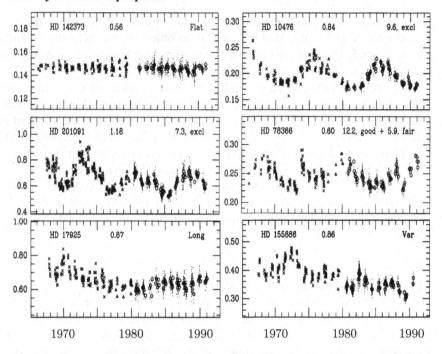

Fig. 11.8. Records of the relative Ca II H+K fluxes of main-sequence stars: ×, Wilson's records (1966–1977); triangles and dots, Ca II HK survey (1977–1992); open circles are 30-day averages. The relative Ca II H+K flux is expressed in S; see Eq. (9.4). The top of each panel shows the stellar identification, color index $B-V$, and a classification of the long-term variability or period(s) in case of cyclic activity; see Section 11.7 (figure from Baliunas *et al.*, 1995).

to the Vaughan–Preston gap (Section 9.3). This gap appears to indicate a transition in the character of the long-term variability of the chromospheric emission.

Among the stars below the Vaughan–Preston gap, almost all early K-type main-sequence stars, with $0.8 < B-V < 1.1$, display a pronounced, smooth, cycle like variation in the Ca II H+K emission, similar to that of the Sun (see HD 10476 in Fig. 11.8). Most of the lesser active F-type stars, with $0.35 < B-V < 0.6$, have nearly flat records (see HD 142373 in Fig. 11.8), or slightly inclined runs of S; only a few are classified as (possibly) cyclic or variable because of noticeable variability over intervals of several years. Among the S records of the G-type stars, with $0.6 < B-V < 0.8$, some display conspicuously cyclic variation, whereas others show a variable S level (some of which are classified as possibly cyclic); there are also flat records. Note that even the least active stars with a flat long-term behavior exhibit an appreciable, intrinsic variation during each observing season. Baliunas *et al.* (1995) suggest that some of the records may correspond to stars that are in a Maunder Minimum state (Section 6.1); they point to HD 3651, which appeared to enter into a Maunder-Minimum state during the survey.

The active stars above the Vaughan–Preston gap exhibit a variety of long-term Ca II H+K emission patterns. Several have been classified as cyclic. Five stars are

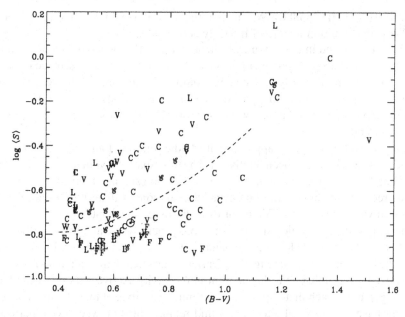

Fig. 11.9. The mean Ca II H+K emission index $\langle S \rangle$ against $B-V$ for main-sequence stars. The long-term variability classification is indicated by the following characters: C, cycle period determined; $\mathcal{C}$, secondary period; L, long-term trend; V, variable without clear period; F, flat; $\odot$, Sun. The dashed curve indicates the Vaughan–Preston gap, separating the very active from less active stars (from Baliunas *et al.*, 1995).

classified as having two cycle periods (including HD 78366 in Fig. 11.8) – these are found close to the Vaughan–Preston gap. In many cases, the activity varies so irregularly that the corresponding stars are simply called variable (HD 155886 in Fig. 11.8). Many among the most active stars are classified as exhibiting a long-term trend (HD 17925 in Fig. 11.8).

The sample of the coolest stars, with $B-V > 1.1$, beyond the end of the Vaughan–Preston gap, is small. These stars also appear to show various long-term behaviors. Some stars display a conspicuous cyclic variability (see HD 201091 in Fig. 11.8) but others show another variability pattern. The Mt. Wilson long-term variability program did not cover the domain of the late K- and M-type dwarfs, because these coolest stars are very weak in the near-ultraviolet. Hence we know very little on the long-term variability of fully convective stars and stars with very deep convection zones. For some BY Dra stars (active Ke and Me type dwarfs; cf. Table 9.2 and Section 14.3), long periods of 50 to 60 yr have been determined from broadband photometric variations by Phillips and Hartmann (1978), Hartmann *et al.* (1981a), and Rodono *et al.* (1983). It is not known, however, how typical these long periods are for the coolest active stars.

We note that the diversity in the classification of stars above and beyond the Vaughan–Preston gap by Baliunas *et al.* (1995) is caused by their classification to one apparently dominant feature. In fact, all these active stars are quite variable (none is classified as flat), and many show a long-term variation under the guise of cyclic or variable (see

HD 201091 and HD 155886). We suggest that *complexity* in the long-term behavior up to 25 yr is the common feature of the truly active stars.

A long-term trend in activity may indicate an underlying activity cycle with a period well in excess of 25 yr. If this were the case, then the frequent occurrence of such long-term trends among the truly active stars would be compatible with cyclic activity with periods longer that 25 yr in many of these stars. The periods of 50 to 60 yr determined from broadband photometric variations of some BY Dra stars, mentioned earlier, would fit such a trend.

At the time of this writing, apparently no studies on truly long-term variation of single G- and K-type giants have been published. The time span of 8 yr in the study by Choi *et al.* (1995) of 12 giants (see Section 9.3) is too short to establish their long-term behavior, but the records do show variations over intervals longer than the rotation period. The long-term variability of RS CVn binaries is discussed in Chapter 14.

The amplitude of a cyclic variation, or, more generally, the amplitude of the long-term variation of the Ca II H+K signal, is small relative to the mean flux level as measured above the basal flux level (see Fig. 9.3). Variable stars below the Vaughan–Preston gap do stay above the basal flux level during minimum activity. Active stars above the Vaughan–Preston gap may reach that gap during minimum activity, but they do not cross that gap.

There have been several attempts to find relationships between cycle periods, cycle amplitudes, and stellar parameters; see Baliunas and Vaughan (1985) for a review of early trials. More recent publications on this matter are by Soon *et al.* (1993) and by Baliunas *et al.* (1996). These attempts have not revealed what determines cycle periods and amplitudes, or the characteristics of a more complex long-term behavior for more active stars.

The Ca II HK project has shown beyond doubt that only a relatively large fraction of the rather slowly rotating G- and K-type stars exhibit fairly smooth cyclic variations (that is, as "smooth" as the solar cycle is; see Section 6.1).

12

Stellar magnetic phenomena

12.1 Outer-atmospheric imaging

The nearest cool star confronted us with the reality that cool stars have extremely inhomogeneous outer atmospheres. This was first confirmed for stars other than the Sun by the modulation of broadband signals, caused by starspots, and later by the discovery of the quasi-periodic variation in the Ca II H+K signal of some cool stars by Vaughan *et al.* (1981) caused by the rotation of an inhomogeneously covered stellar surface. Insight in stellar dynamos requires observational data on the properties of stellar active regions and their emergence patterns. For instance, we would like to know the sizes of stellar active regions and their lifetimes, the details of the structure of starspots, the emission scale height at different temperatures, and so on. In fact, we would like to know the entire three-dimensional geometrical structure of the outer atmospheres of cool stars. For that knowledge to be obtained, stellar surfaces should somehow be imaged by sounding the atmosphere from the photosphere on out. We would like to learn all this not merely for stars with activity levels similar to that of the Sun, but also for other stars, from the extremely active, tidally interacting binary systems for which much of the surface seems to be covered by areas as bright as solar active regions with a small fraction being even brighter, down to the very slowly rotating giant stars whose average coronal brightness is well below that of a solar coronal hole.

With instruments becoming ever more sensitive, and with inversion techniques improving, such three-dimensional mapping of the atmospheres of cool stars is coming within reach. Signal-to-noise ratios are still far below desirable values, however, and multispectral and long-duration observations prove difficult to organize, particularly for satellite observations. The rapidly growing literature on many intrinsically distinct stars provides many snapshots of stellar surfaces, but integrated synoptic information on a significant sample of stars has yet to be obtained. Some emerging trends are collected in this chapter.

At present, there are five distinct means to explore the spatial structure of the atmospheres of cool stars:

1. Direct imaging. If the telescope and stellar source are large enough, some information can be obtained on the structure of the stellar surface by imaging techniques. The number of stars for which this is feasible with current space-based instrumentation is, however, very small. Gilliland and Dupree (1996) discuss Hubble Space Telescope imaging of the M2 Iab supergiant α Ori.

They find evidence for a chromosphere extending out to ~2 stellar radii (see Section 12.5), and a hot spot on a disk that is imaged with only approximately ten resolution elements.

2. Interferometry. In principle, the extent of stellar sources (in particular close binary systems or pre-main sequence stars with extended disks) allows resolution of details smaller than the stellar disk or binary system through interferometry. This method is used most frequently at radio wavelengths. Only a few optical studies have been performed on some cool supergiants to date (Baldwin, 1996), but new instrumentation will boost the potential. The ESO Very large Telescope, for example, promises to resolve several hundred cool stars with at least a dozen pixels across their disks. Low-resolution (relative to the stellar diameters, but ~0.2 milliarcsec on the sky) radio images have been made for a few RS CVn systems (e.g., Lestrade, 1996). Benz *et al.* (1998) present the first VLBI image of a wide dMe binary with a resolution of ~1 milliarcsec.

3. Eclipse mapping. In binary systems, the crossing of one component in front of the other allows some low-resolution recovery of atmospheric structure on the sides of the stars facing each other during eclipse, and off the limbs in favorable conditions. Eclipse mapping is possible for a small subset of close and active binary systems in which the passage of the relatively small companion star behind the (sometimes extended) outer atmosphere of the active star allows pencil-beam mapping studies. The targets suitable for these studies include V471 Tau, a K2 V star with a white dwarf companion, and the eclipsing ζ Aur systems consisting of a bright K (super)giant with a hot B-star companion (see, for example Reimers, 1989; only approximately a dozen such system are known presently).

4. (Zeeman) Doppler imaging. Inhomogeneities that move across the stellar disk because of stellar rotation result in line-profile distortions caused by the Doppler effect. Stellar surfaces can be imaged by modeling these time-dependent distortions. Latitude information can be recovered by the observed velocity amplitude of the signal, and some height information (as in the case of high prominences) from the rapidity with which structures cross the line profile compared to the surface rotation period. The technique is limited to rapid rotators with $v \sin(i)$ values significantly exceeding the intrinsic line widths (that is, in excess of both the thermal and nonthermal broadening). In principle, it can be combined with an analysis of the photospheric magnetic field through the use of Zeeman-sensitive lines.

5. Rotational-modulation mapping. Structures that rotate onto and off the visible hemisphere cause brightness variations that can be used to derive longitude maps of activity. A limiting factor is that the time scale for intrinsic evolution of the source regions should be large in comparison to the stellar rotation period. The use of rotational modulation remains the most accurate way to determine the stellar surface rotation period (Middelkoop *et al.*, 1981; Vaughan *et al.*, 1981). This method is intrinsically free from the projection effects and the uncertainties in stellar radius that are associated with the measurement of $v \sin(i)$ through Doppler line broadening. The Ca II H+K modulation period is a weighted

average for an inhomogeneously covered, differentially rotating sphere; we return to this point in Section 12.7.

Multiwavelength (or "panchromatic," as J. Linsky calls it) observations have the potential of resolving some of the ambiguities inherent to these. The different contrasts, center-to-limb variations, or emission scale heights in different wavelength bands can be used to obtain a more comprehensive picture of stellar atmospheres: disk-occultation of the high corona will lag that for the transition-region lines emitted by lower regions, whereas the light curves for the optically thick chromosphere will show dimming well before occultation (see, Schrijver, 1988, for some sample light curves based on a coarse model of the Sun as a star). These advantages have yet to be exploited.

None of these imaging methods allows high-quality studies of stars that rotate as slowly as the Sun, for which the evolutionary time scale for active regions is typically less than the rotation period. For stars with rotation periods of less than approximately a week, however, useful information can be gathered by using (combinations of) these methods.

12.2 Stellar plages, starspots, and prominences

The flux–flux relationships contain much information about stellar outer atmospheres (Sections 8.4, 9.4, and 9.5.2) but provide this information only as disk integrals. The remarkable tightness of the relationships between disk-integrated fluxes over orders of magnitude in activity, and the agreement with solar data throughout the solar cycle, suggest that the properties of active and quiet regions on stellar surfaces are somehow similar, in the sense of resulting from scaling self-similarities. But that does not mean that the properties are identical, particularly for the most active stars.

Some information on the properties of stellar active regions is obtained by analyzing the rotational modulation of stellar radiative signals, as has been done for a few short-period eclipsing binaries (reviewed by Walter, 1996). With the use of Mg II h and k observations of the RS CVn type binary AR Lac (G2 IV+K0 IV; see Sections 12.3 and 14.3), the plage filling factor was found to vary from 2% to 20% over the years. The associated plage flux density is $F_{\text{Mg II}} \approx 10^7$ erg cm^{-2} s^{-1}, which, when transformed to Ca II H+K [see Eq. (9.6)] somewhat exceeds the very top of the flux range observed for stars as a whole (Fig. 9.4) and is three times larger than the flux from the brightest nonflaring parts of solar plages (see Fig. 8.8). For the areas surrounding these bright regions on AR Lac, a Mg II h+k flux density was inferred of 2×10^6 erg cm^{-2} s^{-1}, which is comparable to the flux density of an average solar active region. Another such extreme example is the W UMa system 44 Boo. Vilhu *et al.* (1989) find a Lyα flux density for plages of 6.5×10^6 erg cm^{-2} s^{-1}, whereas solar active regions reach only about one third of that (Schrijver *et al.*, 1985).

If the calibration relationship Eq. (8.8) were to be applied, magnetic flux densities of 7 kG and 0.5 kG would be inferred for the bright regions and for their dimmer surroundings on AR Lac, respectively. The value for the active region exceeds the field strength that should suppress surface convection (Section 4.1.2), which would be expected to result in large dark spots rather than very bright plages. Hence, it appears that calibration relationships between radiative and magnetic flux densities such as Eq. (8.8) that are based on solar data cannot be applied to extremely active stars, presumably because of the differential saturation of chromospheric emissions, which is not incorporated in that relationship.

These findings on the extreme nature of active regions and their surroundings on very active stars are supported by direct measurements of magnetic fields, which suggest that the most rapidly rotating K- and M-type dwarf stars with periods below ∼3 days are covered with an average magnetic flux density of up to 2,500 G (see Section 9.2.1, compare Figs. 9.1 and 11.5), and thus even more in their active regions. These stars with extremely high magnetic filling factors nevertheless adhere to the relationships between radiative flux densities (Section 9.4).

The correlation between the variations in outer-atmospheric radiative losses from very different temperature domains for stellar active and quiet regions has been studied for only a few stars in some detail. Among the best-studied systems in this respect is the K2–3 V–IV star II Peg (see, e.g., Rodono *et al.*, 1986, for the first extensive study). The ultraviolet fluxes measured by the International Ultraviolet Explorer (*IUE*) in a handful of spectral lines varied in phase, which was interpreted as an active region (complex) moving onto and off the disk (although Butler, 1996, later wondered whether the variability could not have been caused by major flares).

Coronal signals give the largest modulation contrast, but the strong coronal variability obfuscates the sought-after modulation. Moreover, sufficiently long observing intervals with X-ray instruments are rarely granted. Consequently, little headway has been made on the topic of the surface distribution and the evolution of stellar coronal condensations.

Even with the Ca II H+K data base of the Mt. Wilson observatory, spanning more than two decades for some hundred stars, it is difficult to derive properties of stellar plages, because intrinsic variability (ranging from flares to active-region evolution) and calibration uncertainties contribute to the observed variation (see Section 9.4). Solar data provide a gauge for the interpretation of this data base. It turns out, for example, that on the time scale of the rotation period, some 70% of the fluctuations in the solar signal are associated with rotational modulation rather than with active-region evolution (Donahue and Keil, 1995). The variance caused by active-region evolution and rotational modulation leaves part of the observed variations unexplained; this residual variability is associated with other contributions to the variance, such as stellar cycles, active nests, and short-term changes. That such extra variability can be observed holds promise for stellar data.

Unfortunately, in many stars the temporal domains of the various contributions to the total variance appear to overlap. Donahue *et al.* (1997a) attempt to disentangle these contributions by using the pooled variance $\sigma_P^2(\hat{t})$; this is the average value of the variances derived for an entire data set for contiguous time intervals of duration $\hat{t}$. The pooled variance is a monotonically increasing function of $\hat{t}$.

The effects of active-region growth and development can be seen in many fairly active stars from the increase in the pooled variance with $\hat{t}$ for time scales somewhat larger than the stellar rotation period. In some cases, however, active-region evolution either contributes too little to the total variance or its range overlaps with the rotation-period domain so that little increase is seen in $\sigma_P^2(\hat{t})$ between the rotation period and a period of several years. Donahue *et al.* (1997a) interpret the continued increase of $\sigma_P^2(\hat{t})$ on longer time scales as being caused by active-region complexes or (composite) nests, contributing to the signal up to at least several hundred days and on average as long as 2 yr. It is also possible that long-lived patterns in enhanced network associated with very large decayed active regions (see Section 6.3.2.2) contribute to these variations.

Even though the interpretation of the pooled variance in terms of time scales is difficult, Donahue *et al.* (1997b) point out that $\sigma_P^2(\hat{t})$ shows interesting differences from star to star. There is a trend for stars to evolve through three characteristic types of behavior as activity decays through magnetic braking: (a) active stars have plateaus in $\sigma_P^2(\hat{t})$ that generally lie between the rotation period of 2–10 days and $\sim$30 days or more, suggesting that the active-region evolution occurs on the time scale of a month or so; (b) less active stars show a continuous increase of pooled variance with $\hat{t}$, which could indicate a dominance of the evolution of active complexes over rotational modulation; and (c) inactive stars in which the increase in pooled variance is limited to below about 50 days and that show plateaus only at intervals longer than the rotation period. Only one out of every three of the main-sequence stars in the Mt. Wilson HK-project shows some form of plateau in $\sigma_P^2(\hat{t})$.

Stellar active regions are associated with starspots, as indicated by the rotational modulation of the white-light or broadband intensity as well as line-profile distortions caused by Doppler shifts. Starspots are defined as regions of the stellar photosphere that appear dark, but whether such regions consist of single, coherent, dark structures or comprise some number of small, localized darkenings remains unclear. There is a rapidly growing literature on starspots as inferred from intensity variations, with an increasing number of Doppler studies among them. The molecular bands of TiO also provide a diagnostic for starspots. It is beyond the scope of this book to present an extensive review of this literature, so we limit ourselves to a brief summary with some examples.

Butler (1996) reviewed two decades of studies of five RS CVn systems with *IUE* in conjunction with ground-based optical observatories. The emissions from the relatively high-temperature chromosphere and from the transition region generally show a marginal to good anticorrelation between visual and ultraviolet brightness expected from the cospatiality of spots and plages. For Mg II h and k lines, originating at lower temperatures, on the other hand, the signal is ambiguous, sometimes with a good correlation, a good anticorrelation, or no correlation at all. We suggest that this reflects the reduced contrast between plages and their surroundings in Mg II compared to the higher-temperature lines (Sections 8.4.2 and 9.4) and the relatively important role of the magnetic network outside active regions (Section 9.5): if a star is covered by network of a range of strengths covering much of the surface and by active regions with a sufficiently small filling factor, then correlations between various modulation signals can have different magnitudes and even signs, because their signals are weighted differently in the different diagnostics.

The modeling of light curves of spot-studded stars is frequently used in attempts to obtain surface information on stars, but the derived information is often ambiguous. The observed signals may show reasonably large amplitudes, with long-term coherence, so that simple light-curve analyses require only a few, typically two, spot groups to match much of the modulation curve. Yet many more may exist for a similar light curve. Eaton *et al.* (1996) have illustrated this problem for the most active stars, including the RS CVn binaries. They compared observed visible light curves and simulated curves based on the assumption that stars are randomly covered with large spots of different areas. The model spots are given lifetimes of one or a few years, which is long compared to solar spots and probably requires starspot nests that persist for that period of time. The observed variability allows up to some 40 spots to exist on the stellar surface at any one time whereas inversion models of the observed light curves require only one or two. In fact, in their model, more than a dozen or so are needed to avoid overly strong changes in phase and shape as a result of spot evolution.

Doppler modeling frequently suggests that more than two spot regions exist, and some reconstructions suggest that the large spotted areas consist of numerous smaller structures (e.g., Kuerster *et al.*, 1994, for the K0 V star AB Dor). The modeling of stellar photospheres as covered by starspots is becoming more and more sophisticated, as simple models allowing for only a single, homogeneous darkening are being replaced by models that allow multiple spots, some requiring the coexistence of umbrae and penumbrae, but not yet allowing for changes in spot populations. What is clear, however, is that these spots or spot complexes can be very substantial (e.g., Hall, 1996). The weak-line K4 T Tauri star V410 Tau (Strassmeier *et al.*, 1997), the K0 III+? RS CVn system XX Tri (Hampton *et al.*, 1996), and the K2–3 V–IV star II Peg (Doyle *et al.*, 1988) are extreme examples, with amplitudes in the visible of 0.65, 0.6, and 0.5 magnitudes, or factors of 1.8, 1.7, and 1.6, respectively. For II Peg this amounts to a spot coverage of some 60% of one hemisphere – if indeed the other hemisphere is completely unspotted. In contrast, the spot filling factor for the Sun never exceeds 0.2% (Allen, 1972). Note that models assume that the unspotted atmosphere of the star does not vary with stellar activity level, although it is known that for the Sun an increased coverage by faculae increases the integrated brightness of the Sun during the maximum of the sunspot cycle even more than it is decreased by the ensemble of spots (Section 6.1.2); the plage surrounding spots and the enhanced network resulting from plage decay are brighter than the average photosphere. Another complication is that the ratio of facular to spot areas decreases with an increasing activity for the Sun (Foukal, 1998).

The problem of a zero-coverage reference brightness is largely circumvented when TiO bands are observed: the dissociation temperature of this molecule is so low that it does not occur in regular quiet photospheres, with the exception of the cool M-type dwarf stars. Yet, in order for quantitative spot parameters to be derived, a reference spectrum for a supposedly unspotted star of similar fundamental properties or a theoretical spectrum is required in order to have information on other lines overlapping with the TiO bands. This diagnostic was originally used for stars by Ramsey and Nations (1980) to demonstrate the presence of very cool regions in the photosphere of the RS CVn binary HR 1099. Neff *et al.* (1995) use this diagnostic to demonstrate that the unspotted brightness of the star II Peg is ~0.4 magnitudes (or 45%) brighter than its historical maximum; apparently the stellar disk is always covered by starspots to a substantial degree. The TiO diagnostic confirms that spot coverage in active systems can indeed reach up to some 50% of one hemisphere (O'Neal *et al.*, 1998).

Spot temperatures (generally taken to be the same for all spots on a star, averaging over umbral and penumbral regions) appear to lie some 500 K to 2,000 K below the ambient photospheric temperature, with an average temperature difference of the order of 1,000 K. O'Neal *et al.* (1996) find evidence for a weak trend for the difference between starspot temperature and photospheric effective temperature (typically some 700 K to 1,700 K) to increase with increasing surface gravity (for the Sun the contrast is about 1,700 K between the umbrae and the quiet photosphere; see Section 4.2.1).

In two-thirds of the nearly thirty stars for which Doppler maps have been published prior to 1996, indications have been found for spots at high latitude or even at the poles (Strassmeier, 1996). The sample is dominated by tidally interacting RS CVn binaries, but also includes young T Tauri stars, single FK Comae stars, and W UMa contact systems. Polar spots have also been reported on young, rapidly rotating main-sequence stars, such as on AB Dor and some Pleiades. No clear correlation with target class and the occurrence of polar spots has been found. Strassmeier (1996), and other contributors to a workshop dedicated to stellar-surface structure (in particular Byrne, 1996; and Unruh, 1996), address the reality of the polar spots. Whereas many doubts have been raised, there is a preponderance of evidence favoring their existence. The distinguishing signature of polar spots is a dark feature in line profiles that exhibits a Doppler swing with an amplitude that is much smaller than that associated with the equatorial rotation rate, because of their high latitudes.

High-latitude starspots in rapidly rotating stars may be the consequence of strong Coriolis forces on magnetic flux bundles rising from deep within the star. Schüssler (1996) speculates that these forces may deflect the rising flux toward latitudes of up to 60°, which is not high enough to explain all stellar observations. Truly polar spots could result either from a flux eruption originating very deep in stars with relatively small radiative interiors or in stars with a relatively shallow convective envelope by a poleward slip of the previously anchored deep segments of flux rings following an earlier eruption of flux at midlatitudes elsewhere on that ring (in co-rotating binaries, the relevant forces would have to be phrased in terms of tidal forces, but such models have not yet been developed).

As discussed earlier, lifetimes of stellar active complexes are difficult to determine. A lower limit is probably given by the lifetimes of (clusters of) starspots. The most active stars, the tidally interacting RS CVn binaries, have spots reported with radii that exceed 20° in astrocentric coordinates. These appear to live from a few months up to several years. Hall *et al.* (1995), for instance, give lifetimes for spots as determined from photometry ranging from just over 1 yr to more than 4 yr for 12 Cam (K0 III+? V). Strassmeier *et al.* (1994) give an average of 2.2 yr, ranging from a lower limit of 2 months up to 4.5 yr for a 15-yr data set of *V*-band photometry of HR 7275 (K1 IV–III; Strassmeier *et al.*, 1993), with no obvious relationship between spot (group) size and lifetime. They also report significant changes from one rotation to the next, suggesting that small-scale changes to the spot groups, or more likely spot nests, occur on a time scale of no more than weeks.

In these modeling efforts that invoke long-lived spot nests, we again face the fundamental problem posed by the inversion of a Fredholm equation, as in the case of spectroscopic inversion techniques to determine stellar coronal structures (Section 9.6). The study by Eaton *et al.* (1996), described earlier, addresses parts of this problem. They suggest that the apparent longevity of spots could be an artifact of the chance grouping of many spots on a differentially rotating star: randomly evolving spots are subject to rotational shear, which forms and destroys clusters with relatively long coherence times (see also Hall and Busby, 1990). On a positive note, the agreement of the observed and modeled "coherence" times for spot groups on differentially rotating stars could mean that stellar differential rotation (Section 12.7) has been determined with some accuracy.

Prominences in stellar outer atmospheres can in principle be observed through their effects on optical (e.g., Hα; see Section 8.2) and ultraviolet line profiles: when a prominence crosses the stellar disk, it absorbs chromospheric emission from below, resulting in a dark feature in the spectrum that is Doppler-shifted depending on the location of the prominence as projected onto the stellar disk. Evidence for prominences has been seen in a number of stars (reviewed by Collier Cameron, 1996), including the young K0 V star AB Dor, dwarfs in the young α Per cluster, the dMe star HK Aqr, and a number of RS CVn systems. Apparently, prominences are a common phenomenon. In stars with a sufficiently inclined rotation axis, an estimate can be made of their latitude. In AB Dor, for example, they were found to cluster around 60° in latitude, perhaps forming the counterparts of solar polar-crown filaments; in other stars with a range of axial inclinations, their detection suggests that they exist over a range of latitudes. There is some ambiguous evidence that they are preferentially located near spotted areas, but in general they form so high in the stellar atmosphere that they are sensitive to the field over a large area of the stellar photosphere.

Detected stellar prominences are generally positioned quite a bit higher over the photosphere than their solar counterparts, up to several stellar radii. Schroeder (1983) reported a prominencelike feature in the atmosphere of the ζ Aur system (see Table 9.2 for the definition of that class) 32 Cyg (a K4–K5 Ib supergiant with a B6–B7 companion; Eaton, 1993) inferred from an eclipse observation made with the *IUE*: this prominence was suspended at $15R_\odot$ above the star with $R_* \approx 180 R_\odot$ with a linear extent perpendicular to the line of sight of some $30R_\odot$ and a column density of 10^{24} cm^{-2}. Note that the surface gravity in this star is very low, so that the height expressed in pressure scale heights or in terms of kinetic energy for ballistic trajectories is much less extreme than the absolute value suggests. Whether this feature, or the entire chromosphere of this system, is corotating remains a matter of debate (see, e.g., Griffin *et al.*, 1993, on that topic for 22 Vul). Byrne *et al.* (1996) find that prominences on the dMe star HK Aqr, apparently comparable to the stellar disk in size, lie well above the stellar surface, probably somewhere between 0.3 and 2.4 stellar radii. Even in suitable cases, stellar prominences need to be larger than their solar counterparts to be detectable. The column densities reported on so far substantially exceed the range of column densities observed for the Sun ($10^{18} - 10^{19}$ cm^{-2}): V741 Tau (K0 V+WD; $P_{rot} \sim 0.5$ days), for example, exhibited X-ray dips on successive rotations compatible with extinction for a column density of 10^{20} cm^{-2} (Jensen *et al.*, 1986).

On some stars, even the large prominences that are detectable are relatively numerous. Approximately seven prominences were inferred for observations of AB Dor (Collier Cameron, 1996). Each lasts for ~ 1 week, while they form and disappear on a time scale of ~ 1 day. In the rapidly rotating dMe star HK Aqr, Byrne *et al.* (1996) also find evidence for the existence of at least seven relatively cool "clouds."

Large prominences (and other cool material at coronal heights) should also be detectable in the extreme ultraviolet. Their distinguishing property is the wavelength dependence of the absorption in the bound-free continua of H and He; this would have to be disentangled from interstellar extinction, however.

12.3   The extent of stellar coronae

The direct determination of the chromospheric emission scale height is difficult for stars other than the Sun. Some measurements have been made for the extended cool outer atmospheres of giant stars, as is discussed in Section 12.5, but in view of the limited amount of information we restrict ourselves in this section to a discussion of the extent of stellar coronae.

The problem of recovering the coronal structure of a rotating star is mathematically similar to that of recovering the coronal temperature structure (Section 9.6). Indeed, methods developed to derive solar differential emission-measure curves have been used as the basis of light-curve inversions (such as the method used by Siarkowski, 1992, which was originally developed by Withbroe, 1975). Because of this similarity, the same kinds of problems are encountered, particularly the requirement of positive definiteness of the solution, regularization to smooth unrealistic fluctuations, and error analysis. If the sources on the stellar surface evolve significantly during a rotation period, the inversion algorithms fail altogether. But because error analysis is as hard here as in the case of spectroscopic inversions, this often escapes notice and leads to the introduction of spurious features. If we assume that the evolutionary time scale of large (nonflaring) coronal features – such as coronal condensations over active regions – is similar to that for the Sun, then stars with rotation periods of a week or less seem the most promising targets.

The height extent of outer-atmospheric emission can only be measured more or less directly in eclipsing systems, most commonly RS CVn systems. (Note that eclipse mapping is much less sensitive to intrinsic variability than rotation-modulation inversion, because the time interval needed to obtain the required observations is much shorter.) All other methods to measure the emission scale heights involve some semi-empirical physical modeling, be it using the scaling laws for quasi-static loops (with estimates of electron densities) or extensive coronal modeling. Two very well studied such systems are AR Lac and TY Pyx (G5 IV+G5 IV). AR Lac has been the target of coronal imaging studies ever since the pioneering study by Walter *et al.* (1983). The system is an eclipsing RS CVn binary of a $1.54 R_\odot$ G2 IV primary and a $2.81 R_\odot$ K0 IV secondary tidally locked in a 1.983-day orbit, separated by $9.22 R_\odot$. White *et al.* (1990) analyzed a 2-day continuous *EXOSAT* observation of the system. They find a coronal component with a temperature near 7 MK that originates in two distinct regions (it is unclear whether both lie on the G2 IV star) covering approximately one-tenth of the available surface area. One of these regions has an emission scale height of 10^{10} cm, the other of either 10^{10} cm or 10^{11} cm. This component has an inferred electron density of $\sim 5 \times 10^{10}$ cm^{-3}, and if the RTV scaling law [Eq. (8.22)] applies, this implies loop lengths comparable to those in solar X-ray bright points, that is, very much smaller than the inferred heights. This compact component shows an orbital modulation of a factor of ~ 2. A hotter component with a temperature between 15 MK and 40 MK shows no detectable modulation. This suggests the existence of extended loops with lengths comparable to the binary separation. White *et al.* (1990) proposed that this component consists of plasma that is ejected from flaring loops, but that remained trapped in the system's magnetosphere.

Coronal eclipse maps of AR Lac by Siarkowski *et al.* (1996) suggest that there are compact sources of emission with heights less than the stellar radii (and primarily concentrated on the sides of the stars facing each other; see Section 12.6). There appears to be also a "halo" component enveloping the entire system, and there is some evidence of emission linking the two stars. Perhaps not surprisingly, the structure changes dramatically between the three snapshots taken over a period of 9 yr.

If the hot component is indeed extended, that is hard to reconcile with the high electron densities inferred from spectroscopic data of other RS CVn systems at the same coronal temperatures. A fundamental difference between AR Lac and the systems in Table 9.3 is that the orbital separation is smallest for AR Lac. In this respect it is interesting to compare the other well-studied eclipsing system, TY Pyx, with ξ UMa B. TY Pyx is a 3.2-day system of two $1.59 R_\odot$ G2–5 IV to IV–V stars, separated by $12.25 R_\odot$; ξ UMa B is a 3.98-day system of a G0 V star with a dwarf companion also with a separation of $\sim 12 R_\odot$. These systems are roughly comparable, except for the small difference in evolutionary status of the components. For TY Pyx a sharp eclipse is seen in the X-ray light curve observed with *EXOSAT*'s Low-Energy instrument. At the higher energies observed with the Medium-Energy experiment no significant modulation is detected. Although this is similar to what was seen in the case of AR Lac, Culhane *et al.* (1990) conclude that this particular phenomenon can be explained if the high-temperature emission comes from the two hemispheres facing each other in this system of two comparable components. If these emissions originate in compact, high-density structures distributed over a considerable part of these hemispheres, then the results for TY Pyx and ξ UMa B could be compatible. Siarkowski (1996) uses another technique to demonstrate that the same light curve could be described by an extended component between the stars.

The nonuniqueness of the models makes it hard to formulate definitive conclusions about the hot coronal components. It remains unclear whether hot 10–30 MK plasma is contained in extended or in very compact loops, or both, or differently on different stars.

Radio observations have been used for some direct imaging of coronae of cool stars. Benz *et al.* (1998) imaged the system UV Cet (a wide binary of two dM5.5e stars in a 27-yr orbit). They find two separate sources close to the location of one of the components, separated by 4.4 stellar radii, aligned with the rotation axis; the authors suggest that the stellar center of mass is between the two radio sources. The associated loops, with field strengths at the top of at least 15 G, consequently extend at least out to $1.2 R_*$, or 120,000 km, above the photosphere. In comparison, the largest solar loops extend to $0.4 R_\odot$, 280,000 km, that is, much less in terms of the stellar radius, but more in absolute size.

12.4   Stellar flares

Stellar flares offer diagnostics of stellar magnetic activity that supplement diagnostics discussed in previous sections, namely, magnetic line broadening and slowly varying outer-atmospheric emissions. For instance, the cooling times deduced from flares observed in soft X-rays allow estimates on the extent and density of the emitting coronal regions. Moreover, flare properties may reveal aspects of stellar dynamos that cannot be inferred from other disk-integrated diagnostics.

A historic document in the study of stellar flares is the spectrogram of a dwarf M-type star with emission lines, in which Joy and Humason (1949) discovered a brightening by approximately one magnitude both in line emissions and in the continuum emission. Features of this brightening are similar to that of an intense, white-light solar flare, which prompted the search for stellar flares, at first by optical photometry of dKe and dMe stars, which are cool main-sequence stars with emission lines in their spectrum.

In the optical spectrum, stellar flares stand out in the Balmer lines and also in the ultraviolet U band of the UBV photometric system, because of the appreciable brightening of the Balmer continuum during flares. Stellar flares are also observed in radiowaves and microwaves, throughout the ultraviolet, and in X-rays. The similarity between stellar flares and solar flares is demonstrated by the resemblance of the time profiles of the various emissions, and by the coincidences and characteristically different patterns in the various spectral windows (multiwavelength observations are shown in Byrne, 1989). For one thing, the rise time to maximum flare brightness is much shorter than the decay time. Moreover, many stellar flares show a brief, spiky impulsive phase, apparently as a direct response to particle acceleration, and a much more gradual main phase, which is largely determined by the thermal radiation of the plasma heated by the flare, which resembles many solar flares (Section 8.3.3, Fig. 8.6). Some stellar flares are preceded by a precursor, such as a preflare brightening; see Byrne (1983) for time profiles and discussion.

Temperatures in stellar flares cover a broad range over chromospheric, transition-region and coronal temperatures, like in the Sun – apparently the flare temperatures do not depend on the stellar effective temperature. From an analysis of flares of the dM4.5e star YZ CMi observed simultaneously at ultraviolet, optical and radio wavelengths, Van den Oord *et al.* (1996) conclude that these flares can be interpreted in terms of a reconnection model for solar two-ribbon, white-light flares.

Pettersen (1989) explored the occurrence of stellar flares across the H–R diagram. He found that stellar flares are observed in all types of stars that show magnetic activity. Conversely, practically all stars that show flaring possess convective envelopes. Very few flarelike brightenings have been reported for hot stars outside the domain of cool stars, but these may be caused by an unnoticed cool binary companion or a cool star within the source area.

In the paragraphs that follow, we discuss types of stars that stand out by their high frequency and magnitude of flaring. Such stars also show a high level of background magnetic activity, but as yet no rule has been proposed that successfully relates flare frequency and magnitude to some measure for overall magnetic activity.

The dKe and dMe stars frequently show flares; some of these are remarkably strong. These red stars are main-sequence stars, or still-contracting stars that have nearly reached the zero-age main sequence. Such flaring stars are often called UV Ceti stars, but this is not a well-defined subcategory, because apparently all dKe and dMe stars are prone to flaring. Many of these stars are also called BY Dra stars because of their photometric variability attributed to starspots (Section 14.3).

Optical flares on red stars are much more conspicuous than on the Sun, partly because their photospheres are much fainter. More importantly, however, flares on dKe and dMe stars are intrinsically more energetic than solar flares: many stellar flares release an energy of $E \sim 10^{34}$ erg, whereas for the most energetic solar flares values of $E \sim 10^{32}$ erg have been recorded.

Among dMe stars, there is no change in flare productivity and flare magnitude corresponding to the transition from stars with deep convective envelopes to fully convective stars (see Pettersen, 1989). This conclusion is in keeping with the finding that there is no sharp change in outer-atmospheric activity across the boundary where the stars become fully convective (Section 11.5). Apparently the transition from a solar-type boundary-layer dynamo (Chapter 7) to a dynamo operating in a deep convection zone (Section 11.5) is very smooth, and very deep convection zones can support efficient dynamos.

The more energetic flares observed on red main-sequence stars suggest a dynamo action that differs from that in the Sun and other G-type main-sequence stars in some important aspect. Recall that the argument for a boundary-layer dynamo in the Sun is prompted by the observation of the ordered, nearly E–W orientation of solar active regions (Section 6.1). In the Sun, complexity in the photospheric magnetic field favors the occurrence of major flares (Section 8.3). Generalizing this finding, we argue that the occurrence of more intense flares on red main-sequence stars than on the Sun and other G-type main-sequence stars supports the idea that dynamos in the deep convective envelopes of red stars produce rather irregular patterns of bipolar active regions in the stellar photosphere.

Pre-main-sequence stars produce strong flares. In the case of classical T Tauri stars (Section 11.6), flares may occur on the star and in the accretion disk; also the interface between disk and star may be involved. In star-forming regions, many flaring objects have been observed. The boundary between pre-main-sequence and young main-sequence flaring stars is vague; the introduction of terms such as flare stars and flash stars does not help to clarify this situation.

Close binaries containing cool components produce conspicuous flares. In RS CVn systems and in Algol binaries (Section 14.3), flares releasing energies up to $\sim 10^{37}$ erg are observed – these energies are some orders of magnitude larger than in the most intense flares in dMe stars. Moreover, RS CVn flares last longer, up to ~ 7 h in the ultraviolet, and the coronal temperatures are higher, up to $\sim 10^8$ K. Uchida and Sakurai (1983) proposed a model of interacting magnetospheres involving both components. Van den Oord (1988) indicated the possibility that the enormous energies released in such energetic flares can be stored in a magnetic filamentlike configuration in between the two stellar components but connected with one of the components.

In passing we note that active stars display a variety of phenomena in radio waves. The radio events associated with stellar flares carry only a very small fraction of the flare energy, yet they are important as diagnostic tools in the study of the energy release and acceleration mechanisms in flares and other transients. There are radio bursts that are not associated with detectable stellar flares – some of these may be related to ejecta from stellar atmospheres that are not accompanied by flares, such as are observed in the solar atmosphere (Section 8.3.1). In the case of radio phenomena associated with stellar flares, the time profiles of the radio phenomena are not always so nicely correlated with the time profiles in other spectral windows. There are highly polarized radio flares, with extremely high brightness temperatures, that appear to require a coherent emission mechanism. For radio flares we refer to reviews by Kuijpers (1989), by Haisch *et al.* (1991b), and by Bastian (1996).

12.5   Direct evidence for stellar winds

Measuring properties of hot, thermally driven stellar winds, like that of the Sun, is difficult, because the tenuous material can do little to affect the X-ray, ultraviolet, and visible spectra emitted by the central star and its outer atmosphere. Mass loss through hot winds is inferred from the solar example and from its statistical effects on the stellar rotation rates (Chapter 13).

Winds may be detected through radio and near-infrared free-free emission, but only with difficulty even in favorable cases. Only upper limits have been derived for mass loss rates of cool stars on the main sequence or in the early phases of post-main-sequence evolution. Brown *et al.* (1990), for example, used the Very Large Array (*VLA*) to search late-A- and early-F-type stars for free-free microwave emission emitted by stellar winds. They derive upper limits to the mass loss of these stars of 10^{-10}–10^{-9} $M_\odot$/yr. The detection of Procyon (F5 IV–V) as a 3.6-cm radio continuum source allowed Drake *et al.* (1993) to conclude that $\dot{M} < 2 \times 10^{-11}$ $M_\odot$/yr for a warm wind with electron temperatures in the range 10^4–10^5 K. Mullan *et al.* (1992) interpreted emission observed in the millimeter wavelength range of a sample of dMe stars as free-free emission from a hot stellar wind. They estimated mass-loss rates of a few times 10^{-10} $M_\odot$/yr. Van den Oord and Doyle (1997) analyzed the same data with a different model and concluded that the emission does not originate in a wind but rather is associated with the stellar disk; through indirect arguments they place the upper limit to the mass loss for these stars some 2 orders of magnitude lower.

Theoretically, the mass-loss rates through hot winds may vary greatly from star to star. The stellar mass-loss rate $\dot{M}$ is defined as the mass flux through a sphere at some distance from the star. For the distance R_s to the sonic point, where the wind velocity equals the sound speed c_s, we have

$$\dot{M} = 4\pi \mu n_s c_s R_s^2, \tag{12.1}$$

where μ is the mean particle mass and n_s is the particle density at the sonic distance R_s. Mullan (1996) argues that in a hot, thermally driven, isothermal wind (which is discussed for the Sun in Section 8.9), the density in the subsonic part of the wind is essentially exponentially decreasing, with a scale height given by the wind temperature T_w. Hence n_s is expected to be very sensitive to T_w. If T_w would scale with the characteristic coronal (color) temperature (see Section 9.6), then the stellar mass loss in active stars could exceed the solar mass loss by orders of magnitude. The upper limits derived for the hot winds of cool main-sequence stars are 3–4 orders of magnitude higher than the solar mass-loss rate of $\approx 2 \times 10^{-14}$ $M_\odot$/yr, so that the potentially strong dependence of stellar mass loss on the stellar activity level, allowed by Eq. (12.1), cannot be ruled out. Parameters other than the temperature appear to play a less important role in determining $\dot{M}$. If the coronal temperature is fixed, a Weber–Davis model (Section 8.9.2) predicts an increase in $\dot{M}$ of only a factor of 3 when the rotation rate is increased by a factor of 20.

For stars that are substantially different from a Sunlike star, evidence has been found for cool winds. One case for a cool main-sequence star is V471 Tauri, in which a white dwarf orbits around a K2 V star. The separation is large enough so that no significant tidal or magnetic interaction is expected. The white dwarf can be used to probe the wind of the K2 V star as it moves through its wind. Mullan *et al.* (1989) detected signatures of a nonsteady or nonspherically-symmetric outflow from the K dwarf at a velocity of 600–800 km/s and a temperature of no more than a few times 10^4 K. Assuming the mass loss to be steady and spherically symmetric, they estimate the mass-loss rate to be a few times 10^{-11} $M_\odot$/yr, that is, a thousand times the solar mass-loss rate. This suggests that this mass loss may be dominated by coronal mass ejections (compare Section 8.3) rather than by a steady outflow.

For stars well away from the main sequence, the mass loss rates are large, and the winds cool; in late-K- and M-type (super)giants, (much of) the wind is at chromospheric temperatures, which makes detection of the wind easier (for a review based primarily on *IUE* data, see Dupree and Reimers, 1987). For stars that lose mass through intermittent ejections, these result in discrete absorption features within spectral lines or in emission features superposed on (weak) continuum emission. In a steady cool wind of chromospheric temperatures, scattering occurs in strong resonance lines of low ionization stages in the outer layers of the wind, so that the resulting lines reveal expanding circumstellar matter through line asymmetries or broad absorption features on the blue side of the line core. A limitation for velocity measurements is that the absorption features should lie within the line core because too little emission exists outside them in these cool stars. Hence, the velocities should not exceed $\sim$100 km/s for giants and twice that for cool supergiants (see Section 9.3 on the line widths).

For the cool giants and supergiants beyond the Linsky–Haisch dividing line (see Section 11.1) that show no detectable coronal emission, the terminal velocities of the winds range from several kilometers per second up to $\sim$150 km/s, but they are in all but a very few cases well below the surface escape velocities of the stars. Consequently, most of the work provided by the mechanism that drives the wind is done against gravity close to the star.

The wind temperature for (super)giants seems to be no higher than $\sim$20,000 K, decreasing at large distances. Very near the Linsky–Haisch dividing line, mildly active stars are found that also have signatures of cool winds; these are called hybrid stars (Hartmann *et al.*, 1980).

Drake and Linsky (1986) concluded on the following trend in stellar mass loss. For giants the mass loss rate increases from 10^{-10} $M_\odot$/yr for early-K III stars to 10^{-8} $M_\odot$/yr for M5 stars. For bright giants values range from 10^{-9} $M_\odot$/yr for mid-G II to 10^{-7} $M_\odot$/yr for M5 II stars. Supergiants lose mass at rates from 10^{-8} $M_\odot$/yr for mid-G Ib to 10^{-6} $M_\odot$/yr for M2 supergiants (see Dupree, 1982, on the ionization state of Mg in winds and its influence on mass-loss estimates). These mass-loss rates scale roughly with the surface area of the star (compare Fig. 2.10), $\dot{M}_*/M_\odot \approx 2.5 \times 10^{-14}(R_*/R_\odot)^2$. This scaling was first suggested by Goldberg (1979; see Drake and Linsky, 1986 – also for other fits to data on single (super)giants – and the review by Dupree and Reimers, 1987). From a sample of six ζ Aur binary systems, Reimers (1975) proposed a parameterization of the mass-loss rates $\dot{M}_*/M_\odot \propto 5 \times 10^{-13}L'/g'R'$ (primes denote quantities expressed in solar units), but Dupree and Reimers (1987) show that $\dot{M}_*/M_\odot \approx 10^{-12}R'^2$ works just as well. Although the same proportionality with the stellar surface area shows up for these binary systems as for (effectively) single giants, the constant of proportionality is a factor of 40 larger. It is unclear whether this may be caused by the ionization of the wind by the B-type companions of these cool supergiants, causing a substantial misestimate of the mass loss.

For cool giants of a given T_{eff} and g, estimated mass loss rates range over some 2 orders of magnitude. Much of this spread is related to model uncertainties, leaving the intrinsic range unknown.

For stars beyond the Linsky–Haisch dividing line, the power required to drive the wind appears to be approximately one millionth of the luminosity, with little dependence on the evolutionary phase of the stars (Judge and Stencel, 1991). Because of the decrease in the chromospheric radiative power with decreasing effective temperature, this means that beyond about spectral type M5 III, wind losses dominate the outer-atmospheric radiative power.

Magnetic fields are often assumed to play an important role in the driving of mass loss from evolved K- and early-M giants, primarily because Alfvén-wave damping provides a means to drive these winds where other mechanisms seem to fail (e.g., Hartmann and MacGregor, 1980; Charbonneau and MacGregor, 1995; and the review by Harper, 1996). Yet these stars lie in a domain of the Hertzsprung–Russell diagram where magnetic activity is particularly weak, as discussed in Section 11.1. Alternatively, the wind could be driven by (magneto)acoustic waves; this appears problematic because these waves are efficiently damped in the lower domains of the stellar outer atmosphere (although Cuntz, 1997, argues that stochastic shocks might be important after all). Very low degree oscillations (pulsations) may play an important role (e.g., Judge and Stencel, 1991). For the asymptotic giant-branch stars, radiation pressure on dust grains that couples to the gaseous matter through drag may play a role. Broad line profiles of, for instance, singly and doubly ionized carbon and silicon provide evidence for acoustic shocks or Alfvén waves: velocities are required that reach up to 1.6 times the local sound speed (see, e.g., Hartmann *et al.*, 1985, for hybrid stars, and Carpenter and Robinson, 1997, on cool supergiants).

What causes the transition from hot to cool outer atmospheres across the dividing line? It has been proposed that this is related to the stability of coronal loops. In Section 9.6 we described studies concluding that if the loop length L is substantially longer than the pressure scale height $H_{p,5}$ at 10^5 K, only hot loops are possible, but if L is close to $H_{p,5}$ then both hot and cool loops would be possible. Assuming that active-region sizes and thus heights L of the loops scale with the stellar radius R_*, Antiochos *et al.* (1986) determined the value of L/H_p across the H–R diagram. In this way, they found a locus that mimics the X-ray dividing line closely for L/H_p between 5 and 10. Thus they suggested that the variation of the outer-atmospheric emissions across the H–R diagram reflects the response of the atmospheric structure to stellar size and gravity, rather than the variation in the magnetic activity. This is not acceptable, however. As the authors admitted, their model does not explain the marked change in the stellar-wind characteristics across the dividing line. In addition, their model does not explain the difference in the chromospheric emission: merely basal emission on the cool side from the dividing line, and enhanced emission on the warm side. Their model is not compatible with the activity–rotation relation (Section 9.4) either.

The problem of the origin of the cool, slow, and massive stellar wind in relation with the outer-atmospheric dividing lines was taken up by Rosner *et al.* (1991, 1995). Rosner *et al.* (1991) point out that because of the low gravity in giants and supergiants, coronal temperatures equal or exceed the escape temperature. Hence a corona in an open-field configuration would be entirely unconfined and would expand freely into space at the sound speed, so that these stars can only keep detectable amounts of plasma at coronal temperatures if that is confined to closed magnetic loops. They concluded that, wherever the magnetic field is open, the matter must be cool. Then they argue that such cool outer-atmospheric matter experiences strong Alfvén wave reflection at their base, which leads to a cool, slow stellar wind.

Rosner *et al.* (1995) attribute the changes in the outer-atmospheric structure that occur when a star evolves across the dividing line to a fundamental change in the dynamo. To the left of the dividing line, stars are presumed to rotate sufficiently rapidly to maintain a large-scale dynamo that leads to the emergence of a strong magnetic field. In the flux loops of the resulting large active regions, plasma can be confined at transition-region and coronal temperatures. The authors argue that, when a giant crosses the dividing line, the rotation rate drops below the level required to maintain a large-scale dynamo that produces large active regions. Then only a seed field is proposed to remain upon which the convection works to create a small-scale magnetic structure (in their solar illustration, the authors refer to the pepper-and-salt structure, which we take for the weak internetwork field; see Section 4.6). In this small-scale structure, the outer-atmospheric loops are very low, hence $L < H_p$, so according to Antiochos *et al.* (1986) only cool loops are stable in this condition. Rosner *et al.* (1995) pointed out that above the small-scale field, the magnetic field falls off steeply with height, so that the magnetic pressure $B^2(z)/(8\pi)$ rapidly drops below the gas pressure $p(z)$. Hence the magnetic field cannot confine the plasma, so it is expected to open up to form a radial field in which a stellar wind is propelled by Alfvén-wave reflection.

The arguments advanced by Rosner *et al.* (1991, 1995) for the formation of cool loops may also provide an explanation for hybrid giants in the neighborhood of the dividing lines (Section 11.1). A hybrid star could be a star of low activity level that is covered for a small fraction by large-scale active regions and corresponding hot coronal loops, while the rest of the photosphere is covered with cool, small-scale magnetic structure, with a predominantly radial outer-atmospheric field on top.

12.6   Large-scale patterns in surface activity

Stellar rotational modulation sometimes presents remarkably coherent signals over long periods of time. This suggests either very long-lived active regions or starspot clusters, or – more likely – preferred longitudes for the emergence of active regions (the stellar equivalent of activity complexes and sunspot nests). Whereas on many stars only the longitudinal phase information is known, these longitudes may reflect preferred sites of flux emergence, similar to the (composite) activity nests on the Sun (see Section 6.2). In some cases, the persistent coherence could be caused by evolving active-regions that cluster by chance as they are subjected to differential rotation and region evolution – as in the case of starspots discussed in Section 12.2 – but that would require even longer-lived regions, also an indication of the existence of active-region nests.

The location of active longitudes in tidally interacting binaries may well be a function of the system's parameters. Analyses of observations spanning typically half a century of a small set of tidally interacting RS CVn binaries (including RT And, SV Cam, WY Cnc, CG Cyg, and BH Vir) suggest the existence of active-longitude belts centered on one or both of the quadrature directions $90°$ and $270°$ for noncontact systems with periods of less than a day (Zeilik *et al.*, 1994, references therein, and Heckert *et al.*, 1998). The system RS CVn itself, with a period of $P = 4.8$ days, may have migrating active longitudes (Heckert and Ordway, 1995); UX Ari ($P = 6.4$ days) shows a persistent feature for 7 yr at $340°$, opposite from the companion star (see references in Hall, 1991). Systems with orbital periods above 5–10 days, such as RS CVn, SS Boo, and MM Her, show no evidence of such preferred longitudes. The RS CVn system HK Lac ($P = 24$ days) does show preferred longitudes separated by $110°$, but these are not at the quadrature points. This small sample suggests that preferred activity at the quadrature points occurs in the shortest-period systems. It isn't clear whether this is a consequence of the strength of the tidal interaction or whether preferred longitudes at the quadrature points are a property of main-sequence components of binaries, because there is a correlation between stellar and orbital parameters in these systems: the short-period systems contain main sequence stars, whereas the long-period systems contain an evolved primary, with generally a good correlation between the primary radius and the orbital and rotational periods.

The existence of pronounced, persistent active longitudes in short-period binaries suggests that tidal interaction is a prerequisite for the formation of such preferentially active zones. This suggestion is supported by the absence of reports on active longitudes on rapidly rotating cluster stars. There are, however, reports on persistent active zones in single stars that exist in between the more widespread emergence of other active regions on their surface. First of all there is the Sun, which exhibited some preferred areas of activity persisting for up to a few years (Van Driel-Gesztelyi *et al.*, 1992). Soon *et al.* (1999) report on active longitudes (likely active-region nests) that persist for over 10 yr in the G2 V star HD 1835 ($P \approx 7.7$ days, and a cycle period of 9.1 yr, as determined with "fair" accuracy by Baliunas *et al.*, 1995) and in the star HD 82885 (G8 IV–V, but also G8 IIIv, $P \approx 18$ days, with signatures of two periodicities of 7.9 yr, "fair," and 13 yr, "poor;" Baliunas *et al.*, 1995). If these assessments prove correct, we have to face the problem of how an active longitude can survive for a time longer than the cycle period without a companion star to provide a preferential reference frame.

Another preferred zone of activity was discussed in Section 12.2: the most rapidly rotating cool stars frequently show polar spots located near or truly over the rotational pole. Note that whereas planetary dynamos as well as pulsar fields indicate that the dynamo axis need not be aligned with the rotation axis, no indications have thus far been found for a significant inclination of activity belts in cool stars with respect to the equator. The axially symmetric differential rotation may well be instrumental in the alignment of the magnetic and rotation axes.

12.7 Stellar differential rotation

We argued in Section 6.3.2 that differential rotation plays an important role in determining the pattern of the magnetic field in the solar photosphere. In Chapter 7 we described the important role assigned to differential rotation in the dynamo. The solar differential rotation, as measured directly for the surface and indirectly for the interior through helioseismology, is discussed in Chapter 3 for the interior. Unfortunately, the possibilities to measure the differential rotation of other stars are limited. Internal differential rotation cannot be probed by using seismology: although acoustic models of the lowest spherical harmonic degree ($\ell = 1$ and 3) – in principle observable in a disk-integrated stellar signal – contain information on the internal rotation rate, they cannot unveil the details of the expectedly weak differential rotation within a convective envelope (very strong differential rotation can be ruled out by evolutionary arguments; see Chapter 13).

Surface differential rotation modifies disk-integrated line profiles by the differential Doppler broadening at different latitudes. For stars with surface rotation velocities comparable to that of the Sun, however, the velocity dispersion associated with convection (Section 2.5) and waves overwhelms the mean rotational broadening, making it impossible to find second-order effects associated with differential rotation. In rapidly rotating stars, in contrast, some information may be obtained. The extraction of information on differential rotation from line profiles relies on accurate knowledge of the intrinsic line profile, however; the substantial problems in studies of rotational broadening have been discussed in detail by Collins and Truax (1995). Some results on differential rotation of F-type main-sequence stars were published by Gray (1982). When expressed as a ratio of angular velocities Ω or rotation periods P,

$$\frac{\Omega(\theta)}{\Omega(\theta = 0)} = \frac{P(\theta = 0)}{P(\theta)} = 1 - k \sin^2(\theta), \tag{12.2}$$

the parameter k is found to be approximately a factor of 3–4 smaller on average for the F-type stars than for the Sun ($k_\odot \approx 0.15$ for Doppler measurements of the photosphere). Johns–Krull (1996) applies this method to three rapidly rotating T Tauri stars but finds no significant indication of differential rotation.

Two other methods are used to estimate surface differential rotation: Doppler imaging and light-curve analysis. Whereas Doppler imaging provides a direct measurement of the stellar surface differential rotation, its applicability is limited to rapid rotators with substantial inclinations of the rotation axes out of the plane of the sky to allow derivation of latitude information. The analysis of light curves allows a much larger sample of stars to be studied, but this yields no direct information on latitude. Fourier methods result in estimates of the mean rotation period for different epochs. As time passes, the mean latitude of the active phenomena is assumed to change, in analogy to the solar case, which results in a slight change of the inferred rotation period if the star rotates differentially. When observations are available over a sufficiently long interval, the range in observed periods is a measure for the differential rotation. With this method it is impossible to establish whether the equator or the pole rotates most rapidly, whether more complicated torsional patterns exist, or even what the exact magnitude of the differential rotation is, because the latitudinal extent of a possible (multiple) activity belt is unknown. Active-region evolution complicates matters even more, because fortuitous patterns or persistent nests can affect the observed periods markedly. For stars with rotation periods of the order of a few days or at most 1–2 weeks, the effect of active-region evolution on surface differential rotation may be less severe, because for these stars the time scale for active-region evolution is likely to be shorter than the rotation period. In contrast, the larger number of active regions complicates the interpretation of the signal, while it may also dampen the signal's amplitude (e.g., Hempelmann and Donahue, 1997). Nevertheless, the range in rotation periods for a given star is likely to be some quantitative measure of the magnitude of surface differential rotation.

In order to interpret stellar results properly, it is important to first look at the Sun as a star. It turns out that the inferred magnitude of and dependence on cycle phase of the solar differential rotation depend on the details of the subset of the data. Donahue and Keil (1995) use a data base spanning 8 yr of disk-integrated measurements of the solar Ca II K line. They applied a periodogram analysis to selected data with simulated observing window of 200 days, roughly equal to the Mt. Wilson Ca II HK observing season of 120–180 nights. At first glance, their plot of rotation period versus time appears consistent with an equatorward drift of the main activity belt in each solar cycle, jumping to higher rotation periods as a new cycle begins. But if studied in detail, the rotation periods do not match those of the mean longitude of activity, nor does the trend match the actual cycle length (Schrijver, 1996). The more recent study by Hempelmann and Donahue

(1997) uses the entire data set for a wavelet analysis. They recover the solar differential rotation more accurately, but multiple maxima and an ambiguous mean rotation period make even the results based on this excellent data set hard to interpret quantitatively.

The difficulty in recovering the differential rotation of the photosphere from solar data is a warning that the stellar results are to be interpreted with caution. Yet with appropriate methods and data bases spanning more than a full activity cycle with excellent coverage, it seems feasible to derive some quantitative estimate of the surface differential rotation over the width of activity belts.

Starspot records of some stars now span a period of over 80 yr. The measured relative range in periods, $\delta P/P$, can be converted into a coefficient k of differential rotation, as defined in Eq. (12.2), assuming that starspots cover the entire range of starspot latitudes rather than narrow activity belts (Hall and Busby, 1990). The results are not very sensitive to the actual width of the belts, however, as long as these are relatively broad. Hall (1991) plots k versus the mean stellar rotation period P for a sample of 85 single and (close) binary stars. At any value of P, k ranges over approximately an order of magnitude with the Sun very near to the best fit:

$$\log k = (-2.30 \pm 0.06) + (0.85 \pm 0.06) \log P(\theta = 0), \tag{12.3}$$

for P in days.

Hall notes that the relationship between a quasi-Rossby number $\mathcal{R}_o{}^*$ and k is as good as the fit in Eq. (12.3), and that an even somewhat better fit is found by allowing for a dependence on the separation between the binary components. This number $\mathcal{R}_o{}^*$ is a rather artificially derived quantity, however. Hall uses the curve $\mathcal{R}_o$ for main-sequence stars and then multiplies that by a number that increases by a factor of 1.4 for each luminosity subclass, which is by a factor of 5.6 between LC V and LC III–II. With that scaling, the transition from "variable" to "nonvariable" at a photometric precision of 1% matches where $\mathcal{R}_o{}^* \equiv 1$. Whereas this may be esthetically pleasing in view of some dynamo theories (Section 11.3), the interpretation of what $\mathcal{R}_o{}^*$ actually measures is left unaddressed.

The above fit extends over 3 orders of magnitude in k (and even more in P), ranging from three times bigger than the solar value of $k_\odot \approx 0.15$ for slow rotators, to some 200 times less at $k \approx 0.001$ for the most rapid rotators.

Equation (12.3) suggests a small value of k and thus nearly rigid rotation *in relative terms* for rapid rotators. The differential-rotation quantity that is relevant to the dynamo may, however, not be k but rather the time it takes to wind up a field line in a differentially rotating star. From Eq. (12.3), a difference of a full revolution between pole and equator would be realized in the shearing time scale [Eq. (6.11)],

$$\hat{t}_s = \frac{P(\theta = 0)}{k} \approx 200 \times P^{0.15}(\theta = 0), \tag{12.4}$$

for $\hat{t}_s$ and P in days; this is remarkably insensitive to the stellar rotation period. A star with a rotation period of 1 day would wind up field lines in fact 1.7 times *faster* than a star with a rotation period of 30 days, despite the much smaller value of k.

A few stars do not adhere to Eq. (12.3): Hall (1991) discusses three stars for which starspot modulation periods have been determined for many years but which show no systematic trend of these periods with time. Moreover, in some stars the data suggest that the pole rotates faster than the equator (see, e.g., Donahue and Baliunas, 1992, on the G0 V star HD 114710).

Donahue *et al.* (1996) discuss the spread δP in rotation periods observed in a sample of $\sim$100 stars in the Mt. Wilson Ca II HK program. Approximately one-third of these stars showed clear periodicities that are attributed to rotational modulation, and the range in observed periods could be fitted by

$$\frac{\delta P}{P} \propto P^{0.3\pm0.1}, \tag{12.5}$$

implying a weaker dependence of k on rotation rate than in Eq. (12.3), and thus a stronger increase of the shearing time scale in Eq. (12.4) with P. It remains to be seen whether this difference is significant, and if so, whether it reflects a different behavior between relatively slowly rotating single main-sequence stars that make up the sample of Donahue *et al.* (1996) and the RS CVn systems that dominate the sample of Hall (1991).

The theoretical model for stellar differential rotation by Kitchatinov and Ruediger (1995) – discussed in Section 3.3 – predicts a dependence of the differential rotation with depth and latitude on P:

$$\max\left(\frac{\partial\Omega}{\partial r}\right) \propto \Omega^m, \tag{12.6}$$

$$\frac{\Delta P}{P} \propto P^n, \tag{12.7}$$

for ΔP the pole–equator difference; $m \approx -2.4$ and $n \approx 1.6$. A more recent study by Rüdiger *et al.* (1998) resulted in $n \approx 0.3$, showing the sensitivity of the results to the model assumptions. Both gradients decrease with an increasing rotation rate. The relative surface differential rotation in these models is of the same magnitude as the observed values, which is encouraging. The model by Kitchatinov and Ruediger (1995) involves a balance between two opposing forces that drive a meridional circulation, which is, for one thing, sensitive to the anisotropy of turbulent convection. Such sensitivities cannot yet be modeled accurately, so we should allow for a considerable difference between the theoretical and empirical values for n. The agreement for single main-sequence stars in the sample of Donahue *et al.* (1996) appears promising. Many of the more active stars in the sample of Hall (1991) are close binaries in which tidal effects may modify the large-scale circulation, possibly contributing to the difference between the exponents in Eqs. (12.4) and (12.5).

The apparent insensitivity of the shearing time scale to the stellar rotation rate may be an ingredient of the explanation of the tightness of the relationships between stellar activity diagnostics discussed in Section 9.4: the surface pattern of dispersing flux, caught in the differential rotation and meridional flow, appears to be subjected to differential rotation that is only moderately different from that of the Sun. The time scale for differential rotation is, clearly, not the only time scale that is important in establishing these patterns. Let us reconsider the four time scales that are important for the large-scale dispersal of magnetic flux over the solar surface: the time scales for shearing $\hat{t}_s$ [Eq. (6.11)], meridional flow $\hat{t}_m$ [Eq. (6.12)], random-walk diffusion $\hat{t}_d$ [Eq. (6.13)], and the hybrid time scale for shear and diffusion $\hat{t}_h$ [Eq. (6.14)]. In order to maintain the same ordering of these time scales, $\hat{t}_s < (\hat{t}_m, \hat{t}_h) < \hat{t}_d$ for stars other than the Sun requires, with Eq. (12.4),

$$1 \lesssim \left[2.5 \frac{R_*}{R_\odot} \frac{v_{0,\odot}}{v_{0,*}} \left(\frac{P_\odot}{P_*} \right)^{0.15}, 5.4 \frac{R_*}{R_\odot} \left(\frac{D_\odot}{D_*} \right)^{1/2} \left(\frac{P_\odot}{P_*} \right)^{0.075} \right]$$

$$\lesssim 30 \left(\frac{R_*}{R_\odot} \right)^2 \frac{D_\odot}{D_*} \left(\frac{P_\odot}{P_*} \right)^{0.15}, \tag{12.8}$$

for a meridional flow v_0 and a diffusion coefficient D for flux dispersal. For a G2 V star such as the Sun, these inequalities hold for rotation periods up to $500 P_\odot$ if we assume the meridional flow to be independent of rotation rate, in which case the inequalities would hold for all truly Sunlike stars with any significant level of activity. Unfortunately, the meridional flow $v_{0,*}$ and the dispersal coefficient D_* are at present unknown for stars other than the Sun. It is unclear, however, whether the patterns of activity of the relatively active stars are indeed governed by these same parameters: the dynamo in stars much more active than the Sun may well operate in a different spatiotemporal mode (see Chapter 15).

13

Activity and rotation on evolutionary time scales

13.1   The evolution of the stellar moment of inertia

In Section 11.3 we conclude that the stellar rotation rate is the prime parameter determining the level of stellar magnetic activity. The activity level responds to changes in the stellar rotation rate that are the result of the loss of angular momentum through the magnetized stellar wind, and of evolutionary changes in the moment of inertia; both of these effects may also affect the internal differential rotation.

For rigidly rotating bodies, the angular velocity $\Omega(t)$ responds to changes in the angular momentum $\mathcal{L}(t)$ and in the moment of inertia $I(t)$. Provided that the rotation does not significantly affect the moment of inertia, these quantities are related through

$$\Omega(t) = \frac{\mathcal{L}(t)}{I(t)}. \tag{13.1}$$

Because of the evolutionary changes in the stellar interior, reflected in radial expansion or contraction and associated changes in density ρ, the moment of inertia $I(t)$ of the star changes with time:

$$I(t) = \frac{8\pi}{3} \int_{r_a}^{r_b} \rho(r, t)\, r^4 \, dr, \tag{13.2}$$

for an integration over the entire stellar volume, that is, with $r_a = 0$ and $r_b = R_*(t)$. During the main-sequence phase, the change in I is rather small: the moment of inertia for the Sun has changed by only $\sim$7% since its arrival on the main sequence (see Claret and Giménez, 1989; Fig. 13.1). Major changes in the moment of inertia I occur during the pre-main-sequence evolution and during the giant stage. For post-main-sequence stars, the moment of inertia increases by up to 4 orders of magnitude, even prior to reaching the tip of the first ascent near the asymptotic giant branch. The decrease in I during the pre-main-sequence contraction leads to strong spin up.

If there were no magnetic braking, the rotation rate beyond the main sequence would be primarily governed by the stellar expansion, only slightly modified by internal changes that deviate from simple radial scaling (e.g., Endal and Sofia, 1978a; Rutten and Pylyser, 1988). For stars that rotate as rigid bodies, the changes in the moment of inertia caused by internal changes other than homogeneous expansion are quantified through the radius of gyration $\beta_I R_*$:

$$I \equiv M_* \, (\beta_I R_*)^2 \propto \frac{\beta_I^2}{g}. \tag{13.3}$$

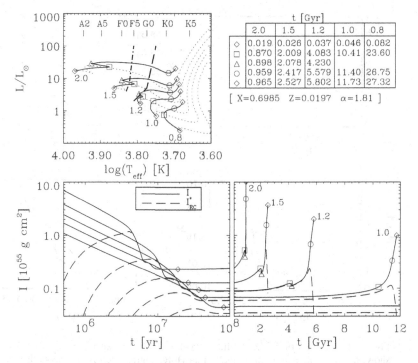

Fig. 13.1. Evolutionary tracks (top panel) for late-type stars of various masses, from the pre-main sequence to main sequence (dotted curves in the top panel), and from there to the base of the giant branch (solid curves in the top panel). The diamonds indicate the zero-age main sequence (at the lower-left end of the solid curves) and the end of model computations. Stellar masses are given in units of the solar mass. The dashed-dotted curve marks the onset of envelope convection. Ages at selected points along the tracks are listed in the table in the top right of this figure for stellar masses indicated in the top row. The evolutionary variations of moments of inertia of the entire star (solid curves) and of the radiative interior (dashed curves) are shown in the lower panels (figure from Charbonneau *et al.*, 1997).

The gyration constant β_{I} at first decreases somewhat beyond the ZAMS, and then increases by a factor of ~ 2.2 to a maximum around the upturn near the asymptotic giant branch, with minor differences for stars with masses from $1.5 M_{\odot}$ up to $3 M_{\odot}$ (Rutten and Pylyser, 1988). In later phases, β_{I} decreases again. The value of β_{I} is substantially smaller than unity because of the concentration of the mass toward the stellar core (compare Fig. 13.2, lower-left panel).

The magnetic brake applies directly to the top of the convective envelope, but the convective flows transport the angular momentum throughout the envelope, thus decelerating its entire volume. The rotational coupling to the radiative interior has been argued to occur on a much longer time scale (as discussed in Section 13.3.2). It is therefore of interest to study the moments of inertia of radiative interior I_{RI} and of the convective envelope I_{CE} separately. The lower-right panel in Figure 13.2 shows the rapid increase in the ratio $I_{\mathrm{CE}}/I_{\mathrm{RI}}$ as stars evolve off the main sequence toward the giant branch.

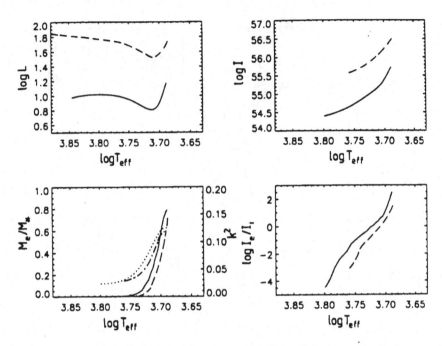

Fig. 13.2. For stars of $1.5M_\odot$ (solid curves) and $2.5M_\odot$ (dashed curves), the four panels show: *top left*, the evolutionary track in an H–R diagram; *top right*, the stellar moment of inertia I (g/cm^2); *lower left*, the fractional mass content of the convective envelope M_e/M_* and the square of the gyration constant $k \equiv \beta_I$ as dotted and dashed-dotted curves; and *lower right*, the ratio of moment of inertia of the convective envelope to that of the radiative interior (figure from Schrijver and Pols, 1993).

13.2   Observed rotational evolution of stars

An early indication of the importance of magnetic braking came from the study of rotational velocities of stars near the main sequence. In his classical work, Kraft (1967) showed that under the assumption of rigid-body rotation, the observed rotation rates for main-sequence stars in the interval $1.5 \lesssim M/M_\odot \lesssim 20$ are consistent with a power law dependence for the total angular momentum $\mathcal{L}$ on stellar mass M_* of the form

$$\mathcal{L} \propto M_*^n, \tag{13.4}$$

with $n = 1.57$, since then commonly referred to as the Kraft relationship (the index n has since been revised to ~ 2; see Kawaler, 1987). The rotation rates of cooler, less massive stars with convective envelopes dip below this relationship (Fig. 13.3). Since Kraft's pioneering study, it has been found that the youngest main-sequence stars lie only slightly below the relationship in Eq. (13.4), but the difference increases rapidly with age, particularly beyond approximately G0. Because the moment of inertia I changes little while on the main sequence, magnetic braking necessarily plays a dominant role in the history of activity for these stars. Skumanich (1972) showed that for G- and K-type main-sequence stars older than $\sim 10^9$ yr, v_{eq} decreases roughly with $t^{-1/2}$ [see Eqs. (13.7) and (13.8)]. This general empirical relationship, revisited by several authors, has stood the test of time, albeit that the equations fitted to the data do differ somewhat from author to author.

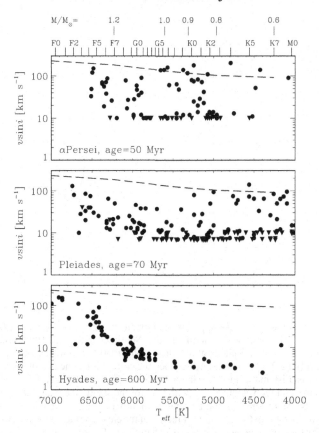

Fig. 13.3. Projected equatorial rotational velocities vs. effective temperature in three young open clusters. Triangles are upper limits. The dashed curve is the run of rotation rate with mass predicted by Eq. (13.4) extrapolated to lower masses, assuming a power-law index $n = 2.02$, as revised by Kawaler (1987; figure from Charbonneau *et al.*, 1997; data courtesy of D. Soderblom).

Figures 13.4 and 13.5 display observed $v \sin i$ values together with the average run of the equatorial velocities as a function of effective temperature for luminosity class (LC) III ($2 \lesssim M_*/M_\odot \lesssim 5$) and LC IV ($1.2 \lesssim M_* \lesssim 2$) stars, respectively. These figures show that the mean rotational velocities, $\bar{v}_{eq}(T_{eff})$, decrease monotonically as stars age, with the exception of the cool giants beyond K0 III (a phenomenon not yet understood). The sample for giants is biased toward lower masses blueward of the giant branch, because the more massive stars evolve more rapidly. Subsequent phases of core-helium burning, not addressed in these studies, add stars of lower masses (the clump giants) to the cool side of Fig. 13.4, which complicates the interpretation of the data. The dashed curves in the figures show the rotational evolution in the absence of magnetic braking predicted by Eq. (13.1) for $M_* = 2.5M_\odot$ and $1.5M_\odot$.

The predicted rotation rates in the absence of magnetic braking lie above the running mean through the empirical data for the relatively cool giants and particularly so for the subgiants (Rutten and Pylyser, 1988). Apparently, the evolutionary increase in the moment of inertia is too small to account for the observed drop in rotation rates, so that

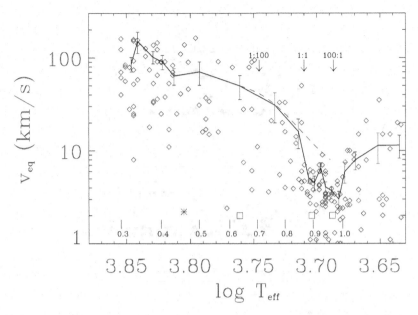

Fig. 13.4. Observed rotation velocity, v_{eq} (corrected for an average projection factor $\langle \sin i \rangle = 4/\pi$), vs. T_{eff} for LC III giant stars (modified after Schrijver and Pols, 1993). Bars near the abscissa mark $B-V$ color. Squares above these color markers show the age in units of 2.5×10^6 yr of evolving $2.5 M_\odot$ stars, starting when the convection zone is four density scale heights deep. The asterisk marks the granulation boundary (Gray and Nagel, 1989). The solid curve represents the average velocity $\frac{4}{\pi} \bar{v}_{eq}$ (with bars for the standard deviation) obtained by ordering the data by decreasing T_{eff} and binning them into adjacent groups of ten stars. The value of T_{eff} where the ratio of the moments of inertia of the envelope and of the interior equals 1:100, 1:1, or 100:1 are marked on the upper right-hand side. The run of $v_{eq}(T_{eff})$ for a model for a rigidly rotating star of $2.5 M_\odot$ without magnetic braking, as described by Eq. (13.1), is indicated by the dashed line.

angular momentum loss must occur as these stars evolve toward the giant branch (e.g., Gray and Nagar, 1985, for LC IV and Gray and Endal, 1982, for LC III). Further support for this conclusion comes from a study of the distribution of rotation rates at different T_{eff}: if the distribution of the initial rotation rates, $H_0(v_{eq})$, and of the orientations of rotation axes with which stars are born do not change with time, the rotational history of stars can be derived from the comparison of distributions $H(v_{eq}, T_{eff})$ at different effective temperatures, because of the monotonically redward evolution during the phases relevant to this problem. In the absence of magnetic braking, $H(v_{eq}, T_{eff})$ would transform as $f(T_{eff}) H^*(v_{eq})$, which is contradicted by the observations (Gray, 1989). Hence, magnetic braking must take place.

Not far redward of the onset of envelope convection, the relative ages of the stars can be estimated fairly easily because age is (almost) uniquely related to the position in the H–R diagram for single (sub)giants (see the evolutionary tracks in Figs. 2.10 and 13.1). Consequently, the observed variation in the mean rotation rates $\bar{v}_{eq}$ along evolutionary tracks in the Hertzsprung–Russell diagram allows the direct study of magnetic braking, at least up to just before the giant branch where the stellar evolutionary phase in the H–R diagram becomes ambiguous. For LC III giants, $\bar{v}_{eq}$ appears to decrease smoothly by a

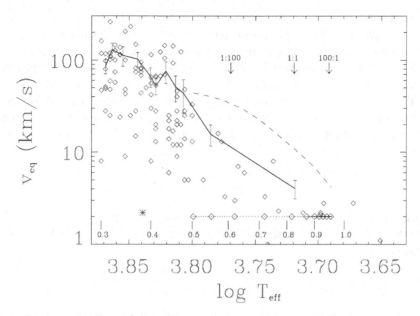

Fig. 13.5. As Fig. 13.4, but for LC IV stars of $\sim 1.5\,M_\odot$. The stellar age is indicated above the $B-V$ markers by diamonds placed 25×10^6 yr apart and dots placed 2.5×10^6 yr apart (from Schrijver and Pols, 1993).

factor of ~ 10 between $B-V = 0.5$ and $B-V = 0.9$. Gray (1991) suggested that this smooth transition is an artifact of the sample and that the decrease is in fact much more drastic (occurring near the thick dashed curve in Fig. 13.1). Rutten and Pylyser (1988) point out that there is no clear discontinuity in activity levels in a much larger sample of Ca II H+K flux densities (see Fig. 2.17; supported by data on evolved stars in open galactic clusters; cf. Beasley and Cram, 1993), so if atmospheric activity is a monotonic measure of rotation rate (Section 11.3), the rotation rate does indeed appear to decrease smoothly as $T_{\rm eff}$ decreases.

For the LC IV stars, the decrease in $\bar{v}_{\rm eq}$ with decreasing $T_{\rm eff}$ is quite pronounced, reaching an order of magnitude already by the time the stars reach $B-V \approx 0.6$ – a decrease so strong that magnetic braking must be invoked. Redward of that point, the run of $\bar{v}_{\rm eq}(T_{\rm eff})$ for the LC IV stars is probably strongly biased toward higher values because of the detection limit at approximately $v \sin(i) \approx 1 - 2$ km/s.

13.3   Magnetic braking and stellar evolution

13.3.1   *Magnetic braking for mature main-sequence stars*

In view of the lack of empirical data on winds of magnetically active stars on the main sequence and the early phases of evolution as giant stars, the properties of the magnetized solar wind as discussed in Section 8.9.2 are frequently used to provide the basis for a study of the evolution of the angular momentum in cool stars. We first concentrate on main-sequence stars, for which the small changes in the moment of inertia I can be ignored in a first analysis of Ω as a function of time.

The Weber–Davis model for angular-momentum loss through a magnetized stellar wind, as expressed in Eq. (8.27), is valid only within the equatorial plane. Some equivalent expression is required to estimate the angular-momentum loss integrated over the entire stellar sphere. Charbonneau *et al.* (1997) summarize the derivation of such an expression: if field lines would bend only in the latitudinal direction, and provided that the mass flux and the Alfvén radius r_A are independent of latitude θ, the equatorial Weber–Davis solution can be extended to a spherical model. The *total* specific angular momentum carried away by the magnetically coupled wind then equals that of a thin, rigidly rotating spherical shell of radius r_A, namely $\mathcal{L}_{\mathrm{sph}} = (2/3)\Omega r_A^2$. Therefore,

$$\frac{d\mathcal{L}}{dt} = 4\pi \rho_A r_A^2 v_{r_A} \left(\frac{2}{3}\Omega r_A^2 \right). \tag{13.5}$$

Conservation of the mass and magnetic flux through the wind base at $r = r_0$ implies that at the Alfvén radius r_A, the following holds: $r_0^2 B_{r_0} = r_A^2 B_{r_A}$ and $B_{r_A}^2 = 4\pi \rho_A v_{r_A}^2$. Consequently,

$$\frac{d\mathcal{L}}{dt} = -\frac{2}{3} B_{r_0}^2 r_0^4 \frac{\Omega}{v_{r_A}}. \tag{13.6}$$

Inserting a relationship between B_{r_0} and Ω allows the computation of the angular-momentum history of the star. For the period range from a few days up to 20 days, the roughly linear dependence in Eq. (11.13) or the steeper dependence in Eq. (11.12) found from direct field measurements can be used. Most studies take $B_{r_0} \propto \Omega$, stimulated by the agreement with a well-known relationship between age and angular velocity for main-sequence stars, as we show now.

Equation (13.6) can be further simplified by noting that for thermally driven winds, $v_{r_A} \simeq c_s$ (see Fig. 8.12). If the wind temperature is taken to be independent of the average magnetic flux density at the surface and of the rotation rate, then Eq. (13.6) transforms into

$$\frac{d\Omega}{dt} \propto -\Omega^3 \tag{13.7}$$

for a constant moment of inertia. In the limit that $t \gg t_0$, and therefore $\Omega(t) \ll \Omega(t_0)$, this integrates to

$$v_{\mathrm{eq}}(t) \propto t^{-1/2}, \tag{13.8}$$

which is the same as the relationship determined empirically by Skumanich (1972); see the discussion below Eq. (13.4). This agreement is encouraging, given the crude approximations made in the above derivation. Note that Eq. (13.7) should only be applied to stars for which the wind is driven thermally, and not by substantial magneto-centrifugal forces associated with the field. This transition in the driving of the wind occurs near $\Omega \gtrsim 5\Omega_\odot$ (Charbonneau *et al.*, 1997). Moreover, stars that rotate substantially faster than that are subject to activity saturation (Section 11.3), which also limits the validity of Eq. (13.7). A number of studies accommodate these effects by using a parameterization proposed by Chaboyer *et al.* (1995):

$$\frac{d\Omega}{dt} = c\,\Omega^3 \quad \text{if} \quad \Omega \le \Omega_c\,; \quad \frac{d\Omega}{dt} = c\Omega_c^2\,\Omega \quad \text{if} \quad \Omega > \Omega_c, \tag{13.9}$$

where Ω_c is determined by a comparison with observational data to be of the order of a day.

13.3.2 *Magnetic braking while the moment of inertia changes*

The rotational evolution subject to magnetic braking is complicated by, among many other things, the possibility that stars may rotate differentially. Fortunately, the one example of a cool star that can be studied in great detail suggests that differential rotation can be neglected to first approximation: the Sun shows only a weak (albeit significant) differential rotation with depth within the convective envelope, and no strong shear between the radiative core and the convective envelope (Chapter 3). In order to allow for a potentially substantial gradient with depth, models are introduced in which the radiative interior and the convective envelope are allowed to rotate as separate solid bodies with different angular velocities Ω_{CE} and Ω_{RI} (a concept initially proposed by Stauffer and Hartmann, 1986).

In such two-component models, the magnetic brake acting on the envelope induces shear between interior and envelope. This shear is counteracted primarily by (magneto) hydrodynamical processes, that by far dominate the gas-kinetic viscosity in the stellar interior. Too little is known about these processes to formulate a consistent ab initio model of their dependence on stellar parameters (see the discussion by Charbonneau *et al.*, 1997). One proposed coupling mechanism is based on a (weak) magnetic field that permeates both domains. Shear winds up this field, until it is strong enough to transport angular momentum efficiently. Charbonneau and MacGregor (1993) conclude that even a modest 1-G field could synchronize the core and the envelope of the Sun within a million years.

Another means to couple interior and envelope is through large-scale circulation. If this circulation is deep enough, it affects the surface abundances of, particularly, lithium, beryllium, and boron, because sufficiently deep transport of these elements through the stellar interior leads to their destruction by nuclear reactions (see, for instance, models by Pinsonneault *et al.*, 1989, 1990, 1991, and a review of observational data on lithium by Pallavicini, 1994).

In view of the incomplete understanding of the transport mechanism(s) of angular momentum through stellar interiors, some studies introduce a *coupling time scale* $\hat{t}_C$ on which the angular momentum between the interior and envelope is exchanged. The coupling time scale may be fixed or it may be a function of stellar properties, and thus of time. Its value can be constrained through a comparison of models and observations.

Given a two-component model and a parameterization of the magnetic brake and of the coupling time scale, the evolution of the angular momenta $\mathcal{L}_{CE}$ and $\mathcal{L}_{RI}$ is described by the following pair of equations:

$$\frac{d\mathcal{L}_{CE}}{dt} = +\frac{\Delta\mathcal{L}}{\hat{t}_C} + \frac{d\mathcal{L}_d}{dt} - \frac{\mathcal{L}_{CE}}{\hat{t}_{\mathcal{L}}}, \qquad (13.10)$$

$$\frac{d\mathcal{L}_{RI}}{dt} = -\frac{\Delta\mathcal{L}}{\hat{t}_C} - \frac{d\mathcal{L}_d}{dt}, \qquad (13.11)$$

with angular momenta $\mathcal{L}_{CE} = I_{CE}\Omega_{CE}$, $\mathcal{L}_{RI} = I_{RI}\Omega_{RI}$, and a difference in the angular momenta between the radiative core and the envelope of $\Delta\mathcal{L} = (\Omega_{RI} - \Omega_{CE})I_{CE}I_{RI}/(I_{RI} +$

I_{CE}). The quantity $d\mathcal{L}_d/dt$ reflects the angular-momentum transfer (postulated to be instantaneous) associated with the changes in the depth of the convective envelope, negligible only for main-sequence evolution. The magnetic brake is assumed to act only on the convective envelope, on a time scale $\hat{t}_C$. Stellar rotational evolution can then be studied by integrating Eqs. (13.10) and (13.11).

13.3.3 Post-main-sequence evolution

Stars with $M_*/M_\odot \lesssim 1.3$ maintain a convective envelope throughout their main-sequence and post-main-sequence evolution up to the giant branch (Section 2.4). In contrast, more massive stars have no convective envelope on the main sequence and develop one only after crossing the boundary line for convective envelopes in the H–R diagram (Fig. 2.10).

Schrijver and Pols (1993) model the angular-momentum loss from the convective envelope by the magnetized wind through a generalization of Eq. (13.7) (continuing initial studies by Endal and Sofia, 1978a; Gray and Endal, 1982; and Rutten and Pylyser, 1988):

$$\frac{d\mathcal{L}_{CE}}{dt} = \alpha E_D(T_{eff})\left(\frac{R_*}{R_\odot}\right)^n \Omega_{CE}^3. \tag{13.12}$$

The function $E_D(T_{eff})$ is introduced to allow for changes in the efficiency of the dynamo and of the magnetic brake as a function of the stellar effective temperature (see Section 11.4). For the particular case of Eq. (13.6), the value of n equals 4, but that value is varied in the simulations to coarsely model, among other effects, variations in the surface area in which the field lines close back on the stellar surface (this region is referred to as the dead zone). The constant of proportionality $\alpha = 2.6 \times 10^{-21} I_0 R_0^2$ is consistent with the Skumanich relationship, Eq. (13.8). The application of Eq. (13.8) requires that it is expressed as a relationship between angular velocity and angular momentum by using $v_{eq} = \mathcal{L} R_0/I_0$. Schrijver and Pols (1993) used I_0 and R_0 for a G0 zero-age main-sequence star.

It is instructive to consider the time scale $\hat{t}_\mathcal{L}$ for magnetic braking that is implied by Eq. (13.12). For $n = 4$,

$$\hat{t}_\mathcal{L} = \frac{\mathcal{L}}{|\dot{\mathcal{L}}|} = \frac{I_j R_\odot^4}{\alpha E_D(T_{eff}) R_*^2 v_{eq}^2}, \tag{13.13}$$

where $I_j = I_*$ if the star rotates rigidly or $I_j = I_{CE}$ if the convective envelope is slowed down separately from the radiative interior. Figure 13.6 shows that if the brake were to reach full efficiency, that is, $E_D(T_{eff}) \equiv 1$, when an evolving star begins to develop a convective envelope as it evolves redward in the Hertzsprung–Russell diagram, $\hat{t}_\mathcal{L}$ is of the order of only a few years for subgiants if the angular momentum is drained from the convective envelope only. For giants $\hat{t}_\mathcal{L}$ is of the order of a millennium, which is still short compared to evolutionary time scales.

The difference in the angular momenta between the interior and envelope depends on the ratio of the time scale $\hat{t}_\mathcal{L}$ for magnetic braking and the time scale $\hat{t}_C$ for the coupling between the envelope and interior: a small ratio $\hat{t}_\mathcal{L}/\hat{t}_C$ would lead to a rapid spin-down of the envelope immediately redward of the onset of convection. This is ruled out by the

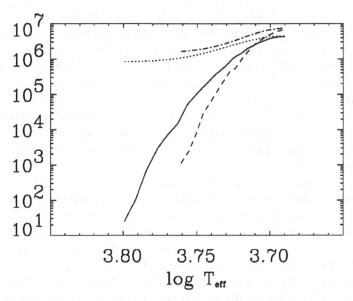

Fig. 13.6. Time scale $\hat{t}_{\mathcal{L}}$ (in years) for magnetic braking as given by Eq. (13.13) for a rigidly rotating convective envelope (solid and dashed curves for $1.5M_\odot$ and $2.5M_\odot$, respectively), and for a rigidly rotating star as a whole (dotted and dashed-dotted curves for $1.5M_\odot$ and $2.5M_\odot$, respectively). The curves are plotted for $v_{eq} = 100$ km/s and assume the dynamo efficiency $E_{\mathrm{D}}(T_{\mathrm{eff}}) \equiv 1$ (figure from Schrijver and Pols, 1993).

observations summarized in Figs. 13.4 and 13.5 for giants and subgiants, respectively. Hence, $\hat{t}_{\mathrm{C}} \lesssim \hat{t}_{\mathcal{L}}$, or the brake's efficiency $E_{\mathrm{D}}(T_{\mathrm{eff}})$ is strongly reduced when convection sets in, or both.

Even if $\hat{t}_{\mathrm{C}}$ were so small that the star would rotate (nearly) rigidly, a magnetic brake at full strength as given by Eq. (13.12) with $n = 4$ would slow subgiants down far too rapidly (compare the evolutionary time scales in Figs. 13.4 and 13.5, with the time scales in Fig. 13.6). It appears that the dynamo efficiency $E_{\mathrm{D}}(T_{\mathrm{eff}})$ should be as small as 10^{-6}, depending on $\hat{t}_{\mathrm{C}}$, upon onset of envelope convection, and increase smoothly to approximately unity later on. This strong reduction of the dynamo efficiency in stars with relatively thin convective envelopes is consistent with the discussion in Section 11.4 of the observed magnetic activity in warm main-sequence stars.

The rotational velocities of subgiants with masses near $1.5M_\odot$ decrease strongly within a period of roughly 5×10^7 yr. During this phase of the stellar evolution, the moment of inertia of the envelope, I_{CE}, is less than 1% of the stellar value, I_*. As the stars evolve, the convective envelope deepens. If the interior would be left rotating substantially faster than the envelope, the deepening envelope would dredge up the angular momentum from the radiative interior and spin up again. Such a resurgence of the rotation rate and associated activity is not observed, and therefore much of the interior's angular momentum must have been removed before I_{CE} reaches a substantial fraction of I_*, that is, well before $\log T_{\mathrm{eff}} \sim 3.72$ (at about G2 III).

Schrijver and Pols (1993) argue that for rigidly rotating giant stars a magnetic brake with $n = 4$ in Eq. (13.12) is too weak to drain sufficient angular momentum from them during their rapid evolution. This holds even if the dynamo efficiency $E_D(T_{eff})$ would equal unity from the moment envelope convection sets in. But the comparison of time scales suggests that $E_D(T_{eff})$ should be much less than unity for the sample of the evolved stars with masses around $2M_\odot$ just after the onset of convection. If the same expression in Eq. (13.12) is to be applicable to both giants and subgiants, which would be pleasing but which is not based on any modeling, then $n \approx 6$ while $E_D(T_{eff})$ should increase rapidly over an interval $\Delta(B-V)$ of at least 0.2. The latter is compatible with the results for main-sequence stars with shallow convective envelopes discussed in Section 11.4. Such a high value of n may also be the reason why braking seems to be inefficient in M-type main-sequence stars (see Fig. 13.3).

A value of $n \approx 6$ and an even steeper increase in $E_D(T_{eff})$ is required for the two-component model in which the envelope and interior rotate as coupled rigid bodies. In this case the coupling time scale $\hat{t}_C$ can be at most one or two million years. In both the one- and two-component models, one cannot uniquely determine the steepness of $E_D(T_{eff})$ or the value of $\hat{t}_C$: a steeper $E_D(T_{eff})$ reduces the braking of the envelope at low values of I_{CE}/I_{RC}, but a lower value of $\hat{t}_C$ would yield the same result.

Schrijver and Pols (1993) argue that whereas the angular velocity of the envelope and interior cannot be allowed to differ by very large factors, the angular velocity of the interior of subgiants can in principle be up to four times as large as that of the envelope in some phases of the evolution. This would place very specific requirements on $E_D(T_{eff})$, the braking strength, and the coupling time scale. In that case, a strong resurgence in the surface rotation rate can be prevented even if the envelope and the interior are only weakly coupled.

The main conclusions from these exploratory studies are that (a) it is likely that the magnetic braking is weak for some time following the onset of convection in stars with masses near $1.5-2M_\odot$ and likely also in more massive stars, and (b) that coupling between the convective envelope and at least some substantial part of the radiative interior must occur for warm main-sequence stars and evolved stars with masses up to $\sim 2M_\odot$ and perhaps more.

13.3.4   *Rotational evolution of very young stars*

We now turn to the evolution of stellar rotation in very young stars. Figure 13.3 summarizes the observed rotational velocities for three young clusters. By the age of the α Persei cluster, some 50 Myr, there is a substantial spread in rotation velocities, which narrows remarkably for the late F- and G-type stars by the age of the Pleiades clusters, at only 70 Myr. The swift change implies a spin-down time scale of the order of the age difference between these two clusters, that is, only ~ 20 Myr. This is a powerful constraint on the magnetic brake. By the age of the Hyades, some 600 Myr, the spread has been reduced strongly, and a reasonably well-defined, monotonically decreasing relationship exists between the rotation rate and effective temperature, extending from approximately spectral type F5 down to late-K.

The still younger T Tauri stars with ages of up to only a few million years appear to have a bimodal distribution in their rotation periods, with a slowly rotating group with periods of 8.5 ± 2.5 days and a faster rotating group with periods of 2.2 ± 1 days (Attridge and Herbst, 1992; also Bouvier *et al.*, 1993). It has been suggested that this bimodality reflects the interaction of these stars with circumstellar disks.

The study of the evolution of very young stars up to the zero-age main sequence phase is complicated by our lack of knowledge of rotation rates at birth and by the potential coupling of the stellar magnetic field to an accretion disk during the early years of the star's existence. Another factor is that for star formation to occur at all, protostars must shed most of their initial magnetic flux and angular momentum (see, for example, Shu *et al.*, 1987). One can, however, include the determination of the initial distribution of velocities, the role of disk coupling, and the consequences of internal differential rotation, by adjusting models of the rotational history of stars to the observed distributions from the T Tauri phase, through the young cluster phases, to the current distribution of main-sequence stars.

Keppens *et al.* (1995) and Krishnamurthi *et al.* (1997) apply such fitting methods, using the Weber–Davis wind model discussed in Section 8.9 and in this chapter, or the parameterization of Eq. (13.7), respectively. Keppens *et al.* argue that, despite the many parameters in the model, the resulting constraints are rather tight and subject to relatively little ambiguity. During most of the pre-main-sequence evolution, the change in the moment of inertia dominates the rotational evolution (Fig. 13.1). They describe the coupling between the stellar radiative interior and the convective envelope using a constant time scale $\hat{t}_C$. They find that during the last phases of pre-main-sequence evolution, the balance between the magnetic braking and rotational coupling between envelope and interior dominate the evolution. The best fit is achieved for (a) a coupling time scale of $\hat{t}_C \approx 10\,\mathrm{Myr}$, (b) a dynamo in which the mean field strength increases linearly with angular velocity, but which saturates to a fixed level for rotation periods less than a little more than 1 day, and (c) a circumstellar disk for stars with initial periods above 5 days that does not persist for more than $\sim$4 Myr. Krishnamurthi *et al.* (1997) argue that the dynamo saturation occurs at a critical angular velocity that is dependent on stellar parameters, possibly occurring at a fixed Rossby number. These results compare reasonably well with simulations of the Pleiades by Soderblom *et al.* (1993) who did not include the very early phases involving disk coupling.

Specific uncertainties that remain concern, among others, the details of disk–star coupling, the temporal change of the coupling time scale $\hat{t}_C$, the possibility that the dynamo strength scales differently with angular velocity for fully convective phases of evolution, and the possible sensitivity of the dynamo efficiency to the magnitude of the shear between interior and envelope.

The convergence of surface rotation rates that occurs for ages beyond $\sim$1 Gyr is an important property of stellar rotational history. For stars of solar age, the details of the initial dispersion in rotation rates have been largely lost. This is true for the directly observable surface rotation rates, but numerical evolutionary models suggest that it is also the case for internal differential rotation. It is this reduction of the initial dispersion that allowed Skumanich to find the age–activity relationship for mature main-sequence stars discussed in Section 13.2. This convergence now appears to be the consequence of the nonlinear response of angular-momentum loss to angular velocity.

14

Activity in binary stars

It has been known for quite some time that cool stars in binaries of short orbital period exhibit strong Ca II H and K emission (see Hiltner, 1947; Gratton, 1950). Observations from space confirmed that all outer-atmospheric emissions typical for magnetic activity are present in the radiation from sufficiently close binary systems. Such binary systems are too close for their components to be observed separately. They betray their binary nature in periodical Doppler shifts; these systems are called spectroscopic binaries.

Whereas for single stars the pattern of activity across the H–R diagram can be understood as the result of stellar evolution and magnetic braking, there is no such simple pattern for active binaries. This chapter discusses the evolution of activity in close binaries and some peculiar features of active binaries. We begin with theoretical considerations of tidal interaction, which, for one thing, tries to synchronize rotational spin and orbital revolution and to maintain that synchronization. Then we discuss how the loss of angular momentum from the active component(s) by stellar wind drives the binary components together while speeding up the rotation and thereby enhancing the activity of the active component(s). Finally, we interpret the observational data on active binaries.

14.1   Tidal interaction and magnetic braking

First we consider the *equipotential surfaces* in a binary system whose components are separated. The potential around a binary system is

$$\Phi = -G \left(\frac{M_1}{r_1} + \frac{M_2}{r_2} \right) - \frac{1}{2} y^2 \omega^2, \tag{14.1}$$

where $M_{1,2}$ and $r_{1,2}$ are the mass of and distance to each component, y is the distance to the axis of revolution, and ω is the angular velocity of the orbital revolution. The first term at the right-hand side of Eq. (14.1) is the potential due to gravity, and the second is the potential due to centrifugal acceleration. On an equipotential surface, given by $\Phi =$ constant, a test particle can move freely without requiring work. Figure 14.1 shows the cross sections of equipotential surfaces with the orbital plane.

In a detached binary system, each component is surrounded by closed potential surfaces low above the star. The first common equipotential surface, named the *Roche surface*, is of particular importance in binary evolution. When one component fills its Roche lobe, the system is called *semi-detached*. If that component expands during its evolution (or

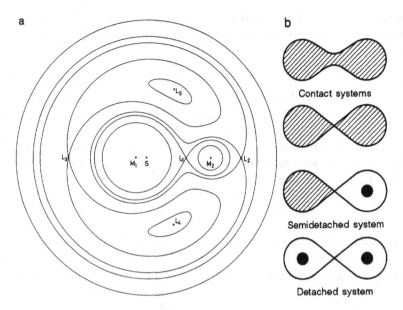

Fig. 14.1. Panel *a*: Geometry of equipotential surfaces in a binary system. The curves of $\Phi =$ constant according to Eq. (14.1) are drawn in the orbital plane for a mass ratio $M_2/M_1 = 0.17$. The Roche surface meets itself at the Lagrange point L_1. S is the center of gravity of the system, through which the axis of revolution passes. Panel *b*: Types of binary systems (from Unsöld and Baschek, 1991).

the distance between the components decreases), gas flows from that component to the other component through the first Lagrange point L_1. In most cases that gas flow does not enter the companion star directly, but rather, because of the conservation of the angular momentum, forms a rotating (accretion) disk around the companion.

If the two components completely fill a common equipotential surface, we speak of a *contact system*; the W Ursa Majoris binaries, mentioned in several preceding sections, belong to this category.

If there were no internal friction in a binary component, its tidal deformation would be symmetric about the plane determined by the axis of revolution and the line joining the centers of the two components. If there is friction and if the spin and orbital motion are not synchronized ($P_{rot} \neq P_{orb}$), or the orbital eccentricity $e \neq 0$, there is a phase shift in the tidal bulge (as for the tidal bulge in the oceans of the Earth) that breaks the symmetry. The tilted mass distribution exerts a torque on the star, leading to an exchange of angular momentum between spin and orbital motions. This torque tries to synchronize orbital motion and spin, and to circularize elliptical orbits; the magnitude of this torque depends strongly on the tidal friction. The friction is not the result of the gas-kinetic viscosity, which is much too small to bring about effects on time scales of stellar evolution, but it is caused by the turbulent viscosity in convective envelopes (Zahn, 1966).

We summarize some results of the theory of tidal interaction developed by Zahn (1977, 1989) for systems with the stellar axes of rotation parallel to the axis of revolution, and for a small difference between orbital and rotation periods.

Zahn relates the turbulent viscosity v_t to the mean velocity of the convective blobs v and their vertical mean path ℓ as computed in the mixing-length approximation (see Section 2.2):

$$v_t = v_{ML}\ell_{ML}/3 \tag{14.2}$$

[compare the turbulent diffusivity in Eq. (7.6)]. This expression is valid only if the convective turnover time ℓ/v is short compared to the tidal period – Zahn (1989) also gives estimates for v_t in the opposite case.

For a star of mass M_1 and moment of inertia I_1 about its rotation axis, with a deep convective envelope, with an orbit close to circular and a uniform rotation, the tidal torque $\mathcal{T}$ can be approximated by

$$I_1\frac{d(\Omega_1 - \omega)}{dt} = \mathcal{T} \simeq 6\frac{k_2}{\hat{t}_f}\left(\frac{M_2}{M_1}\right)^2 M_1 R_1^2 \left(\frac{R_1}{d}\right)^6 (\Omega_1 - \omega), \tag{14.3}$$

where M_2 is the mass of the companion star, R_1 is the radius of the considered star and Ω_1 is its rotational rate, d is the separation of the centers of gravity, ω is the angular velocity for orbital revolution, k_2 is the so-called apsidal constant, and $\hat{t}_f$ is the time scale for convective friction

$$\hat{t}_f = \left(\frac{M_1 R_1^2}{L_1}\right)^{1/3}, \tag{14.4}$$

where L_1 is the stellar luminosity; for the Sun, $\hat{t}_f \simeq 160$ days. This expression is valid wherever convection transports virtually the complete stellar energy flux.

Under the above limiting conditions, the synchronization time scale $\hat{t}_{syn}$ is

$$\frac{1}{\hat{t}_{syn}} \equiv -\frac{1}{\Omega_1 - \omega}\frac{d\Omega_1}{dt} = 6\frac{k_2}{\hat{t}_f}\left(\frac{M_2}{M_1}\right)^2 \frac{M_1 R_1^2}{I_1}\left(\frac{R_1}{d}\right)^6; \tag{14.5}$$

the change in the orbital period and effects of nonzero eccentricity are neglected.

The time scale $\hat{t}_{circ}$ for circularization of the orbit, in the limit of corotation ($\Omega_1 = \omega$), is given by

$$\frac{1}{\hat{t}_{circ}} \equiv -\frac{1}{e}\frac{de}{dt} = \frac{63}{4}\frac{k_2}{\hat{t}_f}\left(\frac{M_2}{M_1}\right)\left(1 + \frac{M_2}{M_1}\right)\left(\frac{R_1}{d}\right)^8, \tag{14.6}$$

where e is the eccentricity of the orbit. For the derivation of the above equations and for more general expressions, we refer to the cited papers by Zahn, to the review by Savonije and Papaloizou (1985), and references given in those papers.

Zahn (1977) estimated from Eq. (14.5) that a binary component of the age of the Sun should be in nearly synchronous rotation if the relative separation (d/R) is less than $\sim$40, with periods shorter than $\sim$25 days. From Eq. (14.6) it follows that a component of the age of the Sun should be in a circular orbit if its relative separation is less than 15, with its orbital period shorter than $\sim$5 days. A comparison of Eq. (14.5) and Eq. (14.6) shows that for detached systems the synchronization time is appreciably shorter than the circularization time. It has been concluded that Zahn's formulas imply that binaries with virtually circular orbits tend to be synchronized.

Plots of observed orbital eccentricity against the orbital period bring out the effect of circularization: main-sequence spectroscopic binaries tend to have nearly circular orbits

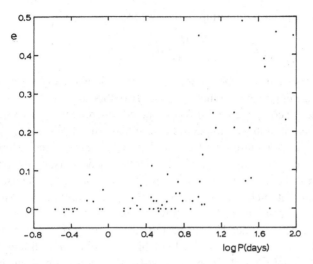

Fig. 14.2. Eccentricity e against orbital period P_{orb} for main-sequence spectroscopic binaries of spectral type F0–G9 (from Middelkoop, 1981).

for periods shorter than 8 days, whereas for longer periods the eccentricities scatter over a wide range (Koch and Hrivnak, 1981; Middelkoop, 1981). For giants, the separation between predominantly circular orbits and a broad range of eccentricities is at $P_{orb} = 120$ days (Middelkoop and Zwaan, 1981). The separations between predominantly circular orbits and the broad range of eccentricities are remarkably well defined (see Fig. 14.2), despite the differences in age of the field binaries.

There are a few exceptions to the rule that synchronization tends to precede circularization: some close binaries have been found with circular orbits that have active components rotating much more rapidly or much more slowly than the orbital rate. Habets and Zwaan (1989) studied four cases and argued that the orbits were circularized during an evolutionary phase quite distinct from the present one. In two of their cases, they suggest that it happened during an early pre-main-sequence phase when the progenitors were much bigger than the present components. During the contraction to the main sequence the circularity of the orbits was maintained, but the tidal interaction became too weak to maintain synchronous rotation. In the other two cases, circularization was reached during the red-giant phase of the present white-dwarf companion in the system when the tidal interaction was strong. After the formation of the white dwarf, circularization was maintained but synchronization was lost. In general, synchronization cannot always be maintained during phases of rapid stellar evolution.

Tidal interaction transfers angular momentum from orbital motion to rotation of the component(s) with a convective envelope, or the other way around, until the rotation of the cool component(s) is synchronized with orbital motion, and the orbits have become circular.

Loss of angular momentum from the binary system by a magnetized stellar wind (Section 13.3) from at least one of its components changes the binary evolution drastically (Mestel, 1968). Because of the tidal coupling, the angular momentum lost with the wind is drawn from the *total* angular momentum in the system,

$$\mathcal{L}_{\text{orb}} = \frac{M_1 M_2}{M_1 + M_2} d^2 \omega, \tag{14.7}$$

which in well-detached systems is some orders of magnitude larger than the spin angular momentum in the active component(s). One consequence of this large reservoir of angular momentum is that an active component with synchronized spin remains active for a much longer time than a single star of the same spectral type and luminosity class. Furthermore, the loss of orbital angular momentum reduces the distance d between the components. From Kepler's third law it follows that $\mathcal{L}_{\text{orb}} \propto P_{\text{orb}}^{1/3}$; hence P_{orb} decreases with a decreasing $\mathcal{L}_{\text{orb}}$. For synchronized component(s) with convective envelopes, P_{rot} consequently decreases with time, so that such components become increasingly active! Moreover, tidal interaction with angular-momentum loss through magnetized winds brings binary components closer together and thus greatly hastens the merging of the two stars. An early application by Verbunt and Zwaan (1981) of binary evolution with angular momentum loss by magnetic braking of a cool component showed the importance of this mechanism in the evolution and powering of low-mass X-ray binaries.

14.2   Properties of active binaries

The high activity level of many spectroscopic binaries and the occurrence of many active binaries (the so-called RS CVn binaries; see Section 14.3) among the otherwise so quiet subgiant domain of the H–R diagram has attracted a lot of attention. This section compares the properties of active binaries with those of active single stars.

Even the most active binaries closely follow the flux–flux relations between coronal, transition-region, and chromospheric emission as established by single stars (see Section 9.4 for illustrations). The active binaries define the upper part of the flux–flux relations, with a long overlap with the part occupied by single stars (see Fig. 3 in Schrijver and Zwaan, 1991). With Oranje *et al.* (1982a) we conclude that the global properties of chromospheres, transition regions and coronae are governed by one activity parameter, irrespective of whether the star is single or a component in a close binary (see Sections 9.4 and 11.3).

The study of the activity–rotation relation for binary components is facilitated by the property that circular orbits tend to indicate a synchronized rotation of the cool components (exceptions to this rule are known; see Section 14.1). The study by Rutten (1987; see Section 11.3) revealed that many active-binary components follow the activity–rotation relation as established for single stars. The relatively high activity level of those components is simply a consequence of their high rotation rates. But this does not explain some of the most active binaries: Rutten (1987) also found stars that are significantly more active than would follow from their rotation rate and the standard activity–rotation relation (as was noticed earlier by Basri *et al.*, 1985); he called these stars overactive. The great majority of these overactive stars are well-known binaries of the RS CVn type.

Attempts to relate overactivity to one binary or stellar parameter have failed. High-resolution spectrograms covering the orbital period indicate that the bulk of the active emission comes from the coolest and most evolved component, or from both components if they are similar, with the exception of Capella in which the warmest, most rapidly rotating component is the most active. Schrijver and Zwaan (1991) investigated the problem of overactivity in a principal-component analysis in a sample of binaries, using the soft X-ray flux density F_X, the orbital period P_{orb}, the binary separation d, the radii R_c and R_h, and the masses M_c and M_h; the index c stands for the active (usually cool) component, and h stands for the less active (usually hot) companion. The data were found to span a two-dimensional subspace covering 87% of the variance. A suitable two-parameter description turned out to be

$$F_X = 7.7 \times 10^7 \left(\frac{R_h}{R_\odot}\right)^{1.1} P_{orb}^{-1.4} \ (\mathrm{erg \, cm^{-2} s^{-1}}), \tag{14.8}$$

with P_{orb} in days. Note that with little loss in the quality of the fit, Eq. (14.8) can be rewritten in many different forms, because of power-law relations between stellar parameters, such as $P_{orb} \propto d^{1.34}$ and $M_h \propto R_h^{0.33}$, and there are three-parameter correlations as well.

The essence of the above outcome is that, whereas the bulk of the emissions comes from the active component, the issued flux densities depend on an *intrinsic* property of the *companion*: the companion affects the activity of the primary. Interpreting overactivity as an enhancement of the level of activity of one active component, one can combine the active single stars and overactive binaries in one relation:

$$F_X \approx 7 \times 10^6 \, P_{orb}^{-0.8} \left[1 + 8 \left(\frac{R_h}{R_\odot}\right) P_{orb}^{-0.5}\right], \tag{14.9}$$

where P is the orbital or rotational period. The factor between brackets is the enhancement factor because of overactivity, with $R_h = 0$ for single stars. The fit (Fig. 14.3) is successful in relating single and binary stars over a large range of (over)activity levels. Note that data points at the top right-hand side in Fig. 14.3 may indicate a decline of outer-atmospheric activity with increasing rotation rate that is similar to the decline of activity with a further increase of rotation rate beyond $P_{rot} = 1$ day mentioned in Section 11.3.

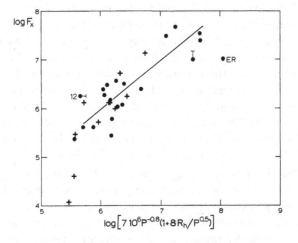

Fig. 14.3. Parameterization of the activity–rotation relationship for active single stars and (over)active binaries. The flux density in soft X-rays F_X (in erg cm^{-2}s^{-1}) is plotted against the best-fit expression in Eq. (14.9). Symbols: +, single stars; •, binaries; ER represents the very close binary ER Vul, with $P_{orb} = 0.69$ days (from Zwaan, 1991).

14.3   Types of particularly active stars and binary systems

Early efforts to find patterns in the photometric and spectroscopic phenomena in close binaries have led to a certain classification of active binaries (see Table 9.2). In this section, we discuss some of the properties of these binary types.

The type of *RS Canum Venaticorum* (RS CVn) *binaries* is not sharply defined. In his review, Hall (1981) distinguished short-period ($P_{orb} < 1$ day), regular, and long-period ($P_{orb} > 2$ weeks) RS CVn binaries. The prototype RS CVn consists of two subgiants (F4 V–IV + K0 IV), with $P_{orb} = 4.8$ days. Among the objects called RS CVn binaries, the active components are of spectral type late F, G, or early K and of luminosity class V, IV, III, or II; the orbital periods range from 0.5 to 140 days. Originally RS CVn variables were discovered by a photometric wave visible in broadband photometry; this wave is attributed to rotation modulation caused by the uneven distribution of starspots across the stellar surface (Section 12.2).

It is clear now that all RS CVn binaries are active because of the relatively rapid rotation of at least one component with a convective envelope and, conversely, that all binaries with a sufficiently active F-, G- or K-type primary component may be called RS CVn binaries.

The great majority of the RS CVn binaries have the rotation of their active components synchronized with the orbital motion (see Hall, 1981). This synchronization is inferred from the equality of the photometric and the orbital periods (except for small differences attributed to the slow migration of starspot complexes). In addition, many of the RS CVn binaries have virtually circular orbits, and a few are synchronized but not yet circularized. These are expected results of tidal interaction (Section 14.1).

The *Algol-type binaries* are eclipsing binaries, with late B-, A- and early F-type main-sequence stars as primaries, and G-, K-, or M-type subgiants or giants as secondaries. Most of these binaries are semidetached; the periods range from ~1 day to a few days. The evolved secondaries appear to be less massive than the main-sequence primaries, which is interpreted as evidence that mass transfer has been in progress already for some time.

The high activity level of the synchronized, rapidly rotating secondary is not conspicuous in the optical spectral region because of the outshining primary, except when the primary is (partly) hidden during an eclipse. The activity manifests itself all the time in intense X-ray and radio emissions, and in a high frequency of intense flaring. Singh *et al.* (1996) compared the X-ray emissions from Algol systems with those from RS CVn binaries, and they tested several possible relationships between X-ray emission parameters and binary parameters. They found a highly significant correlation between the X-ray flux density at the stellar surface of the supposedly active component and rotational (= orbital) period for both types of binaries. The Algol binaries tend to be ~3–4 times less bright compared to RS CVn systems with the same rotation/revolution period. Singh *et al.* (1996) concluded that this refutes a suggestion in the literature that the excess emission from RS CVn systems could be due to ongoing mass transfer, because in that case the Algol systems would be expected to be brighter than RS CVns.

Comparing the X-ray surface fluxes from Algol systems listed by Singh *et al.* (1996) with those of single stars, we find that semidetached systems are markedly overactive, but indeed less so than many of the RS CVn binaries.

The prototype of the *BY Draconis stars* and ~70% of the other members of that group are close binaries, but others are not. The BY Dra variables are chromospherically active K- and M-type main-sequence stars, with broadband light variations attributed to starspots (Vogt, 1975); the implied rotational periods are in the range 1–5 days. Apparently a high rotation rate, with $v_e > 5$ km/s, is the prerequisite for the BY Dra syndrome (Bopp and Fekel, 1977); this high rate suggests that the stars are young or tidally locked in a short-period binary system.

The type of BY Dra variables overlaps with that of the *UV Ceti stars* (or flare stars; see Section 12.4), and, quite in general, with the dKe and dMe emission-line main-sequence stars.

The *W Ursae Majoris* (W UMa) *binaries* are contact binaries consisting of cool main-sequence stars, with orbital periods shorter than 1 day (see Rucinski, 1994 and references given there). Their mass ratios M_2/M_1 range from 0.5 to 0.1. Despite the different masses, the optical spectra of the two components are similar. It is believed that the two components are contained in a common envelope, and that the more massive component generates most of the energy flux, which it then shares with the less massive component through the convective envelope (see Rucinski and Vilhu, 1983).

The W Uma systems are believed to evolve toward lower mass ratios and higher effective temperatures; there is loss of angular momentum through magnetic braking. The final state is probably a rapidly rotating single star as the more massive star engulfs its companion.

Stars of the *FK Comae Berenices* (FK Com) type are extremely rapidly rotating late F-, G- and early K-type giants (Bopp and Stencel, 1981). These are single stars, probably formed by a fusion of binary components, so representing the final phase of the binary fusion that is at work in the W Uma contact binaries. The FK Com stars are among the most active stars; they fit the flux–flux relations for transition-region and chromospheric emissions (Bopp and Stencel, 1981; Zwaan, 1983); their soft X-ray emission appears to deviate from the flux–flux relations. The FK Com stars are the only (apparently) single stars that are significantly overactive (Rutten, 1987).

15

Propositions on stellar dynamos

In this final chapter we present a synopsis of the observational constraints on dynamo processes in stars with convective envelopes that complements our review of studies of the solar dynamo in Chapter 7. We do not try to summarize the rapidly growing literature on mathematical and numerical models of stellar dynamos, but rather we attempt to capture the observational constraints on dynamos in a set of propositions, following Schrijver (1996). You will encounter some speculative links that attempt to bring together different facets of empirical knowledge, but we shall always distinguish conclusions from hypotheses.

Throughout this book, we use the term dynamo in a comprehensive sense, implying the ensemble of processes leading to the existence of a magnetic field in stellar photospheres, which evolves on times scales that are very short compared to any of the time scales for stellar evolution or for large-scale resistive dissipation of magnetic fields. Such a dynamo involves the conversion of kinetic energy in convective flows into magnetic energy.

Solar magnetic activity is epitomized by the existence of small-scale (compared to the stellar surface area), long-lived (compared to the time scale of the convective motions in the photosphere), highly structured magnetic fields in the photosphere, associated with nonthermally heated regions in the outer atmosphere, in which the temperatures significantly exceed that of the photosphere. Other cool stars exhibit similar phenomena, which are collectively referred to as stellar magnetic activity. We define the strength of the dynamo on an empirical basis, as some monotonically increasing function of the outer-atmospheric radiative losses.

> **I.** *A dynamo resulting in magnetic activity appears to operate in all rotating stars with a convection zone directly beneath the photosphere. In (effectively) single stars, the dynamo strength is a monotonic function of rotation rate, at least for stars with rotation periods exceeding about two days. This function appears to be continuous as far as changes are concerned that exceed the intrinsic scatter associated with stellar variability.*

Signatures of magnetic activity are observed in the warm F-type stars from very near the boundary in the Hertzsprung–Russell diagram where spectroscopic signatures of convection are detected (Section 11.4), down to the coolest stars along the main sequence (Section 11.5). In very cool, extremely slowly rotating, single giant stars, the magnetic activity is weak at best (Section 11.1). Some form of a dynamo is suggested to operate

344

even in these stars, however, so that their cool wind can be accelerated by Alfvén waves (Section 12.5).

Some minimal rotation rate may be required to excite a dynamo that can produce appreciable magnetic activity (Section 11.3), but diagrams of, for example, chromospheric radiative losses as a function of stellar effective temperature suggest that this minimal rate lies below the current spectroscopic and photometric detection limits for stellar rotation: the outer-atmospheric activity decreases smoothly to the minimal basal level as stellar rotation slows down (Section 11.3), subject to magnetic braking and to the increase in the moment of inertia as the star evolves (Chapter 13).

Magnetic activity is weak in stars with very shallow convective envelopes, probably because of a weak dynamo rather than a reduced atmospheric response to the magnetic field. Such a reduced atmospheric response might occur because for stars with convective envelopes that are only a few pressure scale heights deep, the convective collapse (Section 4.6), that in other stars leads to the formation of intrinsically strong fields, is probably not fully effective. The rapid, smooth increase in dynamo efficiency with decreasing effective temperature up to stars like the Sun makes it clear, however, that this effect is secondary to the change in dynamo efficiency:

> **II.** *Atmospheric radiative losses associated with magnetic activity observed for main-sequence stars of a given rotation period – and by inference also the dynamo efficiency – increase rapidly with an increasing depth of the convective envelope from the onset of convection up to $B-V \approx 0.6$. A similar increase is inferred for (sub)giants.*

The efficiency function $g(B-V)$ that describes the dynamo efficiency of a given mean surface rotation period (Section 11.4) has a color dependence for main-sequence stars that resembles that which the mixing-length hypothesis predicts for the convective turnover time $\hat{t}_c$ near the bottom of the convective envelope (Section 11.3). However, the results of numerical simulations of convection in strongly stratified media (Section 2.5) appear to be at odds with the intuitive notion of overturning eddies and with the simple dimensional arguments upon which mixing-length theory is based. Hence, the question is what $g(B-V)$ actually measures; it may be sensitive to some other parameter that resembles $\hat{t}_c$ (Section 11.4). Any acceptable dynamo model should be qualitatively consistent with the surprising conclusion that whereas signatures of a magnetic dynamo are observed in the Hertzsprung–Russell diagram just redward of the point where convection becomes the dominant mechanism of energy transport through the stellar envelope (Section 11.4), the dynamo reaches its peak efficiency not until the convective envelope encompasses more than half the stellar volume.

Studies of magnetic braking (Section 13.3) and variability (Section 12.7) suggest that a very similar efficiency function for the dynamo strength is applicable to both main-sequence stars and evolved stars of moderate to high activity levels. For the evolved stars, this efficiency function may have to be scaled by a luminosity-class-dependent function [Section 12.7, below Eq. (12.3)] that increases from unity to no more than 4 from the main sequence to a luminosity class III giant star. This efficiency function is *not* compatible with the convective turnover time $\hat{t}_c(T_{\text{eff}}, g)$ computed for (sub-)giants. No scaling appears to be required for the less active stars, where giants and main-sequence stars lie on the same relationship between radiative flux densities and rotation period (Section 11.3).

The partial success of the Rossby parameter $\mathcal{R}_o$ as the controlling activity parameter for main-sequence stars is puzzling, because after all, the dynamo may well be a truly three-dimensional process – as conceptual models discussed in Chapter 7 have it – rather than being restricted to a thin (boundary) layer characterized by a single number. Perhaps we shouldn't expect it to function as the sole parameter for all cool stars. Hence:

III. *The dynamo strength is a function of the stellar rotation rate (hence for single stars an indirect function of stellar age); it must also depend on some property of the convective envelope, but there is no unequivocal evidence that the dynamo efficiency scales uniquely with the Rossby number for the entire sample of cool stars.*

The occurrence of magnetic activity in stars with very different internal structures and rotation rates does not necessarily require that their dynamos are characterized by similar spatiotemporal patterns. Even stars of the same internal structure and rotation rate may differ considerably in their dynamo patterns, at least temporarily, as the Maunder Minimum in the history of solar activity (Section 6.1.2) demonstrates. The complexity of the dynamo response to nonlinearities in the governing equations has been demonstrated by Jennings and Weiss (1991) by means of an "extremely simple model" of a mean-field dynamo. The dynamo strength S, measured by the magnetic energy density in the mean field, shows a complex behavior as it increases with dynamo number N_D through a series of bifurcations between different modes. The function $S(N_D)$ shows a nonmonotonic behavior, whereas in several ranges of N_D there are multiple values for S that are associated with stable (oscillatory) solutions, and there are several transitions in the symmetries of the solutions, including dipole, quadrupole, and mixed-mode solutions. With increasing nonlinearity, asymmetries become more pronounced, the spatial structure more complicated, and the temporal behavior more irregular. The model allows for asymmetries in which almost all activity can be concentrated on a single hemisphere, as happened during the solar Maunder Minimum (Section 6.1.2).

Other dynamo models also suggest that the dynamo process can exhibit different modes of operation within the full range of governing parameters. Whereas observations do not rule out such transitions from one dynamo mode to another as, for example, the rotation rate changes, the continuity of the rotation–activity relationship does require that these transitions are continuous in the magnitude of atmospheric activity, at least to within the substantial scatter about the mean relationship, as reflected in proposition I.

The dynamo number N_D includes the angular velocity, the differential rotation, and the dispersal of magnetic field. Translating this representation of N_D into stellar properties is not straightforward: Eqs. (11.8) and (11.9) show two different translations, both involving the Rossby number $\mathcal{R}_o$, but differing in the inclusion of the relative depth of the convective envelope. The measure for differential rotation in these models, involving either the radial or latitudinal shear of the differential rotation, or both, cannot currently be reliably derived from observations and requires some a priori parameterization before model properties can be investigated. We point out that the surface differential rotation appears to decrease in relative magnitude as stars rotate more and more rapidly (Section 12.7), but the time it takes a field line to be wrapped once around the star depends only weakly on the stellar rotation period [Eq. (12.4)]. The meridional flow appears to be important as a second (perhaps largely dependent) controlling parameter, leading to potentially complex bifurcation patterns in a parameter space that is at least three dimensional (e.g., Jennings and Weiss, 1991). This flow cannot at present be derived for stars other than the Sun.

There are two groups of stars that are (nearly) completely convective: the late K- and M-type giants, and the M-type dwarfs. A boundary-layer dynamo is not an acceptable model for the dynamo operating in these stars. In Sections 7.3 and 7.4 we consider the theory of dynamo action in deep convective layers, and in Chapters 11, 12, and 13 we discuss the empirical aspects of magnetic activity in such late-K- and M-type stars. The main conclusions are summarized in the next two paragraphs.

In single red giants, magnetic activity is at an extremely low level: the outer atmospheric emissions are reduced to the minimum basal flux densities. The magnetic structure of the atmospheres is believed to consist of small bipoles and a weak radial field, which is invoked to generate Alfvén waves to power the cool stellar winds (Chapter 13).

Among the single, often relatively rapidly rotating, late K- and M-type dwarfs, in contrast, many stars are quite active. The starspots and the large and frequent flares on dMe stars suggest that large, strong active regions occur, although possibly in more complex patterns. The latter may be caused by a smaller degree of ordering in the active regions of dM stars than is the case for the Sun. The boundary-layer dynamo developed for the Sun (Chapter 7) was primarily proposed because of the large degree of order that is observed for the emerging active regions: the Hale–Nicholsen and Joy rules (Table 6.1) require strong fields that have to be amplified and stored below the surface for a substantially longer time than the field's buoyancy in the convective envelope allows, because otherwise the rotational shear would be insufficient to impose the regularity of the E–W orientation of active regions. In (nearly) fully convective stars, this ordering may be much weaker.

The transition in mean activity level when comparing partially convective stars to fully convective stars appears to be smooth, particularly for the fairly rapidly rotating main-sequence stars. This suggests a smooth transition from one dynamo mode to another, or perhaps it is a changing mixture of three processes occurring side by side: the boundary-layer dynamo mode (Sections 7.1 and 7.2), the deep-envelope dynamo mode (Section 7.4), and the turbulent dynamo mode (Section 7.3). In strictly Sunlike stars, the dynamo is likely dominated by a mode that depends on the existence of a boundary layer between the radiative interior and the convective envelope. As the convective envelope deepens, another mode may begin to dominate in which much of the dynamo action occurs in the deep interior, but still relies on (differential) rotation (see Section 7.4, and also Section 11.5), or perhaps – as Durney *et al.* (1993) argued, but which we find less convincing – there is a transition to a turbulent dynamo that relies on the generation of a (chaotic) magnetic field throughout the convective envelope with no need for differential rotation. For evolved stars in which rotation dies away when the convective envelope becomes very deep, the decay of the large-scale dynamo apparently leaves at most a very weak turbulent dynamo beyond the Linsky–Haisch dividing line.

Among the main-sequence stars, there are several patterns of long-term variability in their activity (Section 11.7); only a fraction of these stars show cyclic variations that resemble that of the Sun. The present (1999) picture may be summarized as follows:

IV. *For weakly to moderately active G- and K-type main-sequences stars – which are below the Vaughan–Preston gap – activity tends to be cyclic, but no clear trend of cycle period with other stellar parameters, such as effective temperature T_{eff} or convective turnover time $\hat{t}_c$, has been found. For the most active stars several variability patterns exist, but generally no unambiguous activity cycle is seen.*

It is possible that a cyclic variation with periods of many decades is hidden in the long-term photometric trends shown by some active stars (Section 11.7).

Throughout their activity variations on time scales up to decades, relatively active main-sequence stars never reach low states of activity (Section 11.7), which suggests that the dynamo patterns are more complicated than the (slightly overlapping) sunspot cycle seen on the Sun. Perhaps multiple sunspot cycles are present simultaneously on the surface of these active stars, preventing them from ever appearing inactive.

For the most active stars, the kinematic dynamo description is likely to fail altogether, as significant changes of the large-scale flows and circulations are to be expected because flows are countered by the strong magnetic field that threads the entire convective envelope. This may explain saturation of activity:

V. *The outer-atmospheric radiative losses appear to saturate for stars with rotation periods below approximately 1 or 2 days, but it remains unclear whether this reflects a saturation (a) in the dynamo action (by a backreaction of the field on the flows), (b) in the heating mechanism (either because driving motions are suppressed by the field or because the maximum available power has been exploited), or (c) in the total absolute magnetic flux (which may saturate because of increased flux cancellation rates between opposite polarities with increasing activity).*

The evidence for saturation comes not only from the study of outer-atmospheric radiative losses (with filling factors for the intrinsically strong field approaching unity, Section 12.2) but also from studies of magnetic braking (Section 13.3), particularly in single stars in young clusters. This suggests that it is not simply the heating in the outer atmosphere that limits the radiative losses associated with activity. We cannot rule out the interpretation, however, that the braking saturates because the coronal field is almost completely closed (see the discussion on the magnetospheric dead zone for the wind in Section 13.3) at approximately the same period as the heating mechanism reaches saturation. Note that the reduction of activity in the tidally interacting (contact) W Uma systems (Sections 11.3 and 14.3) may have a different origin than the leveling off and apparent downturn of the rotation–activity relationship for the most rapidly rotating single stars in young clusters, even though the occurrence of this "supersaturation" at about the same rotation period suggests a common cause.

We now turn to the coupling of rotation and activity through stellar evolution.

VI. *Outer-atmospheric activity is accompanied by a loss of stellar angular momentum, which has been demonstrated to be an important process in the deceleration of stellar rotation of stars from the main sequence up to and including at least luminosity class III.*

The less efficient braking of warm stars with shallow envelopes is readily attributed to the low dynamo efficiency for these stars. Moreover, the emerging bipoles are likely to be small because of the shallowness of the convective envelope. Whereas the mass loss rate may be large (Section 11.4), angular momentum is lost less efficiently because of the reduced arm over which the wind exerts its torque.

The relative inefficiency of the magnetic brake in the very cool M-type dwarf stars (Section 13.2) may be attributed to either or both of two effects: (a) it may be the result of the small stellar radius and the consequently strongly reduced efficiency of the brake (Section 13.3), or (b) possibly, but less likely, to the fact that a turbulent dynamo operates in these stars associated with small-scale fields (e.g., Durney *et al.*, 1993). Note that in addition to the reduced rate of angular momentum loss, the rotation rate of these stars is sustained for a relatively long time because of the long phase of contraction toward the main sequence (see Fig. 13.1).

The potential of a transition from a predominantly globally ordered dynamo to a turbulent dynamo in very slow rotators could be linked to the driving of the cool wind (Section 12.5) of evolved stars that lie beyond the corona-wind dividing line (Section 11.1).

VII. *The boundary line in the Hertzsprung–Russell diagram marking the onset of granulation – and thus of convection – in the stellar atmosphere appears to be the only separator of a fundamental nature where the stellar dynamo is concerned; other proposed dividing lines delineate rapid but continuous changes resulting from angular-momentum loss through the stellar magnetosphere, from changes in the moment of inertia associated with evolution, or from differences in the internal stellar structure.*

The overactivity of stars in tidally interacting binaries (Section 14.2) leads to the following proposition:

VIII. *The presence of a companion star enhances the dynamo action in cool components of tidally interacting binaries not only by enforcing rapid rotation through tidal forces but also by some additional interaction. This overactivity may be caused by the influence of the tidal interaction on the (pattern of) differential rotation.*

Whereas in a single star the convective envelope spins down because of magnetic braking, the envelopes of active components in tidally interacting binaries are accelerated by the tidal torque as the components spiral toward each other (Section 14.1). It remains a mystery, however, how the differences between accelerating and decelerating torques in close binaries and effectively single stars affect the differential rotation; we remind the reader that in the case of the Sun, the decelerating torque leads to the counterintuitive pattern in the differential rotation in which the equatorial zone rotates most rapidly.

We conclude with some observations on the geometry of the dynamo:

IX. *In the most rapidly rotating stars, generally members of tidally interacting binaries, starspots occur at high latitudes if not at the poles. In systems with orbital periods below ∼5 days, preferred longitudes exist, often at the quadrature points (i.e., on a line perpendicular to the plane defined by the stellar rotation axis and the line connecting the stars). No such preferred longitudes exist for longer-period systems.*

Dynamo models allow changes in the geometrical properties of the dynamo with rotation rate. A consistent model should also explain the existence of high-latitude starspots in rapidly rotating stars that are likely the result of flux emergence *in situ* near the poles. In stars that are not almost fully convective, this has been proposed to be the consequence of strong Coriolis forces that could deflect rising flux toward high latitudes. In stars with relatively small radiative interiors, in contrast, it could be caused by a poleward slip of deep segments of flux rings (Section 12.2).

Presumably, the field geometry becomes more and more complicated as activity increases. Nevertheless, the stellar outer-atmospheric radiative losses adhere to the flux–flux relationships (Section 9.4), with exceptions for only a few classes of cool stars.

> **X.** *The time-dependent distributions of the magnetic field over the stellar surface and the associated radiative losses always allow the transformation of locally valid relationships to comparable relationships valid for stars as a whole.*

This remarkable property indicates that rather strict rules underlie the surface distributions of magnetic configurations (active regions, network, and so on) over the stellar surface. This suggests that there are similarities in the spatiotemporal patterns in the dynamos of a great variety of stars with convective envelopes.

Appendix I: Unit conversions

Quantity	Symbol	mksa units		Gaussian units (or other)
Length	ℓ	7.25×10^5 m †	1	arcsecond
	$R_\odot$	6.96×10^8 m	1	solar radius
Mass	$M_\odot$	1.99×10^{30} kg	1	solar mass
Gravit. acceler.	$g_\odot$	274 m/s^2	2.7×10^4	cm/s^2
Force	F	1 newton (N)	10^5	dynes (dyn)
Pressure	P	1 pascal (Pa)	10	dyn/cm^2
Temperature	T	1 kelvin (K)	1/11,600	eV
Energy	E	1 joule (J)	10^7	ergs
		1.6×10^{-19} J	1	eV
Power	P	1 watt (W)	10^7	ergs/s
	$L_\odot$	3×10^{26} W	1	solar luminosity
Flux density	F	1 W/m^2	10^3	ergs s^{-1} cm^{-2}
Current	I	1 ampere	3×10^9	statamperes
Electric field	E	1 V/m	$\frac{1}{3} \times 10^{-4}$	statvolt/cm
Conductivity	σ	mho/m	9×10^9	s^{-1}
Resistance	R	ohm	$\frac{1}{9} \times 10^{-11}$	s/cm
Magnetic flux	Φ	1 weber (Wb)	10^8	maxwell (Mx)
Magnetic induction	B	1 tesla (T)	10^4	gauss (G)

† The apparent solar radius changes by about 32 arcsec throughout the year owing to the eccentricity of the Earth's orbit. Consequently, 1 arcsec $= (7.25 \pm 0.12) \times 10^5$ m.

Bibliography

Achmad, L., de Jager, C., and Nieuwenhuijzen, H.: 1991, *Astron. Astrophys.* **250**, 445

Acton, L. W.: 1978, *Astrophys. J.* **225**, 1069

Albregtsen, F. and Maltby, P.: 1978, *Nature* **274**, 41

Allen, C. W.: 1972, *Astrophysical quantities*, Athlone Press, Univ. of London, London, U.K.

Altschuler, M. D. and Newkirk, G.: 1969, *Sol. Phys.* **9**, 131

Anderson, L. S.: 1989, *Astrophys. J.* **339**, 558

Antiochos, S. K. and Dahlburg, R. B.: 1997, *Sol. Phys.* **174**, 5

Antiochos, S. K., Haisch, B. M., and Stern, R. A.: 1986, *Astrophys. J.* **307**, L55

Antiochos, S. K. and Noci, G.: 1986, *Astrophys. J.* **301**, 440

Antiochos, S. K., Shoub, E. C., An, C. H., and Emslie, A. G.: 1985, *Astrophys. J.* **298**, 876

Aschwanden, M. J. and Benz, A. O.: 1997, *Astrophys. J.* **480**, 825

Astronomer Royal: 1925, *Mon. Not. R. Astron. Soc.* **85**, 553

Athay, R. G.: 1976, *The solar chromosphere and corona: quiet Sun*, D. Reidel Publ. Cie., Dordrecht, Holland

Athay, R. G. and Dere, K. P.: 1990, *Astrophys. J.* **358**, 710

Athay, R. G. and White, O. R.: 1979, *Astrophys. J.* **229**, 1147

Attridge, J. M. and Herbst, W.: 1992, *Astrophys. J., Lett.* **398**, 61

Axford, W. I. and McKenzie, J. F.: 1991, in *Solar Wind Seven*, Proceedings of the 3rd COSPAR Colloquium, p. 1

Axford, W. I. and McKenzie, J. F.: 1997, in J. R. Jokipii, C. P. Sonnett, and M. S. Giampapa (Eds.), *Cosmic winds and the heliosphere*, University of Arizona Press, Tucson, Arizona, p. 31

Ayres, T. R.: 1979, *Astrophys. J.* **228**, 509

Ayres, T. R.: 1981, *Astrophys. J.* **244**, 1064

Ayres, T. R.: 1991a, in P. Ulmschneider, E. Priest, and R. Rosner (Eds.), *Mechanisms of Chromospheric and Coronal Heating*, Springer-Verlag, Heidelberg, 228

Ayres, T. R.: 1991b, *Astrophys. J.* **375**, 704

Ayres, T. R., Linsky, J. L., Vaiana, G. S., Golub, L., and Rosner, R.: 1981a, *Astrophys. J.* **250**, 293

Ayres, T. R., Marstad, N. C., and Linsky, J. L.: 1981b, *Astrophys. J.* **247**, 545

Ayres, T. R., Testerman, L., and Brault, J. W.: 1986, *Astrophys. J.* **304**, 542

Babcock, H. W.: 1961, *Astrophys. J.* **133**, 572

Baldwin, J. E.: 1996, in K. G. Strassmeier and J. L. Linsky (Eds.), *Stellar Surface Structure*, IAU Symp. 176, Kluwer Academic Publishers, Dordrecht, The Netherlands, p. 139

Baliunas, S. L., Donahue, R. A., Soon, W. H., Horne, J. H., Frazer, J., Woodard-Eklund, L., Bradford, M., Rao, L. M., Wilson, O. C., Zhang, Q., Bennett, W., Briggs, J., Carroll, S. M., Duncan, D. K., Figueroa, D., Lanning, H. H., Misch, A., Mueller, J., Noyes, R. W., Poppe, D., Porter, A. C., Robinson, C. R., Russell, J., Shelton, J. C., Soyumer, T., Vaughan, A. G., and Whitney, J. H.: 1995, *Astrophys. J.* **438**, 269

Baliunas, S. L., Nesme-Ribes, E., Sokoloff, D., and Soon, W. H.: 1996, *Astrophys. J.* **460**, 848

Baliunas, S. L. and Vaughan, A. H.: 1985, *Ann. Rev. Astron. Astrophys.* **23**, 379

Baliunas, S. L., Vaughan, A. H., Hartmann, L., Middelkoop, F., Mihalas, D., Noyes, R. W., Preston, G. W., Frazer, J., and Lanning, H.: 1983, *Astrophys. J.* **275**, 752

Balke, A. C., Schrijver, C. J., and Zwaan, C.: 1993, *Sol. Phys.* **143**, 215

Balthasar, H., Vázques, M., and Wöhl, H.: 1986, *Astron. Astrophys.* **185**, 87

Barbera, M., Mecela, G., Sciortino, S., Harnden, F. R., and Rosner, R.: 1993, *Astrophys. J.* **414**, 846

Basri, G.: 1987, *Astrophys. J.* **316**, 377
Basri, G. and Bertout, C.: 1989, *Astrophys. J.* **341**, 340
Basri, G., Laurant, R., and Walter, F. M.: 1985, *Astrophys. J.* **298**, 761
Basri, G. and Marcy, G. W.: 1995, *Astron. J.* **109**, 762
Basri, G., Marcy, G. W., and Valenti, J. A.: 1992, *Astrophys. J.* **390**, 622
Basri, G. S., Marcy, G. W., and Valenti, J.: 1990, *Astrophys. J.* **360**, 650
Bastian, T. S.: 1996, in Y. Uchida, T. Kosugi, and H. S. Hudson (Eds.), *Magnetodynamic Phenomena in the Solar Atmosphere*, IAU 153 Colloquium, Kluwer Acad. Publ., Dordrecht, Holland, p. 259
Bastian, T. S., Benz, A. O., and Gary, D. E.: 1998, *Ann. Rev. Astron. Astrophys.* **36**, 131
Beasley, A. J. and Cram, L. E.: 1993, *Astrophys. J.* **417**, 157
Becker, U.: 1955, *Z. Astrophys.* **37**, 47
Beckers, J. M.: 1977, in A. Bruzek and C. J. Durrant (Eds.), *Illustrated glossary for solar and solar-terrestrial physics*, D. Reidel Publ. Cie., Dordrecht, Holland, p. 21
Beckers, J. M. and Morrison, R. A.: 1970, *Sol. Phys.* **14**, 280
Beckers, J. M. and Schröter, E. H.: 1968, *Sol. Phys.* **4**, 142
Belcher, J. W. and MacGregor, K. B.: 1976, *Astrophys. J.* **210**, 498
Benz, A. O., Conway, J., and Güdel, M.: 1998, *Astron. Astrophys.* **331**, 596
Benz, A. O., Kosugi, T., Aschwanden, M. J., Benka, S. G., Chupp, E. L., Enome, S., Garcia, H., Holman, G. D., Kurt, V. G., and Sakao, T.: 1994, *Sol. Phys.* **153**, 33
Berger, M. A.: 1991, in P. Ulmschneider, E. R. Priest, and R. Rosner (Eds.), *Mechanisms of Chromospheric and Coronal Heating*, Springer-Verlag, Berlin, p. 570
Berger, M. A.: 1994, in G. H. J. Van den Oord (Ed.), *Fragmented energy release in Sun and stars—The interface between MHD and plasma physics*, Space Science Rev. **68**, p. 3
Berger, T. E.: 1996, *Ph.D. thesis*, Stanford University, Stanford, Ca.
Berger, T. E., De Pontieu, B., Schrijver, C. J., and Title, A. M.: 1999, *Astrophys. J., Lett.* **519**, 97
Berger, T. E., Löfdahl, M. G., Shine, R. A., and Title, A. M.: 1998, *Astrophys. J.* **495**, 973
Berger, T. E. and Title, A. M.: 1996, *Astrophys. J.* **463**, 365
Bertout, C.: 1989, *Ann. Rev. Astron. Astrophys.* **27**, 351
Bertout, C., Basri, G., and Bouvier, J.: 1988, *Astrophys. J.* **330**, 350
Bhatia, A. K. and Kastner, S. O.: 1999, *Astrophys. J.* **516**, 482
Bidelman, W. P.: 1954, *Astrophys. J., Suppl. Ser.* **1**, 175
Biermann, L.: 1941, *Vierteljahrschrift Astron. Ges.* **76**, 194
Biermann, L.: 1946, *Naturwiss.* **33**, 118
Biermann, L.: 1986, *Max-Planck-Gesellschaft Berichte und Mitteilungen* **2/88**, 63
Bogart, R. S.: 1987, *Sol. Phys.* **110**, 23
Bogdan, T. J., Brown, T. M., Lites, B. W., and Thomas, J. H.: 1993, *Astrophys. J.* **406**, 723
Böhm-Vitense, E.: 1958, *Z. Astrophys.* **46**, 108
Böhm-Vitense, E.: 1989a, *Introduction to stellar astrophysics. I. Basic Stellar Observations and Data*, Cambridge Univ. Press, Cambridge UK
Böhm-Vitense, E.: 1989b, *Introduction to stellar astrophysics. II. Stellar Atmospheres*, Cambridge Univ. Press, Cambridge UK
Böhm-Vitense, E.: 1989c, *Introduction to stellar astrophysics. III. Stellar Structure and Evolution*, Cambridge Univ. Press, Cambridge UK
Bohn, H. U.: 1981, *Ph.D. thesis*, University of Wuerzburg, Germany
Bohn, H. U.: 1984, *Astron. Astrophys.* **136**, 338
Bommier, V., Degl'Inoocnti, E. Landi, Leroy, J. L., and Sahal-Bréchot, S.: 1994, *Sol. Phys.* **154**, 231
Bopp, B. W. and Fekel, F.: 1977, *Astron. J.* **82**, 490
Bopp, B. W. and Stencel, R.: 1981, *Astrophys. J.* **247**, L131
Bouvier, J.: 1990, *Astron. J.* **99**, 946
Bouvier, J., Cabrit, S., Fernandez, M., Martin, E. L., and Matthews, J. M.: 1993, *Astron. Astrophys.* **101**, 485
Bowyer, S. and Malina, R. F.: 1991, in R. F. Malina and S. Bowyer (Eds.), *Extreme Ultraviolet Astronomy*, Pergamon Press, Oxford, 397
Brants, J. J.: 1985a, *Sol. Phys.* **95**, 15
Brants, J. J.: 1985b, *Sol. Phys.* **98**, 197
Brants, J. J.: 1985c, *Observational study of the birth of a solar active region*, PhD Thesis, Utrecht University
Brants, J. J. and Steenbeek, J. C. M.: 1985, *Sol. Phys.* **96**, 229
Brants, J. J. and Zwaan, C.: 1982, *Sol. Phys.* **80**, 251

Braun, D. C. and Fan, Y.: 1998, *Astrophys. J.* **508**, L105
Braun, D. C., LaBonte, B. J., and Duvall, T. L.: 1990, *Astrophys. J.* **354**, 372
Bray, R. J., Cram, L. E., Durrant, C. J., and Loughhead, R. E.: 1991, *Plasma loops in the solar corona*, Cambridge University Press, Cambridge, U.K.
Bray, R. J. and Loughhead, R. E.: 1964, *Sunspots*, Pitman Press, Bath, U.K.
Bray, R. J. and Loughhead, R. E.: 1974, *The solar chromosphere*, Chapman and Hall, London
Bray, R. J., Loughhead, R. E., and Durrant, C. J.: 1984, *The solar granulation (2nd ed.)*, Cambridge University Press, Cambridge
Brekke, P.: 1997, in *Proceedings of the ASPE conference, Preveza, Greece*
Brickhouse, N. S.: 1996, in S. Bowyer and R. F. Malina (Eds.), *Astrophysics in the Extreme Ultraviolet*, Kluwer Academic Publishers, Drodrecht, The Netherlands, p. 105
Brickhouse, N. S., Raymond, J. C., and Smith, B. W.: 1995, *Astrophys. J., Suppl. Ser.* **97**, 551
Brouwer, M. P. and Zwaan, C.: 1990, *Sol. Phys.* **129**, 221
Brown, A., Tjin a Djie, H. R. E., and The, P. S.: 1986, *ESA SP* **263**, 173
Brown, A., Vealé, A., and Judge, Ph.: 1990, *Astrophys. J.* **361**, 220
Brown, J. C., Correia, E., Fárník, F., Garcia, H., Hénouw, J.-C., Rosa, T. N. La, Machado, M. E., Nakajima, H., and Priest, E. R.: 1994, *Sol. Phys.* **153**, 19
Brummell, N. H., Hurlburt, N. E., and Toomre, J.: 1996, *Astrophys. J.* **473**, 494
Brummell, N. R., Cattaneo, F., and Toomre, J.: 1995, *Science* **269**, 1370
Brunner, W.: 1930, *Astron. Mitt. Zürich* **13**, 67
Bruzek, A.: 1967, *Sol. Phys.* **2**, 451
Bruzek, A.: 1969, *Sol. Phys.* **8**, 29
Bruzek, A. and Durrant, C. J.: 1977, *Illustrated glossary for solar and solar-terrestrial physics*, D. Reidel Publ. Cie., Dordrecht, Holland
Buchholz, B. and Ulmschneider, P.: 1994, in J.-P. Caillault (Ed.), *Cool Stars, Stellar Systems, and the Sun*, 8th Cambridge Workshop (Athens, GA), ASP Conf. Series, p. 363
Bumba, V.: 1965, in R. Lüst (Ed.), *Stellar and Solar Magnetic Fields*, IAU Symp. No. 22, North-Holland Publ. Cie, Amsterdam, 305
Bumba, V. and Howard, R.: 1965, *Astrophys. J.* **141**, 1502
Bumba, V. and Suda, J.: 1984, *Bull. Astron. Inst. Czech.* **35**, 28
Burrows, A., Hubbard, W. B., Saumon, D., and Lunine, J. I.: 1993, *Astrophys. J.* **406**, 158
Butler, C. J.: 1996, in K. G. Strassmeier and J. L. Linsky (Eds.), *Stellar Surface Structure*, IAU Symp. 176, Kluwer Academic Publishers, Dordrecht, The Netherlands, p. 423
Buurman, J.: 1973, *Astron. Astrophys.* **29**, 329
Byrne, P. B.: 1983, in P. B. Byrne and M. Rodono (Eds.), *Activity in Red-Dwarf Stars*, IAU Colloq. No. 71, D. Reidel, Dordrecht, The Netherlands, p. 157
Byrne, P. B.: 1989, *Sol. Phys.* **121**, 61
Byrne, P. B.: 1993, *Astron. Astrophys.* **278**, 520
Byrne, P. B.: 1996, in K. G. Strassmeier and J. L. Linsky (Eds.), *Stellar Surface Structure*, IAU Symp. 176, Kluwer Academic Publishers, Dordrecht, The Netherlands, p. 299
Byrne, P. B., Eibe, M. T., and Rolleston, W. R. J.: 1996, *Astron. Astrophys.* **311**, 651
Caligari, P., Moreno-Insertis, F., and Schüssler, M.: 1995, *Astrophys. J.* **441**, 886
Caligari, P., Schüssler, M., and Moreno-Insertis, F.: 1998, *Astrophys. J.* **502**, 481
Callier, A., Chauveau, F., Hugon, M., and Rösch, J.: 1968, *C. R. Acad. Sci. Paris B* **266**, 199
Calvet, N., Hartmann, L., Hewett, R., Valenti, J., Basri, G., and Walter, F.: 1996, in R. Pallavicini and A. K. Dupree (Eds.), *Cool Stars, Stellar Systems and the Sun, 9th Cambridge Workshop*, Astron. Soc. Pacific Conf. Series, San Francisco, p. 419
Canfield, R. C.: 1976, *Sol. Phys.* **50**, 239
Canfield, R. C., Reardon, K. P., Leka, K. D., Shibata, K., Yokoyama, T., and Shimojo, M.: 1996, in Y. Uchida, T. Kosugi, and H. S. Hudson (Eds.), *Magnetodynamic Phenomena in the Solar Atmosphere*, IAU Colloq. 153, Kluwer Acad. Publ., Dordrecht, Holland, p. 49
Cappelli, A., Cerutti–Sola, M., Cheng, C. C., and Pallavicini, R.: 1989, *Astron. Astrophys.* **213**, 226
Carbon, C. F. and Gingerich, O.: 1969, in O. Gingerich (Ed.), *Theory and Observations of Normal Stellar Atmospheres*, MIT Press, Cambridge, Mas., 377
Carlsson, M. and Stein, R. F.: 1992, *Astrophys. J., Lett.* **397**, 59
Carlsson, M. and Stein, R. F.: 1994, in M. Carlsson (Ed.), *Chromospheric dynamics*, Proceedings of a mini-workshop held at the Institute of Theoretical Physics, University of Oslo, Norway, p. 47

Carlsson, M. and Stein, R. F.: 1997, *Astrophys. J.* **481**, 500

Carpenter, K. G. and Robinson, R. D.: 1997, *Astrophys. J.* **479**, 970

Carrington, R. C.: 1858, *Mon. Not. R. Astron. Soc.* **19**, 1

Carrington, R. C.: 1863, *Observations of the Spots on the Sun*, Williams and Norgate, London

Castenmiller, M. J. M., Zwaan, C., and van der Zalm, E. B. J.: 1986, *Sol. Phys.* **105**, 237

Cattaneo, F.: 1999, *Astrophys. J., Lett.* **515**, 39

Cauzzi, G., Canfield, R. C., and Fisher, G. H.: 1996, *Astrophys. J.* **456**, 850

Cayrel de Strobel, G.: 1996, *Astron. Astrophys. Rev.* **7**, 243

Cerruti–Sola, M., Cheng, C. C., and Pallavicini, R.: 1992, *Astron. Astrophys.* **256**, 185

Chaboyer, B., Demarque, P., and Pinsonneault, M. H.: 1995, *Astrophys. J.* **441**, 865

Charbonneau, P. and MacGregor, K. B.: 1993, *Astrophys. J.* **417**, 762

Charbonneau, P. and MacGregor, K. B.: 1995, *Astrophys. J.* **454**, 901

Charbonneau, P. and MacGregor, K. B.: 1996, *Astrophys. J.* **473**, L59

Charbonneau, P. and MacGregor, K. B.: 1997, *Astrophys. J.* **486**, 502

Charbonneau, P., Schrijver, C. J., and MacGregor, K. B.: 1997, in J. R. Jokipii, C. P. Sonnett, and M. S. Giampapa (Eds.), *Cosmic winds and the heliosphere*, University of Arizona Press, Tucson, Arizona, p. 677

Cheng, C.-C., Dere, K. P., Wu, S. T., Hagyard, M. J., and Hiei, E.: 1996, *Adv. Sp. Res.* **17**, 205

Cheng, Q. Q., Envold, O., and Elgarøy, Ø.: 1997, *Astron. Astrophys.* **327**, 1155

Chiuderi, C., Einaudi, G., and Torricelli–Ciamponi, G.: 1981, *Astron. Astrophys.* **97**, 27

Choi, H. J., Soon, W., Donahue, R. A., Baliunas, S. L., and Henry, G. W.: 1995, *Publ. Astron. Soc. Pac.* **107**, 744

Choudhuri, A. R.: 1992, in J. H. Thomas and N. O. Weiss (Eds.), *Sunspots: Theory and Observations*, NATO ASI Series C: Vol. 375, Kluwer Academic Publishers, Dordrecht, p. 243

Choudhuri, A. R. and Gilman, P. A.: 1987, *Astrophys. J.* **316**, 788

Christensen–Dalsgaard, J., Gough, D. O., and Thompson, M. J.: 1991, *Astrophys. J.* **378**, 413

Claret, A. and Giménez, A.: 1989, *Astron. Astrophys. Suppl. Ser.* **81**, 37

Collier Cameron, A.: 1996, in K. G. Strassmeier and J. L. Linsky (Eds.), *Stellar Surface Structure*, IAU Symp. 176, Kluwer Academic Publishers, Dordrecht, The Netherlands, p. 449

Collier Cameron, A., Li, J., and Mestel, L.: 1991, in S. Catalano and J. R. Stauffer (Eds.), *Angular Momentum Evolution of Young Stars*, Kluwer Academic Publishers, Dordrecht, The Netherlands, 297

Collins, G. W. and Truax, R. J.: 1995, *Astrophys. J.* **439**, 860

Cook, J. W., Cheng, C.-C., Jacobs, V. L., and Antiochos, S. K.: 1989, *Astrophys. J.* **338**, 1176

Cook, J. W. and Ewing, J. A.: 1990, *Astrophys. J.* **355**, 719

Cowling, T. G.: 1934, *Mon. Not. R. Astron. Soc.* **94**, 39

Cowling, T. G.: 1946, *Mon. Not. R. Astron. Soc.* **106**, 39

Cowling, T. G.: 1953, in G. P. Kuiper (Ed.), *The Sun*, The University of Chicago Press, Chicago, p. 532

Cowling, T. G.: 1985, *Ann. Rev. Astron. Astrophys.* **23**, 1

Craig, I. J. D. and Brown, J. C.: 1986, *Inverse problems in astronomy*, Adam Hilger Ltd., Bristol, U.K.

Craig, I. J. D., McClymont, A. N., and Underwood, J. H.: 1978, *Astron. Astrophys.* **70**, 1

Culhane, J. L., White, N. E., Parmar, A. N., and Shafer, R. A.: 1990, *Mon. Not. R. Astron. Soc.* **243**, 424

Cuntz, M.: 1997, *Astron. Astrophys.* **325**, 709

Cuntz, M., Rammacher, W., and Ulmschneider, P.: 1994, *Astrophys. J.* **432**, 690

Cuntz, M. and Ulmschneider, P.: 1994, in J.-P. Caillault (Ed.), *Cool Stars, Stellar Systems, and the Sun*, 8th Cambridge Workshop (Athens, GA), ASP Conf. Series, p. 368

Cuntz, M., Ulmschneider, P., and Musielak, Z. E.: 1998, *Astrophys. J., Lett.* **493**, 117

Danielson, R. E.: 1964, *Astrophys. J.* **139**, 45

Daw, A., Deluca, E. E., and Golub, L.: 1995, *Astrophys. J.* **453**, 929

d'Azambuja, M. and d'Azambuja, L.: 1948, *Ann. Obs. Paris, Section Meudon* **VI**, Fasc. VII

De Jager, C.: 1959, in S. Flügge (Ed.), *Astrophysik III: Das Sonnensystem*, Handbuch der Physik Band 52, Springer, Berlin, p. 80

Degenhardt, D. and Lites, B. W.: 1993a, *Astrophys. J.* **404**, 383

Degenhardt, D. and Lites, B. W.: 1993b, *Astrophys. J.* **416**, 875

Delaboudiniere, J.-P., Artzner, G. E., Brunaud, J., Gabriel, A. H., Hochedez, J. F., Millier, F., Song, X. Y., Au, B., Dere, K. P., Howard, R. A., Kreplin, R., Michiels, D. J., Moses, J. D., Defise, J. M., Jamar, C., Rochus, P., Chauvineau, J. P., Marioge, J. P., Catura, R. C., Lemen, J. R., Shing, L., Stern, R. A., Gurman,

J. B., Neupert, W. M., Maucherat, A., Clette, F., Cugnon, P., and van Dessel, E. L.: 1995, *Sol. Phys.* **162**, 291

DeMastus, H. L., Wagner, W. J., and Robinson, R. D.: 1973, *Sol. Phys.* **31**, 449

Dere, K. P.: 1982, *Sol. Phys.* **77**, 77

Dere, K. P.: 1994, *Adv. Sp. Res.* **4**, 13

Dere, K. P., Bartoe, J.-D. F., and Brueckner, G. E.: 1989, *Sol. Phys.* **123**, 41

Dere, K. P., Bartoe, J.-D. F., Brueckner, G. E., Cook, J. W., and Socker, D. G.: 1987, *Sol. Phys.* **114**, 223

Deubner, F.-L.: 1975, *Astron. Astrophys.* **44**, 371

Deubner, F.-L.: 1988, in R. Stailio and L. A. Willson (Eds.), *Pulsation and Mass loss in Stars*, Kluwer Academic Publishers, Dordrecht, 163

Deubner, F.-L.: 1994, in R. K. Ulrich (Ed.), *Helio- and Asteroseismology from the Earth and Space*, Astronomical Society of the Pacific Conference Series, San Francisco

Deubner, F.-L. and Fleck, B.: 1990, *Astron. Astrophys.* **228**, 506

Deubner, F.-L. and Gough, D. O.: 1984, *Ann. Rev. Astron. Astrophys.* **22**, 593

DeVore, C. R.: 1987, *Sol. Phys.* **112**, 17

Ding, M. D., Hénoux, J.-C., and Fang, C.: 1998, *Astron. Astrophys.* **332**, 761

Donahue, R. A. and Baliunas, S. L.: 1992, *Astrophys. J., Lett.* **393**, 63

Donahue, R. A., Dobson, A. K., and Baliunas, S. L.: 1997a, *Sol. Phys.* **171**, 191

Donahue, R. A., Dobson, A. K., and Baliunas, S. L.: 1997b, *Sol. Phys.* **171**, 211

Donahue, R. A. and Keil, S. L.: 1995, *Sol. Phys.* **159**, 53

Donahue, R. A., Saar, S. H., and Baliunas, S. L.: 1996, *Astrophys. J.* **466**, 384

Dowdy, J. F., Rabin, D., and Moore, R. L.: 1986, *Sol. Phys.* **105**, 35

Doyle, J. G., Butler, C. J., Morrison, L. V., and Gibbs, P.: 1988, *Astron. Astrophys.* **192**, 275

Doyle, J. G., Houdebine, E. R., Mathioudakis, M., and Panagi, P. M.: 1994, *Astron. Astrophys.* **285**, 233

Drake, J. J.: 1996, in S. Bowyer and R. F. Malina (Eds.), *Astrophysics in the Extreme Ultraviolet*, IAU Coll. 152, Kluwer Academic Publishers, Dordrecht, The Netherlands, p. 97

Drake, J. J., Laming, J. M., and Widing, K. G.: 1995, *Astrophys. J.* **443**, 393

Drake, S. A. and Linsky, J. L.: 1986, *Astron. J.* **91**, 602

Drake, S. A., Simon, Th., and Brown, A.: 1993, *Astrophys. J.* **406**, 247

Dravins, D. and Nordlund, Å.: 1990a, *Astron. Astrophys.* **228**, 184

Dravins, D. and Nordlund, Å.: 1990b, *Astron. Astrophys.* **228**, 203

Dulk, G. A.: 1985, *Ann. Rev. Astron. Astrophys.* **23**, 169

Dunn, R. B. and Zirker, J. B.: 1973, *Sol. Phys.* **33**, 281

Dunn, R. B., Zirker, J. B., and Beckers, J. M.: 1974, in R. G. Athay (Ed.), *Chromospheric Fine Structure*, IAU Symp. No. 56, D. Reidel Publ. Co., Dordrecht, 45

Dupree, A. K.: 1982, in Y. Kondo, J. Meand, and R. Chapman (Eds.), *Advances in Ultraviolet Astronomy: Four Years of IUE Research*, NASA CP-2238, p. 3

Dupree, A. K.: 1996, in R. Pallavicini and A. K. Dupree (Eds.), *Cool Stars, Stellar Systems and the Sun*, Ninth Cambridge Workshop, Astron. Soc. of the Pac. Conf. Ser. Vol. 109, San Francisco, CA, p. 237

Dupree, A. K., Brickhouse, N. S., and Hanson, G. J.: 1996, in S. Bowyer and B. M. Haisch (Eds.), *Astrophysics in the extreme ultraviolet*, IAU Coll. 152, Kluwer Academic Publishers, Dordrecht, The Netherlands, p. 141

Dupree, A. K., Hartmann, L., and Smith, G. H.: 1990, *Astrophys. J.* **353**, 623

Dupree, A. K. and Reimers, D.: 1987, in Y. Kondo (Ed.), *Exploring the Universe with the IUE Satellite*, (Scientific accomplishments of the IUE), D. Reidel, Dordrecht, p. 321

Durney, B. R.: 1995, *Sol. Phys.* **160**, 213

Durney, B. R.: 1996a, *Sol. Phys.* **166**, 231

Durney, B. R.: 1996b, in A. Sanchez-Ibarra (Ed.), *The Solar Cycle: Recent Progress and Future Resarch*, p. 000

Durney, B. R.: 1997, *Astrophys. J.* **486**, 1065

Durney, B. R., de Young, D. S., and Roxburgh, I. W.: 1993, *Sol. Phys.* **145**, 207

Durney, B. R. and Latour, J.: 1978, *Geophys. Astrophys. Fluid Dyn.* **9**, 241

Durrant, C. W., Mattig, W., Nesis, A., Reiss, G., and Schmidt, W.: 1979, *Sol. Phys.* **61**, 251

Duvall, T. L., Kosovichev, A. G., Scherrer, Ph., Bogart, R. S., Bush, R. I., de Forest, C., Hoeksema, J. T., Schou, J., Saba, J. L. R., Tarbell, T. D., Title, A. M., and Wolfson, C. J.: 1997, *Sol. Phys.* **170**, 63

Eaton, J. A., Henry, G. W., and Fekel, F. C.: 1996, *Astrophys. J.* **462**, 888

Eaton, J. E.: 1993, *Astron. J.* **105**, 1525

Eddy, J. A.: 1976, *Science* **192**, 1189

Eddy, J. A.: 1977, in O. R. White (Ed.), *The Solar Output and its Variation*, Colorado Ass. Univ. Press, Boulder, p. 51

Eddy, J. A.: 1980, in R. Pepin, J. Eddy, and R. Merrill (Eds.), *The Ancient Sun*, Pergamon, 119

Elgarøy, Ø., Engvold, O., and Jonås, P.: 1997, *Astron. Astrophys.* **326**, 165

Ellerman, F.: 1917, *Astrophys. J.* **46**, 298

Elliot, J. R. and Gough, D. O.: 1999, *Astrophys. J.* **516**, 475

Endal, A. S. and Sofia, S.: 1978a, *Astrophys. J.* **210**, 184

Endal, A. S. and Sofia, S.: 1978b, *Astrophys. J.* **220**, 279

Engvold, O. and Elgarøy, Ø.: 1987, in J. L. Linsky and R. E. Stencel (Eds.), *Cool Stars, Stellar Sysems and the Sun*, Springer-Verlag, Berlin, 315

Esser, R. and Habbal, S. R.: 1997, in J. R. Jokipii, C. P. Sonnett, and M. S. Giampapa (Eds.), *Cosmic winds and the heliosphere*, University of Arizona Press, Tucson, Arizona, p. 297

Ewell, M. W.: 1992, *Sol. Phys.* **137**, 215

Ewell, M. W., Zirin, H., Jensen, J. B., and Bastian, T. S.: 1993, *Astrophys. J.* **403**, 426

Faurobert-Scholl, M., Feautrier, N., Machefert, F., Petrovay, K., and Speilfiedel, A.: 1995, *Astron. Astrophys.* **298**, 289

Feldman, U. and Laming, J. M.: 1994, *Astrophys. J.* **434**, 370

Feldman, U., Schühle, U., Widing, K. G., and Laming, J. M.: 1998, *Astrophys. J.* **505**, 999

Firstova, N. M.: 1986, *Sol. Phys.* **103**, 11

Fisher, G., Longcope, D., Metcalf, T., and Pevtsow, A.: 1998, *Astrophys. J.* **508**, 885

Fleck, B. and Schmitz, F.: 1993, *Astron. Astrophys.* **273**, 671

Fleming, T. A., Schmitt, J. H. M.M., and Giampapa, M. S.: 1995, *Astrophys. J.* **450**, 401

Foing, B. and Bonnet, R. M.: 1984, *Astrophys. J.* **279**, 848

Fokker, A. D.: 1977, in A. Bruzek and C. J. Durrant (Eds.), *Illustrated Gossary for Solar and Solar-Terrestrial Physics*, D. Reidel Publ. Cie, Dordrecht, The Netherlans, p. 111

Foukal, P.: 1971, *Sol. Phys.* **20**, 298

Foukal, Peter: 1990, *Solar Astrophysics*, Wiley and Sons, New York

Foukal, P.: 1998, *Astrophys. J.* **500**, 958

Foukal, P. V. and Lean, J.: 1990, *Science* **247**, 556

Fox, Ph.: 1908, *Astrophys. J.* **28**, 253

Fröhlich, C.: 1994, in J. M. Pap, C. Fröhlich, H. S. Hudson, and S. K. Solanki (Eds.), *The Sun as a Variable Star*, IAU Colloq. 143, Cambridge University Press, Cambridge, UK, p. 28

Gahm, G. F., Nordh, H. L., Olofson, S. G., and Carlbourg, N. C.: 1974, *Astron. Astrophys.* **33**, 399

Gaizauskas, V.: 1989, *Sol. Phys.* **121**, 135

Gaizauskas, V.: 1993, in H. Zirin, G. Ai, and H. Wang (Eds.), *The Magnetic and Velocity Fields of Solar Active Regions*, 479

Gaizauskas, V.: 1994, in R. J. Rutten and C. J. Schrijver (Eds.), *Solar Surface Magnetism*, NATO ASI Series C433, Kluwer, Dordrecht, p. 133

Gaizauskas, V.: 1996, *Sol. Phys.* **169**, 357

Gaizauskas, V., Harvey, K. L., Harvey, J. W., and Zwaan, C.: 1983, *Astrophys. J.* **265**, 1056

Gaizauskas, V., Zirker, J., Sweetland, C., and Kovacs, A.: 1997, *Astrophys. J.* **470**, 448

Galsgaard, K. and Nordlund, A.: 1996, *J. Geophys. Res.* **101**, 13445

García de la Rosa, J. I.: 1987, *Sol. Phys.* **112**, 49

Gary, D. E., Zirin, H., and Wang, H.: 1990, *Astrophys. J.* **355**, 321

Giampapa, M. S., Rosner, R., Kashyap, V., and Fleming, T. A.: 1996, *Astrophys. J.* **463**, 707

Giampapa, M. S., Worden, S. P., and Linsky, J. L.: 1982, *Astrophys. J.* **258**, 740

Gilliland, R. L. and Dupree, A. K.: 1996, *Astrophys. J., Lett.* **463**, 29

Gilman, P. A.: 1980, in D. F. Gray and J. L. Linsky (Eds.), *Stellar Turbulence*, Springer-Verlag, Berlin, 19

Giovanelli, R. G.: 1980, *Sol. Phys.* **68**, 49

Giovanelli, R. G.: 1982, *Sol. Phys.* **77**, 27

Giovanelli, R. G. and Jones, H. P.: 1982, *Sol. Phys.* **79**, 267

Godoli, G.: 1967, *Oss. Astrof. Catania Publ.* **115**, 224

Gokhale, M. H. and Zwaan, C.: 1972, *Sol. Phys.* **26**, 52

Goldberg, L.: 1979, *QJRAS* **20**, 361

Goldstein, M. L., Roberts, D. A., and Mattheus, W. H.: 1995, *Ann. Rev. Astron. Astrophys.* **33**, 283

Golub, L.: 1980, *Phil. Trans. R. Soc. Ser. London A* **297**, 595

Golub, L. and Herant, M.: 1989, *SPIE* **1160**, 629

Golub, L., Herant, M., Kalata, K., Lovas, I., Nystrom, G., Pardo, F., Spiller, E., and Wilczynski, J. S.: 1990, *Nature* **344**, 842

Golub, L., Krieger, A. S., Silk, J. K., Timothy, A. F., and Vaiana, G. S.: 1974, *Astrophys. J.* **189**, L93

Golub, L. and Pasachoff, J. M.: 1997, *The Solar Corona*, Cambridge University Press, Cambridge, U.K.

Gomez, M. T., Marmolino, C., Roberti, G., and Severino, G.: 1987, *Astron. Astrophys.* **188**, 169

Goode, Ph. R., Dziembowski, W. A., Korzennik, S. G., and Rhodes, E. J.: 1991, *Astrophys. J.* **367**, 649

Goode, Ph. R., Strous, L. H., Rimmele, T. R., and Stebbins, R. T.: 1998, *Astrophys. J., Lett.* **495**, 27

Gosling, J. T.: 1996, *Ann. Rev. Astron. Astrophys.* **34**, 35

Gough, D. O. and Toomre, J.: 1991, *Ann. Rev. Astron. Astrophys.* **29**, 627

Gratton, L.: 1950, *Astrophys. J.* **111**, 31

Gray, D. F.: 1982, *Astrophys. J.* **258**, 201

Gray, D. F.: 1984, *Astrophys. J.* **281**, 719

Gray, D. F.: 1989, *Astrophys. J.* **347**, 1021

Gray, D. F.: 1991, in S. Catalano and J. R. Stauffer (Eds.), *Angular momentum evolution of young stars*, Kluwer Academic Publishers, Dordrecht, The Netherlands, 183

Gray, D. F. and Endal, A. S.: 1982, *Astrophys. J.* **254**, 162

Gray, D. F. and Nagar, P.: 1985, *Astrophys. J.* **298**, 756

Gray, D. F. and Nagel, T.: 1989, *Astrophys. J.* **341**, 421

Griffin, R. E. M., Huensch, M., Griffin, R. F., and Schroeder, K.-P.: 1993, *Astron. Astrophys.* **274**, 225

Grossmann-Doerth, U., Knölker, M., Schüssler, M., and Weisshaar, E.: 1989, in R. J. Rutten and G. Severino (Eds.), *Solar and Stellar Granulation*, NATO ASI Series C 263, Kluwer, Dordrecht, p. 481

Guedel, M.: 1997, *Astrophys. J., Lett.* **408**, 121

Guenther, E. W. and Emerson, J. P.: 1996, in R. Pallavicini and A. K. Dupree (Eds.), *Cool Stars, Stellar Systems and the Sun, 9th Cambridge Workshop*, Astron. Soc. Pacific Conf. Series, San Francisco, p. 427

Habbal, S. R., Dowdy, J. F., and Withbroe, G. L.: 1990, *Astrophys. J.* **352**, 333

Habbal, S. R. and Grace, E.: 1991, *Astrophys. J.* **382**, 667

Habbal, S. R. and Withbroe, G. L.: 1981, *Sol. Phys.* **69**, 77

Habets, G. M. H. J. and Zwaan, C.: 1989, *Astron. Astrophys.* **211**, 56

Hagenaar, H. J., Schrijver, C. J., and Title, A. M.: 1997, *Astrophys. J.* **481**, 988

Hagenaar, H. J., Schrijver, C. J., Title, A. M., and Shine, R. A.: 1999, *Astrophys. J.* **511**, 932

Haisch, B., Saba, J. L. R., and Meyer, J.-P.: 1996, in S. Bowyer and R. F. Malina (Eds.), *Astrophysics in the Ultraviolet*, IAU Coll. 152, Kluwer Academic Publishers, Dordrecht, The Netherlands, p. 511

Haisch, B. and Schmitt, J. H. M. M.: 1996, *Publ. Astron. Soc. Pac.* **108**, 113

Haisch, B., Schmitt, J. H. M. M., and Rosso, C.: 1991a, *Astrophys. J.* **383**, L15

Haisch, B. M. and Rodonò, M. (Eds.): 1989, *Solar and stellar flares*, Proc. IAU Colloquium 104, Solar Phys. 121, Kluwer, Dordrecht

Haisch, B. M., Strong, K. T., and Rodonò, M.: 1991b, *Ann. Rev. Astron. Astrophys.* **29**, 275

Hale, G. E. and Nicholson, S. B.: 1938, *Publ. Carnegie Inst.* 498

Hall, D. S.: 1981, in R. M. Bonnet and A. K. Dupree (Eds.), *Reidel*, NATO A. S. I., Dordrecht, p. 431

Hall, D. S.: 1991, in I. Tuominen, D. Moss, and G. Ruediger (Eds.), *The Sun and Cool Stars: activity, magnetism, dynamos*, IAU Coll. No. 130, Springer Verlag, Berlin, Germany, p. 352

Hall, D. S.: 1996, in K. G. Strassmeier and J. L. Linsky (Eds.), *Stellar Surface Structure*, IAU Symp. 176, Kluwer Academic Publishers, Dordrecht, The Netherlands, p. 217

Hall, D. S. and Busby, M. R.: 1990, in C. Ibanoglu (Ed.), *Active close binaries*, Kluwer, Dordrecht, The Netherlands, 377

Hall, D. S., Fekel, F. C., Henry, G. W., Eaton, J. A., Barksdale, W. S., Dadonas, V., Eker, Z., Kalv, P., Chambliss, C. R., and Fried, R. E.: 1995, *Astron. J.* **109**, 1277

Hammer, R. and Ulmschneider, P.: 1991, in P. Ulmschneider, E. Priest, and R. Rosner (Eds.), *Mechanisms of Chromospheric and Coronal Heating*, Springer-Verlag, Heidelberg, p. 344

Hampton, M., Henry, G. W., Eaton, J. A., Nolthenius, R. A., and Hall, D. S.: 1996, *Publ. Astron. Soc. Pac.* **108**, 68

Handy, B. N., Acton, L. W., Kankelborg, C. C., Wolfson, C. J., Akin, D. J., Bruner, M. E., Carvalho, R., Catura, R. C., Chevalier, R., Duncan, D. W., Edwards, C. G., Feinstein, C. N., Freeland, S. L., Friedlander, F. M., Hoffman, C. H., Hurlburt, N. E., Jurcevich, B. K., Katz, N. L., Kelly, G. A., Lemen, J. R., Levay, M., Lindgren, R. W., Mathur, D. P., Meyer, S. B., Morrison, S. J., Morrison, M. D.,

Nightingale, R. W., Pope, T. P., Rehse, R. A., Schrijver, C. J., Shine, R. A., Shing, L., Strong, K. T., Tarbell, T. D., Title, A. M., Torgerson, D. D., Golub, L., Bookbinder, J. A., Caldwell, D., Cheimets, P. N., Davis, W. N., Deluca, E. E., McMullen, R. A., Amato, D., Fisher, R., Maldonado, H., and Parkinson, C.: 1999, *Sol. Phys.*, in press

Hansen, C. J., Cox, J. P., and Van Horn, H. M.: 1977, *Astrophys. J.* **217**, 151

Harper, G.: 1996, in R. Pallavicini and A. K. Dupree (Eds.), *Cool Stars, Stellar Systems and the Sun*, Ninth Cambridge Workshop, Astron. Soc. of the Pac. Conf. Ser. Vol. 109, San Francisco, CA, p. 461

Harrison, R. A. and Thompson, A. M.: 1992, *A study in preparation for the SoHO mission*, Technical report, Science and Engineering Research Council, Rutherford Appleton Laboratory, Childon, Didcot, Oxfordshire, England, RAL-91-092

Hart, A. B.: 1954, *Mon. Not. R. Astron. Soc.* **114**, 17

Hart, A. B.: 1956, *Mon. Not. R. Astron. Soc.* **116**, 38

Hartmann, L.: 1990, in G. Wallerstein (Ed.), *Cool Stars, Stellar Sysems and the Sun*, Astron. Soc. Pacific Conf. Series, Vol. 9, San Francisco, 289

Hartmann, L., Bopp, B. W., Dussault, M., Noah, P. V., and Klimke, A.: 1981a, *Astrophys. J.* **249**, 662

Hartmann, L., Dupree, A. K., and Raymond, J. C.: 1980, *Astrophys. J., Lett.* **236**, 143

Hartmann, L., Dupree, A. K., and Raymond, J. C.: 1981b, *Astrophys. J.* **246**, 193

Hartmann, L., Jordan, C., Brown, A., and Dupree, A. K.: 1985, *Astrophys. J.* **296**, 576

Hartmann, L. and MacGregor, K. B.: 1980, *Astrophys. J.* **242**, 260

Hartmann, L. W. and Noyes, R. W.: 1987, *Ann. Rev. Astron. Astrophys.* **25**, 271

Harvey, J.: 1977, *Highlights of Astronomy* **4**, 223, Part II

Harvey, J.: 1985, in M. J. Hagyard (Ed.), *Measurements of Solar Vector Magnetic Fields*, NASA Conf. Pub. 2374, p. 109

Harvey, K. L.: 1989, in R. J. Rutten and G. Severino (Eds.), *Solar and Stellar Granulation*, NATO ASI Series, Kluwer Academic Publishers, Dordrecht, The Netherlands, p. 623

Harvey, K. L.: 1992, in K. L. Harvey (Ed.), *The Solar Cycle*, Ast. Soc. Pacific Conf. Ser. Vol. 27, San Francisco, p. 335

Harvey, K. L.: 1993, *Ph.D. thesis*, Astronomical Institute, Utrecht University

Harvey, K. L. and Harvey, J. W.: 1973, *Sol. Phys.* **28**, 61

Harvey, K. L. and Martin, S. F.: 1973, *Sol. Phys.* **32**, 389

Harvey, K. L., Nitta, N., Strong, K., and Tsuneta, S.: 1994, in Y. Uchida, T. Watanabe, K. Shibata, and H. S. Hudson (Eds.), *X-Ray Solar Physics from Yohkoh*, Universal Academy Press, Tokyo, p. 21

Harvey, K. L. and White, O. R.: 1999, *Astrophys. J.* **515**, 812

Harvey, K. L. and Zwaan, C.: 1993, *Sol. Phys.* **148**, 85

Heasly, J. N., Ridgeway, S. T., Carbon, D. F., Milkey, R. W., and Hall, D. B. N.: 1978, *Astrophys. J.* **219**, 970

Heckert, P. A., Maloney, G. V., Stewart, M. C., Ordway, J. I., Hickman, M. A., and Zeilik, M.: 1998, *Astron. J.* **115**, 1145

Heckert, P. A. and Ordway, J. I.: 1995, *Astron. J.* **109**, 2169

Hempelmann, A. and Donahue, R. A.: 1997, *Astron. Astrophys.* **322**, 835

Hempelmann, A., Schmitt, J. H. M. M., Schulz, M., Ruediger, G., and Stępién, K.: 1995, *Astron. Astrophys.* **294**, 515

Hiltner, W. A.: 1947, *Astrophys. J.* **106**, 481

Hoeksema, J. T.: 1984, *Ph.D. thesis*, Stanford University, Stanford, Ca.

Howard, R.: 1984, *Ann. Rev. Astron. Astrophys.* **22**, 131

Howard, R. and Harvey, J.: 1970, *Sol. Phys.* **12**, 23

Howard, R. and Stenflo, J. O.: 1972, *Sol. Phys.* **22**, 402

Howard, R. A., Sheeley, N. R., Michels, D. J., and Koomen, M. L.: 1985, *J. Geophys. Res.* **90**, 8173

Howard, R. F.: 1959, *Astrophys. J.* **130**, 193

Howard, R. F.: 1972, *Sol. Phys.* **25**, 5

Howard, R. F.: 1991a, *Sol. Phys.* **135**, 43

Howard, R. F.: 1991b, *Sol. Phys.* **134**, 233

Howard, R. F.: 1992, *Sol. Phys.* **137**, 205

Howard, R. F.: 1996, *Ann. Rev. Astron. Astrophys.* **34**, 75

Howard, R. F. and LaBonte, B. J.: 1980, *Astrophys. J.* **239**, L33

Hoyng, P.: 1992, in J. T. Schmelz and J. C. Brown (Eds.), *The Sun, a laboratory for astrophysics*, Kluwer Acad. Publishers, Dordrecht, the Netherlands, p. 99

Hoyng, P.: 1994, in R. J. Rutten and C. J. Schrijver (Eds.), *Solar Surface Magnetism*, Kluwer Acad. Publishers, Dordrecht, the Netherlands, p. 387

Hubbard, W. B., Lunine, J. I., Saumon, D., and Burrows, A.: 1994, *Bull. Am. Astron. Soc.* **185**, 9006

Hudson, H. and Ryan, J.: 1995, *Ann. Rev. Astron. Astrophys.* **33**, 239

Hughes, B. D.: 1995, *Random Walks and Random Environments. Vol. I: Random Walks*, Clarendon Press, Oxford, U.K.

Hughes, B. D.: 1996, *Random Walks and Random Environments. Vol. II: Random Environments*, Clarendon Press, Oxford, U.K.

Hundhausen, A. J.: 1993a, *J. Geophys. Res.* **98**, 13177

Hundhausen, A. J.: 1993b, *J. Geophys. Res.* **98**, 13177

Hundhausen, A. J.: 1994, in M. G. Kivelson and C. T. Russell (Eds.), *Introduction to Space Physics*, Cambridge University Press, Cambridge, UK, p.

Jackson, J. D.: 1975, *Classical Electrodynamics*, John Wiley and Sons, New York

Jahn, K.: 1989, *Astron. Astrophys.* **222**, 264

Jahn, K.: 1992, in J. H. Thomas and N. O. Weiss (Eds.), *Sunspots: Theory and Observations*, NATO ASI Series C: Vol. 375, Kluwer Academic Publishers, Dordrecht, p. 139

Jeffries, R. D.: 1999, in C. J. Butler and J. G. Doyle (Eds.), *Solar and Stellar Activity: Similarities and Differences*, Astron. Soc. Pacific, San Francisco, p. 75

Jennings, R. L. and Weiss, N. O.: 1991, *Mon. Not. R. Astron. Soc.* **252**, 249

Jensen, K. A., Swank, J. H., Petre, R., Guinan, E. F., Sion, E. M., and Shippman, H. L.: 1986, *Astrophys. J., Lett.* **309**, 27

Johns–Krull, C. M.: 1996, *Astron. Astrophys.* **306**, 803

Johns–Krull, C. M. and Valenti, J. A.: 1996, *Astrophys. J., Lett.* **459**, 95

Jordan, C.: 1969, *Mon. Not. R. Astron. Soc.* **142**, 501

Jordan, C.: 1991, in P. Ulmschneider, E. Priest, and R. Rosner (Eds.), *Mechanisms of Chromospheric and Coronal Heating*, Springer-Verlag, Heidelberg, p. 300

Joy, A. H. and Humason, M. L.: 1949, *Publ. Astron. Soc. Pac.* **61**, 133

Joy, A. H.: 1919, *Astrophys. J.* **49**, 167

Joy, A. H.: 1942, *Publ. Astron. Soc. Pac.* **54**, 15

Joy, A. H. and Wilson, R. E.: 1949, *Astrophys. J.* **109**, 231

Judge, P. G. and Stencel, R. E.: 1991, *Astrophys. J.* **371**, 357

Kano, R. and Tsuneta, S.: 1996, *Publ. Astron. Soc. Jpn.* **48**, 535

Kato, T.: 1976, *Astrophys. J., Suppl. Ser.* **30**, 397

Kawaler, S. D.: 1987, *Publ. Astron. Soc. Pac.* **99**, 1322

Keil, S. L. and Canfield, R. C.: 1978, *Astron. Astrophys.* **70**, 169

Keller, C. U.: 1992, *Nature* **359**, 307

Keppens, R., MacGregor, K. B., and Charbonneau, P.: 1995, *Astron. Astrophys.* **294**, 469

Kiepenheuer, K. O.: 1953, in G. P. Kuiper (Ed.), *The Sun*, University of Chicago Press, Chicago, p. 322

Kippenhahn, R. and Weigert, A.: 1990, *Stellar Structure and Evolution*, Springer-Verlag, Berlin

Kitchatinov, L. L. and Ruediger, G.: 1995, *Astron. Astrophys.* **299**, 446

Kjeldseth-Moe, O., Brynildsen, N., Brekke, P., Maltby, P., and Brueckner, G. E.: 1993, *Sol. Phys.* **145**, 257

Klimchuk, J. A., Lemen, J. R., Feldman, U., Tsuneta, S., and Unchida, Y.: 1992, *Publ. Astron. Soc. Jpn.* **44**, L181

Kneer, F.: 1983, *Astron. Astrophys.* **128**, 311

Koch, R. H. and Hrivnak, B. J.: 1981, *Astron. J.* **86**, 438

Komm, R. W., Howard, R. F., and Harvey, J. W.: 1993a, *Sol. Phys.* **143**, 19

Komm, R. W., Howard, R. F., and Harvey, J. W.: 1993b, *Sol. Phys.* **147**, 207

Komm, R. W., Howard, R. F., and Harvey, J. W.: 1995, *Sol. Phys.* **158**, 213

Kopp, G. and Rabin, D.: 1994, in D. M. Rabin, J. T. Jefferies, and C. Lindsey (Eds.), *Infrared Solar Physics*, IAU Symp. No. 154, Kluwer Acad. Publ., Dordrecht, p. 477

Kopp, R. A. and Kuperus, M.: 1968, *Sol. Phys.* **4**, 212

Kosovichev, A. G., Schou, J., Scherrer, P. H., Bogart, R. S., Bush, R. I., Hoeksema, J. T., Aloise, J., Bacon, L., Burnette, A., DeForest, C., Giles, P. M., Leibrand, K., Nigam, R., Rubin, M., Scott, K., Williams, S. D., Basu, S., Christensen–Dalsgaard, J., Daeppen, W., Rhodes, E. J., Duvall, T. L., Howe, R., Thompson, M. J., Gough, D. O., Kekii, T., Toomre, J., Tarbell, T. D., Title, A. M., Mathur, D., Morrison, M., Saba, J. L., Wolfson, C. J., Zayer, I., and Milford, P. N.: 1997, *Sol. Phys.* **170**, 43

Koutchmy, S.: 1977, *Sol. Phys.* **51**, 399

Kraft, R. P.: 1967, *Astrophys. J.* **150**, 551

Krause, F. and Rädler, K. H.: 1980, *Mean-Field Magnetohydrodynamics and Dynamo Theory*, Pergamon Press, London

Krause, F., Rädler, K. H., and Rüdiger, G. (Eds.): 1993, *The Cosmial Dynamo*, IAU Symp. 157, Kluwer Academic Publishers, Dordrecht, the Netherlands

Krishnamurthi, A., Pinsonneault, M. H., Barnes, S., and Sofia, S.: 1997, *Astrophys. J.* **480**, 303

Krucker, S. and Benz, A. O.: 1998, *Astrophys. J., Lett.* **501**, 213

Kucera, T. A., Andretta, V., and Poland, A. I.: 1999, *Sol. Phys.* **183**, 107

Kuerster, M., Schmitt, J. H. M. M., and Gutispoto, G.: 1994, *Astron. Astrophys.* **289**, 899

Kuhn, J. R., Bush, R. I., Scheick, X., and Scherrer, P.: 1998, *Nature* **392**, 155

Kuijpers, J.: 1989, *Sol. Phys.* **121**, 163

Kulaczewski, J.: 1992, *Astron. Astrophys.* **261**, 602

Kumar, P. and Lu, E.: 1991, *Astrophys. J., Lett.* **375**, 35

Kundu, M. R., Raulin, J. P., Nitta, N., Hudson, H. S., Shimojo, M., Shibata, K., and Raoult, A.: 1995, *Astrophys. J.* **447**, L135

Kuperus, M. and Athay, R. G.: 1967, *Sol. Phys.* **1**, 361

Kurokawa, H. and Kawai, G.: 1993, in H. Zirin, G. Ai, and H. Wang (Eds.), *The Magnetic and Velocity Fields of Solar Active Regions*, IAU Colloq. No. 141, Astron. Soc. Pacific, San Francisco, 507

Kurucz, R.: 1979, *Astrophys. J., Suppl. Ser.* **40**, 1

LaBonte, B. J. and Howard, R.: 1980, *Sol. Phys.* **80**, 15

Laming, J. M. and Drake, J. J.: 1999, *Astrophys. J.* **516**, 324

Laming, J. M., Drake, J. J., and Widing, K. G.: 1995, *Astrophys. J., Lett.* **443**, 416

Laming, J. M., Feldman, U., Schuehle, U., Lemaire, P., Curdt, W., and Wilhelm, K.: 1997, *Astrophys. J.* **485**, 911

Landi Degl'Innocenti, E.: 1976, *Astron. Astrophys. Suppl. Ser.* **25**, 369

Landi Degl'Innocenti, E.: 1998, *Nature* **392**, 256

Landini, M. and Monsignori–Fossi, B.: 1991, *Astron. Astrophys. Suppl. Ser.* **91**, 183

Latushko, S.: 1994, *Sol. Phys.* **149**, 231

Lawrence, J. K.: 1991, *Sol. Phys.* **135**, 249

Lawrence, J. K. and Schrijver, C. J.: 1993, *Astrophys. J.* **411**, 402

Leighton, R. B.: 1959, *Astrophys. J.* **130**, 366

Leighton, R. B.: 1964, *Astrophys. J.* **140**, 1547

Leighton, R. B.: 1969, *Astrophys. J.* **156**, 1

Leighton, R. B., Noyes, R. W., and Simon, G. W.: 1962, *Astrophys. J.* **135**, 474

Lemen, J. R., Mewe, R., Schrijver, C. J., and Fludra, A.: 1989, *Astrophys. J.* **341**, 474

Lemmens, A. F. P., Rutten, R. G. M., and Zwaan, C.: 1992, *Astron. Astrophys.* **257**, 671

Lestrade, J.-F.: 1996, in K. G. Strassmeier and J. L. Linsky (Eds.), *Stellar Surface Structure*, IAU Symp. 176, Kluwer Academic Publishers, Dordrecht, The Netherlands, p. 173

Levine, R. H., Altschuler, M. D., Harvey, J. W., and Jackson, B. V.: 1977, *Astrophys. J.* **215**, 636

Liebert, J., Saffer, R. A., Norsworthy, J., Giampapa, M. S., and Stauffer, J. R.: 1992, in M. S. Giampapa and J. A. Bookbinder (Eds.), *Cool Stars, Stellar Systems, and the Sun*, A. S. P. Conference Series, Vol. 26, p. 282

Liggett, M. A. and Zirin, H.: 1985, *Sol. Phys.* **97**, 51

Lighthill, M. J.: 1952, *Proc. R. Soc. London* **211**, 564

Linker, J., Mikic, Z., and Schnack, D. D.: 1996, in K. S. Balasubramanian, S. L. Keil, and R. N. Smartt (Eds.), *Solar Drivers of Interplanetary and Terrestrial Disturbances*, Conf. Series Vol. 95, Astron. Soc. Pacific, San Francisco, 208

Linsky, J. L.: 1986, *Sol. Phys.* **100**, 333

Linsky, J. L. and Haisch, B. M.: 1979, *Astrophys. J.* **229**, L27

Linsky, J. L. and Wood, B. E.: 1996, in R. Pallavicini and A. K. Dupree (Eds.), *Cool Stars, Stellar Systems and the Sun*, Ninth Cambridge Workshop, Astron. Soc. of the Pac. Conf. Ser. Vol. 109, San Francisco, CA, p. 497

Linsky, J. L., Wood, B. E., Brown, A., Giampapa, M. S., and Ambruster, C.: 1995, *Astrophys. J.* **455**, 670

Lites, B. W., Bida, T. A., Johannesson, A., and Scharmer, G. B.: 1991, *Astrophys. J.* **373**, 683

Lites, B. W., Elmore, D. F., Seagraves, P., and Skumanich, A.: 1993a, *Astrophys. J.* **418**, 928

Lites, B. W., Rutten, R. J., and Kalkofen, W.: 1993b, *Astrophys. J.* **414**, 345

Livi, S. H. B., Wang, J., and Martin, S. F.: 1985, *Aust. J. Phys.* **38**, 855

Livingston, W.: 1974, in *Flare-Related Field Dynamics*, NCAR, Boulder, CO, 269

Livingston, W. C. and Harvey, J.: 1971, in R. Howard (Ed.), *Solar Magnetic Fields*, IAU Symp. No. 43, Reidel, Dordrecht, 51

Livingston, W. C. and Orrall, F. Q.: 1974, *Sol. Phys.* **39**, 301

Lockwood, G. W., Skiff, B. A., and Radick, R. R.: 1997, *Astrophys. J.* **485**, 789
Longcope, D. W.: 1996, *Sol. Phys.* **169**, 91
Low, B. C.: 1984, *Astrophys. J.* **281**, 392
Low, B. C.: 1996, *Sol. Phys.* **167**, 217
Low, B. C. and Lou, Y. Q.: 1990, *Astrophys. J.* **352**, 343
Lutz, T. E. and Pagel, B. E. J.: 1982, *Mon. Not. R. Astron. Soc.* **199**, 1101
Lynden-Bell, D. and Pringle, J. E.: 1974, *Mon. Not. R. Astron. Soc.* **168**, 603
MacQueen, R. M., Eddy, J. A., Gosling, J. T., Hildner, E., and et al., R. H. Munro: 1974, *Astrophys. J.* **187**, L85
Maeder, A. and Meynet, G.: 1989, *Astron. Astrophys.* **210**, 155
Magain, P.: 1986, *Astron. Astrophys.* **163**, 135
Maltby, P.: 1994, in R. J. Rutten and C. J. Schrijver (Eds.), *Solar Surface Magnetism*, NATO ASI Series C: Vol. 433, Kluwer Acad. Publishers, Dordrecht, p. 179
Maltby, P., Avrett, E. H., Carlsson, M., Kjeldseth-Moe, O., Kurucz, R. L., and Loeser, R.: 1986, *Astrophys. J.* **306**, 284
Mandelbrot, B.: 1986, *Fractals in physics*, North-Holland, Amsterdam
Mariska, J. T.: 1992, *The Solar Transition Region*, Cambridge University Press, Cambridge, U.K.
Mariska, J. T., Feldman, U., and Doschek, G. A.: 1978, *Astrophys. J.* **226**, 698
Marsden, R. G. and Smith, E. J.: 1996, *Adv. Sp. Res.* **17**, 289
Martens, P. C. H., Van den Oord, G. H. J., and Hoyng, P.: 1985, *Sol. Phys.* **96**, 253
Martin, S. F.: 1984, in S. L. Keil (Ed.), *Proc. Symp. on Small-scale Dynamical Processes in Quiet Stellar Atmospheres*, NSO, Sacramento Peak, NM, 30
Martin, S. F.: 1988, *Sol. Phys.* **117**, 243
Martin, S. F.: 1990a, in V. Ruždjak and E. Tandberg-Hansen (Eds.), *Dynamics of Quiescent Prominences*, IAU Colloq. 117, Springer-Verlag, Berlin, 1
Martin, Sara F.: 1990b, in J.-O. Stenflo (Ed.), *The Solar Photosphere: Structure, Convection and Magnetic Fields*, Proceedings IAU Symposium 138 (Kiev), Kluwer, Dordrecht, p. 129
Martin, S. F., Bilimoria, R., and Tracadas, P. W.: 1994, in R. J. Rutten and C. J. Schrijver (Eds.), *Solar Surface Magnetism*, NATO ASI Series C433, Kluwer Academic Publishers, Dordrecht, p. 303
Martin, S. F., Livi, S. H. B., and Wang, J.: 1985, *Aust. J. Phys.* **38**, 929
Martínez Pillet, V., Moreno-Insertis, F., and Vazquez, M.: 1993, *Astron. Astrophys.* **274**, 521
Mauas, P. J., Avrett, E. H., and Loeser, R.: 1990, *Astrophys. J.* **357**, 279
Maunder, E. W.: 1890, *Mon. Not. R. Astron. Soc.* **50**, 251
Maunder, E. W.: 1919, *Mon. Not. R. Astron. Soc.* **79**, 451
Maunder, E. W.: 1922, *Mon. Not. R. Astron. Soc.* **82**, 534
McClintock, W., Henry, R. C., and Moos, H. W.: 1975, *Astrophys. J.* **202**, 73
McClymont, A. N. and Mikic, Z.: 1994, *Astrophys. J.* **422**, 899
McIntosh, P. S.: 1981, in L. E. Cram and J. H. Thomas (Eds.), *The Physics of Sunspots*, NSO-Sacramento Peak, Sunspot, NM, 7
Mehltretter, J. P.: 1974, *Sol. Phys.* **38**, 43
Mestel, L.: 1968, *Mon. Not. R. Astron. Soc.* **138**, 359
Mestel, L.: 1999, *Stellar Magnetism*, Oxford University Press, Oxford, UK
Mestel, L. and Spruit, H. C.: 1987, *Mon. Not. R. Astron. Soc.* **226**, 57
Metcalf, T. R., Jiao, L., McClymont, A. N., Alexander, N., Canfield, R. C., and Uitenbroek, H.: 1995, *Astrophys. J.* **439**, 474
Mewe, R., Gronenschild, E. H. B. M., and Van den Oord, G. H. J.: 1985, *A & AS* **62**, 1985
Mewe, R., Kaastra, J. S., Schrijver, C. J., Van den Oord, G. H. J., and Alkemade, F. J. M.: 1995, *Astron. Astrophys.* **296**, 477
Meyer, F., Schmidt, H. U., and Weiss, N. O.: 1977, *Mon. Not. R. Astron. Soc.* **179**, 741
Middelkoop, F.: 1981, *Astron. Astrophys.* **101**, 295
Middelkoop, F.: 1982a, *Astron. Astrophys.* **113**, 1
Middelkoop, F.: 1982b, *Astron. Astrophys.* **107**, 31
Middelkoop, F., Vaughan, A. H., and Preston, G. W.: 1981, *Astron. Astrophys.* **96**, 401
Middelkoop, F. and Zwaan, C.: 1981, *Astron. Astrophys.* **101**, 26
Minnaert, M. G. J.: 1946, *Mon. Not. R. Astron. Soc.* **106**, 98
Moffatt, H. K.: 1978, *Magnetic Field Generation in Electrically Conducting Fluids*, Cambridge University Press, Cambridge, UK
Monsignori-Fossi, B. C. and Landini, M.: 1994, *Astron. Astrophys.* **284**, 900

Moreno-Insertis, F.: 1986, *Astron. Astrophys.* **166**, 291
Moreno-Insertis, F.: 1997, in V. G. Hansteen (Ed.), *Solar Magnetic Fields*, University of Oslo, Oslo, p. 3
Mosher, J. M.: 1977, *The Magnetic History of Solar Active Regions.*, CalTech, Pasadena, CA
Muchmore, D. and Ulmschneider, P.: 1985, *Astron. Astrophys.* **142**, 393
Muglach, K. and Solanki, S. K.: 1992, *Astron. Astrophys.* **262**, 301
Muglach, K., Solanki, S. K., and Livingston, W. C.: 1994, in R. J. Rutten and C. J. Schrijver (Eds.), *Solar Surface Magnetism*, NATO ASI Series C Vol. 433, Kluwer Academic Publ., Dordrecht, p. 127
Mullan, D. J.: 1996, in R. Pallavicini and A. K. Dupree (Eds.), *Cool Stars, Stellar Systems and the Sun*, Ninth Cambridge Workshop, Astron. Soc. of the Pac. Conf. Ser. Vol. 109, San Francisco, CA, p. 461
Mullan, D. J., Doyle, J. G., Redman, R. O., and Mathioudakis, M.: 1992, *Astrophys. J.* **397**, 225
Mullan, D. J., Sion, E. M., Bruhweiler, F. C., and Carpenter, K. G.: 1989, *Astrophys. J., Lett.* **339**, 33
Muller, R.: 1992, in J. H. Thomas and N. O. Weiss (Eds.), *Sunspots: Theory and Observations*, NATO ASI Series C: Vol. 375, Kluwer Academic Publishers, Dordrecht, p. 175
Musielak, Z. E., Rosner, R., Stein, R. F., and Ulmschneider, P.: 1994, *Astrophys. J.* **423**, 474
Namba, O. and van Rijsbergen, R.: 1969, in E. A. Spiegel and J.-P. Zahn (Eds.), *Problems of Stellar Convection*, IAU Colloq. 38, Springer-Verlag, Berlin, p. 119
Narain, U. and Ulmschneider, P.: 1990, *Space Sci. Rev.* **54**, 377
Narain, U. and Ulmschneider, P.: 1996, *Space Sci. Rev.* **75**, 453
Neff, J. E., O'Neal, D., and Saar, S. H.: 1995, *Astrophys. J.* **452**, 879
Neidig, D. F.: 1989, *Sol. Phys.* **121**, 261
Nelson, G. D. and Musman, S.: 1978, *Astrophys. J., Lett.* **222**, 69
Newton, H. W. and Nunn, M. L.: 1951, *Mon. Not. R. Astron. Soc.* **111**, 413
Noever, D. A.: 1994, *Astron. Astrophys.* **282**, 252
Nordlund, Å.: 1982, *Astron. Astrophys.* **107**, 1
Nordlund, Å.: 1986, *Sol. Phys.* **100**, 209
Nordlund, Å., Brandenburg, A., Jennings, R. L., Rieutord, M., Ruokolainen, J., Stein, R. F., and Tuominen, I.: 1992, *Astrophys. J.* **392**, 647
Nordlund, Å., Galsgaard, K., and Stein, R. F.: 1994, in R. J. Rutten and C. J. Schrijver (Eds.), *Solar Surface Magnetism*, NATO Advanced Research Workshop, Kluwer Academic Publishers, Dordrecht, The Netherlands, p. 471
Nordlund, Å. and Dravins, D.: 1990, *Astron. Astrophys.* **228**, 155
Nordlund, Å. and Galsgaard, K.: 1997, in G. M. Simnett, C. E. Alissandrakis, and L. Vlahos (Eds.), *Solar and Heliospheric Plasma Physics*, Proceedings of the 8th European Meeting on Solar Physics Held at Halkidiki, near Thessaloniki, Greece, 13–18 May, 1996, Lecture Notes in Physics, Vol. 489, Springer, Heidelberg
Nordlund, Å., Spruit, H. C., Ludwig, H.-G., and Trampedach, R.: 1997, *Astron. Astrophys.* **328**, 229
November, L. J.: 1989, *Astrophys. J.* **344**, 494
November, L. J. (Ed.): 1991, *Solar Polarimetry*, NSO/SP Summer Workshop Series No. 11, Sunspot, New Mexico, USA
November, L. J., Toomre, J., Gebbie, K. B., and Simon, G. W.: 1981, *Astrophys. J., Lett.* **245**, 123
Noyes, R. W.: 1996, in R. Pallavicini and A. K. Dupree (Eds.), *Cool Stars, Stellar Systems, and the Sun*, (Proceedings of the Ninth Cambridge Workshop), ASP Conf. Ser. Vol. 109, San Francisco, p. 3
Noyes, R. W., Hartmann, L., Baliunas, S. L., Duncan, D. K., and Vaughan, A. H.: 1984, *Astrophys. J.* **279**, 763
O'Dell, M. A., Pagani, P., Hendry, M. A., and Collier Cameron, A.: 1995, *Astron. Astrophys.* **294**, 715
Ogawara, Y., Acton, L. W., Bentley, R. D., Bruner, M. E., Culhane, J. L., Hiei, E., Hirayama, T., Hudson, H. S., Kosugi, T., Lemen, J. R., Strong, K. T., Tsuneta, S., Uchida, Y., Watanabe, T., and Yoshimori, M.: 1992, *Publ. Astron. Soc. Jpn.* **44**, L41
Okabe, A., Boots, B., and Sugihara, K.: 1992, *Spatial Tesselations; Concepts and Applications of Voronoi Diagrams*, John Wiley and Sons, Chichester, U.K.
O'Neal, D., Saar, S. H., and Neff, J. E.: 1996, *Astrophys. J.* **463**, 766
O'Neal, D., Saar, S. H., and Neff, J. E.: 1998, *Astrophys. J., Lett.* **501**, 73
Oranje, B. J.: 1983a, *Astron. Astrophys.* **122**, 88
Oranje, B. J.: 1983b, *Astron. Astrophys.* **124**, 43
Oranje, B. J.: 1985, *Astron. Astrophys.* **154**, 185
Oranje, B. J. and Zwaan, C.: 1985, *Astron. Astrophys.* **147**, 265

Oranje, B. J., Zwaan, C., and Middelkoop, F.: 1982a, *Astron. Astrophys.* **110**, 30
Oranje, B. J., Zwaan, C., and Middelkoop, F.: 1982b, *Astron. Astrophys.* **110**, 30
Ossendrijver, A. J. H. and Hoyng, P.: 1996, *Astron. Astrophys.* **313**, 959
Pallavicini, R.: 1994, in J.-P. Caillault (Ed.), *Cool Stars, Stellar Systems, and the Sun*, ASP Conf. Series Vol. 64, San Francisco, CA, p. 244
Pallavicini, R., Serio, S., and Vaiana, G. S.: 1977, *Astrophys. J.* **216**, 108
Pap, J. M., Fröhlich, C., Hudson, H. S., and Solanki, S. K. (Eds.): 1994, *The Sun as a Variable Star*, IAU Colloq. 143, Cambridge University Press, Cambridge, UK
Parker, E. N.: 1955, *Astrophys. J.* **122**, 239
Parker, E. N.: 1958, *Astrophys. J.* **128**, 669
Parker, E. N.: 1963a, *Astrophys. J.* **138**, 552
Parker, E. N.: 1963b, *Interplanetary Dynamical Processes*, Wiley-Interscience, New York
Parker, E. N.: 1966, *Astrophys. J.* **145**, 811
Parker, E. N.: 1972, *Astrophys. J.* **174**, 499
Parker, E. N.: 1975, *Sol. Phys.* **40**, 291
Parker, E. N.: 1979, *Cosmical Magnetic Fields*, Clarendon Press, Oxford
Parker, E. N.: 1984, *Astrophys. J.* **281**, 839
Parker, E. N.: 1988, *Astrophys. J.* **330**, 474
Parker, E. N.: 1993, *Astrophys. J.* **408**, 707
Parker, E. N.: 1994, *Spontaneous Current Sheets in Magnetic Fields*, Oxford University Press, Oxford, U.K.
Parker, E. N.: 1997, in J. R. Jokipii, C. P. Sonnett, and M. S. Giampapa (Eds.), *Cosmic winds and the heliosphere*, University of Arizona Press, Tucson, Arizona, p. 3
Parnell, C. E., Priest, E. R., and Golub, L.: 1994, *Sol. Phys.* **151**, 57
Pasquini, L., Brocato, E., and Pallavicini, R.: 1990, *Astron. Astrophys.* **234**, 277
Patten, B. M. and Simon, Th.: 1996, *Astrophys. J., Suppl. Ser.* **106**, 489
Peres, G.: 1997, in *The corona and solar wind near minimum activity*, European Space Agency, ESA Publications Division, ESTEC, Noordwijk, The Netherlands, p. 55
Pettersen, B. R.: 1989, *Sol. Phys.* **121**, 299
Phillips, K. J. H., Greer, C. J., Bhatia, A. K., and Keenan, F. P.: 1996, *Astrophys. J., Lett.* **469**, 57
Phillips, M. J. and Hartmann, L.: 1978, *Astrophys. J.* **224**, 182
Pierce, A. K. and LoPresto, J. C.: 1984, *Sol. Phys.* **93**, 155
Pinsonneault, M. H., Deliyannis, C. P., and Demarque, P.: 1991, *Astrophys. J.* **367**, 239
Pinsonneault, M. H., Kawaler, S. D., and Demarque, P.: 1990, *Astrophys. J., Suppl. Ser.* **74**, 501
Pinsonneault, M. H., Kawaler, S. D., Sofia, S., and Demarque, P.: 1989, *Astrophys. J.* **338**, 424
Piters, A.: 1995, *Ph.D. thesis*, Astronomical Institute 'Anton Pannekoek', Amsterdam
Pizzo, V., Schwenn, R., March, E., Rosenbauer, H., Muelhaeuser, K.-H., and Neubauer, F. M.: 1983, *Astrophys. J.* **271**, 335
Popper, D. M.: 1980, *Ann. Rev. Astron. Astrophys.* **18**, 115
Porter, L. J. and Klimchuk, J. A.: 1995, *Astrophys. J.* **454**, 499
Press, W. H., Flannery, B. P., Teukolsky, S. A., and Vetterling, W. T.: 1992, *Numerical recipes*, Cambridge University Press, Cambridge, U.K.
Priest, E. R.: 1982, *Solar Magnetohydrodynamics*, Reidel, Dordrecht
Priest, E. R. (Ed.): 1989, *Dynamics and Structure of Quiescent Solar Prominences*, Kluwer, Dordrecht
Priest, E. R.: 1993, in J. L. Linsky (Ed.), *Physics of Solar and Stellar Coronae*, Kluwer Academic Publishers, Dordrecht, The Netherlands, p. 515
Proctor, M. R. E. and Gilbert, A. D. (Eds.): 1994, *Lectures on Solar and Planetary Dynamos*, Cambridge University Press, Cambridge, U.K.
Rabin, D.: 1992, *Astrophys. J.* **391**, 832
Rabin, D., Moore, R., and Hagyard, M. J.: 1984, *Astrophys. J.* **287**, 404
Rammacher, W. and Ulmschneider, P.: 1992, *Astron. Astrophys.* **253**, 586
Ramsey, L. W. and Nations, H. L.: 1980, *Astrophys. J., Lett.* **239**, 121
Rebolo, R., Martin, E. L., Basri, G., Marcy, G. W., and Zapatero-Osorio, M. R.: 1996, *Astrophys. J.* **469**, 53
Rees, D. E.: 1987, in W. Kalkofen (Ed.), *Numerical Radiative Transfer*, Cambridge University Press, Cambridge, p. 213
Reeves, E. M.: 1976, *Sol. Phys.* **46**, 53
Reeves, E. M., Noyes, R. W., and Withbroe, G. L.: 1972, *Sol. Phys.* **27**, 251
Reeves, E. M., Vernazza, J. E., and Withbroe, G. L.: 1976, *Phi. Trans. Roy. Soc. London A* **281**, 319

Reimers, D.: 1975, *Mem. Soc. R. Sci Liège* **8**, 369

Reimers, D.: 1977, *Astron. Astrophys.* **57**, 395

Reimers, D.: 1989, in L. E. Cram and L. V. Kuhi (Eds.), *FGK Stars and T Tauri Stars (NASA SP-502)*, NASA, Washington, p. 53

Renzini, A., Cacciani, C., Ulmschneider, P., and Schmitz, F.: 1977, *Astron. Astrophys.* **61**, 39

Ribes, J. C. and Nesme-Ribes, E.: 1993, *Astron. Astrophys.* **276**, 549

Roberts, P. H. and Stix, M.: 1971, *NCAR-TN/IA-60*, NCAR, Boulder, Colorado, USA

Robinson, R. D.: 1980, *Astrophys. J.* **239**, 961

Robinson, R. D., Worden, S. P., and Harvey, J. W.: 1980, *Astrophys. J.* **236**, L155

Rodono, M., Cutispoto, G., Pazzani, V., Catalano, S., Byrne, P. B., Doyle, J. G., Butler, C. J., Andrews, A. D., Blanco, C., Marilli, E., Linsky, J. L., Scaltriti, F., Busso, M., Cellino, A., Hopkins, J. L., Oakazaki, A., Hayashi, S. S., Zeilik, M., Helston, R., Henson, G., Smith, P., and Simon, T.: 1986, *Astron. Astrophys.* **165**, 135

Rodono, M., Pazzani, U., and Cutispoto, G.: 1983, in P. B. Byrne and M. Rodono (Eds.), *Activity in Red Dwarf Stars*, IAU Colloq. 71, Reidel, Dordrecht, 179

Rosner, R., An, C. H, Musielak, Z. E., Moore, R. L., and Suess, S. T.: 1991, *Astrophys. J.* **372**, L91

Rosner, R.: 1980, in A. K. Dupree (Ed.), *Cool Stars, Stellar Systems, and the Sun*, SAO, Special Report No. 389, 79

Rosner, R.: 1991, in P. Ulmschneider, E. Priest, and R. Rosner (Eds.), *Mechanisms of Chromospheric and Coronal Heating*, Springer-Verlag, Heidelberg, p. 287

Rosner, R., Musielak, Z. E., Cattaneo, F., Moore, R. L., and Suess, S. T.: 1995, *Astrophys. J.* **442**, L25

Rosner, R., Tucker, W. H., and Vaiana, G. S.: 1978, *Astrophys. J.* **220**, 643

Rosner, R. and Weiss, N. O.: 1992, in K. L. Harvey (Ed.), *The Solar Cycle*, Proceedings of the National Solar Observatory / Sacramento Peak 12th summer workshop, Astron. Soc. of the Pacific Conf. Ser. Vol. 27, San Francisco, p. 511

Roudier, Th. and Muller, R.: 1986, *Sol. Phys.* **107**, 11

Roy, J.-R: 1973, *Sol. Phys.* **28**, 91

Rucinski, S. M.: 1994, *Publ. Astron. Soc. Pac.* **106**, 462

Rucinski, S. M. and Vilhu, O.: 1983, *Mon. Not. R. Astron. Soc.* **202**, 1221

Rüdiger, G.: 1994, in M. Schüssler and W. Schmidt (Eds.), *Solar Magnetic Fields*, Cambridge University Press, Cambridge, UK, p. 77

Rüdiger, G., Von Rekowski, B., Donahue, R. A., and Baliunas, S. L.: 1998, *Astrophys. J.* **494**, 691

Rüedi, I., Solanki, S. K., Livingston, W., and Stenlo, J. O.: 1992, *Astron. Astrophys.* **263**, 323

Rugge, H. R. and McKenzie, D. L.: 1985, *Astrophys. J.* **279**, 338

Rust, D. M.: 1968, in K. O. Kiepenheuer (Ed.), *Structure and Development of Solar Active Regions*, IAU Symp. 35, 77

Rutten, R. G. M.: 1984a, *Astron. Astrophys.* **130**, 353

Rutten, R. G. M.: 1984b, *Astron. Astrophys.* **130**, 353

Rutten, R. G. M.: 1987, *Astron. Astrophys.* **177**, 131

Rutten, R. G. M. and Pylyser, E.: 1988, *Astron. Astrophys.* **191**, 227

Rutten, R. G. M. and Schrijver, C. J.: 1986, in M. Zeilik and D. M. Gibson (Eds.), *Cool Stars, Stellar Systems and the Sun*, Springer-Verlag, Berlin, p. 120

Rutten, R. G. M. and Schrijver, C. J.: 1987a, *Astron. Astrophys.* **177**, 155

Rutten, R. G. M. and Schrijver, C. J.: 1987b, *Astron. Astrophys.* **177**, 143

Rutten, R. G. M., Schrijver, C. J., Lemmens, A. F. P., and Zwaan, C.: 1991, *Astron. Astrophys.* **252**, 203

Rutten, R. G. M., Schrijver, C. J., Zwaan, C., Duncan, D. K., and Mewe, R.: 1989, *Astron. Astrophys.* **219**, 239

Rutten, R. J. and Uitenbroek, H.: 1991, *Sol. Phys.* **134**, 15

Ruzmaikin, A., Sokoloff, D., and Tarbell, T. D.: 1991, in I. Tuominen, D. Moss, and G. Ruediger (Eds.), *The Sun and Cool Stars: activity, magnetism, dynamos*, Lecture Notes in Physics, Springer-Verlag, Berlin, p. 140

Saar, S. H.: 1988, *Astrophys. J.* **324**, 441

Saar, S. H.: 1996, in Y. Uchida, T. Kosugi, and H. S. Hudson (Eds.), *Magnetodynamic Phenomena in the Solar Atmosphere*, IAU 153 Colloquium, Kluwer Acad. Publ., Dordrecht, Holland, p. 367

Saba, J. L. R., Schmelz, J. T., Bhatia, A. K., and Strong, K. T.: 1999, *Astrophys. J.* **510**, 1064

Sánchez Almeida, J. and Lites, B. W.: 1992, *Astrophys. J.* **398**, 359

Sánchez Almeida, J. and Martínez Pillet, V.: 1994, *Astrophys. J.* **424**, 1014

Savonije, G. J. and Papaloizou, J. C. B.: 1985, in P. P. Eggleton and J. E. Pringle (Eds.), *Interacting Binaires*, Reidel, 83

Schaller, G., Schaerer, D., Meynet, G., and Maeder, A.: 1992, *Astron. Astrophys. Suppl. Ser.* **96**, 269

Schatten, K. H., Wilcox, J. M., and Ness, N. F.: 1969, *Sol. Phys.* **6**, 442

Schatzman, E. L. and Praderie, F.: 1993, *The Stars*, Springer-Verlag, Berlin, etc.

Schmelz, J. T., Saba, J. L., and Strong, K. T.: 1992, *Astrophys. J., Lett.* **398**, 115

Schmidt, H. U.: 1964, in W. Hess (Ed.), *NASA Symp. on Phys. of Solar Flares*, NASA SP-50, 107

Schmitt, D.: 1993, in F. Krause, K. H. Rädler, and G. Rüdiger (Eds.), *The Cosmic Dynamo*, IAU Symp. 157, Kluwer Academic Publishers, Dordrecht, the Netherlands, 1

Schmitt, J. H. M. M.: 1997, *Astron. Astrophys.* **318**, 215

Schmitt, J. H. M. M., Collura, A., Sciortino, S., Vaiana, G. S., Harnden, F. R., and Rosner, R.: 1990, *Astrophys. J.* **365**, 704

Schmitt, J. H. M. M., Drake, J. J., Haisch, B. M., and Stern, R. A.: 1996a, *Astrophys. J.* **467**, 841

Schmitt, J. H. M. M., Drake, J. J., and Stern, R. A.: 1996b, *Astrophys. J., Lett.* **465**, 51

Schoolman, S. A.: 1973, *Sol. Phys.* **32**, 379

Schrijver, C. J.: 1983, *Astron. Astrophys.* **127**, 289

Schrijver, C. J.: 1987a, *Astron. Astrophys.* **172**, 111

Schrijver, C. J.: 1987b, *Astron. Astrophys.* **180**, 241

Schrijver, C. J.: 1988, *Astron. Astrophys.* **189**, 163

Schrijver, C. J.: 1989, *Sol. Phys.* **122**, 193

Schrijver, C. J.: 1990, *Astron. Astrophys.* **234**, 315

Schrijver, C. J.: 1991, in P. Ulmschneider, E. Priest, and R. Rosner (Eds.), *Mechanisms of Chromospheric and Coronal Heating*, Springer-Verlag, Heidelberg, p. 257

Schrijver, C. J.: 1992, *Astron. Astrophys.* **258**, 507

Schrijver, C. J.: 1993a, *Astron. Astrophys.* **269**, 446

Schrijver, C. J.: 1993b, *Astron. Astrophys.* **269**, 395

Schrijver, C. J.: 1994, in R. J. Rutten and C. J. Schrijver (Eds.), *Solar Surface Magnetism*, NATO ASI Series, Kluwer Academic Publishers, Dordrecht, p. 271

Schrijver, C. J.: 1995, *Astron. Astrophys. Rev.* **6**, 181

Schrijver, C. J.: 1996, in K. G. Strassmeier and J. L. Linsky (Eds.), *Stellar Surface Structure*, IAU Symp. 176, Kluwer Academic Publishers, Dordrecht, The Netherlands, p. 1

Schrijver, C. J., Coté, J., Zwaan, C., and Saar, S. H.: 1989a, *Astrophys. J.* **337**, 964

Schrijver, C. J., Dobson, A. K., and Radick, R. R.: 1989b, *Astrophys. J.* **341**, 1035

Schrijver, C. J., Dobson, A. K., and Radick, R. R.: 1992a, *Astron. Astrophys.* **258**, 432

Schrijver, C. J., Hagenaar, H. J., and Title, A. M.: 1997a, *Astrophys. J.* **475**, 328

Schrijver, C. J. and Haisch, B.: 1996, *Astrophys. J., Lett.* **455**, 55

Schrijver, C. J. and Harvey, K. L.: 1989, *Astrophys. J.* **343**, 481

Schrijver, C. J. and Harvey, K. L.: 1994, *Sol. Phys.* **150**, 1

Schrijver, C. J., Jiménez, A., and Däppen, W.: 1991, *Astron. Astrophys.* **251**, 655

Schrijver, C. J., Lemen, J. R., and Mewe, R.: 1989c, *Astrophys. J.* **341**, 484

Schrijver, C. J. and Martin, S. F.: 1990, *Sol. Phys.* **129**, 95

Schrijver, C. J. and McMullen, R. A.: 1999, *Astrophys. J.*, submitted

Schrijver, C. J., Mewe, R., Van den Oord, G. H. J., and Kaastra, J. S.: 1995, *Astron. Astrophys.* **302**, 438

Schrijver, C. J., Mewe, R., and Walter, F. M.: 1984, *Astron. Astrophys.* **138**, 258

Schrijver, C. J. and Pols, O. R.: 1993, *Astron. Astrophys.* **278**, 51

Schrijver, C. J., Shine, R. A., Hagenaar, H. J., Hurlburt, N. E., Title, A. M., Strous, L. H., Jefferies, S. M., Jones, A. R., Harvey, J. W., and Duvall, Th. L.: 1996a, *Astrophys. J.* **468**, 921

Schrijver, C. J., Shine, R. A., Hurlburt, N. E., Tarbell, T. D., and Lemen, J. R.: 1997b, in O. Kjeldseth-Moe and A. Wilson (Eds.), *Proceedings of the 5th SOHO workshop, Oslo, June 1997*, ESA SP-404, Noordwijk, The Netherlands, p. 669

Schrijver, C. J., Title, A. M., Berger, T. E., Fletcher, L., Hurlburt, N. E., Nightingale, R., Shine, R. A., Tarbell, T. D., Wolfson, J., Golub, L., Bookbinder, J. A., DeLuca, E. E., McMullen, R. A., Warren, H. P., Kankelborg, C. C., Handy, B. N., and De Pontieu, B.: 1999a, *Sol. Phys.*, in press

Schrijver, C. J., Title, A. M., Hagenaar, H. J., and Shine, R. A.: 1997c, *Sol. Phys.* **175**, 329

Schrijver, C. J., Title, A. M., Harvey, K. L., Sheeley, N. R., Wang, Y. M., Van den Oord, G. H. J., Shine, R. A., Tarbell, T. D., and Hurlburt, N. E.: 1998, *Nature* **394**, 152

Schrijver, C. J., Title, A. M., and Ryutova, M. P.: 1999b, *Sol. Phys.*, in press

Schrijver, C. J., Title, A. M., Van Ballegooijen, A. A., Hagenaar, H. J., and Shine, R. A.: 1997d, *Astrophys. J.* **487**, 424

Schrijver, C. J., Van den Oord, G. H. J., and Mewe, R.: 1994, *Astron. Astrophys.* **289**, L23

Schrijver, C. J., Van den Oord, G. H. J., Mewe, R., and Kaastra, J. S.: 1996b, in S. Bowyer and B. M. Haisch (Eds.), *Astrophysics in the extreme ultraviolet*, IAU Coll. 152, Kluwer Academic Publishers, Dordrecht, The Netherlands, p. 121

Schrijver, C. J. and Zwaan, C.: 1991, *Astron. Astrophys.* **251**, 183

Schrijver, C. J., Zwaan, C., Balke, A. C., Tarbell, T. D., and Lawrence, J. K.: 1992b, *Astron. Astrophys.* **253**, L1

Schrijver, C. J., Zwaan, C., Maxson, C. W., and Noyes, R. W.: 1985, *Astrophys. J.* **149**, 123

Schroeder, K.-P.: 1983, *Astron. Astrophys.* **124**, L16

Schröter, E. H.: 1957, *Z. F. Astrophys.* **41**, 141

Schröter, E. H.: 1985, *Sol. Phys.* **100**, 141

Schulz, M.: 1973, *Astrophys. Sp. Sc.* **24**, 371

Schüssler, M.: 1983, in J. O. Stenflo (Ed.), *Solar and stellar magnetic fields*, IAU Symp. No. 102, D. Reidel, Dordrecht, The Netherlands, p. 213

Schüssler, M.: 1990, in J.-O. Stenflo (Ed.), *The Solar Photosphere: Structure, Convection and Magnetic Fields*, Proceedings IAU Symposium 138 (Kiev), Kluwer, Dordrecht, p. 161

Schüssler, M.: 1996, in K. G. Strassmeier and J. L. Linsky (Eds.), *Stellar Surface Structure*, IAU Symp. 176, Kluwer Academic Publishers, Dordrecht, The Netherlands, p. 269

Schüssler, M. and Schmidt, W. (Eds.): 1994, *Solar Magnetic Fields*, Cambridge University Press, Cambridge, U.K.

Schwarzschild, K. and Eberhard, G.: 1913, *Astrophys. J.* **38**, 292

Schwarzschild, M.: 1948, *Astrophys. J.* **107**, 1

Semel, M.: 1985, in R. Muller (Ed.), *High Resolution in Solar Physics*, Springer-Verlag, Berlin, 178

Serio, S., Peres, G., Vaiana, G. S., Golub, L., and Rosner, R.: 1981, *Astrophys. J.* **243**, 288

Sheeley, N. R.: 1969, *Sol. Phys.* **9**, 347

Sheeley, N. R.: 1981, in F. Q. Orrall (Ed.), *Solar Active Regions*, Colorado Assoc. Univ. Press, Boulder, p. 17

Sheeley, N. R.: 1991, *Astrophys. J.* **374**, 386

Sheeley, N. R.: 1992, in K. L. Harvey (Ed.), *The solar cycle*, A. S. P. Conf. Series, Vol. 27, Astron. Soc. of the Pac., San Francisco, p. 1

Sheeley, N. R., Bohlin, J. D., Brueckner, G. E., Purcell, J. D., Scherrer, V. S., and Tousey, R.: 1975, *Sol. Phys.* **40**, 103

Sheeley, N. R., Boris, J. P., Young, T. R., DeVore, C. R., and Harvey, K. L.: 1983, in J. O. Stenflo (Ed.), *Solar and Stellar Magnetic Fields: Origins and Coronal Effects*, IAU Symp. 102, D. Reidel Publishing Company, Dordrecht, The Netherlands, p. 273

Sheeley, N. R., Nash, A. G., and Wang, Y.-M.: 1987, *Astrophys. J.* **319**, 481

Sheeley, N. R., Wang, Y.-M., Hawley, S. H., Brueckner, G. E., Dere, K. P., Howard, R. A., Koomen, M. J., Korendyke, C. M., Michels, D. J., Paswaters, S. E., Socker, D. G., St. Cyr, O. C., Wang, D., Lamy, P. L., Llebaria, A., Schwenn, R., Simnett, G. M., and Biesecker, D. A.: 1997, *Astrophys. J.* **484**, 472

Shibata, K.: 1996, in Y. Uchida, T. Kosugi, and H. S. Hudson (Eds.), *Magnetodynamic Phenomena in the Solar Atmosphere*, IAU Colloq. 153, Kluwer Acad. Publ., Dordrecht, Holland, p. 13

Shimizu, T.: 1995, *Publ. Astron. Soc. Jpn.* **47**, 251

Shimizu, T., Tsuneta, S., Acton, L. W., Lemen, J. R., and Uchida, Y.: 1992, *Publ. Astron. Soc. Jpn.* **44**, L147

Shimojo, M., Hashimoto, T., Shibata, K., Hirayama, T., and Harvey, K. L.: 1996, in Y. Uchida, T. Kosugi, and H. S. Hudson (Eds.), *Magnetodynamic Phenomena in the Solar Atmosphere*, IAU Colloq. 153, Kluwer Acad. Publ., Dordrecht, Holland, p. 449

Shine, R. A., Title, A. M., Tarbell, T. D., Smith, K., Fank, Z. A., and Scharmer, G.: 1994, in R. J. Rutten and C. J. Schrijver (Eds.), *Solar Surface Magnetism*, NATO ASI Series C Vol. 433, Kluwer Academic Publ., Dordrecht, p. 197

Shu, F. H., Adams, F. C., and Lizano, S.: 1987, *Ann. Rev. Astron. Astrophys.* **25**, 23

Siarkowski, M.: 1992, *Mon. Not. R. Astron. Soc.* **259**, 453

Siarkowski, M.: 1996, in K. G. Strassmeier and J. L. Linsky (Eds.), *Stellar Surface Structure*, IAU Symp. 176, Kluwer Academic Publishers, Dordrecht, The Netherlands, p. 469

Siarkowski, M., Pre, P., Drake, S. A., White, N. E., and Singh, K. P.: 1996, *Astrophys. J.* **473**, 470

Similon, Ph. L. and Sudan, R. N.: 1989, *Astrophys. J.* **336**, 442

Simon, G. W., Title, A. M., and Weiss, N. O.: 1995, *Astrophys. J.* **442**, 886

Simon, G. W. and Wilson, P. R.: 1985, *Astrophys. J.* **295**, 241
Simon, T. and Drake, S. A.: 1989, *Astrophys. J.* **346**, 303
Singh, K. P., Drake, S. A., and White, N. E.: 1996, *Astron. J.* **111**, 2415
Skaley, D. and Stix, M.: 1991, *Astron. Astrophys.* **241**, 227
Skumanich, A.: 1972, *Astrophys. J.* **171**, 565
Skumanich, A., Lites, B. W., and Martínez Pillet, V.: 1994, in R. J. Rutten and C. J. Schrijver (Eds.), *Solar Surface Magnetism*, NATO ASI Series C Vol. 433, Kluwer Academic Publ., Dordrecht, p. 99
Skumanich, A., Smythe, C., and Frazier, E. N.: 1975, *Astrophys. J.* **200**, 747
Smith, C. W. and Bieber, J. W.: 1991, *Astrophys. J.* **370**, 435
Smithson, R. C.: 1973a, *Sol. Phys.* **29**, 365
Smithson, R. C.: 1973b, *Sol. Phys.* **29**, 365
Snodgrass, H. B.: 1984, *Sol. Phys.* **94**, 13
Snodgrass, H. B.: 1991, *Astrophys. J.* **383**, L85
Snodgrass, H. B.: 1992, in K. L. Harvey (Ed.), *The Solar Cycle*, Ast. Soc. Pacific Conf. Ser. Vol. 27, San Francisco, p. 205
Snodgrass, H. B. and Dailey, S. B.: 1996, *Sol. Phys.* **163**, 21
Sobotka, M., Bonet, J. A., Vazquez, M., and Hanslmeier, A.: 1995, *Astrophys. J.* **447**, L133
Soderblom, D. R., Stauffer, J. R., MacGregor, K. B., and Jones, B. F.: 1993, *Astrophys. J.* **409**, 624
Solanki, S., Steiner, O., and Uitenbroek, H.: 1991, *Astron. Astrophys.* **250**, 220
Solanki, S. K.: 1993, *Space Sci. Rev.* **63**, 1
Solanki, S. K.: 1997, in G. Simnett, C. E. Alissandrakis, and L. Vlahos (Eds.), *Solar and Heliospheric Plasma Physics*, Springer-Verlag, Berlin, p. 49
Solanki, S. K., Bruls, J. H. M. J., Steiner, O., Ayres, T. R., Livingston, W. C., and Uitenbroek, H.: 1994, in R. J. Rutten and C. J. Schrijver (Eds.), *Solar Surface Magnetism*, (NATO Advanced Study Workshop), Kluwer Academic Publishers, Dordrecht, p. 91
Soon, W., Frick, P., and Baliunas, S.: 1999, *Astrophys. J., Lett.* **510**, 135
Soon, W. H., Baliunas, S. L., and Zhang, Q.: 1993, *Astrophys. J.* **414**, L33
Spiegel, E. A. and Weiss, N. O.: 1980, *Nature* **287**, 616
Spiegel, E. A. and Zahn, J.-P.: 1992, *Astron. Astrophys.* **265**, 106
Spitzer, L.: 1962, *Physics of fully ionized gases*, Interscience, New York
Spruit, H. C.: 1976, *Sol. Phys.* **50**, 269
Spruit, H. C.: 1977a, *Sol. Phys.* **55**, 3
Spruit, H. C.: 1977b, *Magnetic Flux Tubes and Transport of Heat*, Ph.D. Thesis, Utrecht University
Spruit, H. C.: 1979, *Sol. Phys.* **61**, 363
Spruit, H. C.: 1981a, *Astron. Astrophys.* **98**, 155
Spruit, H. C.: 1981b, *Astron. Astrophys.* **102**, 129
Spruit, H. C.: 1981c, in S. D. Jordan (Ed.), *The Sun as a Star*, NASA-CNRS Monograph series on nonthermal phenomena in stellar atmospheres, NASA SP-450, p. 385
Spruit, H. C.: 1983, in J. O. Stenflo (Ed.), *Solar and Stellar Magnetic Fields: Origins and Coronal Effects*, I.A.U. Symposium 102, Reidel, Dordrecht, p. 41
Spruit, H. C.: 1992, in J. H. Thomas and N. O. Weiss (Eds.), *Sunspots: Theory and Observations*, NATO ASI Series C 375, Kluwer Acadmic Publishers, Dordrecht, p. 163
Spruit, H. C.: 1994, in J. M. Pap, C. Fröhlich, H. S. Hudson, and S. K. Solanki (Eds.), *The Sun as a Variable Star*, IAU Colloq. 143, Cambridge University Press, Cambridge, UK, p. 270
Spruit, H. C.: 1997, *Mem. S. A. It.*, in the press
Spruit, H. C., Nordlund, Å., and Title, A. M.: 1990, *Ann. Rev. Astron. Astrophys.* **28**, 263
Spruit, H. C., Title, A. M., and van Ballegooijen, A. A.: 1987, *Sol. Phys.* **110**, 115
Spruit, H. C. and Zwaan, C.: 1981, *Sol. Phys.* **70**, 207
Stauffer, D.: 1985, *Introduction to percolation theory*, Taylor and Francis, London and Philadelphia
Stauffer, J. B. and Hartmann, L.: 1986, *Publ. Astron. Soc. Pac.* **98**, 1233
Stauffer, J. R., Hartmann, L. W., Prosser, C. F., Balachandran, S., Patten, B. M., Simon, Th., and Giampapa, M.: 1997, *Astrophys. J.* **479**, 776
Stępién, K.: 1994, *Astron. Astrophys.* **292**, 191
Steenbeck, M. and Krause, F.: 1969, *Astron. Nachr.* **291**, 49
Stein, R. F.: 1967, *Solar Phys.* **2**, 385
Stein, R. F.: 1968, *Astrophys. J.* **154**, 297
Stein, R. F. and Nordlund, Å: 1998, *Astrophys. J.* **499**, 914

Steiner, O., Grossmann-Doerth, U., Knölker, M., and Schüssler, M.: 1998, *Astrophys. J.* **495**, 468

Stencel, R. E.: 1978, *Astrophys. J.* **223**, L37

Stenflo, J. O.: 1985, *Sol. Phys.* **100**, 189

Stenflo, J. O.: 1994, *Solar Magnetic Fields: Polarized Radiation Diagnostics*, Kluwer Acadmic Publishers, Dordrecht

Stenflo, J. O. and Harvey, J. W.: 1985, *Sol. Phys.* **95**, 99

Stenflo, J. O., Harvey, J. W., Brault, J. W., and Solanki, S.: 1984, *Astron. Astrophys.* **131**, 333

Sterling, A., Hudson, H. S., and Watanabe, T.: 1997, *Astrophys. J., Lett.* **479**, 149

Stix, M.: 1976, in V. Bumba and J. Kleczek (Eds.), *Basic Mechanisms of Solar Activity*, Reidel, Dordrecht, the Nethelands, 367

Stix, Michael: 1989, *The Sun. An Introduction*, Springer, Berlin

Strassmeier, K. G.: 1996, in K. G. Strassmeier and J. L. Linsky (Eds.), *Stellar Surface Structure*, IAU Symp. 176, Kluwer Academic Publishers, Dordrecht, The Netherlands, p. 289

Strassmeier, K. G., Bartus, J., Cutispote, G., and Rodono, M.: 1997, *Astron. Astrophys. Suppl. Ser.* **125**, 11

Strassmeier, K. G., Hall, D. S., and Henry, G. W.: 1994, *Astron. Astrophys.* **282**, 535

Strassmeier, K. G., Rice, J. B., Wehlau, W. H., Hill, G. M., and Mattehws, J. M.: 1993, *Astron. Astrophys.* **268**, 671

Strous, L. H.: 1994, *Ph.D. thesis*, Astronomical Institute, Utrecht University

Strous, L. H., Scharmer, G., Tarbell, T. D., Title, A. M., and Zwaan, C.: 1996, *Astron. Astrophys.* **306**, 947

Strous, L. H. and Zwaan, C.: 1999, *Astrophys. J.* –, in press

Sutmann, G. and Ulmschneider, P.: 1995a, *Astron. Astrophys.* **294**, 232

Sutmann, G. and Ulmschneider, P.: 1995b, *Astron. Astrophys.* **294**, 241

Švestka, Z.: 1976, *Solar Flares*, Reidel Publ. Co., Dordrecht, The Netherlands

Tanaka, K.: 1991, *Sol. Phys.* **136**, 133

Tanaka, Y., Inoue, H., and Holt, S. S.: 1994, *Publ. Astron. Soc. Jpn., Lett.* **46**, 37

Tandberg-Hansen, E.: 1995, *The Nature of Solar Prominences*, Kluwer Acad. Publ., Dordrecht, The Netherlands

Tarbell, T. D., Ferguson, S., Frank, Z., Shine, R. A., Title, A. M., Topka, K. P., and Scharmer, G.: 1990, in J. O. Stenflo (Ed.), *The Solar Photosphere: Structure, Convection and Magnetic Fields*, Kluwer Academic Publishers, Dordrecht, Holland, 147

Thomas, J. H.: 1994, in R. J. Rutten and C. J. Schrijver (Eds.), *Solar Surface Magnetism*, NATO ASI Series C Vol. 433, Kluwer Academic Publ., Dordrecht, the Netherlands, p. 219

Thomas, John H. and Weiss, Nigel O. (Eds.): 1992a, *Sunspots: Theory and Observations*, NATO ASI Series C 375, Kluwer, Dordrecht, the Netherlands

Thomas, J. H. and Weiss, N. O.: 1992b, in J. H. Thomas and N. O. Weiss (Eds.), *Sunspots: Theory and Observations*, NATO ASI Series C: Vol. 375, Kluwer Academic Publishers, Dordrecht, the Netherlands, p. 3

Thompson, B. J., Newmark, J. S., Gurman, J. B., St. Cyr, O. C., and Stetzelberger, S.: 1997, *Bull. Am. Astron. Soc.* **29**, 01.30

Title, A. M. and Berger, T. E.: 1996, *Astrophys. J.* **463**, 797

Title, A. M., Tarbell, T. D., Topka, K. P., Ferguson, S. H., Shine, R. A., and the SOUP team: 1989, *Astrophys. J.* **336**, 475

Title, A. M., Topka, K. P., Tarbell, T. D., Schmidt, W., Balke, A. C., and Scharmer, G.: 1992, *Astrophys. J.* **393**, 782

Topka, K. P., Tarbell, T. D., and Title, A. M.: 1986, *Astrophys. J.* **306**, 304

Tousey, R.: 1973, in M. J. Mycroft and S. K. Runcorn (Eds.), *Space Research XIII*, Akademie-erlag, Berlin, 173

Tuominen, I., Moss, D., and Rüdiger, G. (Eds.): 1991, *The Sun and Cool Stars: activity, magnetism and dynamos*, Springer-Verlag, Berlin

Uchida, Y., Kosugi, T., and Hudson, H. S. (Eds.): 1996, *Magnetodynamic Phenomena in the Solar Atmosphere*, IAU Colloq. 153, Kluwer Acad. Publ., Dordrecht, Holland

Uchida, Y. and Sakurai, T.: 1983, in P. B. Byrne and M. Rodono (Eds.), *Activity in Red Dwarf Stars*, IAU Colloq. 71, Reidel, Dordrecht, 629

Uitenbroek, H.: 1989, *Astron. Astrophys.* **213**, 360

Uitenbroek, H., Noyes, R. W., and Rabin, D.: 1994, *Astrophys. J., Lett.* **432**, 67

Ulmschneider, P.: 1988, *Astron. Astrophys.* **197**, 223

Ulmschneider, P.: 1989, *Astron. Astrophys.* **222**, 171

Ulmschneider, P.: 1990, in G. Wallerstein (Ed.), *Cool Stars, Stellar Systems, and the Sun*, (Proceedings of the Fifth Cambridge Workshop), ASP Conf. Ser. Vol. 9, San Francisco, 116

Ulmschneider, P.: 1991, in P. Ulmschneider, E. Priest, and R. Rosner (Eds.), *Mechanisms of Chromospheric and Coronal Heating*, Springer-Verlag, Heidelberg, p. 328

Ulmschneider, P.: 1996, in R. Pallavicini and A. K. Dupree (Eds.), *Cool Stars, Stellar Systems and the Sun*, Ninth Cambridge Workshop, Astron. Soc. of the Pac. Conf. Ser. Vol. 109, San Francisco, CA, p. 71

Ulrich, R. K.: 1970, *Astrophys. J.* **162**, 993

Ulrich, R. K.: 1993, in W. W. Weiss and A. Baglin (Eds.), *Inside the Stars*, IAU Colloq. 137, Ast. Soc. Pacific Conf. Ser., San Francisco, 25

Unno, W.: 1956, *Publ. Astron. Soc. Jpn.* **8**, 108

Unno, W., Osaki, Y., Saio, H., and Shibahashi, H.: 1989, *Nonradial Oscillations of Stars*, Univ. of Tokyo Press, Tokyo, Japan

Unruh, Y. C.: 1996, in K. G. Strassmeier and J. L. Linsky (Eds.), *Stellar Surface Structure*, IAU Symp. 176, Kluwer Academic Publishers, Dordrecht, The Netherlands, p. 35

Unsöld, A. and Baschek, B.: 1991, *The New Cosmos*, Springer Verlag, Berlin

Vaiana, G. S., Krieger, A. S., and Timothy, A. F.: 1973, *Sol. Phys.* **32**, 81

Vaiana, G. S., Krieger, A. S., Timothy, A. F., and Zombeck, F.: 1976, *Astrophys. Space Sc.* **39**, 75

Valenti, J. A., Marcy, G. W., and Basri, G. S.: 1995, *Astrophys. J.* **439**, 939

Van Ballegooijen, A. A.: 1982a, *Astron. Astrophys.* **113**, 99

Van Ballegooijen, A. A.: 1982b, *Astron. Astrophys.* **106**, 43

Van Ballegooijen, A. A.: 1984a, *Sol. Phys.* **91**, 195

Van Ballegooijen, A. A.: 1984b, in S. L. Keil (Ed.), *Small-scale dynamic processes in quiet stellar atmospheres*, NSO/Sacramento Peak, AURA, Sunspot, p. 260

Van Ballegooijen, A. A.: 1985, in M. J. Hagyard (Ed.), *Measurements of Solar Vector Magneic Fields*, NASA Conf. Pub. 2374, p. 322

Van Ballegooijen, A. A.: 1985, *Astrophys. J.* **298**, 421

Van Ballegooijen, A. A.: 1986, *Astrophys. J.* **311**, 1001

Van den Oord, G. H. J.: 1988, *Astron. Astrophys.* **205**, 167

Van den Oord, G. H. J. and Doyle, J. G.: 1997, *A&A* **319**, 578

Van den Oord, G. H. J., Doyle, J. G., Rodono, M., Gary, D. E., Henry, G. W., Byrne, P. B., Linsky, J. L., Haisch, B. M., Pagano, I., and Leto, G.: 1996, *Astron. Astrophys.* **310**, 908

Van den Oord, G. H. J., Schrijver, C. J., Camphens, M., Mewe, R., and Kaastra, J. S.: 1997, *Astron. Astrophys.* **326**, 1090

Van den Oord, G. H. J. and Zuccarello, F.: 1996, in K. G. Strassmeier and J. L. Linsky (Eds.), *Stellar Surface Structure*, IAU Symp. 176, Kluwer Academic Publishers, Dordrecht, The Netherlands, p. 433

Van Driel-Gesztelyi, L. and Petrovay, K.: 1990, *Sol. Phys.* **126**, 285

Van Driel-Gesztelyi, L., van der Zalm, E. B. J., and Zwaan, C.: 1992, in K. L. Harvey (Ed.), *The Solar Cycle*, NSO/Sacramento Peak, Astron. Soc. Pacific Conf. Series, 27, Sunspot, New Mexico 88349, USA, p. 89

Vaughan, A. H., Baliunas, S. L., Middelkoop, F., Hartmann, L. W., Mihalas, D., Noyes, R. W., and Preston, G. W.: 1981, *Astrophys. J.* **250**, 276

Vaughan, A. H. and Preston, G. W.: 1980, *Publ. Astron. Soc. Pac.* **92**, 38

Vaughan, A. H., Preston, G. W., and Wilson, O. C.: 1978a, *Publ. Astron. Soc. Pac.* **90**, 267

Vaughan, A. H., Preston, G. W., and Wilson, O. C.: 1978b, *Publ. Astron. Soc. Pac.* **90**, 267

Verbunt, F. and Zwaan, C.: 1981, *Astron. Astrophys.* **100**, L7

Vernazza, J. E., Avrett, E. H., and Loeser, R.: 1976, *Astrophys. J., Suppl. Ser.* **30**, 1

Vernazza, J. E., Avrett, E. H., and Loeser, R.: 1981, *Astrophys. J., Suppl. Ser.* **45**, 635

Vernazza, J. E., Foukal, P. V., Huber, M. C. E., Noyes, R. W., Reeves, E. M., Schmahl, E. J., Timothy, J. G., and Withbroe, G. L.: 1975, *Astrophys. J., Lett.* **199**, L123

Vesecky, J. F., Antiochos, S. K., and Underwood, J. H.: 1979, *Astrophys. J.* **233**, 987

Vilhu, O.: 1987, in J. L. Linsky and R. E. Stencel (Eds.), *Cool Stars, Stellar Systems, and the Sun*, Springer-Verlag, Berlin, p. 110

Vilhu, O., Neff, J. E., and Rahunen, T.: 1989, *Astron. Astrophys.* **208**, 201

Vilhu, O. and Rucinki, S. M.: 1983, *Astron. Astrophys.* **127**, 5

Vilhu, O. and Walter, F. M.: 1987, *Astrophys. J.* **321**, 958

Vitense, E.: 1953, *Z. Astrophys.* **32**, 135

Vogt, S. S.: 1975, *Astrophys. J.* **199**, 418

Von Klüber, H.: 1947, *Z. Astrophys.* **24**, 121

Von Steiger, R. and Geiss, J.: 1989, *Astron. Astrophys.* **225**, 222
Von Steiger, R., Geiss, J., and Gloeckler, G.: 1997, in J. R. Jokipii, C. P. Sonnett, and M. S. Giampapa
  (Eds.), *Cosmic winds and the heliosphere*, University of Arizona Press, Tucson, Arizona, p. 581
Von Üxküll, M., Kneer, F., Malherbe, J. M., and Mein, P.: 1989, *Astron. Astrophys.* **208**, 290
Vrabec, D.: 1974, in R. G. Athay (Ed.), *Chromospheric Fine Structure*, IAU Symposium 56, Reidel,
  Dordrecht, 201
Wagner, W. J.: 1984, *Ann. Rev. Astron. Astrophys.* **22**, 267
Waldmeier, M.: 1937, *Z. Astrophys.* **14**, 91
Waldmeier, M.: 1955, *Ergebnisse und Probleme der Sonnenforschung*, Geest & Portig K.-G, Leipzig
Waljeski, K., Moses, D., Dere, K. P., Saba, J. L., Strong, K. T., Webb, D. F., and Zarro, D. M.: 1994,
  Astrophys. J. **429**, 909
Wallenhorst, S. G. and Howard, R.: 1982, *Sol. Phys.* **76**, 203
Wallenhorst, S. G. and Topka, K. P.: 1982, *Sol. Phys.* **81**, 33
Walter, F. M.: 1987, *Publ. Astron. Soc. Pac.* **99**, 31
Walter, F. M.: 1996, in K. G. Strassmeier and J. L. Linsky (Eds.), *Stellar Surface Structure*, IAU Symp. 176,
  Kluwer Academic Publishers, Dordrecht, The Netherlands, p. 355
Walter, F. M., Gibson, D. M., and Basri, G. S.: 1983, *Astrophys. J.* **267**, 665
Wang, H.: 1988, *Sol. Phys.* **116**, 1
Wang, H. and Sakurai, T.: 1998, *Publ. Astron. Soc. Jpn.* **50**, 111
Wang, J., Wang, H., Tang, F., Lee, J. W., and Zirin, H.: 1995, *Sol. Phys.* **160**, 277
Wang, Y.-M., Nash, A. G., and Sheeley, N. R.: 1989, *Science* **245**, 712
Wang, Y. M. and Sheeley, N. R.: 1989, *Sol. Phys.* **124**, 81
Wang, Y.-M. and Sheeley, N. R.: 1993, *Astrophys. J.* **414**, 916
Weart, S. R.: 1970, *Astrophys. J.* **162**, 987
Weart, S. R.: 1972, *Astrophys. J.* **177**, 271
Webb, D. F.: 1998, in D. F. Webb, D. M. Rust, and B. Schmieder (Eds.), *New Perspectives on Solar
  Prominences*, IAU Colloq. 167, 463
Webb, D. F., Martin, S. F., Moses, D., and Harvey, J. W.: 1993, *Sol. Phys.* **144**, 15
Webb, D. F., Rust, D. M., and Schmieder, B. (Eds.): 1998, *New Perspectives on Solar Prominences*, IAU
  Colloq. 167, ASP Conf. Series Vol. 150, San Francisco
Weber, E. J. and Davis, L.: 1967, *Astrophys. J.* **148**, 217
Weiss, N. O.: 1964, *Mon. Not. R. Astron. Soc.* **128**, 225
Weiss, N. O.: 1966, *Proc. Roy. Soc. London Ser.A* **293**, 310
White, N. E., Shafer, R. A., Parmar, A. N., Horne, K., and Culhane, J. L.: 1990, *Astrophys. J.* **350**, 776
White, O. R. and Livingston, W. C.: 1981a, *Astrophys. J.* **249**, 798
White, O. R. and Livingston, W. C.: 1981b, *Astrophys. J.* **249**, 798
White, O. R. and Trotter, D. E.: 1977, *Astrophys. J., Suppl. Ser.* **33**, 391
White, S. M.: 1996, in R. Pallavicini and A. K. Dupree (Eds.), *Cool Stars, Stellar Systems, and the Sun*,
  ASP Conference Series Vol. 109, San Francisco, p. 21
Wikstøl, Ø., Judge, P. G., and Hansteen, V.: 1998, *Astrophys. J.* **501**, 895
Wilcox, J. M.: 1968, *Space Sci. Rev.* **8**, 258
Wild, J. P. and McCready, L. L.: 1950, *Aust. J. Sci.* **A3**, 387
Willson, R. C. and Hudson, H. S.: 1991, *Nature* **351**, 42
Wilson, O. C.: 1966, *Astrophys. J.* **144**, 695
Wilson, O. C.: 1968, *Astrophys. J.* **153**, 221
Wilson, O. C.: 1976, *Astrophys. J.* **205**, 823
Wilson, O. C.: 1978, *Astrophys. J.* **226**, 379
Wilson, O. C. and Bappu, M. K. V.: 1957, *Astrophys. J.* **125**, 661
Withbroe, G. L.: 1975, *Sol. Phys.* **45**, 301
Withbroe, G. L.: 1981, in *Activity and outer atmospheres of the Sun and stars*, 11th Advanced course of the
  Swiss Ntl. Ac. of Sciences, Observatoire de Genève, Sauverny, Switzerland
Withbroe, G. L. and Mariska, J. T.: 1975, *Sol. Phys.* **44**, 55
Withbroe, G. L. and Mariska, J. T.: 1976, *Sol. Phys.* **48**, 21
Woltjer, L.: 1958, *Proc. Nat. Acad. Sci. USA* **44**, 489
Woo, R. and Habbal, S. R.: 1997, *Astrophys. J., Lett.* **474**, 139
Wood, B. E., Brown, A., and Linsky, J. L.: 1995, *Astrophys. J.* **438**, 350
Wood, B. E., Linsky, J. L., and Ayres, T. R.: 1997, *Astrophys. J.* **478**, 745

Yoshida, T. and Tsuneta, S.: 1996, *Astrophys. J.* **459**, 342
Yoshimura, H.: 1975, *Astrophys. J.* **201**, 740
Zahn, J.-P.: 1966, *Ann. Astrophys.* **29**, 489
Zahn, J.-P.: 1977, *Astron. Astrophys.* **57**, 383
Zahn, J. P.: 1989, *Astron. Astrophys.* **220**, 112
Zahn, J.-P.: 1992, *Astron. Astrophys.* **265**, 115
Zahn, J.-P., Talon, S., and Matias, J.: 1997, *Astron. Astrophys.* **322**, 320
Zeilik, M., Gordon, S., Jaderlund, E., Ledlow, M., Summers, D. L., Heckert, P. A., Budding, E., and Banks, T. S.: 1994, *Astrophys. J.* **421**, 303
Zirin, H.: 1972, *Sol. Phys.* **22**, 34
Zirin, H.: 1974, in R. G. Athay (Ed.), *Chromospheric Fine Structure*, IAU Symposium 56, Reidel, Dordrecht, 161
Zirin, H.: 1985a, *Australian J. Phys.* **38**, 961
Zirin, H.: 1985b, *Astrophys. J.* **291**, 858
Zirin, H.: 1988, *Astrophysics of the Sun*, Cambridge Univ. Press, Cambridge UK
Zirin, H. and Liggett, M. A.: 1987, *Sol. Phys.* **113**, 267
Zirker, J. B., Martin, S. F., Harvey, K., and Gaizauskas, V.: 1997, *Sol. Phys.* **175**, 27
Zwaan, C.: 1965, *Ph.D. thesis*, Utrecht University, Utrecht, The Netherlands
Zwaan, C.: 1968, *Ann. Rev. Astron. Astrophys.* **6**, 135
Zwaan, C.: 1978, *Sol. Phys.* **60**, 213
Zwaan, C.: 1981a, in R. M. Bonnet and A. K. Dupree (Eds.), *Solar Phenomena in Stars ad Stellar Systems*, NATO-ASI, Bonas, D. Reidel, Dordrecht, The Netherlands, p. 463
Zwaan, C.: 1981b, in R. M. Bonnet and A. K. Dupree (Eds.), *Solar Phenomena in Stars and Stellar Systems*, NATO-ASI, Bonas, D. Reidel, Dordrecht, The Netherlands, p. 463
Zwaan, C.: 1983, in J. O. Stenflo (Ed.), *Solar and Stellar Magnetic Fields*, IAU Symp. 102, Reidel, Dordrecht, p. 85
Zwaan, C.: 1985, *Sol. Phys.* **100**, 397
Zwaan, C.: 1987, *Ann. Rev. Astron. Astrophys.* **25**, 83
Zwaan, C.: 1991, in P. Ulmschneider, E. R. Priest, and R. Rosner (Eds.), *Mechanisms of Chromospheric and Coronal Heating*, Springer-Verlag, Berlin, 241
Zwaan, C.: 1992, in J. H. Thomas and N. O. Weiss (Eds.), *Sunspots: Theory and Observations*, NATO ASI Series C 375, Kluwer Acadmic Publishers, Dordrecht, p. 75
Zwaan, C.: 1996, *Sol. Phys.* **169**, 265
Zwaan, C., Brants, J. J., and Cram, L. E.: 1985, *Sol. Phys.* **95**, 3
Zwaan, C. and Cram, L. E.: 1989, in L. E. Cram and L. V. Kuhi (Eds.), *FGK stars and T Tauri stars*, NASA SP-502, Washington, p. 215
Zwaan, C. and Harvey, K. L.: 1994, in M. Schüssler and W. Schmidt (Eds.), *Solar Magnetic Fields*, Cambridge University Press, Cambridge, UK, p. 27

Index

Printed in the United States
By Bookmasters